KU-539-811

The Stochastic Finite Element Method

A26129111

The Stochastic Finite Element Method

Basic Perturbation Technique and Computer Implementation

Michael Kleiber

Tran Duong Hien

Polish Academy of Sciences
Warsaw
Poland

JOHN WILEY & SONS

Chichester · New York · Brisbane · Toronto · Singapore

Copyright © 1992 by John Wiley & Sons Ltd,
Baffins Lane, Chichester,
West Sussex PO19 1UD, England

All rights reserved.

No part of this book may be reproduced by any means, or transmitted, or translated into a machine language without the written permission of the publisher.

UNIVERSITY OF STRATHCLYDE
14 JUN 1994
UNIVERSITY LIBRARY

Other Wiley Editorial Offices

John Wiley & Sons, Inc., 605 Third Avenue,
New York, NY 10158-0012, USA

Jacaranda Wiley Ltd, G.P.O. Box 859, Brisbane,
Queensland 4001, Australia

John Wiley & Sons (Canada) Ltd, 22 Worcester Road,
Rexdale, Ontario M9W 1L1, Canada

John Wiley & Sons (SEA) Pte Ltd, 37 Jalan Pemimpin # 05-04,
Black B, Union Industrial Building, Singapore 2057

Library of Congress Cataloging-in-Publication Data

Kleiber, Michal.
The stochastic finite element method : basic perturbation technique and computer implementation / Michael Kleiber, Tran Duong Hien.
p. cm.
Includes bibliographical references and index.
ISBN 0 471 93626 X
1. Finite element method. 2. Stochastic analysis. I. Tran, Duong Hien. II. Title.
TA347.F5K57 1992
620'.001'51535—dc20 92-22814
CIP

British Library Cataloguing in Publication Data

A catalogue record for this book is available from the British Library

ISBN 0 471 93626 X

D
620.0015'15353
KLE

Produced from camera-ready copy supplied by the authors
Printed and bound in Great Britain by Bookcraft (Bath) Ltd.

Contents

Preface ix

Part I Preliminaries 1

1 Some Mathematical Background 3

1.1 Brief Introduction to Probability Theory 3

1.1.1 Random Variables, Probability Density Functions and Cumulative Distribution Functions 3

1.1.2 Expectations and Moments 6

1.1.3 Linear Functions of Random Variables 9

1.1.4 Normal Probability Density Functions 11

1.1.5 Linear Statistical Analysis 12

1.1.6 Random Processes and Random Fields 14

1.2 Discrete Representations of Data 19

1.2.1 Dirac Delta Distribution 19

1.2.2 Discrete Fourier Transform 22

1.2.3 Base-Two Fast Fourier Transform 30

1.3 Calculus of Variation 34

2 FEM in Deterministic Structural Analysis 39

2.1 Introductory Comments and Organization of the Chapter 39

2.2 Equations of Linear Solid Mechanics 40

2.3 Variational Methods 42

2.3.1 Principle of Minimum Potential Energy 42

2.3.2 Multi-Field Principles 45

2.3.3 Hamilton Principle 46

2.4 FEM in Statics 48

2.4.1 Compatible Displacement Model 48

2.4.2 Remarks on Solution Algorithms and Computer Implementation 53

2.5 FEM in Dynamics 55

2.5.1 Compatible Displacement Model 55

2.5.2 Remarks on Solution Algorithms and Computer Implementation 56

2.6 Some Aspects of Nonlinear Analysis 61

2.6.1 Piece-Wise Linearization of Nonlinear Time Response 61

2.6.2 Finite Element Equations 63

2.7 FEM in Linearized Buckling Analysis 65

2.8 Structural Design Sensitivity Analysis 67

2.8.1 Statics 67

2.8.2 Dynamics 71

2.8.2.1 Time Interval Sensitivity 71

2.8.2.2 Time Instant Sensitivity 73

2.8.2.3 Sensitivity of Natural Frequencies and Buckling Loads 75

2.8.2.4 Remarks on Computer Implementation 75

Part II FEM in Stochastic Analysis 81

3 Stochastic Variational Principles 83

3.1 Introductory Remarks 83
3.2 Stochastic Potential Energy Principle 86
3.3 Stochastic Multi-Field Principles 91
3.4 Stochastic Hamilton Principle 93

4 Stochastic Finite Element Analysis 97

4.1 Finite Element Equations: Statics and Dynamics 97
4.2 Probabilistic Distribution Output 104
4.3 Stochastic Free Vibrations 106
4.3.1 Finite Element Equations 106
4.3.2 First and Second Derivatives of Eigenvalues 107
4.3.3 First and Second Derivatives of Eigenvectors 109
4.3.4 First Two Statistical Moments 113
4.4 Stochastic Linearized Buckling Analysis 114
4.4.1 Finite Element Equations 114
4.4.2 First- and Second-Order Eigenvalues 117
4.4.3 First Two Moments for Buckling Load Factors 119
4.4.4 First Two Moments for Buckling Loads 120
4.5 Computational Aspects 123
4.5.1 Notes on Solution Process 123
4.5.2 Random Variable Transformation 125
4.5.2.1 Transformed Uncorrelated Random Variables 125
4.5.2.2 Transformed Finite Element Equations 126
4.5.2.3 Response Probabilistic Distributions 131
4.5.3 Two-Fold Superposition Technique 136
4.5.4 Secularity Elimination 139
4.5.5 Adjoint Variable Technique 142
4.5.6 Concluding Remarks on SFEM Implementation 145
4.6 Numerical Illustrations 146

5 Stochastic Sensitivity: Static Problems 163

5.1 Finite Element Formulation 163
5.2 Computational Aspects 168
5.3 Numerical Illustrations 171

6 Program SFESTA. A Code for the Deterministic and Stochastic Analysis of Statics and Static Sensitivity of 3D Trusses 177

6.1 Data Preparation 177
6.2 Fortran Routines 180
6.2.1 Main Routine `sfesta` 181
6.2.2 Subroutines `constrinp` and `detersens` 186
6.2.3 Subroutines `displcovar` and `displcovar` 192
6.2.4 Subroutine `randsens` 195

7 Stochastic Sensitivity: Dynamic Problems 199
7.1 Finite Element Formulation 199
7.1.1 Time Interval Sensitivity 200
7.1.2 Time Instant Sensitivity 206
7.2 Computational Aspects 210
7.3 Numerical Illustrations 218

8 Program SFEDYN. A Code for the Deterministic and Stochastic Analysis of Dynamics and Dynamic Sensitivity of 3D Frames 223
8.1 Data Preparation 223
8.2 Fortran Routines 228
8.2.1 Subroutine elstiff 228
8.2.2 Subroutine eigsens 234
8.2.3 Subroutine rcovinp 236
8.2.4 Subroutine rhs1ord 240
8.2.5 Subroutines respo1 and rcovprt 242
8.2.6 Subroutine rsecular 249
8.2.7 Subroutines resen1 and resen2 253
8.2.8 Subroutines rsenmean and rsentime 257
8.2.9 Subroutine rsenc1v 262
8.3 Source Files and Overlay Linking 264

9 SFEM in Nonlinear Mechanics
9.1 Finite Element Formulation 271
9.2 Probabilistic Distribution Output 276
9.3 Some Computational Aspects 279

Appendix A SFESTA – User's Manual 283

Appendix B SFEDYN – User's Manual 291

Bibliography 305

Index 313

Glossary of Symbols 319

Acknowledgements

The basic stages of the work on this book were carried out at the Institute of Fundamental Technological Research, Polish Academy of Sciences, Warsaw, in the years 1988 – 1991 in the framework of the government-supported research program CPBP 02.01 on Mechanics of Materials, Structures and Technological Processes.

The final version of the book was written when one of the authors (M. K.) was a Mitsubishi Heavy Industry Visiting Professor at the Large Scale Systems Laboratory, Research Centre for Advanced Science and Technology (RCAST), The University of Tokyo, Japan.

The generous support of all the above institutions is gratefully acknowledged.

Preface

Stochastic finite element methods (SFEM) have recently become an active area of research. As the name suggests, researchers in this field attempt to combine two crucial methodologies developed to deal with complex problems of modern engineering: the finite element analysis and the stochastic analysis.

The finite element method (FEM) is now by far the most effective tool available to analysts interested in computer assisted solving of compound engineering problems. The method has the virtues of simplicity in concept, precision in development and potency in application. As a result of this there exists at present numerous computer programs that can handle, at reasonable cost, very large finite element systems of great engineering significance.

The stochastic analysis in the broadest sense refers to the explicit treatment of uncertainty in any quantity entering the corresponding deterministic analysis. The exact values of these quantities are usually unknown because they cannot be precisely measured. Differences in cross-sectional areas along a beam, eccentricity of mass density and/or of Young's modulus in a disc of a turbomachinery rotor, inaccurate location of a bridge hinge or roller, wind-induced forces acting on a structures, etc. are all examples of inherent characteristics of randomness typical of structural engineering problems. Existing uncertain variations in parameters may have significant effects on such fundamental structural characteristics as static stress distribution, stability and free and forced vibrations; consequently, they must affect the final design.

Useful analytical tools for performing analysis of structures with uncertain properties are provided by the theory of random fields which is an outgrowth of probability theory. However, many newcomers to the field of numerical analysis of such structures may have misgivings about the necessity of entering the arena of advanced probabilistics in order to obtain solutions. In a sense they are right; it is not necessary, only more convenient.

Let us take an example of, say, a material Young's modulus for which we have results of several experiments. By data processing we are able to obtain statistical characteristics of the modulus treated as a random variable. The SFEM will allow us then to determine the statistical characteristics for any response quantity of interest (possibly with limitations resulting from simplifying assumptions adopted while developing the SFEM equations).

An alternative procedure would be to use the series of available Young's modulus values as they stand and to repeatedly perform the FEM calculations for each of them. The results obtained for the response quantities of interest can then be processed by either (a) a formal statistical analysis yielding the statistical parameters required, or (b) an informal approach in which an experienced analyst studies all the structural responses and comes up with his own safety factor to take care of randomness.

Following path (a) should lead to the same results as obtained by adopting the stochastic approach from the beginning, while following path (b) may yield similar results only if the analyst has sufficient insight. But no matter which path is followed it is safe to say generally that both will require more time and effort than the procedure which introduced statistical quantities at the beginning.

In general, FEM is a highly effective method for solving sets of partial differential equations. This book concentrates on equations typical of structural mechanics. The necessity of performing numerical calculations while analysing the stochastic behaviour of structures has now been widely recognized. Since uncertainties appear in the operators of governing equations, even linear (in terms of displacements) systems are in general nonlinear as functions of random variables. The distinguishing feature of SFEM, which is based on the perturbation approach, is treating probabilistic-nonlinear problems with 'deterministic' computational techniques that take full advantage of the mathematical properties of linear operators.

The book offers a thorough treatment of a variety of problems for which we can use the term probabilistic computational mechanics (PCM). Because the PCM equations are so general in terms of the underlying mathematics, they are formally applicable to many other areas of research; thus in this book we shall be able to address a number of quite general issues related to the use of computational techniques in the field of stochastic analysis.

For the purpose of any stochastic analysis uncertainties present in the system need to be distinguished as time-invariant and time-variant. In static analysis it is natural to consider only time-invariant uncertainties (such as experimentally determined material constants). Conversely, forces which induce structural vibration, such as earthquake excitations, wind and wave forces, and jet noise excitation of aircraft panels, are often treated as time-variant uncertainties. The characterization of such uncertainties is presently an area of active research requiring a lot of engineering insight combined with the formal analysis of differential equations with time-variant random excitations. In this book we shall confine ourselves to time-invariant random coefficients and excitations; thus the subject of random vibrations (i.e. vibrations under randomly distributed time-dependent loads) is excluded from this treatment, for instance. Because of the perturbation scheme used in the analysis, the methodology proposed here is fully applicable when uncertainties are not to large and when the probabilistic density functions have decaying tails. Also, even if it believed that the techniques described provide highly useful insights into the manner structural uncertainties propagate, they do not provide adequate information for some more sophisticated questions such as those related to structural reliability.

The first basic goal we set ourselves in this book is getting the reader acquainted with the theoretical fundamentals of SFEM. We must admit, however, that one great difficulty that the would-be user of SFEM formulations will certainly experience is the apparent complexity of programming the method, which may deter people from moving into this important research field. With this thought in mind we decided to include in the text detailed descriptions and listings of a number of specially written SFEM programs. Thus, the second basic goal of ours is that these programs help the reader to take the painful step from theory to program, enabling him to develop (or at least appreciate) programs for his particular applications in his own environment. In the development of the programs the emphasis has been placed on simplicity, ease of understanding and practicality.

The book is organized as follows. It consists of two parts, the first one of which is shorter and just discusses mathematical fundamentals indispensable to properly present the very subject of the book which follows in Part II. No particular prerequisites are required except for the usual engineering undergraduate background in structural mechanics, matrix algebra and computer programming. To better appreciate the material being presented the reader should be familiar with a limited number of basic notions in probability theory; it is also desirable for him to have taken an introductory course in finite elements. Nevertheless, all information crucial to the understanding of the basic derivations contained in the book is summarized in Part I.

Chapter 1 consists of three parts: a brief introduction to probability theory is given in Section 1.1 followed by remarks on the use of the Dirac delta distribution and the presentation of the main idea behind the discrete and so-called fast Fourier transform, Section 1.2. The significance of using variational principles in deriving equations suitable for numerical analysis of differential boundary-value problems is emphasized in Section 1.3.

In Chapter 2 an attempt is made to summarize equations of classical (i.e. deterministic) displacement-type FEM models for linear elastic applications. Conventional problems of linear statics and dynamics are followed by considering problems related to time-independent and time-dependent structural response sensitivity. A brief presentation of a formulation which accounts for different sources of structural nonlinearities is also given. Discussions in this chapter are focused on the variational approach to the problem of formulation of the solid mechanics governing equations, on approximation techniques and on algorithms for solving large systems of algebraic and ordinary differential equations typical of structural systems under static and dynamic loadings. For the structural design sensitivity problems both the direct and adjoint approaches are described and computer implementation issues are discussed.

The reader is encouraged to study Chapters 1 and 2 both as a review and as a reference point for the subsequent development in this book. For readers who are not thoroughly familiar with this background material, these two chapters can also be used as a guide for further reading. References which deal with the contents of these two chapters in more detail are cited therein.

Part II of the book starts with the generalization of the minimum potential energy principle and the Hamilton principle towards inclusion of stochastic effects (Chapter 3) by using the mean-based, second-moment, second-order perturbation technique. These stochastic variational principles are employed in the following chapters to generate finite element equations describing the probabilistic problems of statics, buckling, free and forced vibrations and design sensitivity. The equations are used to obtain the means and covariances of the unknown distributions. We also discuss ways to effectively program solution algorithms devised for these equations.

On the basis of the stochastic minimum potential energy principle and the stochastic Hamilton principle the displacement finite element models of equilibrium equations and equations of motion are introduced in Chapter 4 to solve for the first two moments of the nodal displacements, element strains and stresses. The stochastic finite element formulations for free vibration and linearized buckling analyses are also considered in this chapter. We summarize the essence of stochastic analysis and the advantages of the stochastic finite element approach over statistical techniques. A two-fold superposition technique involving the modal analysis and a random variable transformation is introduced and the description of a secularity elimination procedure based on the fast Fourier analysis and synthesis of complex-valued sequence is given.

In Chapter 5 we incorporate randomness into the problem of static structural sensitivity. Sensitivity of the static response to stochastically described variations in design parameters seems to be an important subject which can be studied using conventional finite element technology. As a result of the analysis we obtain probabilistic distributions (i.e. two statistical moments in the second-moment approach) of sensitivity gradient coefficients. We show that structural design sensitivity and stochastic finite element analysis are similar in terms of both the methodology and computer implementation which greatly facilitates the combined analysis. The computations are much less costly when based on a transformation from the correlated space to an uncorrelated space of random variables. Numerical algorithms worked out are shown to be readily adapted to existing finite element codes whose element matrices are explicitly generated. As an illustration of the above, Chapter 6 presents the microcomputer code SFESTA, a complete Fortran program for deterministic and stochastic analysis of statics and design sensitivity for any medium-scale three-dimensional trusses.

Random and/or design variables can be assumed as cross-sectional area, length or Young's modulus of any structural components. Response functionals can be defined as cost functions and displacement or stress constraint functions. The program SFESTA is fully documented and can be employed for either using it 'as-is' or extending it towards inclusion of a new library of various finite elements, for instance.

In Chapter 7, which devoted to the finite element analysis of stochastic dynamic response sensitivity, we distinguish two different types of sensitivity treatment: the one for evaluating sensitivity of the response taken over any time interval and another one for evaluating sensitivity at a particular time instant. Since both the random and design variables are expressed in a discretized-parameter space, the stochastic sensitivity function can be modelled in a parallel way. A combination of the adjoint variable method and second moment analysis is employed to determine the time response of the first two moments of the sensitivity gradient. It is shown that the effects of finite element mesh density, period elongation and amplitude decay in numerical integration as well as those of secular terms cannot be disregarded. Combined algorithms based on procedures for sensitivity analysis of deterministic systems (Chapter 2) and for displacement-stress response analysis of stochastic systems (Chapter 4) are fully described. The discussion is substantiated in Chapter 8, in which the Fortran code SFEDYN for the deterministic and stochastic analysis of free and forced vibrations and dynamic sensitivity of three-dimensional frames is presented. As with the previous program SFESTA, the code SFEDYN can be directly used for solving a variety of test examples, or it can be further developed to cover other useful structural mechanics situations. Random and design variables can be assumed as cross-sectional area, length, Young's modulus and mass density of any structural components. Numerical algorithms for, and difficulties in, generalizing the stochastic finite element analysis to cover nonlinear structural mechanics problems are discussed in Chapter 9.

The complete description of the input data necessary to run the programs SFESTA and SFEDYN is given in Appendices A and B, respectively.

A few remarks about the mathematics underlying the analysis described in this book should be given. The previous discussion clearly suggests that the problems under scrutiny here can be described in terms of random differential equations. As a matter of fact, after carrying out the FEM-typical spatial discretization of the problem we shall be faced with a system of random ordinary differential equations (or random algebraic equations if problems of statics are considered). It will be assumed that randomness enters the equations only through constant-in-time coefficients at unknown functions and forcing functions (inhomogeneous part). As is known from the theory of random differential equations, this is the simplest case in the sense that it requires a minimum of probabilistic concepts, and allows the treatment to closely follow that for deterministic differential equations.

The book is an outgrowth of the authors' experience in teaching a one-semester course on SFEM to graduate students of civil and mechanical engineering. Depending on the time available, the course has concentrated on selected topics covered in the book, leaving the rest as an extension useful in formulating term projects.

Since the authors are well aware of the difficulties which are often encountered in implementing computer programs from their listings, the source code as well as the compiled IBM PC modules are available from the Publisher at a nominal cost. The reader may also want to contact the authors if he/she is interested in having the full version of the authors' SFEM program which features the following options for any medium-scale complex truss–beam–plate–shell structures: stochastic analysis of statics, stochastic analysis of static sensitivity, stochastic buckling analysis, stochastic analysis of buckling sensitivity, stochastic analysis of free vibrations, stochastic analysis of eigenvalue sensitivity, stochastic analysis of forced vibrations, stochastic analysis of dynamic sensitivity. It goes without saying that the program is capable of the corresponding deterministic analyses as well.

Finally, it should be noted that the book is only an introduction to the important and dynamically developing field of numerical computations for structural systems with load, strength and geometry imperfections. The authors' goal will be met if the reader, after studying the book, is encouraged to employ the formulations and computer programs contained in it in his/her own work, and, possibly, to use them as a starting point for further research.

Part I

PRELIMINARIES

Chapter 1

Some Mathematical Background

1.1 Brief Introduction to Probability Theory

1.1.1 Random Variables, Probability Density Functions and Cumulative Distribution Functions

Probability theory is concerned with the development of a mathematical framework in which to treat *random phenomena*.[1] Thus, inherent to the consideration of random phenomena is a repeated experiment with a set of possible outcomes (or results). If the experiment is performed repeatedly, with all conditions maintained as precisely as possible, and the measured results are identical, then the item which is measured is said to be *deterministic*. Otherwise, i.e. if results are not identical, the item is said to be *random*. Associated with each of the outcomes in the latter experiment is a real number called the *probability of the event*. It seems intuitively reasonable that: (i) the probability is related to the expected relative frequency of occurrence of the event in a long sequence of experiments, (ii) it should lie between zero and unity, and (iii) the sum of probabilities of all possible events in a particular experiment should be unity.

All possible outcomes of the experiment, which can be represented by points called sample points, comprise the sample space of the experiment in question. An event is a collection (or subset) of the sample points in the sample space. If an event is represented as a single outcome, then it is called a simple event; otherwise, the event is compound.

Any real-valued variable defined on a sample space is called a *(real) random variable*. Random variables can be discrete or continuous.

Let X be a continuous random variable which may take on any value on the real line, $-\infty < X < +\infty$. The function $p(x)$ is said to be the *probability density function* (PDF) of the random variable X if the probability that the outcome of the experiment (i.e. the value of X) will fall in the interval $[x - \frac{dx}{2}, x + \frac{dx}{2}]$ along

[1]Only a brief introduction necessary for proper appreciation of later SFEM derivations is given below – for a more thorough treatment of probability theory the reader is referred to [27,28,30,75], for instance.

the real line is $p(x)\mathrm{d}x$. (A lower-case x is used to denote an observed value of the random variable X.) To say this more formally we shall use the notation

$$\Pr\left(x - \frac{\mathrm{d}x}{2} \le X \le x + \frac{\mathrm{d}x}{2}\right) = p(x)\,\mathrm{d}x \tag{1.1}$$

More generally, the notation

$$\Pr(a \le X \le b) = \int_a^b p(x)\,\mathrm{d}x \tag{1.2}$$

is employed to indicate the probability that the random variable X has a value between two points a and b on the real line.

Any function $p(x)$ that satisfies the following three conditions can be chosen as the PDF for a given random variable X:

$$\begin{aligned} &\text{(i)} && p(x) \ge 0 \\ &\text{(ii)} && \int_{-\infty}^{+\infty} p(x)\,\mathrm{d}x = 1 \\ &\text{(iii)} && \Pr(a \le X \le b) = \int_a^b p(x)\,\mathrm{d}x \end{aligned} \tag{1.3}$$

A *cumulative distribution function* (CDF) is the integral of the PDF defined as

$$P(a) = \int_{-\infty}^{a} p(x)\,\mathrm{d}x = \Pr(X \le a) \tag{1.4}$$

In other words, $P(a)$ is the probability that the random variable X will have a value equal to or less than a. The CDF is a monotonically increasing function which possesses the following properties:

$$\begin{aligned} &\text{(i)} && P(-\infty) = 0 \\ &\text{(ii)} && 0 \le P(x) \le 1 \\ &\text{(iii)} && P(+\infty) = 1 \end{aligned} \tag{1.5}$$

Clearly

$$\frac{\mathrm{d}P(x)}{\mathrm{d}x} = \lim_{\Delta x \to 0} \frac{P(x + \Delta x) - P(x)}{\Delta x} = p(x) \tag{1.6}$$

i.e. the PDF $p(x)$ is just the slope of the CDF $P(x)$. Fig. 1.1 depicts the PDF and the CDF of a possible continuous random variable.

Frequently, more than one random variable (say, $X_1, X_2, \ldots, X_n$) need to be associated with an experiment, and a joint behaviour of them is of interest.[1] A *multi-variate* (or *joint*) *probability density function* $p(x_1, x_2, \ldots, x_n)$ must satisfy the following conditions:

---

[1] In a plate structure, for instance, the random thickness may be described at different points, and it is reasonable to anticipate that, although they are random, the plate thickness at one point may not be entirely unrelated to the thickness at other points. The thickness values at different points may then be taken as random variables.

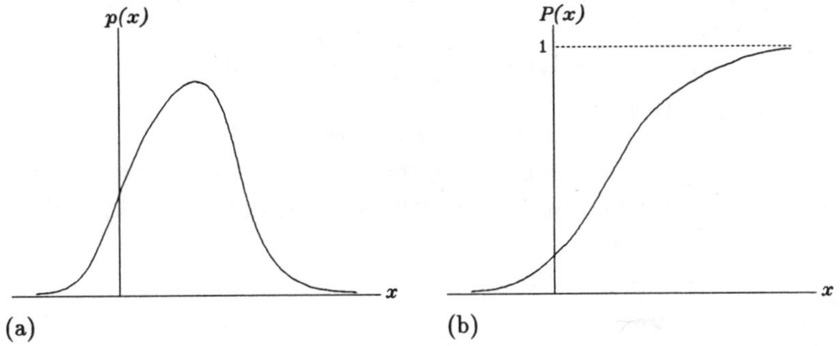

Figure 1.1 Probability density function (a) and cumulative distribution function (b).

$$
\begin{aligned}
&\text{(i)} \quad p(x_1, x_2, \ldots, x_n) \geq 0 \\
&\text{(ii)} \quad \underbrace{\int_{-\infty}^{+\infty}\int_{-\infty}^{+\infty}\cdots\int_{-\infty}^{+\infty}}_{n\text{-fold}} p(x_1, x_2, \ldots, x_n)\, \mathrm{d}x_1 \mathrm{d}x_2 \ldots \mathrm{d}x_n = 1 \\
&\text{(iii)} \quad \Pr(a_1 \leq X_1 \leq b_1, a_2 \leq X_2 \leq b_2, \ldots, a_n \leq X_n \leq b_n) \\
&\qquad = \int_{a_1}^{b_1}\int_{a_2}^{b_2}\cdots\int_{a_n}^{b_n} p(x_1, x_2, \ldots, x_n)\, \mathrm{d}x_1 \mathrm{d}x_2 \ldots \mathrm{d}x_n
\end{aligned}
\tag{1.7}
$$

The symbol $\Pr(a_1 \leq X_1 \leq b_1, a_2 \leq X_2 \leq b_2, \ldots, a_n \leq X_n \leq b_n)$ denotes the probability that the random variables $X_1, X_2, \ldots, X_n$ have values within the prescribed limits.

A *multi-variate* (or *joint*) *cumulative distribution function* is defined by

$$
\begin{aligned}
P(a_1, a_2, \ldots, a_n) &= \int_{-\infty}^{a_1}\int_{-\infty}^{a_2}\cdots\int_{-\infty}^{a_n} p(x_1, x_2, \ldots, x_n)\, \mathrm{d}x_1 \mathrm{d}x_2 \ldots \mathrm{d}x_n \\
&= \Pr(a_1 \leq X_1 \leq b_1, a_2 \leq X_2 \leq b_2, \ldots, a_n \leq X_n \leq b_n)
\end{aligned}
\tag{1.8}
$$

where $a_1, a_2, \ldots, a_n$ are constants. The joint CDF is nondecreasing with respect to each of its arguments. Also

$$
p(x_1, x_2, \ldots, x_n) = \frac{\partial^n P(x_1, x_2, \ldots, x_n)}{\partial x_1 \partial x_2 \ldots \partial x_n}
\tag{1.9}
$$

A joint PDF provides complete probabilistic information about the PDF of any one random variable; the latter function, when obtained from a joint PDF, is called a *marginal probability density function.* A marginal PDF of the i-th random variable is obtained from the joint PDF $p(x_1, x_2, \ldots, x_n)$ by integrating over the remaining variables

$$
\begin{aligned}
p(x_i) = \underbrace{\int_{-\infty}^{+\infty}\int_{-\infty}^{+\infty}\cdots\int_{-\infty}^{+\infty}}_{(n-1)\text{-fold}} & p(x_1, x_2, \ldots, x_{i-1}, x_i, x_{i+1}, \ldots, x_n) \\
& \times \mathrm{d}x_1 \mathrm{d}x_2 \ldots \mathrm{d}x_{i-1} \mathrm{d}x_{i+1} \ldots \mathrm{d}x_n
\end{aligned}
\tag{1.10}
$$

The probability that X_i lies between a_i and b_i may thus be expressed as

$$\Pr(a_i \le X_i \le b_i) = \Pr\Big(a_i \le X_i \le b_i \text{ and } -\infty < X_j < +\infty \text{ for } j = 1, 2, \ldots, i-1, i+1, \ldots n\Big) = \int_{a_i}^{b_i} p(x_i)\,\mathrm{d}x_i \tag{1.11}$$

The joint PDF $p(x_1, x_2, \ldots, x_n)$ contains more probabilistic information than $p(x_1), p(x_2), \ldots, p(x_n)$ separately, since the latter PDFs can be obtained from the former. The *marginal cumulative distribution function* of X_i is defined accordingly as

$$P(a_i) = \int_{-\infty}^{a_i} p(x_i)\,\mathrm{d}x_i \tag{1.12}$$

The random variables $X_1, X_2, \ldots, X_n$ are said to be mutually (statistically) independent if (and only if)

$$p(x_1, x_2, \ldots, x_n) = p(x_1)\,p(x_2)\ldots p(x_n) \tag{1.13}$$

that is, if their joint PDF is equal to the product of the marginal PDFs of each random variable. In terms of CDFs, this definition can be given as

$$P(x_1, x_2, \ldots, x_n) = P(x_1)P(x_2)\ldots P(x_n) \tag{1.14}$$

where $P(x_1), P(x_2), \ldots, P(x_n)$ are the marginal CDFs of the random variables $X_1, X_2, \ldots, X_n$, respectively.

1.1.2 Expectations and Moments

Let $Y = f(X)$ be a real, single-valued continuous function of a random variable X; thus, Y is also a random variable. The expectation (or expected value, or mean value, or statistical average) of the function f is defined as

$$E[Y] = E[f(X)] = \int_{-\infty}^{+\infty} f(x)\,p(x)\,\mathrm{d}x \tag{1.15}$$

where $p(x)$ is the PDF of the random variable X.

In the special case of $f(X) = X$ the expectation $E[f(X)]$ becomes the *expected value*, or *mean*, or *average* of the random variable X

$$E[X] = \int_{-\infty}^{+\infty} x\,p(x)\,\mathrm{d}x \tag{1.16}$$

which will also be denoted by x^0, or μ. Another interesting special case is where $f(X) = X^2$ leads to

$$E[X^2] = \int_{-\infty}^{+\infty} x^2\,p(x)\,\mathrm{d}x \tag{1.17}$$

which is called the *mean square value* of the random variable X. Its square root is called the *root mean square value*. The *variance* of X, denoted by $\mathrm{Var}(X)$, or

σ_X^2, is defined as the mean square value of X about the mean

$$\mathrm{Var}(X) = E[(X - x^0)^2] = \int_{-\infty}^{+\infty} (x - x^0)^2 p(x)\,\mathrm{d}x = E[X^2] - (x^0)^2 \qquad (1.18)$$

The positive square root of the variance is denoted as

$$\sigma_X = +\sqrt{\mathrm{Var}(X)} \qquad (1.19)$$

and is called the *standard deviation* of X. Obviously, if the mean value is zero the standard deviation equals the root mean square value. The *coefficient of variation* of a random variable, denoted by α_X, is expressed as

$$\alpha_X = \left[\frac{\mathrm{Var}(X)}{(E[X])^2}\right]^{\frac{1}{2}} \qquad (1.20)$$

The *m-th moment* of a random variable is defined as

$$\mu_X^m = \int_{-\infty}^{+\infty} x^m p(x)\,\mathrm{d}x \qquad (1.21)$$

Clearly, the first moment ($m = 1$) is just the mean of the random variable X, cf. Eq. (1.16). It is convenient to define a set of moments centred on the mean rather than the origin. The m-th moment about the mean (or the *m-th central moment*) is defined as

$$\bar{\mu}_X^m = E[(X - x^0)^m] = \int_{-\infty}^{+\infty} (x - x^0)^m\, p(x)\,\mathrm{d}x \qquad (1.22)$$

Clearly, the second central moment is just the variance of X, cf. Eq. (1.18). Both the mean and variance give important information about the distribution of the random variable. The mean is a measure of the 'centre of gravity' of the PDF while the variance is a measure of the spread of the density about the mean. The variance corresponds to the moment of inertia of a mass about its centre of gravity.

The notion of expectation applies to a set of random variables as well. Using the same notation as in Section 1.1.1 the expectation of a function $f(X_1, X_2, \ldots, X_n)$ of n random variables $X_1, X_2, \ldots, X_n$ is defined as

$$E[f(X_1, X_2, \ldots, X_n)]$$
$$= \underbrace{\int_{-\infty}^{+\infty}\int_{-\infty}^{+\infty}\cdots\int_{-\infty}^{+\infty}}_{n\text{-fold}} f(x_1, x_2, \ldots, x_n)\, p(x_1, x_2, \ldots, x_n)\,\mathrm{d}x_1\mathrm{d}x_2, \ldots\,\mathrm{d}x_n \qquad (1.23)$$

Certain expectations warrant special names, such as:

(i) The *cross-correlation* between X_k and X_l defined by

$$\mu_{kl} = E[X_k X_l] = \int_{-\infty}^{+\infty}\int_{-\infty}^{+\infty} x_k x_l\, p(x_k, x_l)\,\mathrm{d}x_k\mathrm{d}x_l \qquad (1.24)$$

where $p(x_k, x_l)$ is the joint PDF of X_k and X_l given by

$$p(x_k, x_l) = \underbrace{\int_{-\infty}^{+\infty}\int_{-\infty}^{+\infty}\cdots\int_{-\infty}^{+\infty}}_{(n-2)\text{-fold}} p(x_1, x_2, \ldots, x_n) \times \mathrm{d}x_1 \ldots \mathrm{d}x_{k-1}\mathrm{d}x_{k+1} \ldots \mathrm{d}x_{l-1}\mathrm{d}x_{l+1} \ldots \mathrm{d}x_n \tag{1.25}$$

(ii) The *covariance* between X_k and X_l defined by

$$\begin{aligned} \mathrm{Cov}(X_k, X_l) &= E[(X_k - x_k^0)(X_l - x_l^0)] \\ &= \int_{-\infty}^{+\infty}\int_{-\infty}^{+\infty} (x_k - x_k^0)(x_l - x_l^0)\, p(x_k, x_l)\, \mathrm{d}x_k \mathrm{d}x_l \end{aligned} \tag{1.26}$$

By Eqs. (1.10), (1.16) and (1.24) the covariance may also be expressed as

$$\mathrm{Cov}(X_k, X_l) = E[X_k X_l] - x_k^0 x_l^0 \tag{1.27}$$

We note that if X_k and X_l are statistically independent, then their covariance vanishes, which results directly from the definition of independence of random variables (cf. Eq. (1.13)) and Eq. (1.27).

(iii) The *nondimensional correlation coefficient* between X_k and X_l defined by

$$\varrho_{kl} = \frac{\mathrm{Cov}(X_k, X_l)}{[\mathrm{Var}(X_k)\mathrm{Var}(X_l)]^{\frac{1}{2}}} \tag{1.28}$$

It can be shown that $|\varrho_{kl}| \leq 1$. The random variables X_k and X_l are called *uncorrelated* if $\varrho_{kl} = 0$ (or $\mathrm{Cov}(X_k, X_l) = 0$). Hence, independent random variables are uncorrelated; on the other hand, the zero correlation coefficient (zero covariance) does not necessarily imply statistical independence of the random variables involved.

The mean vector of a set of n random variables $X_1, X_2, \ldots, X_n$ and their covariance matrix are defined respectively as

$$\mathbf{x}^0 = \{\, x_1^0 \;\; x_2^0 \;\; \ldots \;\; x_n^0 \,\} \tag{1.29}$$

$$\mathbf{S_x} = \begin{bmatrix} \mathrm{Var}(X_1) & \mathrm{Cov}(X_1, X_2) & \cdots & \mathrm{Cov}(X_1, X_n) \\ \mathrm{Cov}(X_2, X_1) & \mathrm{Var}(X_2) & \cdots & \mathrm{Cov}(X_2, X_n) \\ \vdots & \vdots & \ddots & \vdots \\ \mathrm{Cov}(X_n, X_1) & \mathrm{Cov}(X_n, X_2) & \cdots & \mathrm{Var}(X_n) \end{bmatrix} \tag{1.30}$$

We conclude this section by quoting the so-called limit theorem known as the Chebyshev inequality. Let X be a random variable with $E(X) = x^0 = \mu_X$ and $\mathrm{Var}(X) = \sigma_X^2$. Then for any real number $\epsilon > 0$,

$$\Pr\Big(|\, X - \mu_X \,| \geq \epsilon \Big) \leq \frac{\sigma_X^2}{\epsilon^2} \tag{1.31}$$

which on writing $\kappa = \frac{\epsilon}{\sigma}$ gives

$$\Pr\Big(|\, X - \mu_X \,| \geq \kappa\, \sigma_X \Big) \leq \frac{1}{\kappa^2} \tag{1.32}$$

This shows clearly the close relationship between the variance of a distribution and the dispersion of that distribution about the mean. The Chebyshev inequality gives a rather weak bound on most distributions. On the other hand, it applies to *all* distributions whose means and variances exist.

1.1.3 Linear Functions of Random Variables

One of most commonly encountered transformations of the random variable X is the linear transformation given by

$$Y = aX + b \tag{1.33}$$

where a and b are constants. Let us see how the means and variances of X and Y are related. We have

$$E[Y] = E[aX + b] = aE[X] + b \qquad \text{or} \qquad y^0 = ax^0 + b \tag{1.34}$$

In a similar way, the variance of Y is found to be

$$\text{Var}(Y) = E[(Y - y^0)^2] = E[(aX + b - ax^0 - b)^2] = a^2 E[(X - x^0)^2] \tag{1.35}$$

or

$$\text{Var}(Y) = a^2 \,\text{Var}(X) \tag{1.36}$$

It is clear from these results that any random variable X with finite mean x^0 and finite variance $\text{Var}(X)$ may be normalized (by using a linear transformation of the type (1.33)) to have zero mean and unit variance. To show this define a new random variable Y such that

$$Y = \frac{X - x^0}{\sigma_X} \tag{1.37}$$

It follows from Eq. (1.35) that

$$y^0 = \frac{E[X] - x^0}{\sigma_X} = 0 \tag{1.38}$$

while the variance of Y is obtained from Eq. (1.36) as

$$\text{Var}(Y) = E[Y^2] = \frac{\text{Var}(X)}{\text{Var}(X)} = 1 \tag{1.39}$$

If we now consider the transformation

$$Y = \sum_{i=1}^{n} a_i X_i \tag{1.40}$$

where X_i are random variables, then it is easy to show that

$$E[Y] = \sum_{i=1}^{n} a_i \, E[X_i] \qquad \text{or} \qquad y^0 = \sum_{i=1}^{n} a_i x_i^0 \tag{1.41}$$

and

$$\text{Var}(Y) = \sum_{i=1}^{n} a_i^2 \,\text{Var}(X_i) + \sum_{i=1}^{n} \sum_{\substack{j=1 \\ j \neq i}}^{n} a_i a_j \text{Cov}(X_i, X_j) \tag{1.42}$$

In particular, if $Y = X_1 \pm X_2$, then

$$E[X_1 \pm X_2] = E[X_1] \pm E[X_2] \tag{1.43}$$

$$\mathrm{Var}(X_1 \pm X_2) = \mathrm{Var}(X_1) + \mathrm{Var}(X_2) \pm 2\mathrm{Cov}(X_1, X_2) \tag{1.44}$$

More generally, if we consider the transformation

$$Y_i = \sum_{j=1}^{n} C_{ij} X_j \qquad i = 1, 2, \ldots, m \tag{1.45}$$

or, in matrix notation,

$$\mathbf{Y}_{m\times 1} = \mathbf{C}_{m\times n} \mathbf{X}_{n\times 1} \tag{1.46}$$

where $\mathbf{Y}_{m\times 1}$ and $\mathbf{X}_{n\times 1}$ are m- and n-dimensional vectors of random variables and $\mathbf{C}_{m\times n}$ is a $m\times n$ matrix of constant coefficients, then

$$E[Y_i] = \sum_{j=1}^{n} C_{ij} E[X_j] \qquad i = 1, 2, \ldots, m \tag{1.47}$$

or, in matrix notation,

$$\begin{aligned} E\left[\{Y_i\}\right] &= \left\{E[Y_1] \;\; E[Y_2] \; \ldots \; E[Y_m]\right\}_{m\times 1} \\ &= \mathbf{C}_{m\times n} \left\{E[X_1] \;\; E[X_2] \; \ldots \; E[X_m]\right\}_{n\times 1} \end{aligned} \tag{1.48}$$

$$\mathbf{y}^0_{m\times 1} = \{y_i^0\}_{m\times 1} = \mathbf{C}_{m\times n} \{x_j^0\}_{n\times 1} = \mathbf{C}_{m\times n} \mathbf{x}^0_{n\times 1} \tag{1.49}$$

The last equation gives the mean values of the m random variables $Y_1, Y_2, \ldots, Y_m$ in terms of the mean values of the n random variables $X_1, X_2, \ldots, X_n$. We may similarly obtain the covariance dependence, which for any two random variables Y_k and Y_l, reads

$$\mathrm{Cov}(Y_k, Y_l) = \sum_{i=1}^{n} \sum_{j=1}^{n} C_{ki}\, C_{lj}\, \mathrm{Cov}(X_i, X_j) \tag{1.50}$$

The above term is the element in the k-th row and l-th column of the covariance matrix. Eq. (1.50) can be presented in matrix notation as

$$\mathbf{S_y}_{m\times m} = \mathbf{C}_{m\times n} \mathbf{S_x}_{n\times n} \mathbf{C}^{\mathrm{T}}_{n\times m} \tag{1.51}$$

where $\mathbf{S_y}$ and $\mathbf{S_x}$ denote the covariance matrix of the random variables $\mathbf{Y}$ and $\mathbf{X}$, respectively.

1.1.4 Normal Probability Density Functions

As an example of many useful probability density functions we shall define below the one which is most commonly employed in scientific and engineering applications.

A random variable X is said to be *Gaussian* (or *normally*) *distributed* if its PDF has the following exponential form:

$$p(x) = \frac{1}{\sigma_X\sqrt{2\pi}} \exp\Big[-\frac{(x-x^0)^2}{2\sigma_X^2}\Big] \tag{1.52}$$

This PDF depends on two parameters: the mean x^0 and the standard deviation σ_X, which thus fully characterize any Gaussian random variable. An equivalent and useful alternative to Eq. (1.52) is obtained by defining a new variable

$$Y = \frac{X - x^0}{\sigma_X} \tag{1.53}$$

which leads to the so-called standardized normal PDF

$$p(y) = \frac{1}{\sqrt{2\pi}} \exp\Big(-\frac{y^2}{2}\Big) \tag{1.54}$$

with

$$y^0 = 0 \qquad\qquad \sigma_Y = 1 \tag{1.55}$$

The Gaussian PDF possesses a number of remarkable mathematical properties. Among them we just mention that linear functions of Gaussian random variables remain Gaussian distributed so that, for instance, if X_1 and X_2 are Gaussian, then $Y = a_1X_1 + a_2X_2$ is Gaussian as well.

The multi-variate normal PDF is given by the expression

$$\begin{aligned} p(x_1, x_2, \ldots, x_n) \quad &= \quad \frac{1}{[(2\pi)^n \det \mathbf{S_X}]^{\frac{1}{2}}} \\ &\quad \times \exp\Big[-\frac{1}{2}\sum_{i,j=1}^{n}(x_i - x_i^0)\,\mathbf{S}^{-1}_{\mathbf{X}_{ij}}(x_j - x_j^0)\Big] \end{aligned} \tag{1.56}$$

where $\mathbf{S_X}$ and $\mathbf{S_X^{-1}}$ are the covariance matrix and its inverse. In the case of two random variables X_1 and X_2 Eq. (1.56) simplifies to

$$\begin{aligned} p(x_1, x_2) = &\frac{1}{2\pi\sigma_{X_1}\sigma_{X_2}(1-\varrho_{12}^2)^{\frac{1}{2}}} \exp\Big\{-\frac{1}{2(1-\varrho_{12}^2)} \\ &\times \Big[\frac{1}{\sigma_{X_1}^2}(x_1 - x_1^0)^2 + \frac{1}{\sigma_{X_2}^2}(x_2 - x_2^0)^2 - \frac{2\varrho_{12}}{\sigma_{X_1}\sigma_{X_2}}(x_1 - x_1^0)(x_2 - x_2^0)\Big]\Big\} \end{aligned} \tag{1.57}$$

where ϱ_{12} is the correlation coefficient between X_1 and X_2, cf. Eq. (1.28).

1.1.5 Linear Statistical Analysis

Let us start this section by considering a function f dependent upon n arguments $x_1, x_2, \ldots, x_n$. Expansion of the function $f(x_1, x_2, \ldots, x_n)$ into the Taylor's series about a particular set of values, say, $x_1^0, x_2^0, \ldots, x_n^0$ of its arguments, reads

$$\begin{aligned} f(x_1, x_2, \ldots, x_n) &= f(x_1^0, x_2^0, \ldots, x_n^0) \\ &+ \sum_{i=1}^{n} \frac{\partial f(x_1^0, x_2^0, \ldots, x_n^0)}{\partial x_i}(x_i - x_i^0) + \cdots \end{aligned} \tag{1.58}$$

where higher-than-linear terms have been neglected. For the $(m\times 1)$-vector function $\mathbf{f}(\mathbf{x}) = \{f_1(x_1, \ldots, x_n), \ldots, f_m(x_1, \ldots, x_n)\}$ the corresponding matrix expression is

$$\mathbf{f}(\mathbf{x})_{m\times 1} = \mathbf{f}(\mathbf{x}^0)_{m\times 1} + \left[\frac{\partial \mathbf{f}(\mathbf{x}^0)}{\partial \mathbf{x}}\right]_{m\times n} (\mathbf{x} - \mathbf{x}^0)_{n\times 1} \tag{1.59}$$

where

$$\frac{\partial \mathbf{f}(\mathbf{x}^0)}{\partial \mathbf{x}} = \begin{bmatrix} \frac{\partial f_1}{\partial x_1} & \frac{\partial f_1}{\partial x_2} & \cdots & \frac{\partial f_1}{\partial x_n} \\ \frac{\partial f_2}{\partial x_1} & \frac{\partial f_2}{\partial x_2} & \cdots & \frac{\partial f_2}{\partial x_n} \\ \vdots & \vdots & & \vdots \\ \frac{\partial f_m}{\partial x_1} & \frac{\partial f_m}{\partial x_2} & \cdots & \frac{\partial f_m}{\partial x_n} \end{bmatrix} \tag{1.60}$$

If now the n arguments $x_1, x_2, \ldots, x_n$ are assumed to be a set of random variables, then the m functions of these variables $f_1, f_2, \ldots, f_n$ are also a set of random variables. If, further, x_i^0 is the mean of the random variable X_i, then Eq. (1.59) is a Taylor's expansion of the vector function $\mathbf{f}$ about the mean values of the random vector variable $\mathbf{X}$.

Taking the expected value of both sides of Eq. (1.59)

$$\begin{aligned} &E[\mathbf{f}(\mathbf{x})] = \left\{ E[f_1] \;\; E[f_2] \;\ldots\; E[f_m] \right\} \\ &= \left\{ f_1(\mathbf{x}^0) \;\; f_2(\mathbf{x}^0) \;\ldots\; f_m(\mathbf{x}^0) \right\} + \left[\frac{\partial \mathbf{f}(\mathbf{x}^0)}{\partial \mathbf{x}}\right] E[(\mathbf{x} - \mathbf{x}^0)] = \mathbf{f}(\mathbf{x}^0) \end{aligned} \tag{1.61}$$

whereas the covariance matrix of the random variables $f_1, f_2, \ldots, f_m$

$$\mathbf{S_f}_{m\times m} = \begin{bmatrix} \mathrm{Var}(f_1) & \mathrm{Cov}(f_1, f_2) & \cdots & \mathrm{Cov}(f_1, f_m) \\ & \mathrm{Var}(f_2) & \cdots & \mathrm{Cov}(f_2, f_m) \\ & \text{symm.} & \ddots & \vdots \\ & & & \mathrm{Var}(f_m) \end{bmatrix} \tag{1.62}$$

is computed as

$$\begin{aligned} \mathbf{S_f}_{m\times m} &= E\left[\left(\mathbf{f}(\mathbf{x}) - \mathbf{f}(\mathbf{x}^0)\right)_{m\times 1} \left(\mathbf{f}(\mathbf{x}) - \mathbf{f}(\mathbf{x}^0)\right)^{\mathrm{T}}_{1\times m}\right] \\ &= \left[\frac{\partial \mathbf{f}(\mathbf{x}^0)}{\partial \mathbf{x}}\right]_{m\times n} \mathbf{S_x}_{n\times n} \left[\frac{\partial \mathbf{f}(\mathbf{x}^0)}{\partial \mathbf{x}}\right]^{\mathrm{T}}_{n\times m} \end{aligned} \tag{1.63}$$

where $\mathbf{S_x}$ is the covariance matrix of the random variables $X_1, X_2, \ldots, X_n$.

It is observed that knowing the mean vector $\mathbf{x}^0$ and the covariance matrix $\mathbf{S_x}$ of the random variables $X_1, X_2, \ldots, X_n$ is all that is required to obtain the mean vector $E[\mathbf{f}]$ and the covariance matrix $\mathbf{S}_f$ of the random variables $f_1, f_2, \ldots, f_n$ (i.e. nothing had to be said of the PDF for $X_1, X_2, \ldots, X_n$). However, if a full probabilistic statement is to be made about the function $\mathbf{f}$, a PDF must be assumed for it; Eq. (1.56) can be used if the normal PDF is assumed.

Let us now try to calculate the mean and covariance matrix of a random vector $\mathbf{q} = \{q_1, q_2, \ldots, q_m\}$ which satisfies the system of algebraic equations

$$\mathbf{A}_{m\times m}\, \mathbf{q}_{m\times 1} = \mathbf{b}_{m\times 1} \tag{1.64}$$

in which the matrix $\mathbf{A}$ and vector $\mathbf{b}$ are explicitly given functions of the random variables $X_1, X_2, \ldots, X_n$ (thus making the vector $\mathbf{q}$ depend on $\mathbf{X}$ implicitly). The mean vector $\mathbf{x}^0_{n\times 1}$ and the covariance matrix $\mathbf{S}_{\mathbf{x}_{n\times n}}$ are assumed given.

Let us expand the vector $\mathbf{q}$ into the Taylor's series as in Eq. (1.59)

$$\mathbf{q}(\mathbf{x})_{m\times 1} = \mathbf{q}(\mathbf{x}^0)_{m\times 1} + \left[\frac{\partial \mathbf{q}(\mathbf{x}^0)}{\partial \mathbf{x}}\right]_{m\times n} (\mathbf{x} - \mathbf{x}^0)_{n\times 1} \tag{1.65}$$

By noting that $\partial \mathbf{q}(\mathbf{x}^0)/\partial \mathbf{x}$ is evaluated at the means of random variables we may use directly the results of Section 1.1.3 which yield

$$\begin{aligned} E[\mathbf{q}]_{m\times 1} &= \left\{ \mathbf{q}_1(\mathbf{x}^0) \;\; \mathbf{q}_2(\mathbf{x}^0) \; \ldots \; \mathbf{q}_m(\mathbf{x}^0) \right\}_{m\times 1} \\ &= \mathbf{A}^{-1}(\mathbf{x}^0)_{m\times m}\, \mathbf{b}(\mathbf{x}^0)_{m\times 1} \end{aligned} \tag{1.66}$$

and

$$\mathbf{S}_{\mathbf{q}_{m\times m}} = \left[\frac{\partial \mathbf{q}(\mathbf{x}^0)}{\partial \mathbf{x}}\right]_{m\times n} \mathbf{S}_{\mathbf{x}_{n\times n}} \left[\frac{\partial \mathbf{q}(\mathbf{x}^0)}{\partial \mathbf{x}}\right]^{\mathrm{T}}_{n\times m} \tag{1.67}$$

However, in the latter equation the matrix $[\partial \mathbf{q}(\mathbf{x}^0)/\partial \mathbf{x}]_{m\times n}$ remains to be computed. To do so we take derivatives of both sides of Eq. (1.64) with respect to the random variables X_i which gives the result

$$\mathbf{A}_{m\times m} \left[\frac{\partial \mathbf{q}}{\partial x_i}\right]_{m\times 1} + \left[\frac{\partial \mathbf{A}}{\partial x_i}\right]_{m\times m} \mathbf{q}_{m\times 1} = \left[\frac{\partial \mathbf{b}}{\partial x_i}\right]_{m\times 1} \qquad i = 1, 2, \ldots, n \tag{1.68}$$

or, when solved for $\partial \mathbf{q}/\partial x_i$, the result

$$\left[\frac{\partial \mathbf{q}}{\partial x_i}\right]_{m\times 1} = \mathbf{A}^{-1}_{m\times m} \left\{ \left[\frac{\partial \mathbf{b}}{\partial x_i}\right]_{m\times 1} - \left[\frac{\partial \mathbf{A}}{\partial x_i}\right]_{m\times m} \mathbf{q}_{m\times 1} \right\} \tag{1.69}$$

Repeating this operation for each of the random variables $X_1, X_2, \ldots, X_n$, and substituting $\mathbf{X} = \mathbf{x}^0$ in the resulting equations of the type (1.69) we obtain the $(m\mathrm{x}n)$-matrix $[\partial \mathbf{q}(\mathbf{x}^0)/\partial \mathbf{x}]$ needed to complete the computation of the covariance matrix $\mathbf{S}_{\mathbf{q}_{m\times m}}$.

1.1.6 Random Processes and Random Fields

The classical theory of probability is concerned with those random variables whose values do not depend upon time, or any other set of parameters. However, in the analysis of many phenomena it is necessary to study random variables which depend upon groups of parameters, most often including time.

Let us define a *random process*[1] $X(\tau)$ as an indexed family of random variables where the index (or parameter) τ belongs to some set T, $\tau \in T$.

The set T is called the parameter set or index set of the process. In many engineering problems the index τ will be time and the underlying intuitive notion will be that of a random variable developing in time.

It is apparent that there are at least two ways to view an arbitrary random process $X(\tau)$:

(i) as a set of random variables in the usual sense of that term defined at each fixed time instant $t \in T$, $X(t)$, $t \in T$,

(ii) as a set of sample functions of τ defined on the domain T, $x = x(\tau)$, $\tau \in T$, where a lower-case $x(.)$ is used to denote a specific realization of the process $X(\tau)$.

The integral function of distribution (i.e. the CDF) of $X(t_1)$ for a fixed $t_1 \in T$ has the form

$$P_1(x_1, t_1) = \Pr\left(X(t_1) \le x_1\right) \tag{1.70}$$

and, consequently, depends upon t_1 as well as upon x_1. The partial derivative of P_1 with respect to x_1

$$p_1(x_1, t_1) = \frac{\partial P_1(x_1, t_1)}{\partial x_1} \tag{1.71}$$

is called a one-dimensional probability density function which has the property, cf. Eq. (1.1)

$$\Pr\left(x_1 - \frac{\mathrm{d}x_1}{2} \le X(t_1) \le x_1 + \frac{\mathrm{d}x_1}{2}\right) = p_1(x_1, t_1)\,\mathrm{d}x_1 \tag{1.72}$$

For problems in which values of random functions at various time instants are considered in isolation from each other, the one-dimensional PDF is an adequate characteristic of the random process. However, it gives no impression of the development of the random process in time. Therefore, a multi-dimensional PDF is needed to characterize the process. The simplest of these – a two-dimensional PDF – is defined as

$$P_2(x_1, t_1; x_2, t_2) = \Pr\left(X(t_1) \le x_1; X(t_2) \le x_2\right) \tag{1.73}$$

[1]In most of the literature, the terms random process, random function and stochastic process are used synonymously.

for which the corresponding PDF is

$$p_2(x_1, t_1; x_2, t_2) = \frac{\partial^2 P_2(x_1, t_1; x_2, t_2)}{\partial x_1 \partial x_2} \tag{1.74}$$

Clearly

$$p_2(x_1, t_1; x_2, t_2) = p_2(x_2, t_2; x_1, t_1) \tag{1.75}$$

and

$$p_1(x_1, t_1) = \int_{-\infty}^{+\infty} p_2(x_1, t_1; x_2, t_2)\, dx_2 \tag{1.76}$$

In general, for an n-dimensional CDF we have

$$P_n(x_1, t_1; \ldots; x_n, t_n) = \Pr\left(X(t_1) \leq x_1; \ldots; X(t_n) \leq x_n\right) \tag{1.77}$$

and

$$p_n(x_1, t_1; \ldots; x_n, t_n) = \frac{\partial^n P_n(x_1, t_1; \ldots; x_n, t_n)}{\partial x_1 \ldots \partial x_n} \tag{1.78}$$

The multi-dimensional PDF is symmetric with respect to the pairs of arguments (x_i, t_i) and (x_j, t_j). If the function p_n is known, then all remaining functions (i.e. the ones with lesser numbers) can be determined. However, the n-dimensional PDF does not in general fully characterize the random process $X(\tau)$, because PDFs with higher number remain unknown.

There exist, however, certain random processes which are completely determined by the assignment of p_n with a finite n.

As in the case of random variables, some of the most important properties of a random process are characterized by its moments, particularly those of the first and the second order.

The m-th moment of a random process $X(\tau)$ at a given $t \in T$ is defined in terms of its first density function $p_1(x_1, t)$ by, cf. Eq. (1.21)

$$\mu_X^m(t) = E[X^m] = \int_{-\infty}^{+\infty} x^m\, p_1(x, t)\, dx \tag{1.79}$$

The first moment, $\mu_X^1(t)$, sometimes denoted by $\mu_X(t)$, is the mean of the stochastic process $X(\tau)$ at t, whereas the second moment, $\mu_X^2(t)$, is the mean square value of $X(\tau)$ at t.

The m-th central moment of a random process $X(\tau)$ at a given t is, cf. Eq. (1.22)

$$\bar{\mu}_X^m(t) = E[(X - x^0)^m] = \int_{-\infty}^{+\infty} (x - \mu_X^1)^m\, p_1(x, t)\, dx \tag{1.80}$$

The variance of $X(\tau)$ at t is $\bar{\mu}_X^2(t)$, which is commonly denoted by $\sigma_X^2(t)$ or $\sigma^2(t)$.

The moments of the process $X(\tau)$ defined in terms of its second density function $p_2(x_1, t_1; x_2, t_2)$ are, in effect, joint moments of two random variables. Thus, the joint moment $\mu_X^{mn}(t_1, t_2)$ of $X(\tau)$ at given t_1 and t_2 is defined by

$$\begin{aligned}\mu_X^{mn}(t_1,t_2) &= E[X^m(t_1)X^n(t_2)] \\ &= \int_{-\infty}^{+\infty}\int_{-\infty}^{+\infty} x_1^m x_2^n\, p_2(x_1,t_1;x_2,t_2)\,\mathrm{d}x_1\mathrm{d}x_2 \qquad (1.81)\end{aligned}$$

The value of $\mu_X^{11}(t_1, t_2)$ is an important measure of independence between $X(t_1)$ and $X(t_2)$ and as such it plays a central role in the theory of stochastic processes. The moment $\mu_X^{11}(t_1, t_2)$ is called the *correlation function* of the stochastic process $X(\tau)$; it clearly depends on both t_1 and t_2.

Instead of $\mu_X^{11}(t_1, t_2)$ the notation $\mu_{XX}(t_1, t_2)$ is sometimes used to emphasize that this function describes the correlation of $X(\tau)$ itself. To distinguish $\mu_{XX}(t_1, t_2)$ from the cross-correlation function

$$\mu_{XY}(t_1,t_2) = E[X(t_1)Y(t_2)] \qquad (1.82)$$

defined for two random variables which belong to two different stochastic processes $X(\tau)$ and $Y(\tau)$, the term of *autocorrelation function* is used for $\mu_{XX}(t_1, t_2)$.

Using a similar reasoning, the autocovariance function of a random process $X(\tau)$ is defined as

$$\begin{aligned}\mathrm{Cov}_{XX}(t_1,t_2) &= E\left[\left(X(t_1)-\mu_X(t_1)\right)\left(X(t_2)-\mu_X(t_2)\right)\right] \\ &= \int_{-\infty}^{+\infty}\int_{-\infty}^{+\infty}\left(x_1-\mu_X(t_1)\right)\left(x_2-\mu_X(t_2)\right)p_2(x_1,t_1;x_2,t_2)\,\mathrm{d}x_1\mathrm{d}x_2 \\ &= \mu_{XX}(t_1,t_2)-\mu_X(t_1)\,\mu_X(t_2) \qquad (1.83)\end{aligned}$$

For $t_1 = t_2 = t$, $\mathrm{Cov}_{XX}(t_1, t_2)$ becomes the variance of $X(\tau)$ at t, i.e.

$$\mathrm{Cov}_{XX}(t,t) = \mathrm{Var}_X(t) = \sigma_X^2(t) = \mu_X^2(t) - \left(\mu_X^1(t)\right)^2 \qquad (1.84)$$

The normalized autocovariance function is called the autocorrelation-coefficient function and is defined as

$$\varrho_{XY}(t_1,t_2) = \frac{\mathrm{Cov}_{XX}(t_1,t_2)}{\sigma_X(t_1)\,\sigma_X(t_2)} \qquad (1.85)$$

A natural generalization of the autocovariance concept to the case of two random processes $X(\tau)$ and $Y(\tau)$ is the *cross-covariance function* given by

$$\mathrm{Cov}_{XY}(t_1,t_2) = E\left[\left(X(t_1)-\mu_X(t_1)\right)\left(Y(t_2)-\mu_Y(t_2)\right)\right] \qquad (1.86)$$

The stochastic quantity $X(\tau)$ may consist of several components $X_i(\tau)$, $i = 1, 2, \ldots, n$, in which case it may be conveniently represented as a vector

$$\mathbf{X}(\tau) = \{\, X_1(\tau) \;\; X_2(\tau) \;\; \ldots \;\; X_n(\tau) \,\}$$

called an *n-dimensional vector random process.* All the definitions above can easily be extended to the case of vector random processes. The mean of a vector stochastic process then takes the form of a mean vector while the correlation and covariance functions become the correlation and covariance function matrices, respectively. For instance, in the correlation matrix

$$\mu_{X_i X_j}(t_1, t_2) = E\left[X_i(t_1)\, X_j(t_2)\right] \tag{1.87}$$

the diagonal entries represent autocorrelations, whereas the off-diagonal entries are cross-correlations.

The first two moment functions of the stochastic process $X(\tau)$ defined above play essentially the same role as the first two moments in describing the behaviour of a random variable. For instance, we have already seen that if a random variable is to have a normal distribution, its mean and variance describe the probabilistic distribution completely. Similarly, a Gaussian random process $X(\tau)$ is completely defined by its mean value function $\mu_X(\tau)$ and its autocovariance function $\mathrm{Cov}_{XX}(\tau, \rho)$, $\tau, \rho \in T$. This process belongs to the group of random processes whose moment functions of an order higher than two can be determined provided the first two moment functions are known. For an arbitrary random process an upper bound of the probability function can be estimated at some instant $\tau = t$, $t \in [t_1, t_2]$, by using the Chebyshev inequality (1.31). Hence, even when the information supplied by the first two moment functions is incomplete to characterize a random process, it is still the most important information about that process.

The random process $X(\tau)$, $\tau \in T$, is called *weakly stationary* (or *weakly homogeneous*) if

$$\begin{aligned}
&\text{(i)} && E[X(\tau)] = \mu_X^1 && \text{for all} \quad \tau \in T \\
&\text{(ii)} && \mathrm{Var}\left(X(\tau)\right) = \sigma_X^2 && \text{for all} \quad \tau \in T \\
&\text{and} && && \\
&\text{(iii)} && \mathrm{Cov}_{XX}(\tau, \tau + h) = \gamma(h) && \text{for all} \quad \tau \in T \\
& && && \text{and arbitrary } h \text{ such that } \tau + h \in T
\end{aligned} \tag{1.88}$$

i.e. these quantities do not depend on time τ (are time-invariant in the sense that their values are not affected by a translation of the time origin).

The random process $X(\tau)$, $\tau \in T$, is called *strongly stationary* (or *strongly homogeneous*) if all its possible moment functions are not affected by a shift in time.

Clearly, the statistical property of strong stationarity implies the property of weak stationary. Conversely, for stationarity in the weak sense to imply stationarity in the strong sense it is necessary that the CDF of $X(\tau)$ depends only on the parameters involved in the definition of the former. In view of the statement made above with regard to the Gaussian distributions it is seen that the stationary Gaussian process is one such example.

Referring to our comment made earlier in this section while discussing the multi-dimensional PDFs we just mention that the stationary Gaussian process may be completely characterized by two PDFs of the following form,[1] cf. Eqs. (1.52) and (1.57)

$$p(x) = \frac{1}{\sigma_X\sqrt{2\pi}} \exp\Big[-\frac{x^2}{2\sigma_X^2}\Big] \tag{1.89}$$

$$p(x_1; x_2, h) = \frac{1}{2\pi\sigma_X^2[1-\varrho_{12}^2(h)]^{\frac{1}{2}}} \exp\Big\{-\frac{x_1^2+x_2^2-2\varrho_{12}(h)\,x_1x_2}{2\sigma_X^2[1-\varrho_{12}^2(h)]}\Big\} \tag{1.90}$$

Unfortunately, the processes we shall be concerned with in this book will not, as a rule, be stationary.

Depending on the nature of the index set T, random processes may be divided into discrete parameter processes and continuous parameter processes. Since the random variable $X(t)$ for a given $t \in T$ can also be either discrete or continuous, we may distinguish the following four classes of random processes:

(i) discrete random processes with discrete parameter (the so-called *discrete random sequence*),

(ii) discrete random processes with continuous parameter (the so-called *discrete random process*),

(iii) continuous random processes with discrete parameter (the so-called *continuous random sequence*),

(iv) continuous random processes with continuous parameter (the so-called *continuous random processes*).

The concept of a *random field*, which we are about to introduce, is a generalization of the concept of a random process, for which the sole independent variable is time. The random field is represented by the symbol $X(\mathbf{x}, \tau)$ which designates a (random) function of the point in space $\mathbf{x}$ and of the time τ. Similarly as before, we understand the expression 'the random function of the argument $(\mathbf{x}, \tau)$' in the sense that at each point $(\mathbf{x}, \tau)$ of the space–time the value $X(\mathbf{x}, \tau)$ is a random variable and, consequently, cannot be exactly predicted.

If the function $X(\mathbf{x}, \tau)$ is defined as a vector function of the form

$$\mathbf{X}(\mathbf{x}, \tau) = \{\, X_1(\mathbf{x}, \tau) \;\; X_2(\mathbf{x}, \tau) \;\; \ldots \;\; X_n(\mathbf{x}, \tau) \,\}$$

then it is called the *n-variate vector random field*. When $\mathbf{x} = \{x_1, x_2, \ldots, x_m\}$ this random vector field is said to be m-dimensional and n-variate [106]. Note that the component functions $X_1(\mathbf{x}, \tau), X_2(\mathbf{x}, \tau), \ldots, X_n(\mathbf{x}, \tau)$ are in general statistically dependent. If both n and m are set equal to unity, then the random function $X(\mathbf{x}, \tau)$ is referred to as the one-dimensional and uni-variate random field.

[1]The zero mean value is assumed for simplicity.

By constructing all the finite-dimensional probability distributions of a random process it is possible to provide a complete description of this field. For practical reasons instead of these distributions one normally considers the corresponding statistical moments at different points. The space–time moments are defined as mean products of the values of the field at different points in space and at different time instants, whereas the space moments correspond to different points in space only.

In view of the formalism involved, any detailed discussion of the moment definitions appears notationally awkward; fortunately, such a discussion is also believed to be fully redundant since the only difference now would be the interpretation of the previously scalar argument τ as the vector $(\mathbf{x}, \tau)$ (for space–time moments) or as the vector $\mathbf{x}$ (for space moments).

To exemplify the notation let us just define the matrix of the covariance functions of two arguments $\xi_1 = (\mathbf{x}_1, \tau_1)$ and $\xi_2 = (\mathbf{x}_2, \tau_2)$ for a multi-dimensional and multi-variate random field $\mathbf{X}(\mathbf{x}, \tau)$ as

$$\mathrm{Cov}_{X_iY_j}(\xi_1, \xi_2) = E\left[\left(X_i(\xi_1) - \mu_{X_i}(\xi_1)\right)\left(X_j(\xi_2) - \mu_{X_j}(\xi_2)\right)\right] \tag{1.91}$$

We may note in closing that the terminology employed above in (i)–(iv) can be extended in the case of random fields by distinguishing random sequences, for which observations are made at discrete (usually equally spaced but not necessarily so) points on a time axis from lattice processes, for which observations are made at the nodes of a lattice in space.

We may also quote a generalization of the notion of random field stationarity to the case of random fields:

> A random field $\mathbf{X}(\xi)$ is called homogeneous if all the joint cumulative distribution functions remain the same when the set of locations $\xi_1, \xi_2, \ldots, \xi_n$ is translated (but not rotated!) in the parameter space.

This implies that all the probabilities depend only on the relative, not the absolute, locations of the points $\xi_1, \xi_2, \ldots, \xi_n$.

1.2 Discrete Representations of Data

1.2.1 Dirac Delta Distribution

To introduce the concept of the Dirac delta distribution let us first consider a function expressed by the equation

$$f(x) = \begin{cases} \frac{1}{w} & \text{for} \quad x \in \left[-\frac{w}{2}, \frac{w}{2}\right] \\ 0 & \text{for} \quad x \notin \left[-\frac{w}{2}, \frac{w}{2}\right] \end{cases} \tag{1.92}$$

The rectangular-shaped function $f(x)$ has the magnitude $1/w$ and base width w; and the area under the function is equal to unity. Let the base of $f(x)$ decrease with a corresponding increase in amplitude so as to maintain the unit area. Taking the limit of $f(x)$ at $w \to 0$ leads to

$$\delta(x) = \lim_{w \to 0} f(x) = \begin{cases} \infty & \text{for} \quad x = 0 \\ 0 & \text{for} \quad x \neq 0 \end{cases} \tag{1.93}$$

but the integral of the function as given by Eq. (1.92) remains unity, that is

$$\int_{-\epsilon}^{+\epsilon} \delta(x)\,\mathrm{d}x = \lim_{w \to 0} \int_{-w/2}^{+w/2} \frac{1}{w}\,\mathrm{d}x = \lim_{w \to 0} \left(\frac{w}{w}\right) = 1 \tag{1.94}$$

where ϵ is an arbitrarily small value, $\epsilon > 0$. With a shift of the variable x Eqs. (1.93) and (1.94) read

$$\delta(x - x_0) = \begin{cases} \infty & \text{for} \quad x = x_0 \\ 0 & \text{for} \quad x \neq x_0 \end{cases} \tag{1.95}$$

$$\int_{x_0-\epsilon}^{x_0+\epsilon} \delta(x - x_0)\,\mathrm{d}x = 1 \tag{1.96}$$

where x_0 is the shift from the origin.

Limiting functions of this type are called *Dirac delta distributions* and are denoted by $\delta(x)$. The Dirac delta measure $\delta(x)$ is not a function in the conventional sense, since its value cannot properly be assigned for every x. Hence, a concept of distribution [42] is required to rigorously define this 'function'. In terms of its properties, $\delta(x)$ is often defined as the one which satisfies Eqs. (1.95) and (1.96) and the condition

$$\begin{aligned} \int_{-\infty}^{+\infty} g(x)\,\delta(x - x_0)\,\mathrm{d}x &= \int_{-\infty}^{+\infty} g(x)\,\delta(x_0 - x)\,\mathrm{d}x \\ &= \int_{-\infty}^{+\infty} g(x_0 - x)\,\delta(x)\,\mathrm{d}x \;=\; g(x_0) \end{aligned} \tag{1.97}$$

Here it is assumed that the test function $g(x)$ is continuous at $x = x_0$. In contrast to the concept of ordinary functions, here we are only interested in the relationship between the values of the integration corresponding to different test functions; and it may not be needed to know all values of $\delta(x)$. (In Eq. (1.97) the value of the integration is simply the value of the test function at $x = x_0$.) It is worth noting that $\delta(x)$ is an even distribution.

On the basis of the above discussion let us now consider another function whose limit exists only in the distribution sense. As an example we take the PDF of a normal random variable X, cf. Eq. (1.52), with zero mean. If the standard deviation σ_X is small, then the probability density is concentrated about the zero mean. As σ_X approaches zero the PDF considered becomes

$$\lim_{\sigma_X \to 0} p(x) = \lim_{\sigma_X \to 0} \frac{1}{\sigma_X \sqrt{2\pi}} \exp\Big[-\frac{x^2}{2\sigma_X^2}\Big] = \delta(x) \tag{1.98}$$

while, by using Eq. (1.3–iii), the probability of $X = 0$ has the limiting value 1

$$\Pr(X=0) = \lim_{\sigma_X \to 0} \Pr(-\sigma_X \leq X \leq +\sigma_X) = \int_{-\epsilon}^{+\epsilon} \delta(x)\, dx = 1 \tag{1.99}$$

The concept of the Dirac delta distribution is very useful in many engineering problems. In structural dynamics the delta distribution is known as the *unit impulse*, for instance; the physical interpretation of this term follows directly from Eq. (1.92).

In the field of computational mechanics, which hinges so much on the use of digital computers, it is natural to consider discrete systems described in terms of sequences of numbers rather than continuous systems described by means of continuous functions. In the terminology of system analysis, input data are often referred to as the *input signal* and the system response as the *output signal.* In discrete systems the signals are defined only for discrete values of an ordering variable ξ denoted by $\xi_{(k)}$, $k = 0, 1, \ldots$. In such cases, ξ is said to be a *discrete variable* and the signals are called *discrete signals.* Discrete signals do not arise naturally but are the result of discretization of continuous signals. Conversion of a continuous signal to a discrete one is carried out at the sampling points ξ_k, which are usually (but not necessarily) taken at equal intervals, so that

$$\xi_{(k)} = k\,\Delta\xi \qquad\qquad k = 0, 1, \ldots \tag{1.100}$$

where $\Delta\xi$ is called the *sampling interval.* Output signals of discrete systems can be regarded as a sequence of sample values $f(\xi_{(k)})$ resulting from the continuous output signal $f(\xi)$. Using the concept of the so-called *zero-order hold* we may assume [43,86]

$$f_{(k)} = f(\xi_{(k)}) = f(k\Delta\xi) \qquad\qquad \xi_{(k)}) \in \left[k\Delta\xi, (k+1)\Delta\xi\right) \tag{1.101}$$

which generates a continuous function in the form of a staircase.

It is worthwhile at this point to introduce the discrete form of the delta distribution known as the Kronecker delta

$$\delta(i-j) = \begin{cases} 1 & \text{for } i = j \\ 0 & \text{for } i \neq j \end{cases} \qquad\qquad i, j = 0, 1, \ldots \tag{1.102}$$

Assuming that the function $f(\xi)$ is continuous at points $\xi_{(k)} = k\Delta\xi$, $k = 0, 1, \ldots$, a sample of $f(\xi)$ evaluated at $\xi_1 = \Delta\xi$ is given by

$$f_{(1)} = f(\xi)\,\delta(\xi - \Delta\xi) = f(\Delta\xi)\,\delta(\xi - \Delta\xi) \tag{1.103}$$

More generally, the discrete signal $f_{(k)}$ can be represented by the series

$$f_k = f(\xi_{(k)}) = \sum_{l=0}^{+\infty} f(l)\,\delta(k-l) \tag{1.104}$$

which may be conveniently employed in the analysis of discrete systems.

1.2.2 Discrete Fourier Transform

A Fourier series may be defined as an expansion of a function $x(\tau)$ in a series of sines and cosines such as

$$x(\tau) = \tfrac{1}{2}a_0 + \sum_{n=1}^{+\infty}\left(a_n \cos n\tau + b_n \sin n\tau\right) \tag{1.105}$$

The condition imposed on $x(\tau)$ to make this representation valid is that $x(\tau)$ is piece-wise regular, i.e. it has only a finite number of finite discontinuities and only a finite number of extreme values.[1] The so called Sturm–Liouville theory [59,68,108] guarantees the validity of Eq. (1.105) and, by use of the orthogonality relations for trigonometric functions on the interval $[0, 2\pi]$, for instance, allows us to compute the expansion coefficients as

$$\begin{aligned} a_n &= \frac{1}{\pi}\int_0^{2\pi} x(\xi)\cos n\xi \,\mathrm{d}\xi \qquad n = 0, 1, \ldots \\ b_n &= \frac{1}{\pi}\int_0^{2\pi} x(\xi)\sin n\xi \,\mathrm{d}\xi \qquad n = 1, 2, \ldots \end{aligned} \tag{1.106}$$

By using Euler's formula, Eq. (1.105) may be rewritten in exponential form as

$$x(\tau) = \sum_{-\infty}^{+\infty} c_n \exp(\mathrm{i}\, n\tau) \tag{1.107}$$

provided

$$\begin{aligned} c_0 &= \tfrac{1}{2}a_0 \\ c_n &= \tfrac{1}{2}(a_n \mp \mathrm{i}\, b_n) \qquad n = \pm 1, \pm 2, \ldots \end{aligned} \tag{1.108}$$

In other words

$$c_n = \frac{1}{2\pi}\int_{-\infty}^{+\infty} x(\xi)\exp(-\mathrm{i}\, n\xi)\,\mathrm{d}\xi \qquad n = 0, \pm 1, \ldots \tag{1.109}$$

Usually, one is concerned with finding the coefficients of the Fourier expansion of a known function; occasionally, one may wish to reverse this process and determine the function represented by a given Fourier series.

We may note that substituting Eqs. (1.106) into Eq. (1.105) leads to the Fourier expansion in the form

$$\begin{aligned} x(\tau) &= \frac{1}{2\pi}\int_0^{2\pi} x(\xi)\,\mathrm{d}\xi + \frac{1}{\pi}\sum_{n=1}^{+\infty}\left[\left(\int_0^{2\pi} x(\xi)\cos n\xi\,\mathrm{d}\xi\right)\cos n\tau\right. \\ &\qquad \left. + \left(\int_0^{2\pi} x(\xi)\sin n\xi\,\mathrm{d}\xi\right)\sin n\tau\right] \\ &= \frac{1}{2\pi}\int_0^{2\pi} x(\xi)\,\mathrm{d}\xi + \frac{1}{\pi}\sum_{n=1}^{+\infty}\int_0^{2\pi} x(\xi)\cos n(\xi - \tau)\,\mathrm{d}\xi \end{aligned} \tag{1.110}$$

[1]These requirements, known as the Dirichlet conditions, are sufficient but not necessary [31,33].

the first (constant) term being the average value of $x(\tau)$ over the interval $[0, 2\pi]$.

The restriction to an interval of length 2π may easily be relaxed since for $x(\tau)$ which is periodic with a period of $2L$ we may write

$$x(\tau) = \frac{1}{2}a_0 + \sum_{n=1}^{+\infty} \left(a_n \cos \frac{n\pi}{L}\tau + b_n \sin \frac{n\pi}{L}\tau \right) \tag{1.111}$$

with

$$\begin{aligned} a_n &= \frac{1}{L}\int_{-L}^{L} x(\xi) \cos \frac{n\pi}{L}\xi \, d\xi \qquad n = 0, 1, \ldots \\ b_n &= \frac{1}{L}\int_{-L}^{L} x(\xi) \sin \frac{n\pi}{L}\xi \, d\xi \qquad n = 1, 2, \ldots \end{aligned} \tag{1.112}$$

The choice of the symmetric interval $[-L, L]$ is not essential since for $x(\tau)$ periodic with a period of $2L$ any interval $[\tau_0, \tau_0 + 2L]$ will do.

The general advantages of the Fourier representation are that it may represent certain, possibly discontinuous, functions over a limited range $[0, 2\pi]$, $[-L, L]$ and so on, or over the infinite interval $(-\infty, +\infty)$ provided the function is periodic.

Furthermore, the Fourier series may also be useful in representing non-periodic functions over the infinite range. In order to see this we note that the Fourier series resulting from Eqs. (1.111) and (1.112) is

$$\begin{aligned} x(\tau) &= \frac{1}{2L}\int_{-L}^{L} x(\xi)\, d\xi + \frac{1}{L}\sum_{n=1}^{+\infty} \left[\left(\int_{-L}^{L} x(\xi) \cos \frac{n\pi}{L}\xi \, d\xi \right) \cos \frac{n\pi}{L}\tau \right. \\ &\quad \left. + \left(\int_{-L}^{L} x(\xi) \sin \frac{n\pi}{L}\xi \, d\xi \right) \sin \frac{n\pi}{L}\tau \right] \\ &= \frac{1}{2L}\int_{-L}^{L} x(\xi)\, d\xi + \frac{1}{L}\sum_{n=1}^{+\infty} \int_{-L}^{L} x(\xi) \cos \frac{n\pi}{L}(\xi - \tau)\, d\xi \end{aligned} \tag{1.113}$$

Let us transform (descriptively rather than rigorously, cf. [19,74,100]) the finite interval $[-L, L]$ into the infinite interval $(-\infty, +\infty)$ by letting the parameter L approach infinity, i.e. by setting [2,98,111]

$$\frac{n\pi}{L} = \omega\,, \qquad \frac{\pi}{L} = \Delta\omega \qquad \text{with} \qquad L \to \infty \tag{1.114}$$

We then obtain from Eq. (1.113)

$$x(\tau) = \frac{1}{\pi}\int_{0}^{+\infty} d\omega \int_{-\infty}^{+\infty} x(\xi) \cos \omega(\xi - \tau)\, d\xi \tag{1.115}$$

which in known as the Fourier integral. It is subjected to the conditions that $x(\tau)$ is (i) piece-wise continuous, (ii) differentiable and (iii) absolutely integrable, i.e.

$$\int_{-\infty}^{+\infty} |\, x(\tau)\,|\, d\tau < \infty \tag{1.116}$$

(On account of the third condition the first term in Eq. (1.113) has vanished while deriving Eq. (1.115).)

Since $\cos\omega(\xi-\tau)$ is an even function of ω, Eq. (1.115) may be rewritten as

$$x(\tau) = \frac{1}{2\pi}\int_{-\infty}^{+\infty}\int_{-\infty}^{+\infty} x(\xi)\,\cos\omega(\xi-\tau)\,\mathrm{d}\xi\mathrm{d}\omega \tag{1.117}$$

By noting that $\sin\omega(\xi-\tau)$ is an odd function of ω, i.e.

$$\frac{1}{2\pi}\int_{-\infty}^{+\infty}\int_{-\infty}^{+\infty} x(\xi)\,\sin\omega(\xi-\tau)\,\mathrm{d}\xi\mathrm{d}\omega = 0 \tag{1.118}$$

multiplying the last equation by the imaginary unit i and subtracting sideways Eq. (1.117) from Eq. (1.118) we obtain the Fourier integral in exponential form:

$$x(\tau) = \frac{1}{2\pi}\int_{-\infty}^{+\infty}\left[\int_{-\infty}^{+\infty} x(\xi)\,\exp(-\mathrm{i}\,\omega\xi)\,\mathrm{d}\xi\right]\exp(\mathrm{i}\,\omega\tau)\,\mathrm{d}\omega \tag{1.119}$$

or

$$x(\tau) = \int_{-\infty}^{+\infty} X(\omega)\,\exp(\mathrm{i}\,\omega\tau)\,\mathrm{d}\omega \tag{1.120}$$

with, cf. Eq. (1.109)

$$X(\omega) = \frac{1}{2\pi}\int_{-\infty}^{+\infty} x(\xi)\,\exp(-\mathrm{i}\,\omega\xi)\,\mathrm{d}\xi \tag{1.121}$$

By changing the order of integration in Eq. (1.119) we arrive at the expression

$$x(\tau) = \int_{-\infty}^{+\infty} x(\xi)\left\{\frac{1}{2\pi}\int_{-\infty}^{+\infty}\exp\left[\mathrm{i}\omega(\tau-\xi)\right]\,\mathrm{d}\omega\right\}\mathrm{d}\xi \tag{1.122}$$

with the integral in brackets 'acting' on $x(\xi)$ exactly as the Dirac delta distribution does

$$\frac{1}{2\pi}\int_{-\infty}^{+\infty}\exp\left[\mathrm{i}\omega(\tau-\xi)\right]\,\mathrm{d}\omega = \delta(\tau-\xi) \tag{1.123}$$

so that we may write, cf. Eq. (1.97)

$$x(\tau) = \int_{-\infty}^{+\infty} x(\xi)\,\delta(\tau-\xi)\,\mathrm{d}\xi \tag{1.124}$$

Some more general comments are now in order. In the field of applied mathematics we frequently encounter pairs of functions related by an expression of the following general form

$$X(f) = \int_{a}^{b} x(\tau)\,K(f,\tau)\,\mathrm{d}\tau \tag{1.125}$$

The function $X(f)$ is called the *integral transform* of the function $x(\tau)$ by the kernel $K(f,\tau)$. The operation may also be described as mapping a function $x(\tau)$ in τ-space into another function $X(f)$ in f-space.[1]

One of the most useful of the infinite number of possible integral transforms is the *direct Fourier transform* (or *Fourier analysis*) given by

$$X(f) = \int_{-\infty}^{+\infty} x(\tau)\,\exp(-\mathrm{i}\,2\pi f\tau)\,\mathrm{d}\tau \tag{1.126}$$

with its *inverse Fourier transform* (or *Fourier synthesis*) defined as

$$x(\tau) = \int_{-\infty}^{+\infty} X(f)\,\exp(\mathrm{i}\,2\pi\tau f)\,\mathrm{d}f \tag{1.127}$$

The independent variable f is as a rule expressed in units of Hertz (Hz). Occasionally it is convenient to use another related independent variable ω defined by

$$\omega = 2\pi f \tag{1.128}$$

so that Eqs. (1.126) and (1.127) become

$$X(\omega) = a_1\int_{-\infty}^{+\infty} x(\tau)\,\exp(-\mathrm{i}\,\omega\tau)\,\mathrm{d}\tau \tag{1.129}$$

$$x(\tau) = a_2\int_{-\infty}^{+\infty} X(\omega)\,\exp(\mathrm{i}\,\omega\tau)\,\mathrm{d}\omega \tag{1.130}$$

Eqs. (1.129) and (1.130) are almost but not quite symmetrical differing in the sign of the imaginary unit i and the factors a_1 and a_2. The specific values of a_1 and a_2 are a matter of choice provided their product equals $1/(2\pi)$. Some authors set $a_1=1/(2\pi)$ and $a_2=1$ [16,20,86], others assign $a_1=a_2=1/(2\pi)^{1/2}$, or use $a_1=1$ and $a_2=1/(2\pi)$, [22]. It will turn out convenient to use at some places in this book Eqs. (1.129) and (1.130) and at some other places Eqs. (1.126) and (1.127), depending on the context.

We may observe that the Fourier transform pair (1.129) and (1.130) is exactly the one we have derived before when considering the Fourier integral, Eqs. (1.120) and (1.121). We further note that once the Fourier transform $X(\omega)$ of a function $x(\tau)$ has been computed, it is easy to compute the Fourier transform $X_1(\omega)$ of the derivative $\mathrm{d}x(\tau)/\mathrm{d}\tau$, since

$$X_1(\omega) = a_1\int_{-\infty}^{+\infty} \frac{\mathrm{d}x(\tau)}{\mathrm{d}\tau}\,\exp(-\mathrm{i}\,\omega\tau)\,\mathrm{d}\tau \tag{1.131}$$

which, upon integrating by parts, results in

$$X_1(\omega) = a_1\exp(-\mathrm{i}\,\omega\tau)x(\tau)\Big|_{\tau=-\infty}^{\tau=+\infty} + a_1\mathrm{i}\,\omega\int_{-\infty}^{+\infty} x(\tau)\exp(-\mathrm{i}\,\omega\tau)\,\mathrm{d}\tau \tag{1.132}$$

[1]In structural dynamics, for instance, this interpretation takes on physical significance as a link between the time and frequency domains.

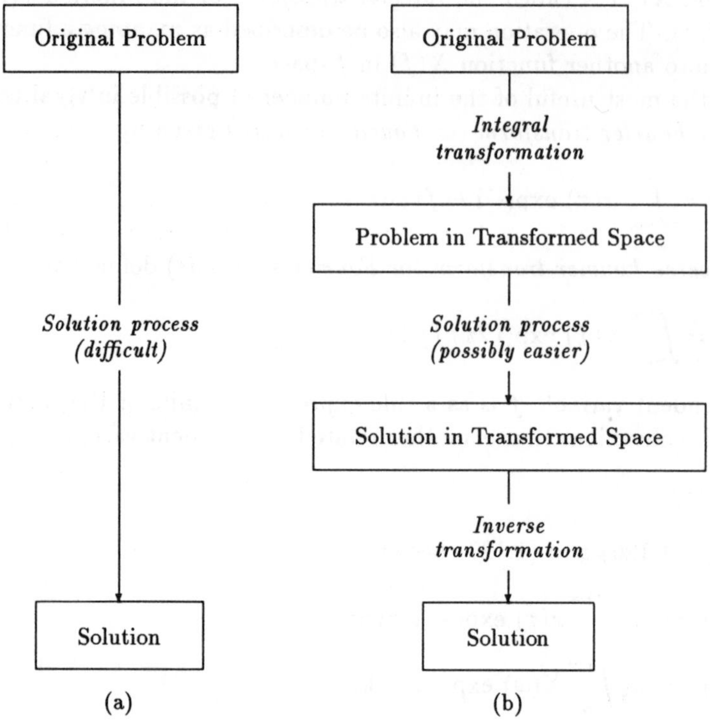

Figure 1.2 (a) Direct solution, (b) solution by using Fourier transform.

Since $x(\tau)$ vanishes as $\tau \to \pm\infty$ (which is necessary for the Fourier transform of $x(\tau)$ to exist) we arrive at

$$X_1(\omega) = \mathrm{i}\,\omega\, X(\omega) \tag{1.133}$$

i.e. the transform of the derivative is $\mathrm{i}\omega$ times the transform of the original function. This result is readily generalized to the case of the n-th derivative of $x(\tau)$, yielding the relation

$$X_n(\omega) = (\mathrm{i}\,\omega)^n\, X(\omega) \tag{1.134}$$

provided all the integrand parts vanish as $\tau \to \pm\infty$. Eq. (1.134) is clearly at the core of the Fourier transform usefulness for solving differential equations as it enables us to replace the direct solution process shown in Fig. 1.2(a) by the sequence of steps shown in Fig. 1.2(b) which are frequently accomplished much more easily.

Fourier transform pairs involve integrations over different domains (the time and the frequency domains in structural dynamics, for instance). Except for some rather simple functions, evaluation of these integrals may cause serious difficulties. Therefore, it is highly desirable to develop a procedure for evaluating

Fourier transforms numerically. Such a procedure for modifying Fourier transform pairs so as to permit their efficient computation on a digital computer is called the *discrete Fourier transform.*

Theoretically, the infinite-range transform $X(f)$ of Eq. (1.126) will not exist for any $x(\tau)$ which represents a random quantity (even when $x(\tau)$ is a stationary signal). This is because $x(\tau)$ may not be absolutely integrable, i.e. may not satisfy the condition (1.116). In practice we cannot record any $x(\tau)$ in the infinite limit. Instead, by restricting the limit to a finite time interval $[0,T]$ the finite-range Fourier transform will generally exist[1] as defined by

$$X(f) = X(f,T) = \int_0^T x(\tau)\exp(-\mathrm{i}\,2\pi f\tau)\,\mathrm{d}\tau \tag{1.135}$$

We note that in this case all the values of $x(\tau)$, $\tau \notin [0,T]$, are presumed to be zero. Next, suppose that the direct transform $X(f,T)$ exists only over the frequency domain $[-F,F]$ and takes on zero values elsewhere[2] so that the inverse transform $x(t,F)$ reads

$$x(\tau,F) = \int_{-F}^{F} X(f)\exp(\mathrm{i}\,2\pi f\tau)\,\mathrm{d}f \tag{1.136}$$

Assume that the discrete signal sample values resulting from a continuous time signal $x(\tau)$, $\tau \in [0,T]$, are recorded at L equally spaced points $t_0, t_1, \ldots, t_{L-1}$, a sampling time interval Δt apart, and all the values outside the sample length $[0,T]$ are equal to zero. That is

$$\begin{aligned} x_k &= \begin{cases} x(t_k) & \text{for} \quad k = 0,1,\ldots,L-1 \\ 0 & \text{for} \quad k < 0 \quad \text{and} \quad k \geq L \end{cases} \\ t_k &= k\,\Delta t \\ T &= L\,\Delta t \\ \Delta t &= t_{k+1} - t_k \end{aligned} \tag{1.137}$$

Since the time domain extends from 0 to T, the smallest (fundamental) frequency corresponding to the largest period that can be considered in the analysis equals $1/T$; and this can be used to define the sampling frequency interval Δf for discretization of $X(f)$, i.e.

$$\Delta f = \frac{1}{T} \tag{1.138}$$

The continuous frequency signal $X(f,T)$, $f \in [-F,F]$, can then be sampled at L equally spaced points $f_0, f_1, \ldots, f_{L-1}$, a sampling frequency interval Δf apart, as

[1]The finite-range Fourier transform $X(f,T)$ *always* exists for finite length records of stationary random data $x(\tau)$, $\tau \in [0,T]$, [15].

[2]In a strict sense, the dual assumption of the simultaneous existence of $x(\tau)$, $\tau \in [0,T]$, and $X(f)$, $f \in [-F,F]$, is not possible because of the uncertainty principle [14,16].

$$
\begin{aligned}
X_j &= \begin{cases} X(f_j, T) & \text{for} \quad j = 0, 1, \ldots, L-1 \\ 0 & \text{for} \quad j < 0 \quad \text{and} \quad j \geq L \end{cases} \\
f_j &= j\,\Delta f = \frac{j}{T} = \frac{j}{L\Delta t} \\
F &= \tfrac{1}{2} L\,\Delta f = \frac{1}{2\Delta t}
\end{aligned}
\tag{1.139}
$$

where the frequency bandwidth F is known as the *Nyquist cutoff frequency*,[1] which is the highest frequency that can be numerically 'measured' corresponding to a given sampling time interval Δt.

By using the time-frequency discretization expressed by Eqs. (1.137) and (1.139) we can approximate the finite-range direct transform (1.135) via a sequence of discrete signals as

$$
\begin{aligned}
X_j &= \sum_{k=0}^{L-1} x(t_k) \exp(-\mathrm{i}\, 2\pi f_j t_k)\, \Delta t = \Delta t \sum_{k=0}^{L-1} x_k \exp(-\mathrm{i}\, 2\pi\, j \Delta f\, k \Delta t) \\
&= \Delta t \sum_{k=0}^{L-1} x_k \exp\Big(-\mathrm{i} \frac{2\pi}{L} jk\Big) \qquad j = 0, 1, \ldots, L-1
\end{aligned}
\tag{1.140}
$$

and the finite-range inverse transform (1.136) as

$$
\begin{aligned}
x_k &= \sum_{j=0}^{L-1} X(f_j) \exp(\mathrm{i}\, 2\pi f_j t_k)\, \Delta f \\
&= \Delta f \sum_{j=0}^{L-1} X_j \exp\Big(\mathrm{i} \frac{2\pi}{L} jk\Big) \qquad k = 0, 1, \ldots, L-1
\end{aligned}
\tag{1.141}
$$

For the sake of computational convenience the factors Δt and Δf in the sums (1.140) and (1.141) may be symmetrized to obtain

$$
X_j := \frac{1}{\Delta t} X_j = \sum_{k=0}^{L-1} x_k \exp\Big(-\mathrm{i} \frac{2\pi}{L} jk\Big) \tag{1.142}
$$

$$
x_k := \frac{1}{\Delta f} x_k = \sum_{j=0}^{L-1} X_j \exp\Big(\mathrm{i} \frac{2\pi}{L} jk\Big) \tag{1.143}
$$

where Δt and Δf have been simply included with X_j and x_k to have the scale factor of unity before the summations are performed. This form of the finite-range Fourier transform enables us to write one single procedure for both the direct and inverse Fourier transform.

It is emphasized that the accuracy of numerical solutions depends to a large degree on the choice of the sampling time interval Δt (in the Fourier analysis)

---

[1]The cutoff frequency may also be determined by the following observation. If the sampling time interval Δt is given, then the sampling rate equals $1/\Delta t$ sample points per unit time. Since at least two sample points per cycle are needed to define a frequency component in the continuous signal $x(\tau)$, the maximum value of the frequency that can be obtained by sampling at a rate of $1/\Delta t$ sample points per unit time is $1/(2\Delta t)$ cycles per unit time.

and of the sampling frequency interval Δf (in the Fourier synthesis). In the context of the conversion from an integral to a finite sum an important problem of digitization is to select an appropriate sampling interval so that a continuous signal can be discretized by a sequence of L equally spaced sample values with no significant loss of information.

In the Fourier analysis, selecting a value of Δt too small leads to redundant data and unnecessarily increases the computational effort. On the other hand, if Δt is too large, confusion between low and high frequency components in the original signal $x(\tau)$, $\tau \in [0,T]$ may occur; frequencies higher than the Nyquist frequency F may be folded back into the frequency band $[0,F]$. This effect is known as *aliasing* or *folding*. In contrast to analog data processing, aliasing in numerical processing is inherent and may constitute serious errors. To be specific, let us consider the exponential function $\exp(-\mathrm{i}\,2\pi\tau f)$, $f \in [0,F]$, at $\tau = \Delta t$. Noting Eq. $(1.139)_3$ we write

$$\begin{aligned}\exp\Big[-\mathrm{i}\,2\pi\Delta t(2nF+f)\Big] &= \exp\Big[-\mathrm{i}\,2\pi\frac{1}{2F}(2nF+f)\Big] \\ &= \exp(-\mathrm{i}\,2n\pi)\,\exp\Big(-\mathrm{i}\frac{\pi}{F}f\Big) = \exp(-\mathrm{i}\,2\pi\Delta t\,f) \\ &\qquad\qquad n = 1,2,\ldots\end{aligned} \tag{1.144}$$

Clearly, when data are sampled at intervals Δt this function does not distinguish between a frequency f and frequencies $2nF + f$, n being an integer number. In other words, data at frequencies $2nF+f$ give the same information as do those at frequency f. To handle this aliasing problem by sampling a continuous time signal $x(\tau)$ defined in the time domain $[0,T]$ and in the frequency domain $[0,F]$ at equally spaced points Δt apart, the number of sampling time intervals required to avoid aliasing is

$$L_{(\Delta t)} \geq \frac{T}{\Delta t} = \frac{T}{1/(2F)} = 2FT \tag{1.145}$$

Similarly, in the Fourier synthesis, by sampling a continuous frequency signal $X(f)$ at equally spaced points Δf apart, the number of sampling frequency intervals required is

$$L_{(\Delta f)} \geq \frac{2F}{\Delta f} = \frac{2F}{1/T} = 2FT \tag{1.146}$$

Thus, the same number of sampling intervals is required for the discretization of continuous signals on the time and frequency scales.

To save low computational cost, the practical approach to eliminate aliasing in digitization is to filter the original data prior to sampling so that information above a maximum frequency of interest is no longer involved. The filtered data can then be used as input data for the discrete Fourier transform problem considered. The reader who is interested in filtering techniques is referred to [16,19,55,60,98,100], for instance.

1.2.3 Base-Two Fast Fourier Transform

If no attention is paid to computational cost, Eqs. (1.140) and (1.141) can be used as the basis for numerical analysis. However, summations in these equations may be considerably simplified if we note that: (i) the exponential functions are harmonic, (ii) only integer values of the discrete variables j and k are employed in the exponentials and the product $j \times k$ is repeated when the sums are carried out, and (iii) the features pointed out in (i) and (ii) are extended over the range L^2. The procedure known as the *fast Fourier transform* (FFT) takes advantage of these facts. The technique is so efficient that it has made the frequency-domain analysis competitive with the traditional time-domain approach. In many areas of computational engineering such as time-series and spectral analyses, signal and system problems, etc. this powerful algorithm has caused a complete change of attitude toward what can be done by using Fourier or Laplace methods.

To see how the FFT procedure works let us define

$$W(n) = \exp\Big(-\mathrm{i}\,\frac{2\pi}{L}\,n\Big) \tag{1.147}$$

where n is an integer number, and note that

$$\begin{aligned} &\text{(i)} \qquad W(L) = 1 \\ &\text{(ii)} \qquad W(n_1 + n_2) = W(n_1)\, W(n_2) \end{aligned} \tag{1.148}$$

Eq. (1.142) then becomes

$$X_j = \sum_{k=0}^{L-1} x_k\, W(jk) \qquad\qquad j = 0, 1, \ldots, L-1 \tag{1.149}$$

Eq. (1.149) is the Fourier transform of x_k when x_k is a discrete signal sequence of sampling length L. Thus the computations require a total of L^2 complex multiply–add operations. (One complex multiply–add is equivalent to four real multiply–adds.) We shall consider the case when the sequence x_k has a sampling length of the form $L = 2^\beta$, β being an integer, since the FFT may then be shown to be the most efficient.

Let us observe that when $L = 2^\beta$ every integer index j, $0 \le j \le L-1$, has a unique representation in the binary form

$$j = (j_{\beta-1}, j_{\beta-2}, \ldots, j_0) = \sum_{\alpha=0}^{\beta-1} 2^\alpha j_\alpha \tag{1.150}$$

where j_α is the remainder in the successive division of j by 2. In some computer languages, Fortran or C for instance, we can write the expression of j_α as

$$j_\alpha = \mathrm{mod}\,(j\, \underbrace{/2/2\ldots/2}_{(\alpha+1)\text{-fold}}) \tag{1.151}$$

Note that j_α takes only the value 0 or 1. As an example, with $L = 16 = 2^4$ every integer number j, $0 \le j \le 15$, can be uniquely represented by

$$\begin{aligned} j_{\text{decimal}} &= 2^3 \times j_3 + 2^2 \times j_2 + 2^1 \times j_1 + 2^0 \times j_0 \\ &= j_3 j_2 j_1 j_{0\,\text{binary}} = (j_3, j_2, j_1, j_0) \end{aligned} \tag{1.152}$$

and when $j = 13$, for instance, we get

$$\begin{aligned} 13_{\text{decimal}} &= 2^3 \times 1 + 2^2 \times 1 + 2^1 \times 0 + 2^0 \times 1 \\ &= 1101_{\text{binary}} = (1, 1, 0, 1) \end{aligned} \tag{1.153}$$

Using Eq. (1.150), Eq. (1.149) can be expressed in the binary form

$$\begin{aligned} X(j_{\beta-1}, j_{\beta-2}, \ldots, j_0) = \sum_{k_0=0}^{1} \sum_{k_1=0}^{1} \cdots \sum_{k_{\beta-1}=0}^{1} x\,(k_{\beta-1}, k_{\beta-2}, \ldots, k_0) \\ \times W\Big(\sum_{\alpha=0}^{\beta-1} 2^\alpha j_\alpha \sum_{\alpha=0}^{\beta-1} 2^\alpha k_\alpha\Big) \end{aligned} \tag{1.154}$$

From Eq. (1.148–ii), $W(.)$ in Eq. (1.154) can be rewritten as

$$\begin{aligned} W\Big(\sum_{\alpha=0}^{\beta-1} 2^\alpha j_\alpha \sum_{\alpha=0}^{\beta-1} 2^\alpha k_\alpha\Big) = W\Big(2^{\beta-1} k_{\beta-1} \sum_{\alpha=0}^{\beta-1} 2^\alpha j_\alpha\Big) \\ \times W\Big(2^{\beta-2} k_{\beta-2} \sum_{\alpha=0}^{\beta-1} 2^\alpha j_\alpha\Big) \ldots W\Big(k_0 \sum_{\alpha=0}^{\beta-1} 2^\alpha j_\alpha\Big) \end{aligned} \tag{1.155}$$

Remembering Eq. (1.148–i) and the assumption that $L = 2^\beta$, the first term on the right-hand side of Eq. (1.155) can be decomposed to give

$$\begin{aligned} W\Big(2^{\beta-1} k_{\beta-1} \sum_{\alpha=0}^{\beta-1} 2^\alpha j_\alpha\Big) = W(2^\beta 2^{\beta-2} j_{\beta-1} k_{\beta-1})\, W(2^\beta 2^{\beta-3} j_{\beta-2} k_{\beta-1}) \ldots \\ \times W(2^\beta j_1 k_{\beta-1})\, W(2^{\beta-1} j_0 k_{\beta-1}) = W(2^{\beta-1} j_0 k_{\beta-1}) \end{aligned} \tag{1.156}$$

since $2^{\beta-2} j_{\beta-1} k_{\beta-1}$, $2^{\beta-3} j_{\beta-2} k_{\beta-1}$, $\ldots$, $j_1 k_{\beta-1}$ are integer-valued expressions. This equation represents the main idea behind the FFT and applies not only to the case of $L = 2^\beta$. Similarly, decomposition of the second term of Eq. (1.155) leads to

$$\begin{aligned} W\Big(2^{\beta-2} k_{\beta-2} \sum_{\alpha=0}^{\beta-1} 2^\alpha j_\alpha\Big) = W(2^\beta 2^{\beta-3} j_{\beta-1} k_{\beta-2})\, W(2^\beta 2^{\beta-4} j_{\beta-2} k_{\beta-2}) \ldots \\ \times W(2^{\beta-1} j_1 k_{\beta-2})\, W(2^{\beta-2} j_0 k_{\beta-2}) = W\Big((2j_1 + j_0) 2^{\beta-2} k_{\beta-2}\Big) \end{aligned} \tag{1.157}$$

As the procedure proceeds over the terms (except for the last one) of Eq. (1.155), at each step a new term which does not vanish by the condition of Eq. (1.148–i) is added. This decomposition process continues until the last term is reached so that Eq. (1.154) takes the form

$$\begin{aligned} X(j_{\beta-1}, j_{\beta-2}, \ldots, j_0) = \sum_{k_0=0}^{1} \sum_{k_1=0}^{1} \cdots \sum_{k_{\beta-1}=0}^{1} x_0(k_{\beta-1}, k_{\beta-2}, \ldots, k_0)\, W(2^{\beta-1} j_0 k_{\beta-1}) \\ \times W\Big((2j_1 + j_0) 2^{\beta-2} k_{\beta-2}\Big) \ldots W\Big((2^{\beta-1} j_{\beta-1} + 2^{\beta-2} j_{\beta-2} + \ldots + j_0) k_0\Big) \end{aligned} \tag{1.158}$$

Carrying out each of the sums corresponding to the indices $k_{\beta-1}$, $k_{\beta-2}, \ldots, k_0$ separately and labelling the intermediate results leads to the recursive equations

$$\begin{aligned}
x_1(j_0, k_{\beta-2}, \ldots, k_0) &= \sum_{k_{\beta-1}=0}^{1} x_0(k_{\beta-1}, k_{\beta-2}, \ldots, k_0) W(2^{\beta-1} j_0 k_{\beta-1}) \\
x_2(j_0, j_1, k_{\beta-3}, \ldots, k_0) &= \sum_{k_{\beta-2}=0}^{1} x_1(j_0, k_{\beta-2}, \ldots, k_0) W\big((2j_1 + j_0) 2^{\beta-2} k_{\beta-2}\big) \\
\ldots &\qquad \ldots \\
x_\beta(j_0, j_1, \ldots, j_{\beta-1}) &= \sum_{k_0=0}^{1} x_{\beta-1}(j_0, j_1, \ldots, k_0)\, W\Big(k_0 \sum_{\alpha=0}^{\beta-1} 2^\alpha j_\alpha\Big) \\
X(j_{\beta-1}, j_{\beta-2}, \ldots, j_0) &= x_\beta(j_0, j_1, \ldots, j_{\beta-1})
\end{aligned} \tag{1.159}$$

The numerical procedure based on these equations is referred to as the Cooley–Tukey algorithm of the base-two FFT [26]. The extensions to the case of L chosen according to criteria other than $L = 2^\beta$ can be found in [16,20,29,32], for instance. It is seen that the summation in the i-th equation of Eq. (1.159) contains only one complex multiply–add operation with the summation index $k_{\beta-i} = 1$, since if $k_{\beta-i} = 0$, then $W(nk_{\beta-i}) = 1$. Furthermore, it can be shown in the discrete Fourier transform that the relation $W(n) = -W(n + L/2)$ holds true. The total number of complex multiply–add operations is then equal to $L\beta/2$. Recalling that $\beta = \log_2 L$ and the total number of complex multiply–add operations required in the traditional Fourier transform is equal to L^2, the base-two FFT reduces the computation cost $\mathrm{O}(L/\log_2 L)$ times.[1] In the general case when $L = n_1 n_2 \ldots n_j \neq 2^\beta$ this factor is equal to $\mathrm{O}[L/(n_1 + n_2 + \ldots + n_j)]$ [29].

As an example let us now consider a simple case of FFT with $L = 4$, i.e. $\beta = 2$. The objective is to compute coefficients X_j of the Fourier analysis expressed by Eq. (1.142) on the basis of the procedures described above. Eq. (1.142) becomes in this case

$$X_j = \sum_{k=0}^{3} x_k \exp\Big(-\mathrm{i}\,\frac{\pi}{2} jk\Big) \qquad j = 1, 2, 3 \tag{1.160}$$

By using Eq. (1.150) the indices j and k involved in Eq. (1.160) can be written in the binary form as

$$\begin{aligned}
j &= (j_1, j_0) = 2j_1 + j_0 \\
k &= (k_1, k_0) = 2k_1 + k_0
\end{aligned} \tag{1.161}$$

and Eq. (1.160) takes the form

$$X(j_1, j_0) = \sum_{k_0=0}^{1} \sum_{k_1=0}^{1} x_0(k_1, k_0) \exp\Big[-\mathrm{i}\,\frac{\pi}{2}(2j_1 + j_0)(2k_1 + k_0)\Big] \tag{1.162}$$

[1] O(.) stands for 'the order of'.

Decomposition of the exponent function term in Eq. (1.162) gives

$$\begin{aligned}&\exp\Big[-\mathrm{i}\,\frac{\pi}{2}(2j_1+j_0)(2k_1+k_0)\Big]\\&\quad=\exp\Big[-\mathrm{i}\,\pi k_1(2j_1+j_0)\Big]\exp\Big[-\mathrm{i}\,\frac{\pi}{2}k_0(2j_1+j_0)\Big]\end{aligned}\tag{1.163}$$

In this equation the first exponent function on the right-hand side can be shown to be

$$\begin{aligned}\exp\Big[-\mathrm{i}\,\pi k_1(2j_1+j_0)\Big] &= \exp(-\mathrm{i}\,2\pi k_1 j_1)\exp(-\mathrm{i}\,\pi k_1 j_0)\\&= \exp(-\mathrm{i}\,\pi k_1 j_0)\end{aligned}\tag{1.164}$$

since $\exp(-\mathrm{i}\,2\pi k_1 j_1)=1$. Substituting Eq. (1.164) into Eq. (1.163) and inserting the resulting equation into Eq. (1.162) gives

$$X(j_1,j_0)=\sum_{k_0=0}^{1}\exp\Big[-\mathrm{i}\frac{\pi}{2}k_0(2j_1+j_0)\Big]\sum_{k_1=0}^{1}x_0(k_1,k_0)\exp(-\mathrm{i}\,\pi k_1 j_0)\tag{1.165}$$

Carrying out the inner sum over k_1 and denoting the result by $x_1(j_0,k_0)$ yields

$$\begin{aligned}x_1(j_0,k_0) &= \sum_{k_1=0}^{1}x_0(k_1,k_0)\exp(-\mathrm{i}\,\pi k_1 j_0)\\&= x_0(0,k_0)+x_0(1,k_0)\exp(-\mathrm{i}\,\pi j_0)\end{aligned}\tag{1.166}$$

Similarly, the sum over k_0 in Eq. (1.165) labelled by $x_2(j_0,j_1)$ becomes

$$\begin{aligned}X(j_1,j_0)=x_2(j_0,j_1) &= \sum_{k_0=0}^{1}x_1(j_0,k_0)\exp\Big[-\mathrm{i}\frac{\pi}{2}k_0(2j_1+j_0)\Big]\\&= x_1(j_0,0)+x_1(j_0,1)\exp\Big[-\mathrm{i}\frac{\pi}{2}(2j_1+j_0)\Big]\end{aligned}\tag{1.167}$$

Thus, evaluation of the FFT components for the case $L=4$ proceeds 'computationally' as follows:

for $j_0=0$

$$\begin{aligned}x_1(0,0) &= x_0(0,0)+x_0(1,0)\\x_1(0,1) &= x_0(0,1)+x_0(1,1)\\x_2(0,0) &= x_1(0,0)+x_1(0,1)=X(0,0)=x_0\\A &= \exp(-\mathrm{i}\,\pi)\\x_2(0,1) &= x_1(0,0)+x_1(0,1)A=X(1,0)=x_2\end{aligned}$$

for $j_0=1$

$$\begin{aligned}x_1(1,0) &= x_0(0,0)+x_0(1,0)\,A\\x_1(1,1) &= x_0(0,1)+x_0(1,1)\,A\\x_2(1,0) &= x_1(1,0)+x_1(1,1)\exp\Big(-\mathrm{i}\,\frac{\pi}{2}\Big)=X(0,1)=x_1\\x_2(1,1) &= x_1(1,0)+x_1(1,1)\exp\Big(-\mathrm{i}\,\frac{3\pi}{2}\Big)=X(1,1)=x_3\end{aligned}$$

It is seen that the FFT algorithm requires only three complex operations to evaluate the exponent function at $\pi/2$, π and $3\pi/2$, whereas in the conventional Fourier transform nine complex operations would be needed.

It is emphasized in closing that the discussion of Section 1.2 has necessarily been rather sketchy and should be understood merely as an attempt to introduce fundamental ideas of numerically oriented Fourier transform problems. Nevertheless, it is believed that the material covered should turn out sufficient for a successful computer implementation of the concepts involved; for a more complete treatment the reader is referred to [16,19,20,30,68,74,97], for instance.

1.3 Calculus of Variations

The mathematical problem of minimizing (or maximizing) an integral is dealt with in a special branch of the calculus, called *calculus of variations.* The mathematical theory shows that the final result can be established without taking into account the infinity of tentatively possible paths. A tentative path which differs from the actual path in an arbitrary but infinitesimal degree is called a *variation* of the actual path, and the calculus of variations investigates the changes in the value of an integral caused by such infinitesimal variations of the path.

To be specific let us start with a simple case of

$$\mathcal{J}[u] = \int_{x_1}^{x_2} f\left[x, u(x), \frac{\mathrm{d}u(x)}{\mathrm{d}x}\right] \mathrm{d}x \tag{1.168}$$

where $\mathcal{J}$ is the quantity that is to take on an extreme value, f is a known function of the variables indicated but the dependence of u on x is not fixed; that is, $u(x)$ is unknown. In other words, we are to choose the path $u(x)$ passing

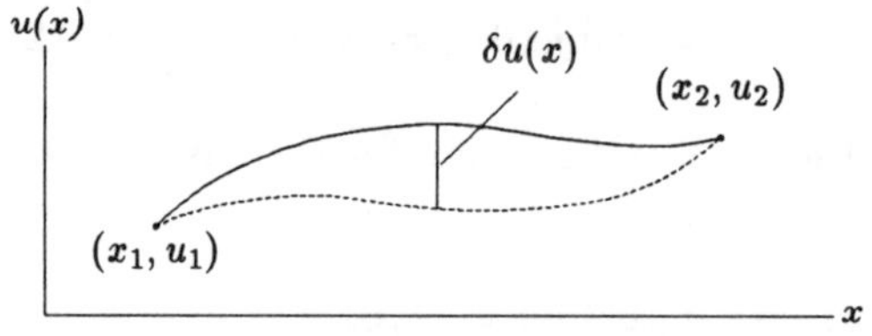

Figure 1.3 Optimum path of $u(x)$ and variation $\delta u(x)$

through points (x_1, u_1) and (x_2, u_2), cf. Fig. 1.3, which renders the functional $\mathcal{J}$ stationary. In this book, and almost always in cases of physical significance, the stationary value of $\mathcal{J}$ will be a minimum. Let us assume that an optimum path exists and compare the value of $\mathcal{J}$ for our (unknown) path with that obtained from neighbouring paths. The difference between these two for a given x is called the variation of $u(x)$ and it is denoted by $\delta u(x)$; for convenience the following notation is adopted:

$$\delta u(x) = \epsilon\,\eta(x) \tag{1.169}$$

where a scalar factor ϵ gives the magnitude of the variation. The function $\eta(x)$ must be differentiable and such that

$$\eta(x_1) = \eta(x_2) = 0 \tag{1.170}$$

i.e. all varied paths must pass through the fixed end points. Let us further define

$$u(x;\epsilon) = u(x) + \epsilon\,\eta(x) \tag{1.171}$$

and assume that $u(x;0) = u(x)$ is the unknown path that will minimize $\mathcal{J}$. The functional $\mathcal{J}$ may now be treated as a function of the parameter ϵ

$$\mathcal{J}[\epsilon] = \int_{x_1}^{x_2} f\left[x, u(x;\epsilon), \frac{\mathrm{d}u(x;\epsilon)}{\mathrm{d}x}\right] \mathrm{d}x \tag{1.172}$$

and the condition for an extreme value is that

$$\left.\frac{\mathrm{d}\mathcal{J}[\epsilon]}{\mathrm{d}\epsilon}\right|_{\epsilon=0} = 0 \tag{1.173}$$

The dependence of the integral on x is contained in $u(x;\epsilon)$ and $\mathrm{d}u(x;\epsilon)/\mathrm{d}x$, therefore

$$\frac{\mathrm{d}\mathcal{J}[\epsilon]}{\mathrm{d}\epsilon} = \int_{x_1}^{x_2} \left[\frac{\partial f}{\partial u}\frac{\partial u}{\partial \epsilon} + \frac{\partial f}{\partial u_{,x}}\frac{\partial u_{,x}}{\partial \epsilon}\right] \mathrm{d}x \tag{1.174}$$

where $u_{,x}(x;\epsilon) = \mathrm{d}u(x;\epsilon)/\mathrm{d}x$. Noting that by Eq. (1.171)

$$\frac{\partial u(x;\epsilon)}{\partial \epsilon} = \eta(x) \qquad \frac{\partial u_{,x}(x;\epsilon)}{\partial \epsilon} = \frac{\mathrm{d}\eta(x)}{\mathrm{d}x} \tag{1.175}$$

Eq. (1.174) reads

$$\frac{\mathrm{d}\mathcal{J}[\epsilon]}{\mathrm{d}\epsilon} = \int_{x_1}^{x_2} \left[\frac{\partial f}{\partial u}\eta(x) + \frac{\partial f}{\partial u_{,x}}\frac{\mathrm{d}\eta(x)}{\mathrm{d}x}\right] \mathrm{d}x \tag{1.176}$$

Integration of the second term by parts leads to the expression

$$\frac{\mathrm{d}\mathcal{J}[\epsilon]}{\mathrm{d}\epsilon} = \int_{x_1}^{x_2} \left[\frac{\partial f}{\partial u}\eta(x) - \frac{\mathrm{d}}{\mathrm{d}x}\left(\frac{\partial f}{\partial u_{,x}}\right)\eta(x)\right] \mathrm{d}x + \eta(x)\left.\frac{\partial f}{\partial u_{,x}}\right|_{x_1}^{x_2} \tag{1.177}$$

which, by Eqs. (1.170) and (1.173), becomes

$$\int_{x_1}^{x_2} \left[\frac{\partial f}{\partial u} - \frac{\mathrm{d}}{\mathrm{d}x}\left(\frac{\partial f}{\partial u_{,x}}\right)\right] \eta(x)\,\mathrm{d}x = 0 \tag{1.178}$$

We note that in this form ϵ has been set to zero and is no longer part of the problem. Moreover, it is sometimes more convenient to consider Eq. (1.178) multiplied by ϵ, which gives

$$\int_{x_1}^{x_2} \left[\frac{\partial f}{\partial u} - \frac{\mathrm{d}}{\mathrm{d}x}\left(\frac{\partial f}{\partial u_{,x}}\right)\right] \epsilon\,\eta(x)\,\mathrm{d}x = \left.\frac{\mathrm{d}\mathcal{J}[\epsilon]}{\mathrm{d}\epsilon}\right|_{\epsilon=0} \epsilon \;=\; \delta\mathcal{J} \;=\; 0 \tag{1.179}$$

where $\delta\mathcal{J}$ is known as the first variation of the functional $\mathcal{J}$; its significance can also be seen by considering the Taylor expansion for the function

$$\Delta\mathcal{J}(\epsilon) = \mathcal{J}(\epsilon) - \mathcal{J}(0) \tag{1.180}$$

which has the form

$$\Delta\mathcal{J}(\epsilon) = \left.\frac{\mathrm{d}\mathcal{J}}{\mathrm{d}\epsilon}\right|_{\epsilon=0}\epsilon + \frac{1}{2!}\left.\frac{\mathrm{d}^2\mathcal{J}}{\mathrm{d}\epsilon^2}\right|_{\epsilon=0}\epsilon^2 + \cdots = \delta\mathcal{J} + \frac{1}{2!}\delta^2\mathcal{J} + \cdots \tag{1.181}$$

For some 'technical' reasons it is useful to note that

$$\delta\left(\frac{\mathrm{d}u}{\mathrm{d}x}\right) = \frac{\mathrm{d}}{\mathrm{d}x}(\delta u) \tag{1.182}$$

because

$$\delta\left(\frac{\mathrm{d}u}{\mathrm{d}x}\right) = \epsilon\,\eta'(x) = \epsilon\,\frac{\mathrm{d}\eta}{\mathrm{d}x} = \frac{\mathrm{d}}{\mathrm{d}x}(\epsilon\,\eta) = \frac{\mathrm{d}}{\mathrm{d}x}(\delta u) \tag{1.183}$$

In other words, the derivative of the variation is equal to the variation of the derivative.

We can similarly prove that variations performed on the sums, products and ratios of two functions are defined quite analogously to the corresponding rules of differentiation so that

$$\begin{aligned}
u(x) &= u_1(x) + u_2(x) && \rightarrow \quad \delta u(x) = \delta u_1(x) + \delta u_2(x) \\
u(x) &= u_1(x)u_2(x) && \rightarrow \quad \delta u(x) = \delta u_1(x)\,u_2(x) + u_1(x)\delta u_2(x) \\
u(x) &= \frac{u_1(x)}{u_2(x)} && \rightarrow \quad \delta u(x) = \frac{\delta u_1(x)\,u_2(x) - u_1(x)\delta u_2(x)}{\delta u_2^2(x)}
\end{aligned} \tag{1.184}$$

Also, the variation of a definite integral is equal to the definite integral of the variation, i.e.

$$\delta\int_{x_1}^{x_2} u(x)\,\mathrm{d}x = \int_{x_1}^{x_2} \delta u(x)\,\mathrm{d}x \tag{1.185}$$

Since $\eta(x)$ is arbitrary, condition (1.173) for the existence of a stationary value of $\mathcal{J}$ can only be satisfied if the bracketed term in Eq. (1.178) is identically zero, i.e.

$$\frac{\partial f}{\partial u} - \frac{\mathrm{d}}{\mathrm{d}x}\left(\frac{\partial f}{\partial u_{,x}}\right) = 0 \tag{1.186}$$

This is a partial differential equation for the unknown function $u(x)$, referred to as the *Euler equation* for the problem (1.168). It can readily be seen that Eq. (1.186) is the necessary and sufficient condition for the integral (1.168) to be stationary. In other words:

> The necessary and sufficient condition for the integral (1.168) to be stationary, with the boundary conditions $u(x_1) = u_1$ and $u(x_2) = u_2$, is that the differential equation of Euler (1.186) be satisfied.

The simple variational problem considered so far may be generalized in several respects. For instance, we may consider a functional typical of static three-dimensional solid mechanics applications, which has a general form as

$$\mathcal{J}[u_i] = \int_{\Omega} f[x_i, u_i(x_k), u_{i,j}(x_k)]\,\mathrm{d}\Omega \tag{1.187}$$

where i, j, k run over the sequence 1,2,3; x_i and u_i represent the vectors $\mathbf{u} = \{u_1, u_2, u_3\}$ and $\mathbf{x} = \{x_1, x_2, x_3\}$; a comma indicates partial differentiation; and the boundary conditions are given as

$$u_i(x_k) = \hat{u}_i \qquad x_k \in \partial\Omega \tag{1.188}$$

$\partial\Omega$ being the boundary of a three-dimensional region Ω. In other words, we have now the integrand in the form of a function of three independent variables $u_i(x_k)$, which involves the functions themselves and their first partial derivatives. Without presenting tedious derivations [41,64,70,76,120], which basically follow those described earlier in this section, we simply cite the final result in terms of the Euler equations as

$$\frac{\partial f}{\partial u_i} - \sum_{j=1}^{3} \frac{\mathrm{d}}{\mathrm{d}x_j}\left(\frac{\partial f}{\partial u_{i,j}}\right) = 0 \qquad i = 1, 2, 3 \tag{1.189}$$

Thus:

> Integral (1.187), with boundary condition (1.188), will be stationary if the set of three simultaneous partial equations of the second order (1.189) (i.e. the Euler equations for the integral) are satisfied.

Chapter 2

FEM in Deterministic Structural Analysis

2.1 Introductory Comments and Organization of the Chapter

There have been so many textbooks on different aspects of the 'classical'[1] finite element method published in the literature during the last two decades or so that any attempt to repeat at some length the basic derivations typical of the method would seem superfluous at best. On the other hand, this text is a FEM-based book itself, dealing with a special aspect of the method as applied to problems of solid mechanics which involve some parameter uncertainties. Therefore, we at least need to set up the FEM background in such a way as to make it possible to elaborate on probabilistic issues. In an attempt to compromise on these two conflicting aspects we have decided to summarize in this chapter the description of the classical FEM on the structure level as this turns out to be essential for consistent derivations of stochastic FEM equations, and to be very sketchy on all issues related to the element level. Since all the finite elements used in our programs have already been standardized in the literature, we shall just briefly characterize them at places where we report on test calculations.

In this introductory exposition of Chapter 2 we shall confine ourselves to three-dimensional problems only – it is assumed that analogous derivations needed to obtain equations describing linear elasticity problems in one and two dimensions (plane stress, plane strain, axisymmetry, trusses, Euler beams, Kirchhoff plates, etc.) are straightforward and known to the reader. Also, discussions of computational aspects will basically remain unchanged for such simplified theories. It is in that sense that referring to 'structural' analysis (as will repeatedly be done in this book) and explicitly presenting only three-dimensional equations appears to be admissible.

[1]Meaning 'deterministic' in the context of this book.

In accordance with the above philosophy the presentation of the FEM as applied to statics and dynamics will be limited to just quoting essential equations, Sections 2.2–2.5. As the next step, in Section 2.6 we shall devote some space to the discussion of a significantly more advanced problem of nonlinear structural analysis. This will enable us to smoothly pass on to discuss the so-called linearized buckling problem in Section 2.7. The material presented in Section 2.6 will also be indispensable for a consistent presentation of stochastic analysis concepts for nonlinear structural problems in Chapter 9.

Compared with the compact presentations of Sections 2.2–2.5 we shall be much more explicit while describing less conventional aspects of FEM applications to a newly developed field of structural sensitivity analysis. Both static and dynamic sensitivity analysis using FEM will be reviewed and the background set up for later generalizations towards inclusion of probabilistic aspects which will follow in Chapters 5–8.

The notation employed throughout this chapter is an outcome of our attempt to follow the established literature standards to the highest degree possible.

2.2 Equations of Linear Solid Mechanics

All considerations in this book will be referred to a fixed, rectangular, cartesian coordinate system $\{x_k\}$, $k = 1,2,3$. Indicial notation will be employed with indices repeated twice, implying summation. The latin indices i, j, k, l, m, n run over the sequence 1,2,3 (unless indicated otherwise).

Governing equations for a linear elastic body which takes up a region Ω[1] in the space parameterized by $\{x_k\}$ may be summarized as follows:

(a) Displacement of a point of the body with coordinates x_k, $x \in \Omega$, at time $\tau \in [0, \infty)$ is described by the vector $\mathbf{u} = \{u_i\} = \{u_1, u_2, u_3\}$. This vector is assumed to be a sufficiently smooth function of x_k and τ, $u_i = u_i(x_k, \tau)$.

(b) The state of strain at $x_k \in \Omega$ is defined by means of a symmetric strain tensor $\boldsymbol{\varepsilon} = \{\varepsilon_{ij}\}$ in which ε_{11}, ε_{22}, ε_{33} and $\varepsilon_{12} = \varepsilon_{21}$, $\varepsilon_{13} = \varepsilon_{31}$, $\varepsilon_{23} = \varepsilon_{32}$ are normal and shear strains, respectively. The strain tensor is defined by means of the strain-displacement relationship given as

$$\varepsilon_{ij} = \frac{1}{2}(u_{i,j} + u_{j,i}) \tag{2.1}$$

where a comma followed by, say, 'i' stands for partial differentiation with respect to the coordinate x_i. Clearly, $\varepsilon_{ij} = \varepsilon_{ij}(x_k, \tau)$.

(c) Stresses are described by means of the symmetric stress tensor $\boldsymbol{\sigma} = \{\sigma_{ij}\}$. The components σ_{11}, σ_{22} and σ_{33} are called the normal stresses while the

[1]For the sake of simplicity, we shall be talking of a body Ω instead of a body which takes up a region Ω. In the parlance of continuum mechanics we thus identify for pragmatic reasons the body with its reference configuration.

components $\sigma_{12} = \sigma_{21}$, $\sigma_{13} = \sigma_{31}$ and $\sigma_{23} = \sigma_{32}$ are called the shear stresses. The stress tensor depends on both space and time coordinates, $\sigma_{ij} = \sigma_{ij}(x_k, \tau)$. Stresses are related to strains by means of the constitutive relations which for linear elasticity have a linear, homogeneous form of a generalized Hooke's law as

$$\sigma_{ij} = C_{ijkl}\, \varepsilon_{kl} \tag{2.2}$$

The tensor of elastic moduli has the following symmetries: $C_{ijkl} = C_{jikl}$ and $C_{ijkl} = C_{klij}$ (which implies also $C_{ijkl} = C_{ijlk}$). For an isotropic material all the elastic moduli may be expressed in terms of two independent elastic constants as

$$C_{ijkl} = E\left[\frac{1}{(1+\nu)(1-2\nu)}\delta_{ij}\delta_{kl} + \frac{1}{2(1+\nu)}(\delta_{ik}\delta_{jl} + \delta_{il}\delta_{jk})\right]$$

where E is Young's modulus and ν is Poisson's ratio while δ_{ij} is the Kronecker delta.

(d) The motion of the body is governed by the equations of motion which have the form

$$\sigma_{ij,j} + \varrho f_i = \varrho\, \ddot{u}_i \qquad x_k \in \Omega\,;\;\; \tau \in [0, \infty) \tag{2.3}$$

in which ϱ is the material density, ϱf_i is the body force per unit volume while a dot over the kernel letter indicates time differentiation.

(e) From the viewpoint of boundary conditions the surface $\partial\Omega$ of the body Ω can be divided into two parts: the part $\partial\Omega_u$ over which boundary conditions are prescribed in terms of displacements $\hat{u}_i$

$$u_i = \hat{u}_i \qquad x_k \in \partial\Omega_u\,;\;\; \tau \in [0, \infty) \tag{2.4}$$

and the part $\partial\Omega_\sigma$ over which the boundary conditions are prescribed in terms of given boundary tractions $\hat{t}_i$

$$\sigma_{ij}n_j = \hat{t}_i \qquad x_k \in \partial\Omega_\sigma\,;\;\; \tau \in [0, \infty) \tag{2.5}$$

where n_i is the unit vector normal to the boundary $\partial\Omega$.

(f) The initial conditions have the form

$$\begin{aligned} u_i &= u_i^0 \qquad & x_k \in \Omega\,;\;\; \tau = 0 \\ \dot{u}_i &= \dot{u}_i^0 \qquad & x_k \in \Omega\,;\;\; \tau = 0 \end{aligned} \tag{2.6}$$

This completes the specification of equations describing problems of linear elasticity. Thus, the whole system of equations consists of:

- strain-displacement relations (2.1),
- constitutive relations (2.2),
- equations of motion (2.3),
- boundary conditions (2.4) and (2.5),
- initial conditions (2.6).

In this system we have 15 unknowns functions: 3 displacement components $u_i(x_k, \tau)$, 6 strain components $\varepsilon_{ij}(x_k, \tau)$ and 6 stress components $\sigma_{ij}(x_k, \tau)$ which have to be solved for using 15 field equations (2.1)–(2.3) together with appropriate boundary and initial conditions.

By substituting Eq. (2.1) into (2.2) and Eq. (2.2) into (2.3) and (2.5) we may express our initial-boundary value problem in terms of three displacement components only

$$\begin{aligned}
&(C_{ijkl}\, u_{k,l})_{,j} + \varrho f_i = \varrho\, \ddot{u}_i && x_k \in \Omega\,; && \tau \in [0, \infty) \\
&C_{ijkl}\, u_{k,l}\, n_j = \hat{t}_i && x_k \in \partial\Omega_\sigma\,; && \tau \in [0, \infty) \\
&u_i = \hat{u}_i && x_k \in \partial\Omega_u\,; && \tau \in [0, \infty) \\
&u_i = u_i^0 && x_k \in \Omega\,; && \tau = 0 \\
&\dot{u}_i = \dot{u}_i^0 && x_k \in \Omega\,; && \tau = 0
\end{aligned} \tag{2.7}$$

So far, we have dealt with the problem of linear dynamics of an elastic body. The static behaviour may be described by assuming that the problem is time-independent, implying that all the functions are dependent on spatial variables x_k only. The equations of motion simplify then to the equations of equilibrium in which the tern $\varrho\ddot{u}_i$ is nonexistent, and the initial conditions (2.6) are no longer necessary.

2.3 Variational Methods

2.3.1 Principle of Minimum Potential Energy

The equations quoted in Section 2.2 form the basis of the so-called local formulation for linear elasticity problems. An exact analytical solution to most practical problems described in such a manner (even after introducing approximations leading to simpler theories like plane stress, plane strain or shell theory) can hardly be expected. That is why an alternative formulation, referred to as the variational formulation, has been extensively investigated in the literature and is briefly reviewed below. The importance of variational statements of physical laws goes far beyond their use as simply an alternative to the local formulation. In fact, variational or weak forms of the laws of mechanics are often considered to be the only natural and rigorously correct way to think about them. However, aside from this basic observation, it is sufficient for us to note that the use of variational statements makes it possible to concentrate in a single functional all of the intrinsic features of the problem at hand: the governing field equations as well as the boundary and initial conditions. This ensures a simultaneous and consistent approximation to all the equations in case an approximate solution is sought. Also, using an approximate type of variational formulations allows us to relax the smoothness requirements imposed on the functions appearing in the functional which may be of great significance in constructing approximate solutions.

Without going into theoretical details of variational formulations which can be found elsewhere [4,5,64,70,102,120], let us now first illustrate the approach by considering the principle of minimum potential energy. Only static deformation processes are considered in this section.

The total potential energy of the body can be written as

$$\mathcal{J}_{\mathrm{P}}[u_i] = \int_{\Omega} \left(\frac{1}{2} C_{ijkl}\, \varepsilon_{ij}\, \varepsilon_{kl} - \varrho f_i\, u_i \right) \mathrm{d}\Omega - \int_{\partial\Omega_\sigma} \hat{t}_i\, u_i\, \mathrm{d}(\partial\Omega) \tag{2.8}$$

where the strain ε_{ij} is treated as a function of u_i according to Eq. (2.1). Let us assume that the body forces ϱf_i, the surface tractions $\hat{t}_i$ and the surface displacements $\hat{u}_i$ are prescribed, and kept unchanged in magnitude and direction during the variation procedure to follow. Let us also define an admissible infinitesimal displacement function, which is here understood as any arbitrary continuous and piece-wise differentiable vector function $u_i = u_i(x_k)$ satisfying the kinematic boundary condition (2.4). The principle of minimum total potential energy states:

> Among all the admissible displacement functions the actual displacements make the total potential energy (2.8) an absolute minimum.

Thus, for the true solution u_i we have

$$\delta\mathcal{J}_{\mathrm{P}}[u_i] = 0 \qquad\qquad \delta^2\mathcal{J}_{\mathrm{P}}[u_i] \geq 0 \tag{2.9}$$

and the equality sign in Eq. (2.9) holds only if all the components of the strain tensor ε_{ij} computed from the variation δu_i vanish (i.e. if this particular choice of δu_i implies a rigid body motion).

As we shall see below, the principle provides a means of formulating the equilibrium equations in the interior Ω, and the static boundary equations on the surface $\partial\Omega_\sigma$ of the body but makes no contribution to the difficult task of solving the boundary-value problem. The latter issue will be undertaken in Section 2.4 of this chapter.

Let us explicitly calculate the first variation of the total potential energy. By using the symmetry of the tensor C_{ijkl}, noting that

$$\delta\mathcal{J}_{\mathrm{P}}[u_i] = \frac{\partial \mathcal{J}_{\mathrm{P}}}{\partial u_i}\delta u_i \tag{2.10}$$

and employing Eq. (2.1) to express ε_{ij} in terms of u_i, we obtain

$$\delta\mathcal{J}_{\mathrm{P}}[u_i] = \int_\Omega C_{ijkl}\, u_{i,j}\, \delta u_{k,l}\, \mathrm{d}\Omega - \int_\Omega \varrho f_i\, \delta u_i\, \mathrm{d}\Omega - \int_{\partial\Omega_\sigma} \hat{t}_i\, \delta u_i\, \mathrm{d}(\partial\Omega) \tag{2.11}$$

Integrating the first term by parts yields the expression

$$\begin{aligned}\delta\mathcal{J}_{\mathrm{P}}[u_i] &= \int_\Omega (C_{ijkl}\, u_{i,j}\, \delta u_k)_{,l}\, \mathrm{d}\Omega - \int_\Omega (C_{ijkl}\, u_{i,j})_{,l}\, \delta u_k\, \mathrm{d}\Omega \\ &\quad - \int_\Omega \varrho f_i\, \delta u_i\, \mathrm{d}\Omega - \int_{\partial\Omega_\sigma} \hat{t}_i\, \delta u_i\, \mathrm{d}(\partial\Omega)\end{aligned} \tag{2.12}$$

which, by employing the Gauss–Ostrogradski theorem,[1] using Eq. (2.2) and noting that $\delta u_i = 0$ on $\partial\Omega_u$, becomes

$$\delta\mathcal{J}_{\mathrm{P}}[u_i] = \int_\Omega (\sigma_{ij,j} + \varrho f_i)\, \delta u_i\, \mathrm{d}\Omega + \int_{\partial\Omega_\sigma} (\sigma_{ij}\, n_j - \hat{t}_i)\, \delta u_i\, \mathrm{d}(\partial\Omega) \tag{2.13}$$

Since the stress components belong to the actual solution and both the integrands vanish on account of Eq. (2.3) (with no inertial force) and Eq. (2.5), we find that the first variation of $\mathcal{J}_{\mathrm{P}}$ vanishes.

Using now Eq. (2.11) to compute the second variation we obtain

$$\delta^2\mathcal{J}_{\mathrm{P}}[u_i] = \int_\Omega C_{ijkl}\, \delta u_{i,j}\, \delta u_{k,l}\, \mathrm{d}\Omega \tag{2.14}$$

which is non-negative, $\delta^2\mathcal{J}_{\mathrm{P}} \geq 0$, because $C_{ijkl}\varepsilon_{ij}\varepsilon_{kl}$ is by the definition of the material strain energy a positive definite quadratic form (i.e. $C_{ijkl}\varepsilon_{ij}\varepsilon_{kl} > 0$ for any $\varepsilon_{ij} \neq 0$).

Since no restrictions have been imposed on δu_i in the above proof (except that they are admissible and infinitesimal) we conclude that the total potential energy is minimum for the actual solution.

We note that Eq.(2.11), when rewritten using the condition $\delta\mathcal{J}_{\mathrm{P}} = 0$ as

$$\int_\Omega C_{ijkl}\varepsilon_{ij}\, \delta\varepsilon_{kl}\, \mathrm{d}\Omega = \int_\Omega \varrho f_i\, \delta u_i\, \mathrm{d}\Omega + \int_{\partial\Omega_\sigma} \hat{t}_i\, \delta u_i\, \mathrm{d}(\partial\Omega) \tag{2.15}$$

where (cf. Eq. (2.1))

$$\delta\varepsilon_{ij} = \frac{1}{2}(\delta u_{i,j} + \delta u_{j,i}) \tag{2.16}$$

expresses the principle of virtual work. It states:

> Virtual work done by the stress on virtual strains owing to arbitrary virtual displacements satisfying the prescribed kinematic boundary conditions is equal to virtual work done by the external forces on the same virtual displacements.

We note in passing that it turns out sufficient in many applications to take advantage of the stationary property of the potential energy functional $\mathcal{J}_{\mathrm{P}}$ without referring to its minimum property. Therefore, we often speak of the principle of stationary potential energy.

---

[1]This theorem reads:

$$\int_\Omega A_{,i}\, \mathrm{d}\Omega = \int_{\partial\Omega_\sigma} A\, n_i\, \mathrm{d}(\partial\Omega)$$

for any sufficiently smooth scalar-, vector- or tensor-valued function $A(x_k)$.

2.3.2 Multi-Field Principles

In the preceding section the so-called one-field variational statement of the displacement type was presented. By the assumption that the only independent field entering the system functional is the displacement field, the principle of minimum potential energy gives the equilibrium equations when considering the functional subject to variation. However, in many applications the use of the variational philosophy cannot be so straightforward and the development of more sophisticated formulations, wherein various independent fields may be involved in a functional, is needed. An example of such a multi-field variational statement will be given below.

Let us consider the static problem described by Eqs. (2.1)–(2.5) assuming that $\varrho\ddot{u}_i = 0$ in Eq. (2.3). A complete solution of the problem consists in finding fields of displacements u_i, strains ε_{ij} and stresses σ_{ij} which satisfy the field equations (2.1)–(2.3) in Ω and the boundary conditions (2.4) and (2.5) on $\partial\Omega$. Let us consider the so-called Hu–Washizu functional of the form

$$\begin{aligned}\mathcal{J}_{\text{H-W}}[u_i,\varepsilon_{ij},\sigma_{ij},t_i] &= \int_\Omega \Big(\tfrac{1}{2} C_{ijkl}\,\varepsilon_{ij}\,\varepsilon_{kl} - \varrho f_i u_i\Big)\,\mathrm{d}\Omega - \int_{\partial\Omega_\sigma} \hat{t}_i\,u_i\,\mathrm{d}(\partial\Omega)\\ &\quad - \int_\Omega \sigma_{ij}\left[\varepsilon_{ij} - \tfrac{1}{2}(u_{i,j}+u_{j,i})\right]\mathrm{d}\Omega - \int_{\partial\Omega_u} t_i(u_i - \hat{u}_i)\,\mathrm{d}(\partial\Omega) \qquad (2.17)\end{aligned}$$

where t_i is a stress vector defined on $\partial\Omega_u$. In this expression the functions u_i, ε_{ij}, σ_{ij} and t_i are assumed to be independent, and no subsidiary conditions are additionally considered. Clearly, the fields σ_{ij} and t_i can be interpreted as Lagrange multipliers.

The first variation of the functional $\mathcal{J}_{\text{H-W}}$ reads

$$\begin{aligned}&\delta\mathcal{J}_{\text{H-W}}[u_i,\varepsilon_{ij},\sigma_{ij},t_i]\\ &\quad = \frac{\partial\mathcal{J}_{\text{H-W}}}{\partial u_i}\delta u_i + \frac{\partial\mathcal{J}_{\text{H-W}}}{\partial\varepsilon_{ij}}\delta\varepsilon_{ij} + \frac{\partial\mathcal{J}_{\text{H-W}}}{\partial\sigma_{ij}}\delta\sigma_{ij} + \frac{\partial\mathcal{J}_{\text{H-W}}}{\partial t_i}\delta t_i \qquad (2.18)\end{aligned}$$

By employing the Gauss–Ostrogradski theorem and performing calculations similar to those made in the previous case, cf. Eq. (2.13), the terms on the right-hand side of Eq. (2.18) can be successively expressed as

$$\begin{aligned}\frac{\partial\mathcal{J}_{\text{H-W}}}{\partial u_i}\delta u_i &= \int_\Omega(\sigma_{ij}\,\delta u_{i,j} - \varrho f_i\,\delta u_i)\,\mathrm{d}\Omega - \int_{\partial\Omega_\sigma}\hat{t}_i\,\delta u_i\,\mathrm{d}(\partial\Omega) - \int_{\partial\Omega_u} t_i\,\delta u_i\,\mathrm{d}(\partial\Omega)\\ &= \int_\Omega\Big[(\sigma_{ij}\,\delta u_i)_{,j} - (\sigma_{ij,j}+\varrho f_i)\,\delta u_i\Big]\mathrm{d}\Omega - \int_{\partial\Omega_\sigma}\hat{t}_i\,\delta u_i\,\mathrm{d}(\partial\Omega) - \int_{\partial\Omega_u} t_i\,\delta u_i\,\mathrm{d}(\partial\Omega)\\ &= -\int_\Omega(\sigma_{ij,j}+\varrho f_i)\,\delta u_i\,\mathrm{d}\Omega + \int_{\partial\Omega_\sigma}\sigma_{ij}\,n_j\,\delta u_i\,\mathrm{d}(\partial\Omega)\\ &\quad + \int_{\partial\Omega_u}\sigma_{ij}\,n_j\,\delta u_i\,\mathrm{d}(\partial\Omega) - \int_{\partial\Omega_\sigma}\hat{t}_i\,\delta u_i\,\mathrm{d}(\partial\Omega) - \int_{\partial\Omega_u} t_i\,\delta u_i\,\mathrm{d}(\partial\Omega)\\ &= -\int_\Omega(\sigma_{ij,j}+\varrho f_i)\delta u_i\,\mathrm{d}\Omega + \int_{\partial\Omega_\sigma}(\sigma_{ij}n_j - \hat{t}_i)\delta u_i\,\mathrm{d}(\partial\Omega) + \int_{\partial\Omega_u}(\sigma_{ij}n_j - t_i)\delta u_i\,\mathrm{d}(\partial\Omega)\end{aligned} \qquad (2.19)$$

$$\frac{\partial \mathcal{J}_{\text{H-W}}}{\partial \varepsilon_{ij}} \delta\varepsilon_{ij} = \int_{\Omega} (C_{ijkl}\, \varepsilon_{kl} - \sigma_{ij})\, \delta\varepsilon_{ij}\, \mathrm{d}\Omega \tag{2.20}$$

$$\frac{\partial \mathcal{J}_{\text{H-W}}}{\partial \sigma_{ij}} \delta\sigma_{ij} = \int_{\Omega} \left[\varepsilon_{ij} - \frac{1}{2}(u_{i,j} + u_{j,i}) \right] \delta\sigma_{ij}\, \mathrm{d}\Omega \tag{2.21}$$

$$\frac{\partial \mathcal{J}_{\text{H-W}}}{\partial t_i} \delta t_i = \int_{\partial\Omega_u} (u_i - \hat{u}_i)\, \delta t_i\, \mathrm{d}(\partial\Omega) \tag{2.22}$$

Since the variations δu_i, $\delta\varepsilon_{ij}$, $\delta\sigma_{ij}$ and δt_i are arbitrary and independent, the stationary condition leads to

$$\begin{aligned} \frac{\partial \mathcal{J}_{\text{H-W}}}{\partial u_i} &= 0 & \qquad \frac{\partial \mathcal{J}_{\text{H-W}}}{\partial \varepsilon_{ij}} &= 0 \\ \frac{\partial \mathcal{J}_{\text{H-W}}}{\partial \sigma_{ij}} &= 0 & \qquad \frac{\partial \mathcal{J}_{\text{H-W}}}{\partial t_i} &= 0 \end{aligned} \tag{2.23}$$

In other words, the requirement $\mathcal{J}_{\text{H-W}} = 0$ yields as the Euler conditions for the functional $\mathcal{J}_{\text{H-W}}$ the following conditions:

$$\begin{aligned} &\sigma_{ij,j} + \varrho f_i = 0 & \qquad & x_k \in \Omega \\ &\sigma_{ij}\, n_j = \hat{t}_i & & x_k \in \partial\Omega_\sigma \\ &\sigma_{ij}\, n_j = t_i & & x_k \in \partial\Omega_u \\ &\sigma_{ij} = C_{ijkl}\varepsilon_{kl} & & x_k \in \Omega \\ &\varepsilon_{ij} = \frac{1}{2}(u_{i,j} + u_{j,i}) & & x_k \in \Omega \\ &u_i = \hat{u}_i & & x_k \in \partial\Omega_u \end{aligned} \tag{2.24}$$

which are the equilibrium equations, static boundary conditions, compatibility of the reaction forces, constitutive law, strain-displacement relations and kinematic boundary conditions, respectively.

A whole variety of variational principles can be constructed on the basis of Eq. (2.17) by satisfying a priori some of the field and boundary equations and by introducing appropriate subsidiary conditions [64,120]. Such variational principles usually lose their minimum character, which means that the second variation of the corresponding functionals does not preserve its sign. However, the conditions of functional stationarity still generate the governing differential equations for the problem considered, which is often enough for the development of effective approximate solution methods for problems of elasticity.

2.3.3 Hamilton Principle

In this section we shall consider the dynamic problem of linear elasticity formulated in Section 2.2. In other words, the inertial terms in Eq. (2.3) are no longer negligible, and all the quantities (displacements, strains, stresses) are functions of the space and time variables x_k and τ.

Let us assume $\delta u_i(x_k, t)$ to be a variation of the function $u_i(x_k, \tau)$ at a given time instant $\tau = t$. In view of Eqs. (2.3) and (2.5) the following identity holds:

$$-\int_{\Omega}(\sigma_{ij,j} + \varrho f_i - \varrho\, \ddot{u}_i)\, \delta u_i \,\mathrm{d}\Omega + \int_{\partial\Omega_\sigma}(\sigma_{ij}n_j - \hat{t}_i)\, \delta u_i \,\mathrm{d}(\partial\Omega) = 0 \tag{2.25}$$

By integrating it with respect to time between $\tau = t_1$ and $\tau = t_2$, employing the convention that the values of $u_i(x_k, \tau)$ at $\tau = t_1$ and $\tau = t_2$ are prescribed (so that $\delta u_i(x_k, t_1) = 0$ and $\delta u_i(x_k, t_2) = 0$) and carrying out the integration by parts with respect to each of the independent variables x_k and τ, we obtain

$$\int_{t_1}^{t_2}\Big(\delta\mathcal{T} - \int_{\Omega}\sigma_{ij}\, \delta\varepsilon_{ij} \,\mathrm{d}\Omega + \int_{\Omega}\varrho f_i\, \delta u_i \,\mathrm{d}\Omega + \int_{\partial\Omega_\sigma}\hat{t}_i\, \delta u_i \,\mathrm{d}(\partial\Omega)\Big)\,\mathrm{d}\tau = 0 \tag{2.26}$$

where

$$\mathcal{T} = \int_{\Omega}\frac{1}{2}\varrho\, \dot{u}_i\, \dot{u}_i \,\mathrm{d}\Omega \tag{2.27}$$

is the kinetic energy of the body. As in the case of statics, Eq. (2.26) should be used with two additional conditions:

$$\begin{aligned} &\delta\varepsilon_{ij} = \tfrac{1}{2}(\delta u_{i,j} + \delta u_{j,i}) \qquad && x_k \in \Omega\,; \quad && \tau \in [t_1, t_2] \\ &\delta u_i = 0 && x_k \in \partial\Omega_u\,; && \tau \in [t_1, t_2] \end{aligned} \tag{2.28}$$

If we now assume that the body forces ϱf_i and the boundary tractions t_i are independent of u_i (i.e. that their variations vanish), Eq. (2.26) may be reduced to

$$\delta\int_{t_1}^{t_2}(\mathcal{T} - \mathcal{J}_{\mathrm{P}})\,\mathrm{d}\tau = 0 \tag{2.29}$$

where, as in the previous section

$$\mathcal{J}_{\mathrm{P}}[u_i] = \mathcal{U} - \int_{\Omega}\varrho f_i\, u_i \,\mathrm{d}\Omega - \int_{\partial\Omega_\sigma}\hat{t}_i\, u_i \,\mathrm{d}(\partial\Omega) \tag{2.30}$$

and the strain energy of the linear elastic body is taken to be

$$\mathcal{U} = \int_{\Omega}\frac{1}{2}C_{ijkl}\,\varepsilon_{ij}\,\varepsilon_{kl}\,\mathrm{d}\Omega \tag{2.31}$$

Eq. (2.29) is the Hamilton principle applied to the dynamic problem of the elastic body. It states:

> Among all the admissible displacements which satisfy the prescribed kinematic boundary conditions $u_i = \hat{u}_i$ on $\partial\Omega_u$ and the prescribed conditions at the limits $\tau = t_1$ and $\tau = t_2$, the actual solution renders the functional in Eq. (2.29) stationary.

The Hamilton principle is clearly an extension of the principle of minimum potential energy, cf. Section 2.3.1, to problems of dynamics.

The principle (2.29) is frequently employed in a slightly modified form. In order to obtain this form let us carry out explicitly the variation of the first term in Eq. (2.29) to obtain the expression

$$\int_{t_1}^{t_2} \Big(\int_\Omega \varrho\, \ddot{u}_i\, \delta u_i \,\mathrm{d}\Omega + \delta \mathcal{J}_{\mathrm{P}} \Big)\, \mathrm{d}\tau = 0 \tag{2.32}$$

because

$$\begin{aligned}
\delta \int_{t_1}^{t_2}\int_\Omega \frac{1}{2} \varrho\, \dot{u}_i\, \dot{u}_i \,\mathrm{d}\Omega\,\mathrm{d}\tau &= \int_{t_1}^{t_2}\int_\Omega \varrho\, \dot{u}_i\, \delta\dot{u}_i \,\mathrm{d}\Omega\,\mathrm{d}\tau \\
&= -\int_{t_1}^{t_2}\int_\Omega \varrho\, \ddot{u}_i\, \delta u_i \,\mathrm{d}\Omega\,\mathrm{d}\tau + \int_{t_1}^{t_2}\int_\Omega \varrho \frac{\mathrm{d}}{\mathrm{d}t}(\dot{u}_i\, \delta u_i) \,\mathrm{d}\Omega\,\mathrm{d}\tau \\
&= -\int_{t_1}^{t_2}\int_\Omega \varrho\, \ddot{u}_i\, \delta u_i \,\mathrm{d}\Omega\,\mathrm{d}\tau + \int_\Omega \varrho \Big[\dot{u}_i\, \delta u_i \Big]_{t_1}^{t_2} \,\mathrm{d}\Omega\,\mathrm{d}\tau \\
&= -\int_{t_1}^{t_2}\int_\Omega \varrho\, \ddot{u}_i\, \delta u_i \,\mathrm{d}\Omega\,\mathrm{d}\tau
\end{aligned} \tag{2.33}$$

It follows that

$$\int_\Omega \varrho\, \ddot{u}_i\, \delta u_i \,\mathrm{d}\Omega + \delta \mathcal{J}_{\mathrm{P}} = 0 \tag{2.34}$$

or, more explicitly

$$\int_\Omega \varrho\, \ddot{u}_i\, \delta u_i \,\mathrm{d}\Omega + \int_\Omega C_{ijkl}\, \varepsilon_{ij}\, \delta\varepsilon_{kl} \,\mathrm{d}\Omega = \int_\Omega \varrho f_i\, \delta u_i \,\mathrm{d}\Omega + \int_{\partial\Omega_\sigma} \hat{t}_i\, \delta u_i \,\mathrm{d}(\partial\Omega) \tag{2.35}$$

It is seen that Eq. (2.35) represents the principle of virtual work (Eq. (2.15)) generalized to cover problems of linear dynamics. It is known as the d'Alembert principle, which states that a dynamic system may be set in a state of equilibrium by adding to the external forces a fictitious force commonly known as the inertial force. This force is equal to the mass multiplied by the acceleration, and should always be directed negatively with respect to the corresponding acceleration. The application of the d'Alembert principle allows us to use the equations of equilibrium in obtaining the equations of motion, thus representing a powerful tool of dynamic structural analysis.

2.4 FEM in Statics

2.4.1 Compatible Displacement Model

We have seen in Section 2.3.1 that applying the formal process of the calculus of variation to determine the stationary conditions for a functional yields differential field equations and appropriate boundary conditions for the problem considered. In contrast, substituting into the functional some assumed solutions, involving adjustable parameters and determining the stationary conditions with respect to these parameters provides direct methods to obtain approximate solutions to otherwise intractable problems. The main advantage of such an approach is the replacement of an infinite number of degrees of freedom, which characterize the original problem of solid mechanics, by a finite number of degrees of freedom which appear in the approximate solution.

The Rayleigh–Ritz method consists in approximating the unknown exact solution in the whole region Ω by a linear combination of, say, N suitably chosen coordinate functions which satisfy the kinematic boundary conditions. Substituting the adopted form of the solution in the potential energy functional and finding its stationary value with respect to the coefficients in the linear combination leads to N algebraic equations for the N coefficients. The solution to the problem is thus reduced to an algebraic process. In other word, the Rayleigh–Ritz procedure is as follows: instead of testing all possible admissible functions (infinite in number) to find the true one which gives a stationary value of the functional, we restrict the range of functions compared with the finite family generated by the coordinate functions.

The fundamental difficulty in applying the Rayleigh–Ritz methodology is the choice of the coordinate functions. Generating suitable approximating functions to satisfy the prescribed boundary conditions and to conform with the general deformation mode anticipated in the specific problem can be done fairly readily when the geometry of the solid in question and the corresponding loading are fairly simple. However, as the region becomes more complex in shape, the task becomes increasingly difficult until, for most situations of practical interest, it becomes quite impossible.

With the development of computer methods it has become possible to construct the approximating functions in a localized manner. Let us imagine a solid divided into a number of finite sized and conveniently shaped sub-regions or elements. In the finite element method the distribution of the required quantities (displacements in the case of the potential energy-based approach) is as a rule relatively simple within these elements, but by using a sufficient number of them an acceptable representation of the overall situation is obtained. The basic premise here is that each element is small, and so the variation of the test functions over its entire area will be small as well. The net effect is again to generate a substitutive system with a finite number of degrees of freedom in the form of nodal (generalized) displacements. However, in contrast to the classical Rayleigh–Ritz method, all the integrations required to define the appropriate functional must be evaluated in a piece-wise manner from element to element.

In accordance with our initial assumptions made in Section 2.1, no reference to the problem of selecting the approximating (or test, or shape) functions will be made in what follows. Nor will there be a thorough discussion of algebraic details related to forming the finite element equations on the structure level from the element contributions. Instead, merely a necessary finite element notation will be introduced and a procedure to derive the finite element equations will be described in general terms. The reader wishing to pursue these topics which are no doubt crucial to the FEM success, is referred to the books cited at the end of the book in the Reference list.

The simplest and most widely used finite element model is the so-called compatible displacement model based on the principle of stationary potential energy. The functional $\mathcal{J}_{\mathrm{P}}$ is recast as

$$\mathcal{J}_{\mathrm{P}}[u_i] = \sum_{e=1}^{E} \mathcal{J}_{\mathrm{P}}^{(e)}[u_i^{(e)}] \tag{2.36}$$

where the summation extends over a total of E discrete elements taking up the regions Ω_e, respectively, the index 'e' refers to a typical, or e-th, finite element and $u_i^{(e)}$ are localized displacement functions which vanish everywhere outside of the e-th element. The form (2.36) may be taken as equivalent to the one employed in the original statement of the principle of stationary potential energy provided that displacements satisfy the following requirements:[1] (i) they are continuous and differentiable in each element, (ii) they are continuous across the element boundaries, and (iii) they satisfy the kinematic boundary conditions. These conditions are clearly sufficient for the admissibility of displacements in the variational formulation since the first derivatives have, at worst, a discontinuity between the elements, and their energy is finite.

Assume now that the function $u_i(x_k)$ can be approximated within each element (say, the e-th element) by means of shape functions $\varphi_{i\xi}^{(e)}(x_k)$ as

$$u_i^{(e)}(x_k) = \varphi_{i\xi}^{(e)}(x_k)\, q_\xi^{(e)} \qquad x_k \in \Omega_e\,;\;\; \xi = 1,2,\ldots,N^{(e)} \tag{2.37}$$

where $q_\xi^{(e)}$ is the vector of element nodal (generalized) displacements and $N^{(e)}$ denotes the number of degrees of freedom assumed in the e-th element considered. Summation over repeated indices is implied. Using appropriate transformation the $N^{(e)}$-dimensional vector of elemental degrees of freedom $q_\xi^{(e)}$, $e = 1,2,\ldots,E$, $\xi = 1,2,\ldots,N^{(e)}$, may be derived from the N-dimensional vector of structural degrees of freedom q_α, $\alpha = 1,2,\ldots,N$; N being the total number of degrees of freedom in the discretized system. Let such a transformation be denoted by[2]

$$q_\xi^{(e)} = a_{\xi\alpha}^{(e)}\, q_\alpha \qquad x_k \in \Omega_e\,;\;\; \alpha = 1,2,\ldots,N \tag{2.38}$$

Eqs. (2.37) and (2.38) yield

$$u_i^{(e)}(x_k) = \varphi_{i\xi}^{(e)}(x_k)\, a_{\xi\alpha}^{(e)}\, q_\alpha \qquad x_k \in \Omega_e \tag{2.39}$$

or, more compactly

$$u_i^{(e)}(x_k) = \varphi_{i\alpha}^{(e)}(x_k)\, q_\alpha \qquad x_k \in \Omega_e \tag{2.40}$$

where the 'structure level' shape functions of the e-th element read

$$\varphi_{i\alpha}^{(e)}(x_k) = \varphi_{i\xi}^{(e)}(x_k)\, a_{\xi\alpha}^{(e)} \qquad x_k \in \Omega_e \tag{2.41}$$

---

[1]The conditions (i) and (ii) are slightly different for such structural mechanics problems as beams, plates and shells (i.e. problems involving bending).

[2]Detailed descriptions of the transformation (2.38) are given in every finite element textbook, cf. [3,7,21,25,38,50,58,65,84,94,117,126], for instance.

Instead of the expression (2.37) defined for each finite element separately it will now be convenient to symbolically represent the displacement vector $u_i(x_k)$ defined on the whole region Ω as

$$\begin{aligned} u_i(x_k) &= \sum_{e=1}^{E} u_i^{(e)}(x_k) \\ &= \Big[\sum_{e=1}^{E} \varphi_{i\alpha}^{(e)}(x_k)\Big]\, q_\alpha \;=\; \varphi_{i\alpha}(x_k)\, q_\alpha \qquad\qquad x_k \in \Omega \end{aligned} \tag{2.42}$$

which is a compact and useful expression for further finite element discussions.

The 'element level' expressions for the displacement gradient and the strain tensor have the form

$$\begin{aligned} u_{i,j}^{(e)}(x_k) &= \varphi_{i\xi,j}^{(e)}(x_k)\, q_\xi^{(e)} \\ \varepsilon_{ij}^{(e)}(x_k) &= \tfrac{1}{2}\Big[\varphi_{i\xi,j}^{(e)}(x_k) + \varphi_{j\xi,i}^{(e)}(x_k)\Big]\, q_\xi^{(e)} \;=\; B_{ij\xi}^{(e)}(x_k)\, q_\xi^{(e)} \\ & x_k \in \Omega_e\,;\;\; \xi = 1,2,\ldots,N^{(e)}\,;\;\; e = 1,2,\ldots,E \end{aligned} \tag{2.43}$$

whereas the corresponding 'structure level' expressions are

$$\begin{aligned} u_{i,j}(x_k) &= \varphi_{i\alpha,j}(x_k)\, q_\alpha \\ \varepsilon_{ij}(x_k) &= \tfrac{1}{2}\Big[\varphi_{i\alpha,j}(x_k) + \varphi_{j\alpha,i}(x_k)\Big]\, q_\alpha \;=\; B_{ij\alpha}(x_k)\, q_\alpha \\ & x_k \in \Omega\,;\;\; \alpha = 1,2,\ldots,N \end{aligned} \tag{2.44}$$

In what follows we shall be using the notation (2.44) rather than (2.43) as the former is more compact. However, it has to be always kept in mind that the structure of the $3\times N$ global shape function matrix $\varphi_{i\alpha}$ as defined in Eq. (2.42) is an essential ingredient of the finite element technique.

Let us now present the total potential energy of the body based on the approximation solution assumed in the form (2.42). Using Eq. (2.8) and Eq. (2.42) we arrive at the expression

$$\begin{aligned} \mathcal{J}_{\mathrm{P}}[q_\alpha] &= \Big[\int_\Omega \tfrac{1}{2}\, C_{ijkl}\, B_{ij\alpha}\, B_{kl\beta}\, \mathrm{d}\Omega\Big]\, q_\alpha\, q_\beta \\ &\quad - \Big[\int_\Omega \varrho f_i\, \varphi_{i\alpha}\, \mathrm{d}\Omega + \int_{\partial\Omega_\sigma} \hat{t}_i\, \varphi_{i\alpha}\, \mathrm{d}(\partial\Omega)\Big]\, q_\alpha \end{aligned} \tag{2.45}$$

or, employing the notation

$$\begin{aligned} \int_\Omega C_{ijkl}\, B_{ij\alpha}\, B_{kl\beta}\, \mathrm{d}\Omega &= K_{\alpha\beta} \\ \int_\Omega \varrho f_i\, \varphi_{i\alpha}\, \mathrm{d}\Omega + \int_{\partial\Omega_\sigma} \hat{t}_i\, \varphi_{i\alpha}\, \mathrm{d}(\partial\Omega) &= Q_\alpha \end{aligned} \tag{2.46}$$

at the expression

$$\mathcal{J}_{\mathrm{P}}[q_\alpha] = \tfrac{1}{2} K_{\alpha\beta}\, q_\alpha\, q_\beta - Q_\alpha\, q_\alpha \tag{2.47}$$

The form of Eq. (2.47) is a good illustration supporting the observation made just below Eq. (2.44). Even if no summation over finite elements is explicitly stated in Eq. (2.47) the real solution algorithm for establishing numerically the $N \times N$ stiffness matrix $K_{\alpha\beta}$ and the $N \times 1$ external load vector takes advantage of the localized nature of the shape functions and proceeds element-wise thus greatly facilitating the computational process. A more specific definition of the global stiffness would be

$$K_{\alpha\beta} = \sum_{e=1}^{E} k_{\xi\zeta}^{(e)} \, a_{\alpha\xi}^{(e)} \, a_{\beta\zeta}^{(e)} \tag{2.48}$$

where

$$k_{\xi\zeta}^{(e)} = \int_{\Omega_e} C_{ijkl} \, B_{ij\xi}^{(e)} \, B_{kl\zeta}^{(e)} \, \mathrm{d}\Omega \tag{2.49}$$

is the $N^{(e)} \times N^{(e)}$ element stiffness matrix. Similarly[1]

$$Q_{\alpha} = \sum_{e=1}^{E} Q_{\xi}^{(e)} a_{\alpha\xi}^{(e)} \tag{2.50}$$

where

$$Q_{\xi}^{(e)} = \int_{\Omega_e} \varrho f_i \, \varphi_{i\xi}^{(e)} \, \mathrm{d}\Omega + \int_{\partial\Omega_{\sigma,e}} \hat{t}_i \, \varphi_{i\xi}^{(e)} \, \mathrm{d}(\partial\Omega) \tag{2.51}$$

$\partial\Omega_{\sigma,e}$ being the part of the e-th element boundary which belongs to $\partial\Omega_{\sigma}$. We additionally note that the expression $-\sum_{e=1}^{E} Q_{\xi}^{(e)} a_{\alpha\xi}^{(e)}$ groups all the resultant vectors of forces acting upon particular nodes from their surrounding elements so that Eq. (2.50) simply expresses the equilibrium conditions at all the nodes of the discretized system.

The stationary conditions for the functional (2.47) are obtained by taking the variations of Eq. (2.47) with respect to each particular q_{α} and making them equal to zero, i.e.

$$\frac{\partial \mathcal{J}_{\mathrm{P}}}{\partial q_{\alpha}} = K_{\alpha\beta} \, q_{\beta} - Q_{\alpha} = 0 \tag{2.52}$$

or[2]

$$\begin{aligned} K_{\alpha\beta} \, q_{\beta} &= Q_{\alpha} \\ \mathbf{K} \, \mathbf{q} &= \mathbf{Q} \end{aligned} \tag{2.53}$$

[1]Issues related to the problems involved in the numerical evaluation of the integrals (2.49) and (2.51) are not taken up in this text, cf. [10,25,38,50,104,117,124].

[2]From now on, absolute matrix notation will be used in places to simplify the presentation. Thus, for instance, instead of $K_{\alpha\beta}$, q_{β} and $K_{\alpha\beta} \, q_{\beta} = Q_{\beta}$ we may write $\mathbf{K}$, $\mathbf{q}$ and $\mathbf{Kq} = \mathbf{Q}$, respectively.

Relationship (2.53) is referred to as the fundamental algebraic equation describing the problem of linear statics within the framework of the finite element compatible displacement model.

We note that in the notation subsequently followed in this book the stiffness matrix $K_{\alpha\beta}$ in Eq. (2.53) is the so-called reduced stiffness matrix. This means that the prescribed kinematic boundary conditions are assumed to have been imposed prior to the solution phase and no rigid-body degrees of freedom exist in the system described by Eq. (2.53). Thus, $K_{\alpha\beta}$ is positive definite.[1]

2.4.2 Remarks on Solution Algorithms and Computer Implementation

The effectiveness of the finite element analysis depends to a large degree on the numerical algorithms used to solve the large system of simultaneous, linear algebraic equations of the type (2.53). Since the stiffness matrix **K** is symmetric and positive definite, it is nonsingular and Eq. (2.53) has a unique solution. Although we can obtain the solution to Eq. (2.53) by the well-known Cramer's rule, the large amount of computation required to evaluate determinants makes the use of Cramer's rule impractical. Therefore, we need to select a different method.

In general, there are two types of numerical methods: direct and iterative. Direct methods give the exact solution (neglecting round-off errors) in a finite number of arithmetic operations. Iterative methods give a sequence of approximate solutions that generally converge to the exact solution as the number of iterations tends to infinity. In the application of the finite element method it is the direct techniques which are nearly always used; therefore, we shall confine ourselves to this class of methods in the brief discussion to follow.

Most direct methods are variations of Gaussian elimination. In particular, versions of Gaussian elimination, often referred to as the Cholesky methods, are most widely used in structural analysis [35,54]. All of them are based on reducing the original systems of equations to triangular systems, which are easy to solve.

There are two basic forms of Gaussian elimination: the standard form and the compact form. In the former Eq. (2.53) is reduced to the upper triangular system

$$\mathbf{U}\,\mathbf{q} = \mathbf{Q}^* \qquad (2.54)$$

which is solved by back substitution. The latter algorithm is based on the factorization of **K** into the product

$$\mathbf{K} = \mathbf{L}\,\mathbf{U} \qquad (2.55)$$

where **L** is a unit lower triangular matrix and **U** is an upper triangular matrix.

---

[1]This has not always to be the case when the finite element formulation is based on a multi-field variational principle.

Thus, Eq. (2.53) becomes

$$\mathbf{L}\,\mathbf{U}\,\mathbf{q} = \mathbf{Q} \tag{2.56}$$

which can be divided into two triangular systems

$$\mathbf{L}\,\mathbf{r} = \mathbf{Q} \tag{2.57}$$

and

$$\mathbf{U}\,\mathbf{q} = \mathbf{r} \tag{2.58}$$

Thus, Eq. (2.57) can be solved for **r** by forward substitution, and Eq. (2.58) can be solved for **q** by back substitution. Once we have obtained the **LU** decomposition of **K**, we can solve Eq. (2.53) for any load vector. In particular, the load vector **Q** need not be known at the time of the **LU** decomposition, which is in contrast to standard Gaussian elimination where **K** and **Q** are reduced simultaneously to **U** and $\mathbf{Q}^*$, Eq. (2.54). The decomposition of **K**, Eq. (2.55), can also be expressed as

$$\mathbf{K} = \mathbf{L}\,\mathbf{D}\,\mathbf{L}^{\mathrm{T}} \tag{2.59}$$

where

$$\mathbf{U} = \mathbf{D}\,\mathbf{L}^{\mathrm{T}} \tag{2.60}$$

and **D** is a diagonal matrix. In this case Eq. (2.58) is replaced by

$$\mathbf{D}\,\mathbf{L}^{\mathrm{T}}\mathbf{q} = \mathbf{r} \tag{2.61}$$

In the Cholesky method the matrix **K** is decomposed as

$$\mathbf{K} = \mathbf{U}^{\mathrm{T}}\,\mathbf{U} \tag{2.62}$$

where **U** is an upper triangular matrix with positive diagonal elements.

The advantage of the decomposition (2.59) is that it can be used for non positive definite systems with no complex arithmetic operations required. However, for most situations of practical interest, in which the system matrix is positive definite and all the diagonal elements have positive values, the Cholesky factorization (2.62) is much more suitable.

Apart from decompositions of the type (2.59) and (2.62), there exist various other decomposition schemes based on Gauss elimination, out of which we can mention the so-called static condensation [22,121,122], the substructure analysis [103,104] and the frontal solution [126]. The techniques are specially effective for problems of repetitive-shape type. However, since the order of the system matrix is changed during the solution process it is difficult to adapt these algorithms to fit into a standard numerical code available for problems including different types of finite elements.

In the finite element implementation we frequently have to deal with large coefficient matrices and multi-column right-hand sides. To reduce the high-speed memory required to solve such large-scale systems, some effective storage schemes for the coefficient matrix are used which allow us to save only the nonzero part of the upper triangular portion of the matrix in a multi-block form. For each block, the columns below the skyline of the matrix profile are saved column-wise in a one-dimensional array using the so-called *from-diagonals storage mode* [7,35] or the *to-diagonals storage mode* [124]. The skyline-storage techniques have definite advantages over the banded-storage approach [11,121], since (i) they always require less operating memory, (ii) they are not severely affected by a few very long columns; and the *mean*-bandwidth of the stiffness matrix is of interest rather than the *max*-bandwidth as in the banded-storage case, and (iii) only the dot product is used to perform the factorization and forward–backward substitutions which is important in view of vector-oriented computers.

As will be shown in the following section, the close relationship between the solution of the equilibrium (algebraic) equations and that of the differential equations of motion suggests that the solvers used for the former equations may also be employed in procedures of the modal analysis or the direct step-by-step integration analysis. In this context, the Cholesky procedure seems to be most effective; in particular, it is so when the procedure is employed to transform a generalized eigenproblem to the standard form. For the stochastic finite elements this technique is efficiently applied to the transformation of correlated random variables to a set of uncorrelated ones, cf. Part II.

2.5 FEM in Dynamics

2.5.1 Compatible Displacement Model

The equations describing the response of a body to dynamic excitations within the framework of the finite element displacement model may be obtained by using any of the variational statements (2.29) or (2.35), and the finite element interpolation procedure as presented in Section 2.4.1. We shall use here the d'Alembert principle slightly generalized to account for viscous (or damping) effects. Let us assume first that a part of such effects has the nature of the body forces and is proportional to the velocities of the body particles. Thus, the term $\alpha \int_\Omega \varrho \dot{u}_i \delta u_i \mathrm{d}\Omega$, where α is the proportionality factor, must be added to the left-hand side of Eq. (2.35). Furthermore, let us assume that the remaining part of the damping forces is due to the so-called viscous stresses defined as $\beta C_{ijkl} \dot{\varepsilon}_{kl}$, where β is the proportionality factor. Eq. (2.35) becomes

$$\int_\Omega \varrho\, \ddot{u}_i\, \delta u_i\, \mathrm{d}\Omega + \int_\Omega \left(\alpha \varrho\, \dot{u}_i\, \delta u_i + \beta\, C_{ijkl}\, \dot{\varepsilon}_{ij}\, \delta\varepsilon_{kl} \right) \mathrm{d}\Omega + \int_\Omega C_{ijkl}\, \varepsilon_{ij}\, \delta\varepsilon_{kl}\, \mathrm{d}\Omega = \int_\Omega \varrho f_i\, \delta u_i\, \mathrm{d}\Omega + \int_{\partial\Omega_\sigma} \hat{t}_i\, \delta u_i\, \mathrm{d}(\partial\Omega) \tag{2.63}$$

where the strain variation $\delta\varepsilon_{kl}$ is understood as before, Eq. $(2.28)_1$.

Remembering the considerations of Section 2.4.1, Eq. (2.42) with respect to the adopted finite element notation, we assume now that

$$u_i(x_k, \tau) = \varphi_{i\alpha}(x_k)\, q_\alpha(\tau) \tag{2.64}$$

where the generalized coordinates q_α, $\alpha = 1, 2, \ldots, N$, are now functions of time. By using Eqs. (2.64) and $(2.28)_1$, Eq. (2.63) leads directly to the equations of motion for the spatially discretized system in the form

$$\begin{aligned} M_{\alpha\beta}\, \ddot{q}_\beta(\tau) + C_{\alpha\beta}\, \dot{q}_\beta(\tau) + K_{\alpha\beta}\, q_\beta(\tau) &= Q_\alpha(\tau) \\ \mathbf{M}\,\ddot{\mathbf{q}}(\tau) + \mathbf{C}\,\dot{\mathbf{q}}(\tau) + \mathbf{K}\,\mathbf{q}(\tau) &= \mathbf{Q}(\tau) \end{aligned} \tag{2.65}$$

where, apart from the definitions adopted in Section 2.4.1, Eq. (2.46), we have denoted the system mass and damping matrices as

$$M_{\alpha\beta} = \int_\Omega \varrho\, \varphi_{i\alpha}\, \varphi_{i\beta}\, \mathrm{d}\Omega \tag{2.66}$$

$$C_{\alpha\beta} = \int_\Omega \Big(\alpha \varrho\, \varphi_{i\alpha}\, \varphi_{i\beta} + \beta\, C_{ijkl}\, B_{ij\alpha}\, B_{kl\beta} \Big) \mathrm{d}\Omega = \alpha M_{\alpha\beta} + \beta K_{\alpha\beta} \tag{2.67}$$

These matrices can be obtained from the corresponding elemental matrices by using the same assembly process as employed in generating the system stiffness matrix.

2.5.2 Remarks on Solution Algorithms and Computer Implementation

Eq. (2.65) represents a system of linear ordinary differential equations of second order with constant coefficients. The system can be integrated in the time domain (or in the frequency domain) by using some standard time-stepping procedures (or frequency-stepping procedures via the Laplace or Fourier transform). However, the procedures developed for the solution of general systems of ordinary differential equations can be quite ineffective when used for systems consisting of many equations (say, $10^2 \rightarrow 10^4$ or more) unless specific advantage is taken of the special characteristics of the coefficient matrices. In the finite element context one of the two basic time-stepping algorithms is alternatively considered: the direct integration technique or mode superposition technique. Although both techniques are apparently quite different, they can be shown to be closely related [7,10,22].

(a) Direct (Step-by-Step) Integration Methods

In the direct integration schemes no transformation of Eq. (2.65) into a simpler form is carried out and the equations are time-integrated using a time-stepping procedure. Two basic assumptions are at the root of the direct integration method: (i) Eq. (2.65) is enforced to hold at the time instants $0, \Delta t, 2\Delta t \ldots$ only and need not be satisfied at any other instant, and (ii) the accelerations, velocities and, consequently, displacements are prescribed as functions of time within each time interval considered. Adopting these assumptions implies that the original system (2.65) has to be converted at each selected time instant into a corresponding system of linear algebraic equations with an effective 'stiffness' matrix and an effective 'right-hand side' vector, which can be repeatedly solved by using any of the algorithms described in Section 2.4.2, for instance.

Below we shall briefly describe some aspects of two so-called implicit integration techniques which are widely used in structural dynamics. The techniques are known as the Wilson θ-method and the Newmark method, respectively. Both methods are essentially an extension of the so-called linear acceleration strategy [7,22,64] in which the acceleration is assumed to change linearly in the time interval considered.

The Wilson θ-method is based on the assumption that

$$\ddot{\mathbf{q}}(t+\tau) = \ddot{\mathbf{q}}(t) + \frac{\tau}{\theta\Delta t}\left[\ddot{\mathbf{q}}(t+\theta\Delta t) - \ddot{\mathbf{q}}(t)\right] \qquad \tau \in [0, \theta\Delta t] \tag{2.68}$$

where $\theta \geq 1$ is a parameter of the method. Integrating this equation twice with respect to time τ leads to

$$\begin{aligned}\dot{\mathbf{q}}(t+\tau) &= \dot{\mathbf{q}}(t) + \tau\,\ddot{\mathbf{q}}(t) + \frac{\tau^2}{2\theta\Delta t}\left[\ddot{\mathbf{q}}(t+\theta\Delta t) - \ddot{\mathbf{q}}(t)\right] \\ \mathbf{q}(t+\tau) &= \mathbf{q}(t) + \tau\,\dot{\mathbf{q}}(t) + \frac{\tau^2}{2}\ddot{\mathbf{q}}(t) + \frac{\tau^3}{6\,\theta\Delta t}\left[\ddot{\mathbf{q}}(t+\theta\Delta t) - \ddot{\mathbf{q}}(t)\right]\end{aligned} \tag{2.69}$$

Setting $\tau = \theta\Delta t$ and solving the resulting equations for $\ddot{\mathbf{q}}(t+\theta\Delta t)$ and $\dot{\mathbf{q}}(t+\theta\Delta t)$ gives

$$\begin{aligned}\ddot{\mathbf{q}}(t+\theta\Delta t) &= \frac{6}{(\theta\Delta t)^2}\left[\mathbf{q}(t+\theta\Delta t) - \mathbf{q}(t)\right] - \frac{6}{\theta\Delta t}\dot{\mathbf{q}}(t) - 2\ddot{\mathbf{q}}(t) \\ \dot{\mathbf{q}}(t+\theta\Delta t) &= \frac{3}{\theta\Delta t}\left[\mathbf{q}(t+\theta\Delta t) - \mathbf{q}(t)\right] - 2\dot{\mathbf{q}}(\mathbf{t}) - \frac{\theta\Delta t}{2}\ddot{\mathbf{q}}(t)\end{aligned} \tag{2.70}$$

By substituting Eqs. (2.70) into the equation of motion (2.65) specified for the time instant $t+\theta\Delta t$ we arrive at the linear algebraic system for $\mathbf{q}(t+\theta\Delta t)$

$$\mathbf{K}^{(\text{eff})}\mathbf{q}(t+\theta\Delta t) = \mathbf{Q}^{(\text{eff})}(t+\theta\Delta t) \tag{2.71}$$

with $\mathbf{K}^{(\text{eff})}$ and $\mathbf{Q}^{(\text{eff})}$ representing the effective 'stiffness' matrix and the effective 'load' vector and expressed as

$$\mathbf{K}^{(\text{eff})} = \frac{6}{(\theta\Delta t)^2}\mathbf{M} + \frac{3}{\theta\Delta t}\mathbf{C} + \mathbf{K} \tag{2.72}$$

$$\begin{aligned}\mathbf{Q}^{(\text{eff})} &= \mathbf{Q}(t) + \theta\left[\mathbf{Q}(t+\Delta t) - \mathbf{Q}(t)\right] \\ &\quad + \mathbf{M}\left[\frac{6}{(\theta\Delta t)^2}\mathbf{q}(t) + \frac{6}{\theta\Delta t}\dot{\mathbf{q}}(t) + 2\ddot{\mathbf{q}}(t)\right] \\ &\quad + \mathbf{C}\left[\frac{3}{\theta\Delta t}\mathbf{q}(t) + 2\dot{\mathbf{q}}(t) + \frac{\theta\Delta t}{2}\ddot{\mathbf{q}}(t)\right]\end{aligned} \tag{2.73}$$

Having solved Eq. (2.71) for $\mathbf{q}(t+\theta\Delta t)$ (the right-hand side vector $\mathbf{Q}^{(\text{eff})}(t+\theta\Delta t)$ is known) we may proceed by getting the vector $\ddot{\mathbf{q}}(t+\theta\Delta t)$ from Eq. $(2.70)_1$

followed by the evaluation of $\ddot{\mathbf{q}}(t+\Delta t)$, $\dot{\mathbf{q}}(t+\Delta t)$ and $\mathbf{q}(t+\Delta t)$ (using Eqs. (2.68) and (2.69) at $\tau = \Delta t$) which completes the solution at the time step considered.

The Newmark method is based on the time expansions for the nodal acceleration and velocity vectors in the form

$$\begin{aligned} \ddot{\mathbf{q}}(t+\Delta t) &= \frac{1}{\alpha \Delta t^2}\Big[\mathbf{q}(t+\Delta t) - \mathbf{q}(t)\Big] - \frac{1}{\alpha \Delta t}\dot{\mathbf{q}}(t) - \Big(\frac{1}{2\alpha} - 1\Big)\ddot{\mathbf{q}}(t) \\ \dot{\mathbf{q}}(t+\Delta t) &= \dot{\mathbf{q}}(t) + \Delta t\Big[(1-\delta)\,\ddot{\mathbf{q}}(t) + \delta\,\ddot{\mathbf{q}}(t+\Delta t)\Big] \end{aligned} \tag{2.74}$$

where α and δ are constants which can be selected to obtain the best stability and accuracy characteristics of the integration scheme. As a result of derivations which are analogous to the ones described above we end up with a system of linear algebraic equations of the type similar to Eqs. (2.71)–(2.73), which by the repeated solution for subsequent time instants serves as a means to obtain the overall time response of the structure in the time interval required.

It should be noted that with $\delta \geq 0.5$, $\alpha \geq 0.25(\delta + 0.5)^2$ and $\theta \geq 1.37$ both the Newmark method and Wilson θ-method are unconditionally stable. Also, for a constant value of Δt each of the methods requires only one factorization of the effective 'stiffness' matrix which is then used repeatedly to compute subsequent displacement vectors for changing right-hand side vectors. The similarity of the both methods makes it possible to conveniently implement them in a single finite element code. The methods set no limits on the form of the damping matrix.

The shortcomings of the both methods, and of the integration techniques as a whole, are:

- they always explicitly deal with all the degrees of freedom of the discretized system,
- the computational cost depends strongly on the bandwidth of the coefficient matrices,
- the number of arithmetic operations required is directly proportional to the number of time intervals used in the analysis; thus the techniques are expected to be effective only when the response may be accurately approximated by using a small number (say, 10^2 or so) of time intervals.

(b) Mode Superposition Method

Computation costs of the direct integration strategy may become unacceptably high for the large system response analysis of a relatively long duration (a large number of time intervals required). Therefore, it may often be more effective to first transform the equations of motion into a form for which the step-by-step solution is less costly. The objective of the transformation is to obtain new system mass, damping and stiffness matrices that have a much smaller bandwidth than the original ones.

As a matter of fact, there exist many different orthogonal transformations which would reduce the bandwidth of the system matrices [7,22,75,86]. However, in structural dynamics probably the most effective transformation is accomplished by the change of basis from finite element coordinates to normalized modal coordinates. Using the mode shape matrix as the transformation matrix the system of coupled equations of motion (2.65) is converted into a decoupled system of equations which is then much easier to integrate in time.

The mode shape matrix is obtained as follows. If we substitute a trial function

$$\mathbf{q}(\tau) = \boldsymbol{\Phi} \exp\left[\mathrm{i}\,\omega(\tau - t_0)\right] \tag{2.75}$$

into the homogeneous undamped equation

$$\mathbf{M}\,\ddot{\mathbf{q}}(\tau) + \mathbf{K}\,\mathbf{q}(\tau) = \mathbf{0} \tag{2.76}$$

we obtain the system of simultaneous equations describing the free vibration problem in the form

$$\left(\mathbf{K} - \omega^2\,\mathbf{M}\right)\boldsymbol{\Phi} = \mathbf{0} \tag{2.77}$$

Eq. (2.77) is a generalized algebraic eigenproblem which may be solved for N solutions pairs $(\omega^2_{(1)}, \boldsymbol{\Phi}_{(1)})$, $(\omega^2_{(2)}, \boldsymbol{\Phi}_{(2)})$, ..., $(\omega^2_{(N)}, \boldsymbol{\Phi}_{(N)})$ in which $\omega_{(\alpha)}$ and $\boldsymbol{\Phi}_{(\alpha)}$ are called the α-th system frequency and vibration mode, respectively.

Both matrices $\mathbf{K}$ and $\mathbf{M}$ may be assumed symmetric and positive definite; the so-called consistent mass matrix resulting directly from definition (2.66) is always nonsingular and similarly populated (banded) as the stiffness matrix, whereas a lumped mass matrix with some zero-diagonal entries may always be transformed to become nonsingular by using any technique to eliminate the massless degrees of freedom [7].

The generalized eigenvalue problem (2.77) may easily be transformed to the standard form by decomposing the mass matrix as, cf. Eq. (2.62)

$$\mathbf{M} = \mathbf{U}^{\mathrm{T}}\,\mathbf{U} \tag{2.78}$$

$\mathbf{U}$ being a nonsingular matrix. Substituting Eq. (2.78) into Eq. (2.77) and pre- and post-multiplying the latter by $\mathbf{U}^{-\mathrm{T}}$ and $\mathbf{U}^{-1}$, respectively, the standard eigenproblem is obtained as

$$\left(\tilde{\mathbf{K}} - \omega^2\,\mathbf{I}\right)\tilde{\boldsymbol{\Phi}} = \mathbf{0} \tag{2.79}$$

where

$$\tilde{\mathbf{K}} = \mathbf{U}^{-\mathrm{T}}\,\mathbf{K}\,\mathbf{U}^{-1} \qquad \tilde{\boldsymbol{\Phi}} = \mathbf{U}\,\boldsymbol{\Phi} \tag{2.80}$$

and $\mathbf{I}$ is the unit matrix. Since for symmetric matrices eigenvectors corresponding to different eigenvalues are orthogonal and the eigenvectors are defined up to a scalar multiplier only, we may assume

$$\tilde{\boldsymbol{\Phi}}^{\mathrm{T}}_{(\alpha)}\,\tilde{\boldsymbol{\Phi}}_{(\beta)} = \delta_{\alpha\beta} \tag{2.81}$$

which by Eq. (2.79) implies

$$\tilde{\Phi}^{\mathrm{T}}_{(\alpha)}\,\tilde{\mathbf{K}}\,\tilde{\Phi}_{(\beta)} = \omega^2\,\delta_{\alpha\beta} \tag{2.82}$$

so that by Eqs. (2.78) and (2.80) we also have

$$\Phi^{\mathrm{T}}_{(\alpha)}\,\mathbf{M}\,\Phi_{(\beta)} = \delta_{\alpha\beta} \qquad \Phi^{\mathrm{T}}_{(\alpha)}\,\mathbf{K}\,\Phi_{(\beta)} = \omega^2\,\delta_{\alpha\beta} \tag{2.83}$$

The mode shape matrix $\Phi_{N\times N}$ and the diagonal matrix $\Omega_{N\times N}$ of natural frequencies squared are defined as

$$\begin{aligned} \Phi_{N\times N} &= \phi_{\alpha\beta} = \left[\ \phi_{\alpha(1)}\ \ \phi_{\alpha(2)}\ \ \ldots\ \ \phi_{\alpha(N)}\ \right]_{N\times N}, \qquad \alpha = 1,2,\ldots,N \\ \Omega_{N\times N} &= \mathrm{diag}\left[\ \omega^2_{(1)}\ \ \omega^2_{(2)}\ \ \ldots\ \ \omega^2_{(N)}\ \right]_{N\times N} \end{aligned} \tag{2.84}$$

Let us now present the finite element solution $q_\alpha(x_k,\tau)$ to Eq. (2.65) as

$$q_\alpha(x_k,\tau) = \phi_{\alpha\bar{\alpha}}(x_k)\,r_{\bar{\alpha}}(\tau) \qquad \bar{\alpha} = 1,2,\ldots,\bar{N} \tag{2.85}$$

where $r_{\bar{\alpha}}$ is the vector of modal coordinates, $\bar{N}$ is much smaller than N and $\phi_{\alpha\bar{\alpha}}$ is the $N\times\bar{N}$ submatrix of Φ consisting of $\bar{N}$ columns. The matrix $\phi_{\alpha\bar{\alpha}}$ satisfies the equation

$$K_{\alpha\beta}\,\phi_{\beta\bar{\alpha}} = \Omega_{(\bar{\alpha})}\,M_{\alpha\beta}\,\phi_{\beta\bar{\alpha}} \tag{2.86}$$

where $\Omega_{(\bar{\alpha})}$ is the diagonal matrix of the first $\bar{N}$ natural frequencies squared. Inserting Eq. (2.85) into Eq. (2.65), pre-multiplying the result by $\Phi_{\alpha\bar{\alpha}}$ transposed, imposing the M-orthonormality and K-orthogonality conditions

$$\begin{aligned} \phi^{\mathrm{T}}_{\bar{\alpha}\alpha}\,M_{\alpha\beta}\,\phi_{\beta\bar{\beta}} &= \delta_{\bar{\alpha}\bar{\beta}} \\ \phi^{\mathrm{T}}_{\bar{\alpha}\alpha}\,K_{\alpha\beta}\,\phi_{\beta\bar{\beta}} &= \Omega_{(\bar{\alpha})}\,\delta_{\bar{\alpha}\bar{\beta}} \qquad \text{(no sum on } \bar{\alpha}) \end{aligned} \tag{2.87}$$

and assuming that (so-called modal damping)[1]

$$\phi^{\mathrm{T}}_{\bar{\alpha}\alpha}\,C_{\alpha\beta}\,\phi_{\beta\bar{\beta}} = 2\,\xi_{(\bar{\alpha})}\,\omega_{(\bar{\alpha})}\,\delta_{\bar{\alpha}\bar{\beta}} \tag{2.88}$$

we obtain the system of $\bar{N}$ uncoupled equations for $r_{\bar{\alpha}}(\tau)$ as

$$\ddot{r}_{\bar{\alpha}}(\tau) + 2\,\xi_{(\bar{\alpha})}\,\omega_{(\bar{\alpha})}\,\dot{r}_{\bar{\alpha}}(\tau) + \omega^2_{(\bar{\alpha})}\,r_{\bar{\alpha}}(\tau) = R_{\bar{\alpha}}(\tau) \qquad \begin{array}{c} \bar{\alpha} = 1,2,\ldots,\bar{N} \\ \text{(no sum on } \bar{\alpha}) \end{array} \tag{2.89}$$

in which $\xi_{(\bar{\alpha})}$ is a damping factor associated with the $\bar{\alpha}$-th mode, $\omega_{(\bar{\alpha})}$ is the $\bar{\alpha}$-th frequency of the undamped system and $R_{\bar{\alpha}}$ is the $\bar{\alpha}$-th normalized force defined by

$$R_{\bar{\alpha}}(\tau) = \phi^{\mathrm{T}}_{\bar{\alpha}\beta}\,Q_\beta(\tau) \tag{2.90}$$

Each of the $\bar{N}$ equations in the system (2.89) can be solved independently for the $\bar{N}$ modal coordinates $r_{\bar{\alpha}}(\tau)$ by using any time-stepping scheme or by employing the so-called Duhamel's convolution [22,86]. The final solution, $q_\alpha(\tau)$, $\alpha = 1,2,\ldots,N$, is then obtained from the mode superposition according to Eq. (2.85).

Since it frequently turns out in practice that it is sufficient to take $\bar{N}$ much smaller than N ($\bar{N} \sim 10^1$, for instance), the mode superposition method is a powerful tool of structural analysis.

---

[1] This assumption can be significantly relaxed, the mode superposition method still being applicable. In particular, any damping matrix of the form $\mathbf{C} = \alpha\mathbf{M} + \beta\mathbf{K}$, α and β arbitrary coefficients, may be considered as well.

2.6 Some Aspects of Nonlinear Analysis

2.6.1 Piece-Wise Linearization of Nonlinear Time Response

This book is essentially about stochastic analysis of linear structural systems. However, before going on to discuss the very subject of the book in Part II it will be instructive to pause for a moment to consider some aspects of nonlinear structural analysis. Again, it will be sufficient for the presentation of main ideas involved in such an analysis if we just confine ourselves to the three-dimensional case.

It is now widely acknowledged that for an effective numerical treatment of problems involving large deformations and/or material nonlinearities, it is necessary to use an incremental formulation. A simple (in fact, perhaps the simplest possible) version of such an analysis will be briefly presented below.

As before, we consider the motion of a body with respect to the fixed Cartesian coordinates $\{x_k\}$. Our aim now is to evaluate the subsequent positions of the body at the discrete time points $0, \Delta t, 2\Delta t, \ldots$.[1] We assume that the solution for the kinematic and static variables for all time steps from time 0 to time t, inclusive, has been obtained, and we now seek the solution for the next required equilibrium position. This solution procedure is typical and if applied repeatedly permits us to find the complete solution path. At the end of each step one clearly has to perform the accumulation procedure summing up the values of particular functions at time t and the increments just computed.

We shall adopt the last known configuration (i.e. the one at time t) as the reference state; the region taken up by the body at this instant will be denoted by Ω^t. Assuming that the incremental process from t to $t+\Delta t$ is 'infinitesimal' the problem to be solved may be thought of as the linear mechanics problem for the body subjected to 'initial' stresses (and perhaps some other 'initial' fields) existing at time t. Such a viewpoint corresponds roughly to what is known in the literature as the 'explicit updated Lagrangian description'. The equations describing the incremental problem may be presented in the time interval $t, t+\Delta t$ as follows [9,45,67]:

(a) incremental strain–incremental displacement relations

$$\Delta\varepsilon_{ij} = \frac{1}{2}(\Delta u_{i,j} + \Delta u_{j,i}) \qquad x_k \in \Omega^t \tag{2.91}$$

(b) incremental constitutive equations

$$\Delta\sigma^{\mathrm{II}}_{ij} = C_{ijkl}\,\Delta\varepsilon_{kl} \qquad x_k \in \Omega^t \tag{2.92}$$

(c) incremental equations of motion

$$\Delta\sigma^{\mathrm{I}}_{ij,j} + \varrho\,\Delta f_i = \varrho\,\Delta\ddot{u}_i \qquad x_k \in \Omega^t \tag{2.93}$$

[1] More generally, the time variable τ can run over the sequence $0, t_1, t_2, \ldots$ corresponding to time intervals which are not necessarily equal. Our assumption greatly simplifies the notation without any real loss of generality.

(d) stress-type boundary conditions

$$\Delta\sigma^{\mathrm{I}}_{ij}\, n_j = \Delta\hat{t}_i \qquad x_k \in \partial\Omega^t_\sigma \tag{2.94}$$

(e) kinematic boundary conditions

$$\Delta u_i = \Delta\hat{u}_i \qquad x_k \in \partial\Omega^t_u \tag{2.95}$$

The constitutive moduli are assumed to possibly be functions of 'initial' fields such as stresses and/or internal state variables.

The only symbols which are not self-explanatory in the light of linear mechanics equations (2.91)–(2.95) and the above discussion are the stress increments $\Delta\sigma^{\mathrm{I}}_{ij}$ and $\Delta\sigma^{\mathrm{II}}_{ij}$, referred to as the increments of the first and second Piola–Kirchhoff stress tensor based on the current configuration, respectively. We note that within the description adopted here the total stresses σ^{I}_{ij} and $\sigma^{\mathrm{II}}_{ij}$ are equal to each other at the beginning of the step and, moreover, equal to the usual Cauchy stress tensor σ_{ij}. The incremental stresses $\Delta\sigma^{\mathrm{I}}_{ij}$ and $\Delta\sigma^{\mathrm{II}}_{ij}$ (of which only $\Delta\sigma^{\mathrm{II}}_{ij}$ is symmetric!) are related by the equation

$$\Delta\sigma^{\mathrm{I}}_{ij} = \Delta\sigma^{\mathrm{II}}_{ij} + \Delta u_{i,k}\,\sigma_{kj} \tag{2.96}$$

which completes the field equations for the incremental problem at hand. Eq. (2.96) may be used to eliminate $\Delta\sigma^{\mathrm{I}}_{ij}$ from Eqs. (2.93) and (2.94) thus making the structure of the governing equation system in the nonlinear case very similar to the linear one we considered in Section 2.2.

We note that at the end of the step the updated Lagrangian approach requires transformation of stresses from the reference configuration at time t to the next one just determined; transformation of the coordinate system may turn out convenient as well.

We further note that in the limiting case when $\Delta t \to 0$ the so-called rate-formulation is obtained from the above system as

$$\left.\begin{array}{ll}
\left.\begin{array}{l}
\dot{\sigma}^{\mathrm{I}}_{ij,j} + \varrho\dot{f}_i = \varrho\,\ddot{\dot{u}}_i \\
\dot{\varepsilon}_{ij} = \frac{1}{2}(\dot{u}_{i,j} + \dot{u}_{j,i}) \\
\dot{\sigma}^{\mathrm{II}}_{ij} = C_{ijkl}\,\dot{\varepsilon}_{kl}
\end{array}\right\} & x_k \in \Omega^t\ ; \\
\dot{\sigma}^{\mathrm{I}}_{ij}\, n_j = \dot{\hat{t}}_i & x_k \in \partial\Omega^t_\sigma\ ; \\
\dot{u}_i = \dot{\hat{u}}_i & x_k \in \partial\Omega^t_u\ ; \\
\dot{\sigma}^{\mathrm{I}}_{ij} = \dot{\sigma}^{\mathrm{II}}_{ij} + \dot{u}_{i,k}\,\sigma_{kj} & x_k \in \Omega^t\ ;
\end{array}\right\} \quad t \in [0,\infty) \tag{2.97}$$

Similarly as in the linear case we may eliminate from the above set the strain- and stress-rates which leads to the displacement-rate formulation of the problem in the following form:

$$\left.\begin{array}{ll}
\left(C_{ijkl}\,\dot{u}_{k,l} + \dot{u}_{i,k}\,\sigma_{kj}\right)_{,j} + \varrho\dot{f}_i - \varrho\,\ddot{\dot{u}}_i = 0\ , & x_k \in \Omega^t\ ; \\
\dot{u}_i = \dot{\hat{u}}_i & x_k \in \partial\Omega^t_u\ ; \\
\left(C_{ijkl}\,\dot{u}_{k,l} + \dot{u}_{i,k}\,\sigma_{kj}\right) n_j = \dot{\hat{t}}_i & x_k \in \partial\Omega^t_\sigma\ ;
\end{array}\right\} \quad t \in [0,\infty) \tag{2.98}$$

where the index 'II' on σ_{ij} has been dropped for simplicity.

2.6.2 Finite Element Equations

Returning to the incremental problem described by Eqs. (2.91)–(2.96) it is noted that for a typical incremental step the virtual (incremental) work equation takes the form, cf. Eq. (2.63)

$$\begin{aligned}\int_\Omega \varrho\, \Delta\ddot{u}_i\, \delta(\Delta u_i)\, \mathrm{d}\Omega + \int_\Omega \left[\alpha\, \varrho\, \Delta\dot{u}_i\, \delta(\Delta u_i) + \beta\, C_{ijkl}\, \Delta\dot{\varepsilon}_{ij}\, \delta(\Delta\varepsilon_{kl})\right] \mathrm{d}\Omega \\ + \int_\Omega \left[C_{ijkl}\, \Delta\varepsilon_{ij}\, \delta(\Delta\varepsilon_{kl}) + \sigma_{kl}\, \Delta u_{i,k}\, \delta(\Delta u_{i,l})\right] \mathrm{d}\Omega \\ = \int_\Omega \varrho\, \Delta f_i\, \delta(\Delta u_i)\, \mathrm{d}\Omega + \int_{\partial\Omega_\sigma} \Delta\hat{t}_i\, \delta(\Delta u_i)\, \mathrm{d}(\partial\Omega)\end{aligned} \tag{2.99}$$

This equation describes incrementally the nonlinear problem of structural dynamics with respect to a fixed reference configuration at the beginning of the step. Apart from the incremental quantities involved and the necessary updating procedures Eq. (2.99) differs formally from Eq. (2.63) describing the linear elastic problem just by the additional term representing the incremental virtual work owing to the existence of nonzero stresses σ_{ij} at the beginning of the step.

The derivation of the finite element equations is also formally very similar; adopting Eq. (2.64) as the displacement expansion, introducing it into Eq. (2.99) and using the fact that the generalized displacements $q_\alpha(\tau)$ are variated independently leads to an equation of the form

$$M_{\alpha\beta}\, \Delta\ddot{q}_\beta(\tau) + C_{\alpha\beta}\, \Delta\dot{q}_\beta(\tau) + \left(K_{\alpha\beta}^{(\mathrm{con})} + K_{\alpha\beta}^{(\sigma)}\right) \Delta q_\beta(\tau) = \Delta Q_\alpha(\tau) \tag{2.100}$$

in which

$$K_{\alpha\beta}^{(\mathrm{con})} = \int_\Omega C_{ijkl}\, B_{ij\alpha}\, B_{kl\beta}\, \mathrm{d}\Omega \tag{2.101}$$

is the constitutive stiffness matrix while

$$K_{\alpha\beta}^{(\sigma)} = \int_\Omega \sigma_{ij}\, \varphi_{k\alpha,i}\, \varphi_{k\beta,j}\, \mathrm{d}\Omega \tag{2.102}$$

is the initial stress (or geometric) stiffness matrix. It is noted that the latter matrix is a linear and homogeneous function of the stress components σ_{ij}. The tangent stiffness matrix is more compactly written as

$$K_{\alpha\beta}^{(\mathrm{T})} = K_{\alpha\beta}^{(\mathrm{con})} + K_{\alpha\beta}^{(\sigma)} \tag{2.103}$$

It is sometimes more convenient to use the equation of motion (2.100) in a modified form which reads

$$M_{\alpha\beta}\, \ddot{q}_\beta(t+\Delta t) + C_{\alpha\beta}\, \dot{q}_\beta(t+\Delta t) + K_{\alpha\beta}^{(\mathrm{T})}\, \Delta q_\beta = Q_\alpha(t+\Delta t) - F_\alpha(t) \tag{2.104}$$

which can readily be obtained from Eq. (2.100) by noting that

$$\begin{aligned}\Delta\ddot{q}_\alpha &= \ddot{q}_\alpha(t+\Delta t) - \ddot{q}_\alpha(t) \\ \Delta\dot{q}_\alpha &= \dot{q}_\alpha(t+\Delta t) - \dot{q}_\alpha(t)\end{aligned} \tag{2.105}$$

The symbol $F_\alpha(t)$ in Eq. (2.104) stands for the so-called internal nodal force vector based on (i.e. statically equivalent to) stresses at the beginning of the step and can be calculated as

$$F_\alpha = \int_\Omega \sigma_{ij}\, B_{ij\alpha}\, \mathrm{d}\Omega \tag{2.106}$$

In contrast to linear elasticity problems, it is impossible to develop one single method that could be efficiently used for any nonlinear finite element application. Thus, only an examplary algorithm typical of a broader class of time integration procedures is briefly described below, which is believed to be instructive in further studies; to this aim the reader may consult [7,64,99,125], for instance.

Let us assume that Eq. (2.100) holds in the time interval $[t, t+\theta\Delta t]$. Eq. (2.69) can then be rewritten in incremental form as

$$\begin{aligned} \Delta\dot{\mathbf{q}}(\tau) &= \tau\,\ddot{\mathbf{q}}(t) + \frac{\tau^2}{2\theta\Delta t}\Delta\ddot{\mathbf{q}}(\tau) \\ \Delta\mathbf{q}(\tau) &= \tau\,\dot{\mathbf{q}}(t) + \frac{\tau^2}{2}\ddot{\mathbf{q}}(t) + \frac{\tau^3}{6\,\theta\Delta t}\Delta\ddot{\mathbf{q}}(\tau) \end{aligned} \qquad \tau \in [0, \theta\Delta t] \tag{2.107}$$

The vectors $\Delta\ddot{\mathbf{q}}$ and $\Delta\dot{\mathbf{q}}$ can be expressed in terms of $\Delta\mathbf{q}$ by solving Eq. $(2.107)_2$ for $\Delta\ddot{\mathbf{q}}$ and substituting the result into Eq. $(2.107)_1$. Setting $\tau = \theta\Delta t$ leads to

$$\begin{aligned} \Delta\ddot{\mathbf{q}}^* &= \frac{6}{(\theta\Delta t)^2}\Delta\mathbf{q}^* - \frac{6}{\theta\Delta t}\dot{\mathbf{q}}(t) - 3\ddot{\mathbf{q}}(t) \\ \Delta\dot{\mathbf{q}}^* &= \frac{3}{\theta\Delta t}\Delta\mathbf{q}^* - 3\dot{\mathbf{q}}(t) - \frac{\theta\Delta t}{2}\ddot{\mathbf{q}}(t) \end{aligned} \tag{2.108}$$

where $\Delta\ddot{\mathbf{q}}^*$, $\Delta\dot{\mathbf{q}}^*$ and $\Delta\mathbf{q}^*$ are increments of the nodal accelerations, velocities and displacements, respectively, associated with the extended time interval $\theta\Delta t$. By introducing Eqs. (2.108) into Eq. (2.100) specified at the time instant $t+\theta\Delta t$ we obtain the algebraic system for $\Delta\mathbf{q}^*$ as

$$\mathbf{K}^{(\mathrm{eff})}\Delta\mathbf{q}^* = \Delta\mathbf{Q}^{(\mathrm{eff})} \tag{2.109}$$

where

$$\mathbf{K}^{(\mathrm{eff})} = \frac{6}{(\theta\Delta t)^2}\mathbf{M} + \frac{3}{\theta\Delta t}\mathbf{C} + \mathbf{K}^{(\mathrm{T})} \tag{2.110}$$

$$\Delta\mathbf{Q}^{(\mathrm{eff})} = \theta\Delta\mathbf{Q} + \mathbf{M}\Big[\frac{6}{\theta\Delta t}\dot{\mathbf{q}}(t) + 3\ddot{\mathbf{q}}(t)\Big] + \mathbf{C}\Big[3\dot{\mathbf{q}}(t) + \frac{\theta\Delta t}{2}\ddot{\mathbf{q}}(t)\Big] \tag{2.111}$$

An additional comment on the application of the Wilson θ-method (and of the implicit integration approach as a whole) should be made at this point. The configuration at time $t+\Delta t$ (and, consequently, at $t+\theta\Delta t$), for which the equilibrium conditions are established, is unknown. This usually makes it necessary to carry out additional iterations in order to find a more accurate solution within each time interval. Such a solution can be obtained by using the Newton-type iteration techniques, among which the Newton–Raphson scheme is defined as

$$\mathbf{K}^{(\text{eff})}_{(m)}(t+\theta\Delta t)\,\Delta\mathbf{q}^*_{(m)} = \mathbf{Q}(t+\theta\Delta t) - \mathbf{F}^{(\text{eff})}_{(m-1)}(t+\theta\Delta t) \qquad m = 1,2,\ldots \tag{2.112}$$

where the vector $\mathbf{F}^{(\text{eff})}_{(m-1)}$ stands for the internal nodal forces corresponding to the displacements

$$\mathbf{q}_{(m-1)}(t+\theta\Delta t) = \mathbf{q}_{(m-2)}(t+\theta\Delta t) + \Delta\mathbf{q}^*_{(m-1)} \tag{2.113}$$

$\Delta\mathbf{q}^*_{(m)}$ is the m-th correction to the incremental displacement vector $\Delta\mathbf{q}^*$ and m denotes the iteration number. Since the updating and factorizing of the effective 'stiffness' matrix take place anew at each iteration, the computation cost of the method may be high. It could, therefore, turn out better to use a modified iteration scheme given by

$$\mathbf{K}^{(\text{eff})}(t)\,\Delta\mathbf{q}^*_{(m)} = \mathbf{Q}(t+\theta\Delta t) - \mathbf{F}^{(\text{eff})}_{(m-1)}(t+\theta\Delta t) \tag{2.114}$$

where the effective 'stiffness' matrix has to be factorized only once at the beginning of each time interval.[1] Note that the use of any Newton-type algorithm should lead to the same results provided all the procedures are convergent.

It should also be pointed out that apart from implicit and explicit integration techniques a number of so-called mixed (implicit–explicit) algorithms have recently been developed [12,56]; they are based essentially on breaking down the region Ω into two subregions, in which explicit and implicit time integrations are carried out. If the integration must be performed for many time steps, it may be more effective to use the mode superposition method [8,67,88,92]. Although future work is required to overcome the numerical limitations of both the mixed integration and mode superposition approaches applied to nonlinear problems, they have a great potential, particularly in finite element stochastic and nonlinear structural sensitivity analyses. Appropriate discussion is, however, beyond the scope of this book.

2.7 FEM in Linearized Buckling Analysis

Structural buckling is essentially a nonlinear mechanism. As is known, however, and we shall see in a moment, under some simplifying assumptions this problem may be reduced to a linear analysis. It is the combination of this feature together with the potential importance of the information provided that makes the linearized buckling approach a useful tool in the hands of the structural analyst. We shall confine ourselves to one-parameter external loads and to the elastic behaviour of the material and pose the following question:

[1]It sometimes proves advantageous to choose a slightly different scheme

$$\mathbf{K}^{(\text{eff})}(\bar{t})\,\Delta\mathbf{q}^*_{(m)} = \mathbf{Q}(t+\theta\Delta t) - \mathbf{F}^{(\text{eff})}_{(m-1)}(t+\theta\Delta t)$$

in which the index $\bar{t}$ takes on some a priori selected value from the set $\{0, \Delta t, 2\Delta t, \ldots\}$.

Assume that we know the fundamental[1] elastic solution to a given structural mechanics problem. Is there any possibility, and for what value of load, of another solution into which the structure could bifurcate if it were slightly perturbed from its primary equilibrium path?

The first load level at which there is solution nonuniqueness is called the buckling (or bifurcation) load. The equation governing the nonlinear static response of any elastic structure has the form

$$\left[K^{(e)}_{\alpha\beta} + K^{(\sigma)}_{\alpha\beta}(\sigma_{ij})\right] \Delta q_\beta = \Delta Q_\alpha \tag{2.115}$$

It is assumed here that the system is characterized by the existing nodal displacements q_α and the element stresses σ_{ij} (monitored at the integration points), all being the result of the previous history of the load up to the value Q_α. Assume now that:

(a) prebuckling deformations are so small that the stiffness matrices $K^{(e)}_{\alpha\beta}$ and $K^{(\sigma)}_{\alpha\beta}$ can be formed on the initial undeformed configuration,

(b) the whole value of load Q_α has been applied proportionally to a simple parameter λ such that

$$Q_\alpha = \lambda\, Q^*_\alpha \tag{2.116}$$

where Q^*_α is a reference load vector,

(c) the structure behaves in a linearly elastic fashion up to buckling so that the following relation holds approximately at every integration point:

$$\sigma_{ij} = \lambda\, \sigma^*_{ij} \tag{2.117}$$

where σ^*_{ij} is the stress corresponding to the load value Q^*_α.

Under these assumptions we look now for a load parameter λ such that the system stiffness matrix $K^{(e)}_{\alpha\beta} + K^{(\sigma)}_{\alpha\beta}(\lambda \sigma^*_{ij})$ becomes singular, in which case more than one incremental displacement solution Δq_α is possible. Using the fact that the initial stress matrix is a linear and homogeneous function of stresses we obtain the relation

$$\left[K^{(e)}_{\alpha\beta} + \lambda\, K^{(\sigma)}_{\alpha\beta}(\sigma^*_{ij})\right] \Delta q_\beta = \Delta\lambda\, Q^*_\alpha \tag{2.118}$$

If this equation is to have two different solutions $\Delta q_{\alpha(1)}$ and $\Delta q_{\alpha(2)}$ for the given parameter λ, then

$$\begin{aligned} \left[K^{(e)}_{\alpha\beta} + \lambda\, K^{(\sigma)}_{\alpha\beta}(\sigma^*_{ij})\right] \Delta q_{\beta(1)} &= \Delta\lambda\, Q^*_\alpha \\ \left[K^{(e)}_{\alpha\beta} + \lambda\, K^{(\sigma)}_{\alpha\beta}(\sigma^*_{ij})\right] \Delta q_{\beta(2)} &= \Delta\lambda\, Q^*_\alpha \end{aligned} \tag{2.119}$$

[1] 'Fundamental' here means the one obtained for the ideal system, with no material, geometry or load imperfections considered, no mater how likely they are to occur in practice.

By subtracting sideways we arrive at the condition of existence of a bifurcation point in the form

$$\left[K^{(e)}_{\alpha\beta} + \lambda\, K^{(\sigma)}_{\alpha\beta}(\sigma^*_{ij})\right] v_\beta = 0 \tag{2.120}$$

where $v_\alpha = \Delta q_{\alpha(1)} - \Delta q_{\alpha(2)}$. If Eq. (2.120) has a nonzero solution with respect to v_α, i.e. if $\det[K^{(e)}_{\alpha\beta} + \lambda K^{(\sigma)}_{\alpha\beta}(\sigma^*_{ij})] = 0$, then the resulting value of the parameter λ, $\lambda = \lambda_{(cr)}$, ensures that the external load $Q_{\alpha(cr)} = \lambda_{(cr)} Q^*_\alpha$ corresponds to the structural buckling.

In terms of linear algebra, Eq. (2.120) represents the so-called generalized eigenvalue problem. The solution to it consists of N nontrivial eigenpairs $(\lambda_{(1)}, v_{\alpha(1)})$, $(\lambda_{(2)}, v_{\alpha(2)}), \ldots, (\lambda_{(N)}, v_{\alpha(N)})$ with $\lambda_{(1)}, \lambda_{(2)}, \ldots, \lambda_{(N)}$ (ordered so that $\lambda_{(1)} \leq \lambda_{(2)} \leq \ldots \leq \lambda_{(N)}$) and $v_{\alpha(1)}, v_{\alpha(2)}, \ldots, v_{\alpha(N)}$ being called the eigenvalues and the eigenvectors, respectively. In the buckling analysis we are generally interested in the smallest eigenvalue $\lambda_{(1)}$ only (and possibly in the corresponding buckling mode $v_{\alpha(1)}$).

To sum up the findings of this section, the following steps have to be undertaken in order to find a buckling load for a given elastic structural system:

(a) a linear elastic solution has to be found for a reference load Q^*_α resulting in the stress distribution σ^*_{ij} by using the relation, cf. Eq. (2.53)

$$K^{(e)}_{\alpha\beta}\, q^*_\beta = Q^*_\alpha \tag{2.121}$$

and, for finding stresses, the relation, cf. Eqs.(2.2) and $(2.44)_2$

$$\sigma^*_{ij} = C_{ijkl}\, B_{kl\alpha}\, q^*_\alpha \tag{2.122}$$

(b) the eigenvalue problem (2.120) has to be solved for the smallest eigenvalue $\lambda_{(1)} = \lambda_{(cr)}$ (and possibly the corresponding eigenvector $v_{\alpha(1)}$ which turns out to be helpful in establishing the post-buckling shape of the structure).

The value $\lambda_{(cr)} Q^*_\alpha$ is the buckling load for the structure and load pattern considered.

Some further details on numerical treatment of the buckling problem will be given in Sections 4.4 and 4.5.

2.8 Structural Design Sensitivity Analysis

2.8.1 Statics

The objective of *structural design sensitivity* (SDS) analysis is to consider the relationship between structural response (or state variables) as determined by the solution of the boundary-value problem and design variables available to the analyst and used in the input phase of the solution process. As structural response measures we may take displacements, stresses, buckling loads or natural frequencies, whereas as design variables we usually have at our disposal such parameters as truss and beam cross-sectional areas, plate and shell thicknesses, etc. We note that even for linear elastic problems the system equations may generally be nonlinear functions of the state and design variables.

The gradients of the structural response functionals with respect to the design variables characterize trends of the structural response variation under change of design. Thus, the gradients provide a useful database for choosing search directions to come up with an improved new design.

SDS analysis turns out to be particularly useful when employed in the framework of the finite element formulation. Consider the static structural response of a linear elastic system with N degrees of freedom (i.e. of a structure spatially discretized by FEM) defined by the functional[1]

$$\mathcal{G}(h^d) = G\left[q_\alpha(h^d), h^d\right] \qquad \alpha = 1, 2, \ldots, N\,; \;\; d = 1, 2, \ldots, D \tag{2.123}$$

where G is a given function of its arguments, h^d is a D-dimensional vector of design variables and q_α, as before, is the vector of nodal displacement-type parameters. The displacement vector satisfies the equilibrium equations, cf. Eq. (2.53)

$$K_{\alpha\beta}(h^d)\, q_\beta(h^d) = Q_\alpha(h^d) \tag{2.124}$$

Since the stiffness $K_{\alpha\beta}$ and the load vector Q_α are functions of the design variables, the displacement vector is assumed to be an implicit function of these variables as well.

The objective of the SDS analysis is to determine changes in the structural response functional with variations in design parameters; in other words we are looking for the so-called *sensitivity gradient* $\partial\mathcal{G}/\partial h^d$. By using the chain rule of differentiation we readily arrive at

$$\mathcal{G}^{,d} = G^{,d} + G_{,\alpha}\, q_\alpha^{,d} \qquad \alpha = 1, 2, \ldots, N\,; \;\; d = 1, 2, \ldots, D \tag{2.125}$$

where the symbols $(.)^{,d}$ and $(.)_{,\alpha}$ describe the first partial derivatives with respect to the d-th design variable and the α-th nodal displacement, respectively. Since the design variables h^d are indicated here as the only arguments in the functions $\mathcal{G}$, $K_{\alpha\beta}$, q_α and Q_α, the partial derivatives of these functions with respect to h^d are in fact equal to the corresponding total derivatives. This is not so in the case of G, for which $dG/dh^d = \mathcal{G}^{,d}$. Since G is an explicitly given function of h^d and q_α, the derivatives $G^{,d}$ and $G_{,\alpha}$ may be computed directly while $q_\alpha^{,d}$ remains to be calculated.

Assume that $K_{\alpha\beta}(h^d)$ and $Q_\alpha(h^d)$ are continuously differentiable with respect to the design variables h^d. According to the implicit function theorem [40] the vector $q_\alpha(h^d)$ is then also continuously differentiable. Differentiating both sides of Eq. (2.124) with respect to h^d leads to

$$K_{\alpha\beta}\, q_\beta^{,d} = Q_\alpha^{,d} - K_{\alpha\beta}^{,d}\, q_\beta \tag{2.126}$$

[1] The index 'd' has been positioned as a superscript just for the sake of notational convenience.

where the explicit notation indicating dependence of all the functions on h^d has been suppressed for simplicity. Since the stiffness matrix $K_{\alpha\beta}$ is assumed to be nonsingular, Eq. (2.126) can be solved for $q_\alpha^{\cdot d}$ which, when substituted into Eq. (2.125), results in the relationship

$$\mathcal{G}^{\cdot d} = G^{\cdot d} + G_{.\beta}\, K_{\alpha\beta}^{-1} \left(Q_\alpha^{\cdot d} - K_{\alpha\gamma}^{\cdot d}\, q_\gamma\right) \tag{2.127}$$

The above technique for computing the sensitivity gradients is referred to as the *direct differentiation method* (DDM). This method has been extensively employed in structural optimization.

An alternative strategy, known as the *adjoint variable method* (AVM), can be presented by defining an adjoint variable vector λ_α, $\alpha = 1, 2, \ldots, N$, in such a way that

$$\lambda_\alpha = G_{.\beta}\, K_{\alpha\beta}^{-1} \tag{2.128}$$

which leads to the adjoint equations for the vector λ_α in the form

$$K_{\alpha\beta}\, \lambda_\beta = G_{.\alpha} \tag{2.129}$$

Having solved Eq. (2.129) for the adjoint variables λ_β the sensitivity gradient coefficients may be obtained as

$$\mathcal{G}^{\cdot d} = G^{\cdot d} + \lambda_\alpha \left(Q_\alpha^{\cdot d} - K_{\alpha\beta}^{\cdot d}\, q_\beta\right) \tag{2.130}$$

Formally, the main ideas behind DDM and AVM seem to be identical. However, for realistic design problems their computer performance is considerably different. In the real design process the designer must frequently evaluate the response of systems with a large number of degrees of freedom, and account for numerous loading conditions and numerous performance constraints (involving functional of the type (2.123)) at a given trial design.

Let us assume that instead of a single load vector and a single constraint considered so far there are L load cases and C active performance constraints to be accounted for in the static SDS analysis. Assume also that for each load case at least one constraint is active – otherwise this load case could be suppressed for purposes of the SDS analysis. Furthermore, suppose that the derivatives of the element stiffness matrices and load vectors with respect to D design variables have been computed and assembled, and the global stiffness matrix has been triangularized. These data are constant for linear elastic problems and are assumed to have been saved on a back-up storage memory.

To compute the sensitivity gradient coefficients for each of C constraints using DDM, Eq.(2.124) must be solved for L load conditions. Next, each of the L vectors of the calculated nodal displacements is used to determine the D right-hand sides of Eq. (2.126). Having obtained a solution for $q_\alpha^{\cdot d}$ the sensitivity gradient coefficients may be directly evaluated for C constraints. It is seen that in this approach there are $L \times D$ sets of equations to be solved. More precisely, the forward-reductions (2.57) and backward-substitutions (2.58) in the compact scheme of Gauss elimination procedure must be carried out $L \times D$ times.

On the other hand, assuming in AVM that each of the C constraints is active for all the L loading cases, there are C sets of equations (2.129) to be solved for the adjoint variables λ_α. Design sensitivity coefficients of the C constraints for the L loads are then calculated directly from Eq. (2.130) for the D design variables. It is seen that if $L \times D < C$, then DDM is preferred. This may likely occur in preliminary design problems, when a small number of design variables and a large number of constraints are accounted for in the analysis. In structural optimization problems, for which usually $C < D$, AVM is as a rule more efficient (it is so even for a single loading condition!). Moreover, the equation solution phase of AVM is often less costly since $L+C$ algebraic equations may be solved simultaneously (provided the structural response functional is linear with respect to the displacements, which is frequently assumed in design practice). On the contrary, employing DDM implies two separate solution steps and an additional in-between procedure for the calculation of the right-hand sides in Eq. (2.126).

Similarly to the assembly procedure for the global stiffness matrix, the matrix of derivatives of the global stiffness matrix with respect to design variables is obtained simply by adding derivatives of elemental stiffnesses expressed in the global coordinate system. It is important to note that in contrast to the global stiffness matrix profile almost all entries in the matrices of their derivatives with respect to the particular design variables are equal to zero. This greatly simplifies computation of the sensitivity gradients since almost all operations can be carried out at the element level.

Another aspect which has to be taken into account in developing SDS software is the effective computation of stiffness derivatives with respect to design variables for elements which have stiffness matrices implicitly generated within the finite element program [38,69,124]. In order to generate the stiffness matrices, in most up-to-date finite element codes numerical integration is employed instead of using the closed form expressions in terms of design variables. For such numerically generated element matrices differentiation with respect to design variables can be performed through a sequence of computations used to generate these matrices, leading to implicit design derivative procedures. The element matrices of the design derivatives can also be obtained by using a finite difference scheme. We may take for the e-th element, for instance,

$$\frac{\partial k^{(e)}_{\xi\zeta}}{\partial h^{\bar{d}}} \approx \frac{1}{\epsilon}\left[k^{(e)}_{\xi\zeta}(\mathbf{h} + \mathbf{1}_{(\bar{d})}\epsilon) - k^{(e)}_{\xi\zeta}(\mathbf{h})\right] \qquad \bar{d} = 1, 2, \ldots, D \tag{2.131}$$

where $k^{(e)}_{\xi\zeta}$, as in Eq. (2.49), stands for the e-th element stiffness matrix, $h^{\bar{d}}$ is the $\bar{d}$-th component of the D-dimensional design variable vector $\mathbf{h}$, ϵ represents a small perturbation and the D-dimensional vector $\mathbf{1}_{(\bar{d})}$ has a one at the $\bar{d}$-th position and zeros elsewhere.

2.8.2 Dynamics

2.8.2.1 Time Interval Sensitivity

In the preceding section we have briefly considered some basic aspects of the SDS analysis for systems whose inertial and damping effects are negligible. We now move on to discuss SDS in the context of dynamically loaded finite element systems.

Consider the structural response of a linear elastic system with N degrees of freedom described by an integral functional

$$\mathcal{G}(h^d) = \int_0^{T(h^d)} G\left[q_\alpha(h^d;\tau), h^d\right] \mathrm{d}\tau \qquad \begin{array}{l} \alpha = 1,2,\ldots,N \\ d = 1,2,\ldots,D \end{array} \tag{2.132}$$

where the vector of generalized displacements q_α is, as before, an implicit function of the design variables h^d. The symbol τ, $\tau \in [0, T(h^d)]$, denotes time with $T(h^d)$ representing a terminal time determined uniquely by the so-called terminal-time condition [44]

$$\mathcal{T}[q_\alpha(h^d;T), \dot{q}_\alpha(h^d;T), h^d] = 0 \tag{2.133}$$

The system is assumed to satisfy the finite element equations of motion, cf. Eq. (2.65)

$$M_{\alpha\beta}(h^d)\,\ddot{q}_\beta(h^d;\tau) + C_{\alpha\beta}(h^d)\,\dot{q}_\beta(h^d;\tau) + K_{\alpha\beta}(h^d)\,q_\beta(h^d;\tau) = Q_\alpha(h^d;\tau) \tag{2.134}$$

together with the homogeneous initial conditions[1]

$$q_\alpha(h^d, 0) = 0 \qquad\qquad \dot{q}_\alpha(h^d, 0) = 0 \tag{2.135}$$

The objective of dynamic structural design sensitivity (DSDS) analysis is to evaluate the change in the response functional (2.132) subject to constraints (2.133)–(2.135) with respect to design variations, i.e. to determine the sensitivity gradient of this functional with respect to design variables.

DSDS based on a functional of the type (2.132) will be referred to as *sensitivity over a time interval* or *time interval sensitivity.* Another type of DSDS will be considered later in this chapter.

It may be shown that if the mass, damping and stiffness matrices as well as load vector are continuously differentiable with respect to h^d and if the mass matrix is nonsingular[2], then the displacements, velocities and accelerations are also continuously differentiable with respect to h^d [23]. This guarantees that the dynamic response of the structural system will be as smooth as the dependence

[1]As we shall comment upon later in this chapter, the assumption of homogeneity of initial conditions is not restrictive.

[2]This can always be achieved by elimination of the so-called massless degrees of freedom.

on the design variables in the equations of motion. Differentiation of Eq. (2.132) with respect to the design variables using Leibnitz rule leads to[1]

$$\mathcal{G}^{\cdot d} = G(T)\,T^{\cdot d} + \int_0^T \left[G^{\cdot d}(\tau) + G_{.\alpha}(\tau)\,q_\alpha^{\cdot d}(\tau)\right] \mathrm{d}\tau \tag{2.136}$$

where the symbols $(.)^{\cdot d}$ and $(.)_{.\alpha}$, as defined in Section 2.8.1, are the first partial derivatives with respect to design variables and nodal displacements, respectively. The first term on the right-hand side of Eq. (2.136) involves the derivative of T with respect to the design variables since the terminal-time condition (2.133) used to evaluate T is design-dependent. In order to eliminate the term $T^{\cdot d}$ from Eq. (2.136) we differentiate Eq. (2.133) with respect to h^d to obtain

$$\mathcal{T}^{\cdot d} + \mathcal{T}_{.\alpha}\left[q_\alpha^{\cdot d}(T) + \dot{q}_\alpha(T)\,T^{\cdot d}\right] + \frac{\partial \mathcal{T}}{\partial \dot{q}_\alpha}\left[\dot{q}_\alpha^{\cdot d}(T) + \ddot{q}_\alpha(T)\,T^{\cdot d}\right] = 0 \tag{2.137}$$

Noting that from the terminal-time condition (2.133) the terminal time T is determined uniquely only if the following condition holds true

$$\dot{\mathcal{T}} = \mathcal{T}_{.\alpha}\,\dot{q}_\alpha(T) + \frac{\partial \mathcal{T}}{\partial \dot{q}_\alpha}\ddot{q}_\alpha(T) \neq 0 \tag{2.138}$$

Eq. (2.137) can be rewritten as

$$T^{\cdot d} = -\frac{1}{\dot{\mathcal{T}}}\left[\mathcal{T}^{\cdot d} + \mathcal{T}_{.\alpha}\,q_\alpha^{\cdot d}(T) + \frac{\partial \mathcal{T}}{\partial \dot{q}_\alpha}\dot{q}_\alpha^{\cdot d}(T)\right] \tag{2.139}$$

Using the last expression for $T^{\cdot d}$, Eq. (2.136) becomes

$$\begin{aligned}\mathcal{G}^{\cdot d} &= -\frac{G(T)}{\dot{\mathcal{T}}}\left[\mathcal{T}^{\cdot d} + \mathcal{T}_{.\alpha}\,q_\alpha^{\cdot d}(T) + \frac{\partial \mathcal{T}}{\partial \dot{q}_\alpha}\dot{q}_\alpha^{\cdot d}(T)\right] \\ &\quad + \int_0^T \left[G^{\cdot d}(\tau) + G_{.\alpha}(\tau)\,q_\alpha^{\cdot d}(\tau)\right] \mathrm{d}\tau\end{aligned} \tag{2.140}$$

Recall that $\mathcal{T}$ and its derivatives with respect to either time or design variables are functions of the terminal time T.

To express the sensitivity gradient $\mathcal{G}^{\cdot d}$ explicitly in terms of design variable variations, i.e. to eliminate from Eq. (2.140) the terms involving $q_\alpha^{\cdot d}$ and $\dot{q}_\alpha^{\cdot d}$, the adjoint variable technique can be conveniently used. To this aim we define an adjoint vector $\lambda_\alpha(\tau)$, $\alpha = 1, 2, \ldots, N$, assumed independent of the design variables. Differentiating Eq. (2.134) with respect to h^d, pre-multiplying the result by the adjoint vector λ_α, integrating it over the time interval from 0 to T,

---

[1]From now on, the argument h^d will frequently be suppressed to make the presentation more compact; thus, instead of $K_{\alpha\beta}(h^d)$, $q_\alpha(h^d;\tau)$ and $G[q_\alpha(h^d;T), h^d]$ we may in places write $K_{\alpha\beta}$, $q_\alpha(\tau)$ and $G(T)$, respectively.

integrating the first two terms involving $\dot{q}_{\alpha}^{\cdot d}$ and $\ddot{q}_{\alpha}^{\cdot d}$ by parts with respect to time and noting the symmetry of the mass, damping and stiffness matrices gives

$$\begin{aligned}
&M_{\alpha\beta}\,\lambda_\alpha(T)\,\dot{q}_\beta^{\cdot d}(T) - \left[M_{\alpha\beta}\,\dot{\lambda}_\alpha(T) - C_{\alpha\beta}\,\lambda_\alpha(T)\right] q_\beta^{\cdot d}(T) \\
&\quad + \int_0^T \left[M_{\alpha\beta}\,\ddot{\lambda}_\alpha(\tau) - C_{\alpha\beta}\,\dot{\lambda}_\alpha(\tau) + K_{\alpha\beta}\,\lambda_\alpha(\tau)\right] q_\beta^{\cdot d}(\tau)\,\mathrm{d}\tau \\
&\quad = \int_0^T \lambda_\alpha(\tau)\left[Q_\alpha^{\cdot d}(\tau) - M_{\alpha\beta}^{\cdot d}\,\ddot{q}_\beta(\tau) - C_{\alpha\beta}^{\cdot d}\,\dot{q}_\beta(\tau) - K_{\alpha\beta}^{\cdot d}\,q_\beta(\tau)\right] \mathrm{d}\tau \qquad (2.141)
\end{aligned}$$

Since this equation must hold for arbitrary $\lambda_\alpha(\tau)$, the adjoint variables may be selected so that the coefficients of terms involving $q_\alpha^{\cdot d}$ and $\dot{q}_\alpha^{\cdot d}$ in Eq. (2.140) and Eq. (2.141) are equal, leading to the differential equations of motion for the adjoint system in the form

$$\begin{aligned}
&M_{\alpha\beta}(h^d)\,\ddot{\lambda}_\beta(\tau) - C_{\alpha\beta}(h^d)\,\dot{\lambda}_\beta(\tau) + K_{\alpha\beta}(h^d)\,\lambda_\beta(\tau) \\
&\quad = G_{.\alpha}[q_\beta(h^d;\tau), h^d] \qquad \tau \in [0, T] \qquad (2.142)
\end{aligned}$$

with the terminal conditions for the adjoint displacements and velocities at time T in the form

$$\lambda_\alpha(T) = -\frac{G(T)}{\dot{\mathcal{T}}}\,M_{\alpha\beta}^{-1}\,\frac{\partial \mathcal{T}}{\partial \dot{q}_\beta} \qquad (2.143)$$

$$\dot{\lambda}_\alpha(T) = M_{\alpha\beta}^{-1}\left[C_{\beta\gamma}\,\lambda_\gamma(T) + \frac{G(T)}{\dot{\mathcal{T}}}\,\mathcal{T}_{.\beta}\right] \qquad (2.144)$$

Equations (2.142)–(2.144) can be called the *terminal-value problem.* We note that once the original initial-value problem (2.134),(2.135) is solved by forward time integration and Eq. (2.133) is used to determine T, the right-hand sides of Eqs. (2.142)–(2.144) are known and the terminal-value problem may be solved as well by using backward time integration. In other words, the initial–terminal-value problem described by Eqs. (2.134),(2.135),(2.142)–(2.144) yields a unique solution for the displacement vector function $q_\alpha(\tau)$ and the adjoint vector function $\lambda_\alpha(\tau)$ in the time interval [0,T]. Introducing Eqs. (2.142)–(2.144) into Eq. (2.141) and substituting the result into Eq. (2.140) allows us to rewrite the d-th component of the sensitivity gradient in the final form as

$$\begin{aligned}
\mathcal{G}^{\cdot d} &= -\frac{G(T)}{\dot{\mathcal{T}}}\,\mathcal{T}^{\cdot d} + \int_0^T G^{\cdot d}(\tau)\,\mathrm{d}\tau \\
&\quad + \int_0^T \lambda_\alpha(\tau)\left[Q_\alpha^{\cdot d}(\tau) - M_{\alpha\beta}^{\cdot d}\,\ddot{q}_\beta(\tau) - C_{\alpha\beta}^{\cdot d}\,\dot{q}_\beta(\tau) - K_{\alpha\beta}^{\cdot d}\,q_\beta(\tau)\right] \mathrm{d}\tau \qquad (2.145)
\end{aligned}$$

2.8.2.2 Time Instant Sensitivity

In the preceding section the DSDS problem was formulated as the 'time interval sensitivity' and the sensitivity gradient was not evaluated at any particular time instant within the time interval considered. We now consider the time behaviour of the sensitivity gradient in the time interval $[0,T]$. In other words, the sensitivity of a time integral of the function G is not considered, cf. Eq. (2.136), but rather the sensitivity of an instantaneous value of this function. The analysis will be referred to as the ***sensitivity at an arbitrary time instant*** or, briefly, ***time instant sensitivity***. With no loss of generality, let us assume for simplicity of presentation that the terminal-time condition (2.133) can be expressed in an explicit form with respect to the terminal time T.

By using the concept of the sampling time signals introduced in Section 1.2.1 the structural response functional can be defined in terms of an infinite sequence of adjacent impulses as their sampling intervals tend to zero, that is

$$\mathcal{G}(h^d;t) = \int_0^t G\,[q_\alpha(h^d;\tau),h^d]\,\delta(t-\tau)\,\mathrm{d}\tau \qquad t\in[0,T] \tag{2.146}$$

$$\alpha=1,2,\ldots,N\,;\;\; d=1,2,\ldots,D$$

where t denotes the running terminal time and $\delta(t-\tau)$ is the Dirac delta measure (which is an even distribution). The function $G[q_\alpha(h^d;\tau),h^d]$ is assumed to be continuous in the entire time interval considered, $\tau\in[0,T]$. Eq. (2.146) can be interpreted as a linear combination of shifted unit impulses, wherein the weight of the unit impulse $\delta(t-\tau)$ is $G(\tau)\mathrm{d}\tau$. Since the running terminal time t is given (equal to some a priori selected value in $[0,T]$), differentiation of Eq. (2.146) with respect to h^d and noting Eq. (1.97) yields

$$\mathcal{G}^{,d}(t) = G^{,d}(t) + \int_0^t G_{,\alpha}(\tau)\,q_\alpha^{,d}(\tau)\,\delta(t-\tau)\,\mathrm{d}\tau \tag{2.147}$$

Observing that, cf. Eq. (1.103)

$$G_{,\alpha}(\tau)\,\delta(t-\tau) = G_{,\alpha}(t)\,\delta(t-\tau) \tag{2.148}$$

and using a similar derivation as in the previous section, the terminal problem is obtained as

$$\begin{aligned} M_{\alpha\beta}(h^d)\,\ddot{\lambda}_\beta(\tau) - C_{\alpha\beta}(h^d)\,\dot{\lambda}_\beta(\tau) + K_{\alpha\beta}(h^d)\,\lambda_\beta(\tau) & \\ = G_{,\alpha}[q_\beta(h^d;t),h^d]\,\delta(t-\tau) \qquad & \tau\in[0,t]\,;\;\; t\in[0,T] \end{aligned} \tag{2.149}$$

with the running terminal conditions

$$\lambda_\alpha(t)=0 \qquad \dot{\lambda}_\alpha(t)=0 \tag{2.150}$$

Having solved Eqs. (2.134) and (2.149) the term involving $\dot{q}_\alpha^{,d}$ in Eq. (2.147) can be eliminated to yield the sensitivity gradient in the form

$$\begin{aligned} \mathcal{G}^{,d}(t) = {} & G^{,d}(t) \\ & + \int_0^t \lambda_\alpha(\tau)\Big[f_\alpha^{,d}(\tau) - M_{\alpha\beta}^{,d}\,\ddot{q}_\beta(\tau) - C_{\alpha\beta}^{,d}\,\dot{q}_\beta(\tau) - K_{\alpha\beta}^{,d}\,q_\beta(\tau)\Big]\mathrm{d}\tau \\ & \tau\in[0,t]\,;\;\; t\in[0,T] \end{aligned} \tag{2.151}$$

2.8.2.3 Sensitivity of Natural Frequencies and Buckling Loads

Natural frequencies of free vibrations and buckling loads can be treated as eigenvalues of some generalized eigenproblems. Since the mass matrix as well as the initial stress and elastic stiffness matrices are all design-dependent, the eigenvalues and associated eigenvectors in the appropriate eigenproblems are also implicit functions of design variables. In this section we shall briefly describe the problem of natural frequency design sensitivity[1] assuming for simplicity that the system is conservative and no repeated eigenvalues may occur in it.

For the free vibration problem we thus consider a specific form of the structural response functional defined as

$$\mathcal{G}(h^d) = \omega_{(\alpha)}(h^d) \tag{2.152}$$

where $\omega_{(\alpha)}$ is a fixed, α-th eigenfrequency of the system, $\alpha = 1, 2, \ldots, N$. The equation describing the free vibration problem for the system is considered as the constraint

$$K_{\alpha\beta}(h^d)\, \phi_{\beta\gamma}(h^d) = \omega_{(\alpha)}(h^d)\, M_{\alpha\beta}(h^d)\, \phi_{\beta\gamma}(h^d) \qquad \text{(no sum on } \alpha) \tag{2.153}$$

It can be shown that if the mass and stiffness matrices are positive definite, symmetric and continuously differentiable with respect to the design variables h^d, and if the eigenvalue $\omega_{(\alpha)}$ is not repeated, then the eigenvalue and the corresponding eigenvector of Eq. (2.153) are continuously differentiable with respect to the design variables.

Differentiating Eq. (2.153) with respect to the design variables and premultiplying the resulting equation by the eigenvector matrix $\phi_{\alpha\beta}$ transposed we obtain

$$\begin{aligned} \omega_{(\alpha)}^{,d}\, \phi_{\gamma\alpha}^{\mathrm{T}}\, M_{\alpha\beta}\, \phi_{\beta\delta} &= \phi_{\gamma\alpha}^{\mathrm{T}} \left(K_{\alpha\beta}^{,d} - \omega_{(\alpha)}\, M_{\alpha\beta}^{,d} \right) \phi_{\beta\delta} \\ &\quad + \phi_{\gamma\alpha}^{\mathrm{T}} \left(K_{\alpha\beta}\, \phi_{\beta\delta} - \omega_{(\alpha)}\, M_{\alpha\beta}\, \phi_{\beta\delta} \right) \end{aligned} \tag{2.154}$$

The second term on the right-hand side of Eq. (2.154) is zero according to Eq. (2.153). Using the orthonormalization condition $(2.87)_1$ the final expression for the design sensitivity gradient of the α-th eigenvalue reads

$$\omega_{(\alpha)}^{,d} = \phi_{\gamma\alpha}^{\mathrm{T}} \left(K_{\alpha\beta}^{,d} - \omega_{(\alpha)}\, M_{\alpha\beta}^{,d} \right) \phi_{\beta\delta} \tag{2.155}$$

2.8.2.4 Remarks on Computer Implementation

Let us start the discussion by noting again that the left-hand sides in the equations of motion (2.134) and in the adjoint equations (2.142) and (2.149) differ only by the algebraic sign at the damping term. Defining a *backward time* $\bar{\tau}$ by

$$\bar{\tau} = T - \tau \qquad \text{so that} \qquad \frac{\partial(.)}{\partial \tau} = -\frac{\partial(.)}{\partial \bar{\tau}}$$

[1]The buckling load design sensitivity problem is similar and will not be explicitly treated here.

we observe that the left-hand sides become formally identical provided the change of time variables from τ to $\bar{\tau}$ has been employed in the adjoint equations.

The computations of the time interval sensitivity analysis proceed as follows:

(i) Given a vector of design variables h^d, Eq. (2.134) is integrated forward with respect to time τ, the value of $\mathcal{T}[q_\alpha(h^d;T), \dot{q}_\alpha(h^d;T), h^d]$ is monitored and the time instant $\tau = T$ at which $\mathcal{T} = 0$ is defined as the terminal time. The quantities q_α, $\dot{q}_\alpha$, $\partial\mathcal{T}/\partial q_\alpha$ and $\partial\mathcal{T}/\partial\dot{q}_\alpha$ are calculated at the terminal time.

(ii) Equations (2.143) and (2.144) are solved so that the terminal conditions for the adjoint displacements $\lambda_\alpha(\tau)$ and adjoint velocities $\dot{\lambda}_\alpha(\tau)$ become explicitly given. The adjoint terminal problem (2.142) can be integrated over the *forward time* τ from $\tau = T$ to $\tau = 0$ (or over the backward time from $\bar{\tau} = 0$ to $\bar{\tau} = T$), yielding a unique solution $\lambda_\alpha(\tau)$.

(iii) Having solved the initial–terminal problem (by either the step-by-step direct integration method or by the mode superposition method), the design sensitivity gradient coefficients as given by Eq. (2.145) can be computed using any numerical integration scheme.

The above approach becomes particularly efficient when the structural damping effects are such that the effective use of the mode superposition method can be made, cf. Section 2.5.2. This is so because owing to the similarity of both parts of the initial–terminal problem the eigenproblem has to be solved only once and the same eigenpairs may be used for either the structural or adjoint system.

In the case of the time instant sensitivity analysis the following computational steps have to be performed:

(i) Eq. (2.134) is integrated forward in time from $\tau = 0$ to $\tau = t$, where t is a given time instant at which the dynamic response sensitivity is sought.

(ii) The right-hand side of Eq. (2.149) is formed and the equation is integrated forward from $\tau = t$ to $\tau = 0$ (or backward from $\bar{\tau} = 0$ to $\bar{\tau} = t$).

(iii) The design sensitivity gradient coefficients at time t are evaluated using Eq. (2.151).

(iv) Steps (i), (ii) and (iii) are repeated for every running time instant t at which the sensitivity gradient is to be determined.

Let us assume for the sake of generality that we are interested in the time instant sensitivity results at every discrete time instant in the interval $[0, T]$, T being given. The cost of computations carried out in accordance with the above algorithm would appear unacceptably high, particularly when the direct integration technique would be used. This is so because the terminal problem (2.149),(2.150) would have to be solved repeatedly for every running time instant $t \in [0, T]$ which effectively amounts to $n(n+1)/2$ time steps, where $n = T/\Delta t$ and Δt is the time increment (sampling interval) considered. However, we may note that the operators in the initial–terminal system (2.134),(2.149) are linear, the initial conditions for the structural equations are homogeneous and the (running) terminal conditions for the adjoint equations are homogeneous at every t as well. This observation enables one to develop an effective numerical procedure such that the adjoint system (2.149) can be integrated only once from T to 0 over n time increments. As will be shown below, the response of the adjoint system can be treated through the unit impulse response of a normalized structural system whose operators, being functions of the structural mass, damping and stiffness matrices, are independent of $G[q_\alpha(h^d;t),h^d]$ and t.

In general, the multi-degree-of-freedom system (2.149) cannot be solved in a straightforward manner. This is so because the right-hand side is not an ordinary function of time τ as the Dirac delta distribution is defined in terms of its properties rather than its values, and the generalized coordinates on the left-hand side are coupled. To overcome these difficulties let us first use in Eq. (2.149) (and similarly in Eq. (2.134)) a transformation from the generalized coordinates $\lambda_\alpha(\tau)$ to the normalized coordinates $\vartheta_{\bar\alpha}(\tau)$, $\bar\alpha = 1,\ldots,\bar N \ll N$ as described in Section 2.5.2. Using the normalized mode shape matrix $\phi_{\alpha\bar\alpha}$ obtained from the corresponding eigenproblem we arrive at the system of $\bar N$ uncoupled equations:

$$\begin{aligned}&\ddot\vartheta_{\bar\alpha}(\tau)-2\xi_{(\bar\alpha)}\,\omega_{(\bar\alpha)}\,\dot\vartheta_{\bar\alpha}(\tau)+\omega^2_{(\bar\alpha)}\,\vartheta_{\bar\alpha}(\tau)=\phi^{\mathrm T}_{\bar\alpha\alpha}G_{.\alpha}(t)\,\delta(t-\tau) \qquad \tau\in[0,t]\\ &\vartheta_{\bar\alpha}(t)=0 \qquad \dot\vartheta_{\bar\alpha}(t)=0 \qquad\qquad (\text{no sum on } \bar\alpha)\end{aligned} \tag{2.156}$$

where

$$\vartheta_{\bar\alpha}(\tau)=\phi^{\mathrm T}_{\bar\alpha\alpha}\,\lambda_\alpha(\tau) \qquad \bar\alpha=1,2,\ldots,\bar N\,;\quad \alpha=1,2,\ldots,N \tag{2.157}$$

The $\bar\alpha$-th independent adjoint equation (2.156) may be interpreted as a linear, single-degree-of-freedom system excited by an impulse of fixed magnitude $\phi^{\mathrm T}_{\bar\alpha\alpha}G_{.\alpha}$ acting at $\tau = t$.

Let us next consider the response of the following system:

$$\begin{aligned}&\ddot\vartheta_{\bar\alpha}(\tau)-2\xi_{(\bar\alpha)}\,\omega_{(\bar\alpha)}\,\dot\vartheta_{\bar\alpha}(\tau)+\omega^2_{(\bar\alpha)}\,\vartheta_{\bar\alpha}(\tau)=\delta(t-\tau) \qquad \tau\in[0,t]\\ &\vartheta_{\bar\alpha}(t)=0 \qquad \dot\vartheta_{\bar\alpha}(t)=0 \qquad\qquad (\text{no sum on } \bar\alpha)\end{aligned} \tag{2.158}$$

Integrating Eq. (2.158) in time over a small time interval ranging from t to $t-\epsilon$ at $\epsilon\to 0$ we have

$$\lim_{\epsilon\to 0}\int_t^{t-\epsilon}\Big[\ddot\vartheta_{\bar\alpha}(\tau)-2\xi_{(\bar\alpha)}\omega_{(\bar\alpha)}\,\dot\vartheta_{\bar\alpha}(\tau)+\omega^2_{(\bar\alpha)}\vartheta_{\bar\alpha}(\tau)\Big]\mathrm d\tau=\lim_{\epsilon\to 0}\int_t^{t-\epsilon}\delta(t-\tau)\,\mathrm d\tau=1 \tag{2.159}$$

where

$$\begin{aligned}\lim_{\epsilon\to 0}\int_t^{t-\epsilon}\ddot\vartheta_{\bar\alpha}(\tau)\,\mathrm d\tau &= \lim_{\epsilon\to 0}\Big[\dot\vartheta_{\bar\alpha}(t-\epsilon)-\dot\vartheta_{\bar\alpha}(t)\Big] = \dot\vartheta_{\bar\alpha}(t^-)\\ \lim_{\epsilon\to 0}\int_t^{t-\epsilon}\dot\vartheta_{\bar\alpha}(\tau)\,\mathrm d\tau &= \lim_{\epsilon\to 0}\Big[\vartheta_{\bar\alpha}(t-\epsilon)-\vartheta_{\bar\alpha}(t)\Big] = 0\\ \lim_{\epsilon\to 0}\int_t^{t-\epsilon}\vartheta_{\bar\alpha}(\tau)\,\mathrm d\tau &= 0\end{aligned} \tag{2.160}$$

and consequently, we arrive at the following identity:

$$\dot{\vartheta}_{\bar{\alpha}}(t^-) = 1 \tag{2.161}$$

The notation $\dot{\vartheta}_{\bar{\alpha}}(t^-)$ is meant to indicate the change in velocity at the end (going backward) of the time interval $[t-\epsilon, t]$. Eq. (2.161) can be given the following interpretation. Immediately after a unit impulse excitation the unit mass gains a momentum of unit magnitude, but there is still no time for an instantaneous change in the normalized displacement; therefore, the system is left unexcited. Since the normalized system is at rest immediately prior to the excitation, an instantaneous increment in the normalized velocity appears so that the unit impulse excitation applied at t can be regarded as the equivalent of the unit terminal velocity. In fact this is the way in which initial velocities are imparted to systems possessing inertia [75,97].

In view of this interpretation the response to a unit impulse excitation described by Eq. (2.158) can be expressed as

$$\begin{aligned} &\ddot{\vartheta}_{\bar{\alpha}}(\tau) - 2\xi_{(\alpha)}\,\omega_{(\alpha)}\,\dot{\vartheta}_{\bar{\alpha}}(\tau) + \omega^2_{(\alpha)}\,\vartheta_{\bar{\alpha}}(\tau) = 0 \qquad \tau \in [0,t] \\ &\vartheta_{\bar{\alpha}}(t) = 0 \qquad \dot{\vartheta}_{\bar{\alpha}}(t) = 1 \qquad \text{(no sum on } \bar{\alpha}) \end{aligned} \tag{2.162}$$

Owing to the linearity of the system the normalized adjoint response described by Eq. (2.156) is proportional to the unit impulse response described by Eq. (2.162) by the amplitude $\phi^{\mathrm{T}}_{\bar{\alpha}\alpha} G_{.\alpha}$ evaluated at t. Finally, the adjoint response described by Eq. (2.149) can be obtained by superposition of the modal responses

$$\lambda_\alpha(\tau) = \sum_{\bar{\alpha}=1}^{\bar{N}} \phi_{\alpha\bar{\alpha}}\,\vartheta_{\bar{\alpha}}(\tau)\,\phi^{\mathrm{T}}_{\bar{\alpha}\beta} G_{.\beta} \qquad \tau \in [0,t]\ ;\ \ t \in [0,T] \tag{2.163}$$

The nonhomogeneous initial conditions (2.135) do not obscure the computer implementation. To obtain a homogeneous initial problem from the nonhomogeneous system

$$\begin{aligned} &M_{\alpha\beta}(h^d)\ddot{q}_\beta(h^d;\tau) + C_{\alpha\beta}(h^d)\dot{q}_\beta(h^d;\tau) + K_{\alpha\beta}(h^d)q_\beta(h^d;\tau) = Q_\alpha(h^d;\tau) \\ &q_\alpha(h^d,0) = q^0(h^d) \qquad \dot{q}_\alpha(h^d,0) = \dot{q}^0(h^d) \end{aligned} \tag{2.164}$$

we define a particular solution

$$q^*_\alpha(h^d;\tau) = q^0_\alpha(h^d) - \dot{q}^0_\alpha(h^d)\,\tau \tag{2.165}$$

use a new variable

$$v_\alpha(h^d;\tau) = q_\alpha(h^d;\tau) - q^*_\alpha(h^d;\tau) \tag{2.166}$$

and substitute Eq. (2.166) into Eqs. (2.164) to obtain

$$\begin{aligned} &M_{\alpha\beta}(h^d)\,\ddot{v}_\beta(h^d;\tau) + C_{\alpha\beta}(h^d)\,\dot{v}_\beta(h^d;\tau) + K_{\alpha\beta}(h^d)\,v_\beta(h^d;\tau) \\ &\quad = Q_\alpha(h^d;\tau) - C_{\alpha\beta}(h^d)\,\dot{q}^*_\beta(h^d;\tau) - K_{\alpha\beta}(h^d)\,q^*_\beta(h^d;\tau) \end{aligned} \tag{2.167}$$

$$v_\alpha(h^d,0) = 0 \qquad \dot{v}_\alpha(h^d,0) = 0 \tag{2.168}$$

The eigenvalue sensitivity problem differs from the problems of static or forced-vibration sensitivity by the fact that derivatives of eigenvalues with respect to design variables can be determined without solving an adjoint problem or derivatives of eigenvectors with respect to the design variables. This feature of eigenvalue sensitivity allows direct computation of the sensitivity values once the eigenproblem has been solved. Recall that the problem is formulated assuming that the system is conservative; for nonconservative systems the situation is different [89].

It can be shown that if repeated eigenvalues appear in the free vibration and buckling analysis, then some computational difficulties may arise leading to singular or nondifferentiable problems. Sensitivity analysis formulations for such cases go beyond the framework of this text; they can be found in [44,85,95,101, 114,115], for instance.

Part II

FEM IN STOCHASTIC ANALYSIS

Chapter 3

Stochastic Variational Principles

3.1 Introductory Remarks

It has been common practice in engineering to analyse structural systems by assuming that system parameters are exactly determined. However, since such ideal situations are rarely encountered in engineering reality, the need to address uncertainties in the design is now clearly recognized. Traditionally, designers have used safety factors to provide increased confidence. This approach is questionable, however, as it does not as a rule take into account the underlying, more sophisticated probability characteristics.

In this book we are basically concerned with one major source of uncertainties in the design parameters, which is the so-called *inherent variability*, often referred to as *randomness*.[1] Randomness may exist in the characteristics of the structure itself (e.g. material properties, member sizes) or in the environment to which the structure is exposed (e.g. load, support conditions). Since it is now widely recognized that in some structures the response is sensitive to both the material and geometric properties of the structure as well as to the applied loads, even small uncertainties in these characteristics can adversely affect the structural performance. Moreover, since such uncertainties are usually spatially distributed over the region of the structure and must be modelled as random fields, and the structures themselves are frequently so complex as to exclude their analytical analysis even in the deterministic case, the need for an effective numerical tool to deal with a broad class of stochastic structural problems becomes evident.

Three basic methodologies can be adopted to quantify structural response uncertainties. The first one uses a Taylor series expansion to formulate linear relationships between some characteristics of the random response and the random structural parameters on the basis of the perturbation approach. The second methodology, known as Monte Carlo simulation, relies on the direct use of a computer and simulates an experiment. In such a simulated experiment a set of random numbers is generated first to represent the statistical uncertainties

[1] Other sources were identified in [57,61] as estimation error, model imperfection error and human error.

in the structural parameters. These random numbers are then substituted into the response equation to obtain a set of random numbers which reflect the uncertainty in the structural response, and which has to be analysed using special techniques [6,28,30,96,105,116] to produce a qualification of the uncertainty. The third methodology is known as the Neumann expansion [1,17,87], but has been rarely used in the field of structural mechanics so far [13,107]. The method has been shown to be effective when coupled with the Monte Carlo simulation [123]. The so-called Karkunen–Loeve decomposition has been recently employed yielding a mathematically elegant and computationally tractable algorithm when coupled with either a Neumann expansion scheme [110], or a polynomial chaos expansion and a Galerkin projection [39].

In this book the *second-order* version of the first methodology is exclusively used as it seems both theoretically sound and computationally feasible to apply even for very large structural systems by using techniques typical of contemporary computational mechanics. In particular, we shall present a version of the finite element method which accounts for uncertainties in both the geometry and/or material properties of the structure, as well as in the applied load. The approach will be referred to as the *stochastic finite element method* (SFEM).

As mentioned in the Preface, this book is about dealing numerically with time-invariant uncertainties. The load uncertainties which fall into the scope of this analysis are those described by using a statistical analysis of data obtained from measurement of load intensity without regard for the time frequency of occurrence. The weight of the materials used in construction (dead loads) or the weight of the contents of a building (live loads) are examples of loads of this type.

It is clear that in many real situations accounting for time-varying randomness in the load is much more important than any attempt to model the time-invariant uncertainties mentioned above; the frequently encountered need for random vibration analysis is a good example in this context. The intentional limitation of this book to problems of time-invariant uncertainties is due to significant mathematical simplification resulting from such an assumption; mathematically speaking the problem becomes reducible to the analysis of random differential equations with constant-in-time coefficients and no random process inhomogeneous part. The simplification is warranted by the character of the book which merely attempts to set up a transparent methodology and to present introductory computer implementation of the underlying theory on which to build further developments.

The mathematical essence of our structural mechanics considerations below is the analysis of stochastic partial differential equations with coefficients in the form of constant-in-time, varying-in-space random fields, say $\mathbf{b}(\mathbf{x})$. Some general remarks seem to be in order on what is here meant by a solution $\mathbf{u}(\mathbf{x}, \tau)$ of the random differential equations and what properties of the solution process does one seek in a given situation.

The FEM approach requires first that the random field $\mathbf{b}(\mathbf{x})$ be spatially discretized, i.e. represented by its nodal value vector, say $\tilde{\mathbf{b}}$. The finite element discretization of the displacement-type equations of motion for the linear structural dynamics problem leads then to a system of ordinary differential equations which are linear in the unknown nodal displacement vector $\mathbf{q}(\tau)$ (a vector random process) and nonlinear in the random coefficients $\tilde{\mathbf{b}}$. If now certain joint statistical behaviour of the random variables $\tilde{\mathbf{b}}$ is specified by giving their statistical properties, one is required to determine in terms of these properties certain statistical characteristics of the solution $\mathbf{q}(\tau)$.

Similarly as in the deterministic theory of differential equations, both the quantitative and qualitative analyses of so obtained random ordinary differential equations[1] are relevant and important. Any qualitative analysis would be out of place in this book; some mathematical background may be found in [109], for instance.

In the quantitative approach the complete solution is constituted by giving the joint probability distributions of all the random variables $\mathbf{q}(\tau)$ for all finite sets of time instants; knowledge of this family of distribution functions allows us to determine the probabilities of any useful event associated with the process. In practice this goal is impossible to achieve and, in view of usually limited information on joint statistical behaviour[2] of $\tilde{\mathbf{b}}$, it is not feasible to strive for it either. Thus, solutions of random differential equations considered here mean as a rule much less than joint CDFs; being able to determine a limited number of statistical properties associated with the solution process is very often considered a fully satisfactory situation provided the properties are recognized as pertinent to the problem at hand. Among the most useful properties of a random solution process are certainly the statistical moments, particularly the mean and the correlation function.

Another remark that seems relevant at this point refers to the fact that ordinary differential equations with constant random coefficients can be transformed to the form in which randomness enters only through the initial conditions [109]. Since the solution process for the latter class of differential equations has a simple structure, relatively straightforward mathematical tools can in principle be employed to determine the statistical properties of the solution vector in terms of those associated with the initial data vector. In particular, the solution presents no conceptual difficulty if the solution of the corresponding deterministic differential equations can be found. Thus, it is as a rule effective for systems consisting of a few equations, for instance. For large systems, bearing additionally in mind future generalizations toward the inclusion of nonlinear effects, the techniques advocated in this book and based on the second moment perturbation approach appear to be much more effective, though.

---

[1]As a matter of fact, we implicitly assume that the so-called mean-square treatment of random differential equations is considered in this book, cf. [109].

[2]The random vector $\tilde{\mathbf{b}}$ is usually defined by its first two moments, i.e. expected value and covariance matrix; for simplicity, higher order moments are not considered. This assumption is fully legitimate if $\tilde{\mathbf{b}}$ is stationary; otherwise it has to be treated as an approximation which, in view of the central limit theorem [71,83], is relevant in many cases.

Part II of the book addresses basic theoretical issues and implementation techniques necessary for successful SFEM applications. Confining ourselves basically (except for Chapter 9) to linear systems we shall be concerned with a whole variety of issues ranging from stochastic static analysis via linearized stability and dynamic analyses to sensitivity investigations of stochastic static and dynamic response. The theoretical derivations will closely follow the considerations recently put forward in [51]–[53] and [78]–[82]; the concepts have been reworked, extended and implemented by the authors as documented in [46]–[49]. In order to increase the transparency of this text no direct reference is made below to the source publications when presenting specific concepts and equations.

3.2 Stochastic Potential Energy Principle

The stochastic potential energy principle as adopted in this book is a combination of the principle of minimum potential energy presented in Section 2.3.1 and the second-order perturbation technique.

Let us consider a linear structural system under static load and assume

$$\mathbf{b}(x_k) = \{\; b_1(x_k) \;\; b_2(x_k) \;\; \ldots \;\; b_R(x_k) \;\} \qquad k = 1,2,3$$

to be a set of R random fields which can represent randomness in the cross-sectional area and length of truss and beam members, thickness of plate and shell elements, Young's modulus and mass density of the material, etc., as well as time-invariant randomness in the external load. The first two statistical moments for the random fields $b_r(x_k)$, $r = 1,2,\ldots,R$, are defined as, cf. Eqs. (1.16), (1.26) and (1.91)

$$E[b_r] \;=\; b_r^0 \;=\; \int_{-\infty}^{+\infty} b_r\, p_1(b_r)\, \mathrm{d}b_r \tag{3.1}$$

$$\mathrm{Cov}(b_r, b_s) \;=\; S_{\mathrm{b}}^{rs} \;=\; \int_{-\infty}^{+\infty}\int_{-\infty}^{+\infty} (b_r - b_r^0)(b_s - b_s^0)\, p_2(b_r, b_s)\, \mathrm{d}b_r \mathrm{d}b_s \tag{3.2}$$

$$r, s = 1,2,\ldots,R$$

The latter definition can be replaced by

$$S_{\mathrm{b}}^{rs} = \alpha_{b_r} \alpha_{b_s} b_r^0\, b_s^0\, \mu_{b_r b_s} \tag{3.3}$$

with

$$\begin{aligned} \alpha_{b_r} &= \left[\frac{\mathrm{Var}(b_r)}{b_r^0} \right]^{\frac{1}{2}} \\ \mu_{b_r b_s} &= \int_{-\infty}^{+\infty}\int_{-\infty}^{+\infty} b_r b_s\, p_2(b_r, b_s)\, \mathrm{d}b_r \mathrm{d}b_s \end{aligned} \tag{3.4}$$

where, as in Chapter 1, $E[b_r]$, $\mathrm{Cov}(b_r, b_s)$, $\mathrm{Var}(b_r)$, $\mu_{b_r b_s}$, α_{b_r}, $p_1(b_r)$ and $p_2(b_r, b_s)$ denote the spatial expectation, covariance, variance, correlation functions, the coefficients of variation, PDF and the joint PDF, respectively.

The main idea behind the second-order perturbation approach to the stochastic version of the potential energy principle involves expanding all the random field variables in the problem at hand, i.e. elastic moduli $C_{ijkl}[\mathbf{b}(x_k);x_k]$, mass density $\varrho[\mathbf{b}(x_k);x_k]$, body forces $f_i[\mathbf{b}(x_k);x_k]$, boundary tractions $\hat{t}_i[\mathbf{b}(x_k);x_k]$ and displacements $u_i[\mathbf{b}(x_k);x_k]$ about the spatial expectations of the random field variables $\mathbf{b}(x_k)=\{b_r(x_k)\}$, denoted by $\mathbf{b}^0(x_k)=\{b_r^0(x_k)\}$, via Taylor series with a given small parameter ϵ and retaining terms up to second order. The expansions are explicitly written as[1]

$$
\begin{aligned}
C_{ijkl}[\mathbf{b}(x_k);x_k] &= C^0_{ijkl}[\mathbf{b}^0(x_k);x_k] + \epsilon\, C^{,r}_{ijkl}[\mathbf{b}^0(x_k);x_k]\,\Delta b_r(x_k) \\
&\quad + \tfrac{1}{2}\epsilon^2 C^{,rs}_{ijkl}[\mathbf{b}^0(x_k);x_k]\,\Delta b_r(x_k)\Delta b_s(x_k) \\
\varrho[\mathbf{b}(x_k);x_k] &= \varrho^0[\mathbf{b}^0(x_k);x_k] + \epsilon\,\varrho^{,r}[\mathbf{b}^0(x_k);x_k]\,\Delta b_r(x_k) \\
&\quad + \tfrac{1}{2}\epsilon^2 \varrho^{,rs}[\mathbf{b}^0(x_k);x_k]\,\Delta b_r(x_k)\Delta b_s(x_k) \\
f_i[\mathbf{b}(x_k);x_k] &= f_i^0[\mathbf{b}^0(x_k);x_k] + \epsilon\, f_i^{,r}[\mathbf{b}^0(x_k);x_k]\,\Delta b_r(x_k) \\
&\quad + \tfrac{1}{2}\epsilon^2 f_i^{,rs}[\mathbf{b}^0(x_k);x_k]\,\Delta b_r(x_k)\Delta b_s(x_k) \\
\hat{t}_i[\mathbf{b}(x_k);x_k] &= \hat{t}_i^0[\mathbf{b}^0(x_k);x_k] + \epsilon\, \hat{t}_i^{,r}[\mathbf{b}^0(x_k);x_k]\,\Delta b_r(x_k) \\
&\quad + \tfrac{1}{2}\epsilon^2 \hat{t}_i^{,rs}[\mathbf{b}^0(x_k);x_k]\,\Delta b_r(x_k)\Delta b_s(x_k) \\
u_i[\mathbf{b}(x_k);x_k] &= u_i^0[\mathbf{b}^0(x_k);x_k] + \epsilon\, u_i^{,r}[\mathbf{b}^0(x_k);x_k]\,\Delta b_r(x_k) \\
&\quad + \tfrac{1}{2}\epsilon^2 u_i^{,rs}[\mathbf{b}^0(x_k);x_k]\,\Delta b_r(x_k)\Delta b_s(x_k)
\end{aligned}
\tag{3.5}
$$

where

$$\epsilon\,\Delta b_r(x_k) = \delta b_r(x_k) = \epsilon\left[b_r(x_k) - b_r^0(x_k)\right] \tag{3.6}$$

is the first-order variation of $b_r(x_k)$ about $b_r^0(x_k)$, and

$$
\begin{aligned}
\epsilon^2\Delta b_r(x_k)\Delta b_s(x_k) &= \delta b_r(x_k)\,\delta b_s(x_k) \\
&= \epsilon^2\left[b_r(x_k) - b_r^0(x_k)\right]\left[b_s(x_k) - b_s^0(x_k)\right]
\end{aligned}
\tag{3.7}
$$

denotes the second-order variation of $b_r(x_k)$ and $b_s(x_k)$ about $b_r^0(x_k)$ and $b_s^0(x_k)$, respectively. The symbol $(.)^0$ represents the value of the functions taken at b_r^0 while $(.)^{,r}$ and $(.)^{,rs}$ stand for the first and second (mixed) partial derivatives with respect to the random field variables $b_r(x_k)$ evaluated at their expectations, respectively.

In accordance with the philosophy of the perturbation approach[2] the expansions (3.5) are now substituted into the principle of minimum potential energy,

---

[1]Clearly, repetition of the subscript k in the arguments x_k (and $x_k^{(1)}$ and $x_k^{(2)}$, below) does not imply summation.

[2]Despite the versatility enjoyed by the perturbation method the reader should be aware that certain subtle questions about this method remain unanswered. In particular, the power expansion (3.5) is valid only if the response is analytic in ϵ and the series converges. Criteria for convergence must include the magnitude of the perturbation parameter ϵ – no such criteria have been established in the present context.

Eq. (2.11). Equating terms of equal orders we arrive at the following zeroth-, first- and second-order potential energy variational principles:

- Zeroth-order (ϵ^0 terms, one equation)

$$\int_\Omega C^0_{ijkl} u^0_{i,j}\, \delta u_{k,l}\, \mathrm{d}\Omega = \int_\Omega \varrho^0 f^0_i\, \delta u_i\, \mathrm{d}\Omega + \int_{\partial\Omega_\sigma} \hat{t}^0_i\, \delta u_i\, \mathrm{d}(\partial\Omega) \tag{3.8}$$

- First-order (ϵ^1 terms, R equations)

$$\begin{aligned}\int_\Omega C^0_{ijkl} u^{,r}_{i,j}\, \delta u_{k,l}\, \mathrm{d}\Omega &= \int_\Omega \left(\varrho^{,r} f^0_i + \varrho^0 f^{,r}_i\right) \delta u_i\, \mathrm{d}\Omega \\ &+ \int_{\partial\Omega_\sigma} \hat{t}^{,r}_i\, \delta u_i\, \mathrm{d}(\partial\Omega) - \int_\Omega C^{,r}_{ijkl} u^0_{i,j}\, \delta u_{k,l}\, \mathrm{d}\Omega\end{aligned} \tag{3.9}$$

- Second-order (ϵ^2 terms, one equation)

$$\begin{aligned}\int_\Omega C^0_{ijkl} u^{,rs}_{i,j} S^{rs}_{\mathrm{b}}\, \delta u_{k,l}\, \mathrm{d}\Omega &= \int_\Omega \left(\varrho^0 f^{,rs}_i + 2\varrho^{,r} f^{,s}_i + \varrho^{,rs} f^0_i\right) S^{rs}_{\mathrm{b}}\, \delta u_i\, \mathrm{d}\Omega \\ + \int_{\partial\Omega_\sigma} \hat{t}^{,rs}_i S^{rs}_{\mathrm{b}}\, \delta u_i\, \mathrm{d}(\partial\Omega) &- \int_\Omega \left(2C^{,r}_{ijkl} u^{,s}_{i,j} + C^{,rs}_{ijkl} u^0_{i,j}\right) S^{rs}_{\mathrm{b}}\, \delta u_{k,l}\, \mathrm{d}\Omega\end{aligned} \tag{3.10}$$

The test function δu_i should satisfy the kinematic boundary conditions (2.4). We note that the second-order equation has been obtained here by multiplying the R-variate probability density function $p_R(b_1, b_2, \ldots, b_R)$ by the ϵ^2-terms and integrating over the domain of the random field variables $\mathbf{b}(x_k)$. For instance, the ϵ^2-term involving $\hat{t}^{,rs}_i[\mathbf{b}^0(x_k); x_k]$ reads, cf. Eqs. (1.26) and $(3.7)_1$

$$\begin{aligned}&\int_{-\infty}^{+\infty} \left[\int_{\partial\Omega_\sigma} \epsilon^2\, \hat{t}^{,rs}_i[\mathbf{b}^0(x_k); x_k] \Delta b_r(x_k) \Delta b_s(x_k)\, \delta u_i\, \mathrm{d}(\partial\Omega)\right] p_R\big(\mathbf{b}(x_k)\big)\, \mathrm{d}\mathbf{b} \\ &= \epsilon^2 \int_{\partial\Omega_\sigma} \hat{t}^{,rs}_i[\mathbf{b}^0(x_k); x_k] \underbrace{\left[\int_{-\infty}^{+\infty} \Delta b_r(x_k) \Delta b_s(x_k)\, p_R\big(\mathbf{b}(x_k)\big)\, \mathrm{d}\mathbf{b}\right]}_{S^{rs}_{\mathrm{b}}} \delta u_i\, \mathrm{d}(\partial\Omega) \\ &= \epsilon^2 \int_{\partial\Omega_\sigma} \hat{t}^{,rs}_i\, S^{rs}_{\mathrm{b}}\, \delta u_i\, \mathrm{d}(\partial\Omega)\end{aligned} \tag{3.11}$$

Thus, the occurrence of the double sums $(.)^{,rs} S^{rs}_{\mathrm{b}}$ and $(.)^{,r}(.)^{,s} S^{rs}_{\mathrm{b}}$ in the formulation enables one to deal with only one equation (3.10), instead of $R(R+1)/2$ second-order equations apparently required (Eqs. (3.5) are symmetric with respect to r and s). From a computational standpoint this is particularly important, as shown in the following chapters.

Eq. (3.8) is seen to be identical to the deterministic minimum potential energy principle (2.11); as such it can serve as the basis to obtain the zeroth-order displacements $u^0_i(x_k)$.[1] Once this has been done, the higher-order terms $u^{,r}_i(x_k)$ and $u^{,rs}_i(x_k)$ can be evaluated from Eqs. (3.9) and (3.10), sequentially. We note that except for the unknown functions all terms involved on the left-hand side of Eqs. (3.8)–(3.10) are identical, the probabilistic characteristics of the problem being translated entirely into the effective forces on the right-hand sides.

[1]The function argument $\mathbf{b}^0(x_k)$ will be suppressed at places to clarify the presentation; for instance, instead of $u^0_i[\mathbf{b}^0(x_k); x_k]$ we may write $u^0_i(x_k)$.

Having solved Eqs. (3.8)–(3.10) for $u_i^0(x_k)$, $u_i^{,r}(x_k)$ and $u_i^{,rs}(x_k)$ (to be exact, for $u_i^{,rs}(x_k)S_{\mathrm{b}}^{rs}$) probabilistic distributions of the random displacement field $u_i[\mathbf{b}(x_k);x_k]$ may, for a given ϵ, be calculated. Clearly, setting $\epsilon=0$ yields the deterministic solution. In our case, the formal solution is obtained by setting $\epsilon=1$ which, of course, stipulates that the fluctuation of the random field variables $\mathbf{b}(x_k)$ is small. By introducing the expanded equation $(3.5)_5$ into the expression for the mean value of the random displacement field $u_i[\mathbf{b}(x_k);x_k]$

$$E\Big[u_i[\mathbf{b}(x_k);x_k]\Big] = \int_{-\infty}^{+\infty} u_i[\mathbf{b}(x_k);x_k]\, p_R\big(\mathbf{b}(x_k)\big)\,\mathrm{d}\mathbf{b} \tag{3.12}$$

its second-order estimate is obtained as

$$E\Big[u_i[\mathbf{b}(x_k);x_k]\Big] = u_i^0(x_k) + \frac{1}{2}u_i^{,rs}(x_k)\, S_{\mathrm{b}}^{rs} \tag{3.13}$$

since, cf. Eqs. (1.7–ii), (1.23), (1.26) and Eq. (3.7)

$$\begin{aligned} E[u_i] &= \int_{-\infty}^{+\infty}\Big\{u_i^0[\mathbf{b}^0(x_k);x_k] + u_i^{,r}[\mathbf{b}^0(x_k);x_k]\,\Delta b_r(x_k) \\ &\quad + \frac{1}{2}u_i^{,rs}[\mathbf{b}^0(x_k);x_k]\,\Delta b_r(x_k)\Delta b_s(x_k)\Big\}\, p_R\big(\mathbf{b}(x_k)\big)\,\mathrm{d}\mathbf{b} \\ &= u_i^0[\mathbf{b}^0(x_k);x_k]\underbrace{\int_{-\infty}^{+\infty} p_R\big(\mathbf{b}(x_k)\big)\,\mathrm{d}\mathbf{b}}_{=1} \\ &\quad + u_i^{,r}[\mathbf{b}^0(x_k);x_k]\underbrace{\int_{-\infty}^{+\infty}\Delta b_r(x_k)\, p_R\big(\mathbf{b}(x_k)\big)\,\mathrm{d}\mathbf{b}}_{=0} \\ &\quad + \frac{1}{2}u_i^{,rs}[\mathbf{b}^0(x_k);x_k]\int_{-\infty}^{+\infty}\Delta b_r(x_k)\Delta b_s(x_k)\, p_R\big(\mathbf{b}(x_k)\big)\,\mathrm{d}\mathbf{b} \end{aligned} \tag{3.14}$$

Clearly, if only the first-order accuracy of the displacement estimation is required, then Eq. (3.13) reduces to

$$E\Big[u_i[\mathbf{b}(x_k);x_k]\Big] = u_i^0(x_k) \tag{3.15}$$

The first-order accurate cross-covariances[1] of $u_i[\mathbf{b}(x_k^{(1)});x_k^{(1)}]$ and $u_j[\mathbf{b}(x_k^{(2)});x_k^{(2)}]$ can be evaluated by substituting the second-order expansion of the random displacement field $u_i[\mathbf{b}(x_k);x_k]$ into the expression for the cross-covariance

$$\begin{aligned} &\mathrm{Cov}\Big(u_i[\mathbf{b}(x_k^{(1)});x_k^{(1)}],u_j[\mathbf{b}(x_k^{(2)});x_k^{(2)}]\Big) = S_{\mathbf{u}}^{ij}(x_k^{(1)},x_k^{(2)}) \\ &= \int_{-\infty}^{+\infty}\Big\{u_i[\mathbf{b}(x_k^{(1)});x_k^{(1)}] - E\Big[u_i[\mathbf{b}(x_k^{(1)});x_k^{(1)}]\Big]\Big\} \\ &\quad\times\Big\{u_j[\mathbf{b}(x_k^{(2)});x_k^{(2)}] - E\Big[u_j[\mathbf{b}(x_k^{(2)});x_k^{(2)}]\Big]\Big\}\, p_R\big(\mathbf{b}(x_k)\big)\,\mathrm{d}\mathbf{b} \end{aligned} \tag{3.16}$$

to get

$$S_{\mathbf{u}}^{ij}(x_k^{(1)},x_k^{(2)}) = u_i^{,r}(x_k^{(1)})\,u_j^{,s}(x_k^{(2)})\,S_{\mathrm{b}}^{rs} \tag{3.17}$$

[1]The first-order estimate of cross-covariances is consistent with the second-moment analysis.

The strain probabilistic characteristics then follow as:

- Second-order accurate mean value

$$\begin{aligned} E\Big[\varepsilon_{ij}[\mathbf{b}(x_k);x_k]\Big] &= E\Big[\tfrac{1}{2}\left(u_{i,j}[\mathbf{b}(x_k);x_k]+u_{j,i}[\mathbf{b}(x_k);x_k]\right)\Big] \\ &= \tfrac{1}{2}\left[u^0_{i,j}(x_k)+u^0_{j,i}(x_k)+\tfrac{1}{2}\left(u^{,rs}_{i,j}(x_k)+u^{,rs}_{j,i}(x_k)\right)S^{rs}_{\mathrm{b}}\right] \\ &= \varepsilon^0_{ij}(x_k)+\tfrac{1}{2}\varepsilon^{,rs}_{ij}(x_k)\,S^{rs}_{\mathrm{b}} \end{aligned} \tag{3.18}$$

in which the following expansion has been employed

$$\begin{aligned} \varepsilon_{ij}[\mathbf{b}(x_k);x_k] &= \varepsilon^0_{ij}[\mathbf{b}^0(x_k);x_k]+\varepsilon^{,r}_{ij}[\mathbf{b}^0(x_k);x_k]\,\Delta b_r(x_k) \\ &\quad+\tfrac{1}{2}\varepsilon^{,rs}_{ij}[\mathbf{b}^0(x_k);x_k]\,\Delta b_r(x_k)\Delta b_s(x_k) \end{aligned} \tag{3.19}$$

- First-order accurate cross-covariance

$$\begin{aligned} \mathrm{Cov}\Big(\varepsilon_{ij}[\mathbf{b}(x^{(1)}_k);x^{(1)}_k],\varepsilon_{kl}[\mathbf{b}(x^{(2)}_k);x^{(2)}_k]\Big) &= S^{ijkl}_{\varepsilon}(x^{(1)}_k,x^{(2)}_k) \\ = \tfrac{1}{4}\left[u^{,r}_{i,j}(x^{(1)}_k)+u^{,r}_{j,i}(x^{(1)}_k)\right]\left[u^{,s}_{k,l}(x^{(2)}_k)+u^{,s}_{l,k}(x^{(2)}_k)\right]S^{rs}_{\mathrm{b}} \\ = \varepsilon^{,r}_{ij}(x^{(1)}_k)\,\varepsilon^{,s}_{kl}(x^{(2)}_k)\,S^{rs}_{\mathrm{b}} \end{aligned} \tag{3.20}$$

To determine the first two moments for the stresses the expression for σ_{ij} is written as

$$\begin{aligned} \sigma_{ij} &= C_{ijkl}\,\varepsilon_{kl} \\ &= \left(C^0_{ijkl}+C^{,r}_{ijkl}\Delta b_r+\tfrac{1}{2}C^{,rs}_{ijkl}\,\Delta b_r\Delta b_s\right)\left(\varepsilon^0_{kl}+\varepsilon^{,u}_{kl}\Delta b_u+\tfrac{1}{2}\varepsilon^{,uv}_{kl}\,\Delta b_u\Delta b_v\right) \end{aligned} \tag{3.21}$$

By employing Eq. (3.21) in stress equations similar to Eqs. (3.12) and (3.16) and neglecting the variations of an order higher than two we arrive at:

- Second-order accurate mean value

$$\begin{aligned} E\Big[\sigma_{ij}[\mathbf{b}(x_k);x_k]\Big] &= C^0_{ijkl}(x_k)\,\varepsilon^0_{kl}(x_k)+\tfrac{1}{2}\Big[C^{,rs}_{ijkl}(x_k)\,\varepsilon^0_{kl}(x_k) \\ &\quad+2C^{,r}_{ijkl}(x_k)\,\varepsilon^{,s}_{kl}(x_k)+C^0_{ijkl}(x_k)\,\varepsilon^{,rs}_{kl}(x_k)\Big]S^{rs}_{\mathrm{b}} \end{aligned} \tag{3.22}$$

- First-order accurate cross-covariance

$$\begin{aligned} \mathrm{Cov}\Big(\sigma_{ij}[\mathbf{b}(x^{(1)}_k);x^{(1)}_k],\sigma_{kl}[\mathbf{b}(x^{(2)}_k);x^{(2)}_k]\Big) &= S^{ijkl}_{\sigma}(x^{(1)}_k,x^{(2)}_k) \\ = \Big[C^{,r}_{ijmn}(x^{(1)}_k)\,C^{,s}_{kl\bar{m}\bar{n}}(x^{(2)}_k)\,\varepsilon^0_{mn}(x^{(1)}_k)\,\varepsilon^0_{\bar{m}\bar{n}}(x^{(2)}_k) \\ +\,C^{,r}_{ijmn}(x^{(1)}_k)\,C^0_{kl\bar{m}\bar{n}}(x^{(2)}_k)\,\varepsilon^0_{mn}(x^{(1)}_k)\,\varepsilon^{,s}_{\bar{m}\bar{n}}(x^{(2)}_k) \\ +\,C^0_{ijmn}(x^{(1)}_k)\,C^{,r}_{kl\bar{m}\bar{n}}(x^{(2)}_k)\,\varepsilon^{,s}_{mn}(x^{(1)}_k)\,\varepsilon^0_{\bar{m}\bar{n}}(x^{(2)}_k) \\ +\,C^0_{ijmn}(x^{(1)}_k)\,C^0_{kl\bar{m}\bar{n}}(x^{(2)}_k)\,\varepsilon^{,r}_{mn}(x^{(1)}_k)\,\varepsilon^{,s}_{\bar{m}\bar{n}}(x^{(2)}_k)\Big]S^{rs}_{\mathrm{b}} \end{aligned} \tag{3.23}$$

3.3 Stochastic Multi-Field Principles

In the preceding section a stochastic one-field variational statement with an independent displacement field was introduced and the stochastic potential energy principle was formulated using the second-order perturbation technique. A combination of the second-moment analysis and a multi-field variational principle may be employed as well with the result offering greater flexibility in dealing with randomness in problems of linear elasticity. For example, if the multi-field variational principle described in Section 2.3.2 is employed, then the equilibrium and compatibility conditions, constitutive law, static and kinematic boundary conditions can be simultaneously incorporated into the formulation, leading to a stochastic multi-field principle. The stationary conditions for this principle will generate the zeroth-, first- and second-order equations for the displacement field $u_i[b_r(x_k); x_k]$, strain field $\varepsilon_{ij}[b_r(x_k); x_k]$, stress field $\sigma_{ij}[b_r(x_k); x_k]$ and the reaction force field $t_i[b_r(x_k); x_k]$, respectively. The probabilistic distributions (i.e. the first two moments) for the random field variables $u_i[b_r(x_k); x_k]$, $\varepsilon_{ij}[b_r(x_k); x_k]$ and $\sigma_{ij}[b_r(x_k); x_k]$ can then be evaluated; and the probabilistic description of the structural response is obtained with no subsidiary prescribed conditions. Below we just sketch the pertinent procedure.

Let us consider the functional $\mathcal{J}_{\text{H-W}}$ given by Eq. (2.17) in the case of a structural system whose material and local (i.e. those not referring to the overall shape of the structure) geometric properties, boundary conditions and applied loads are described stochastically. The functions $u_i[b_r(x_k); x_k]$, $\varepsilon_{ij}[b_r(x_k); x_k]$, $\sigma_{ij}[b_r(x_k); x_k]$ and $t_i[b_r(x_k); x_k]$ are assumed to be independently subjected to variations. Similarly as in the case of the stochastic potential energy principle, the second-order expansion about the spatial expectations $\mathbf{b}^0 = b_r^0(x_k)$ of the random variables $\mathbf{b}(x_k) = \{b_r(x_k)\}$ is made for all the random field variables involved in Eq. (2.17). The expanded expressions for C_{ijkl}, ϱ, f_i, $\hat{t}_i$ and u_i are given in Eqs. (3.5) while those for ε_{ij}, σ_{ij}, t_i and $\hat{u}_i$ read

$$
\begin{aligned}
\varepsilon_{ij}[\mathbf{b}(x_k); x_k] &= \varepsilon_{ij}^0[\mathbf{b}^0(x_k); x_k] + \epsilon\, \varepsilon_{ij}^{,r}[\mathbf{b}^0(x_k); x_k]\, \Delta b_r(x_k) \\
&\quad + \frac{1}{2}\epsilon^2 \varepsilon_{ij}^{,rs}[\mathbf{b}^0(x_k); x_k]\, \Delta b_r(x_k) \Delta b_s(x_k) \\
\sigma_{ij}[\mathbf{b}(x_k); x_k] &= \sigma_{ij}^0[\mathbf{b}^0(x_k); x_k] + \epsilon\, \sigma_{ij}^{,r}[\mathbf{b}^0(x_k); x_k]\, \Delta b_r(x_k) \\
&\quad + \frac{1}{2}\epsilon^2 \sigma_{ij}^{,rs}[\mathbf{b}^0(x_k); x_k]\, \Delta b_r(x_k) \Delta b_s(x_k) \\
t_i[\mathbf{b}(x_k); x_k] &= t_i^0[\mathbf{b}^0(x_k); x_k] + \epsilon\, t_i^{,r}[\mathbf{b}^0(x_k); x_k]\, \Delta b_r(x_k) \\
&\quad + \frac{1}{2}\epsilon^2\, t_i^{,rs}[\mathbf{b}^0(x_k); x_k]\, \Delta b_r(x_k) \Delta b_s(x_k) \\
\hat{u}_i[\mathbf{b}(x_k); x_k] &= \hat{u}_i^0[\mathbf{b}^0(x_k); x_k] + \epsilon\, \hat{u}_i^{,r}[\mathbf{b}^0(x_k); x_k]\, \Delta b_r(x_k) \\
&\quad + \frac{1}{2}\epsilon^2\, \hat{u}_i^{,rs}[\mathbf{b}^0(x_k); x_k]\, \Delta b_r(x_k) \Delta b_s(x_k)
\end{aligned}
\tag{3.24}
$$

The second-order expansions (3.5) and (3.24) are now substituted into Eq. (2.18) and the Gauss–Ostrogradski theorem is employed in the resulting equation to obtain the explicit expression for the first variation of the functional $\mathcal{J}_{\text{H-W}}$. Noting that the test functions δu_i, $\delta \varepsilon_{ij}$, $\delta \sigma_{ij}$ and δt_i are arbitrary and independent, by selecting coefficients of like power of ϵ we obtain the expressions for the zeroth-, first- and second-order multi-field variational principle as:

- Zeroth-order (ϵ^0 terms, one equation)

$$
\begin{aligned}
&-\int_\Omega \left(\sigma^0_{ij,j} + \varrho^0 f^0_i\right)\delta u_i \,\mathrm{d}\Omega + \int_{\partial\Omega_\sigma}\left(\sigma^0_{ij} n_j - \hat{t}^0_i\right)\delta u_i \,\mathrm{d}(\partial\Omega) \\
&+ \int_{\partial\Omega_u}\left(\sigma^0_{ij} n_j - t^0_i\right)\delta u_i \,\mathrm{d}(\partial\Omega) + \int_\Omega \left(C^0_{ijkl}\varepsilon^0_{kl} - \sigma^0_{ij}\right)\delta\varepsilon_{ij}\,\mathrm{d}\Omega \\
&+ \int_\Omega \left[\varepsilon^0_{ij} - \frac{1}{2}\left(u^0_{i,j} + u^0_{j,i}\right)\right]\delta\sigma_{ij}\,\mathrm{d}\Omega + \int_{\partial\Omega_u}\left(u^0_i - \hat{u}^0_i\right)\delta t_i \,\mathrm{d}(\partial\Omega) = 0 \qquad (3.25)
\end{aligned}
$$

- First-order (ϵ^1 terms, R equations)

$$
\begin{aligned}
&-\int_\Omega \left(\sigma^{,r}_{ij,j} + \varrho^{,r} f^0_i + \varrho^0 f^{,r}_i\right)\delta u_i \,\mathrm{d}\Omega + \int_{\partial\Omega_\sigma}\left(\sigma^{,r}_{ij} n_j - \hat{t}^{,r}_i\right)\delta u_i \,\mathrm{d}(\partial\Omega) \\
&+ \int_{\partial\Omega_u}\left(\sigma^{,r}_{ij} n_j - t^{,r}_i\right)\delta u_i \,\mathrm{d}(\partial\Omega) + \int_\Omega \left(C^{,r}_{ijkl}\varepsilon^0_{kl} + C^0_{ijkl}\varepsilon^{,r}_{kl} - \sigma^{,r}_{ij}\right)\delta\varepsilon_{ij}\,\mathrm{d}\Omega \\
&+ \int_\Omega \left[\varepsilon^{,r}_{ij} - \frac{1}{2}\left(u^{,r}_{i,j} + u^{,r}_{j,i}\right)\right]\delta\sigma_{ij}\,\mathrm{d}\Omega + \int_{\partial\Omega_u}\left(u^{,r}_i - \hat{u}^{,r}_i\right)\delta t_i \,\mathrm{d}(\partial\Omega) = 0 \qquad (3.26)
\end{aligned}
$$

- Second-order (ϵ^2 terms, one equation)

$$
\begin{aligned}
&-\int_\Omega \left(\sigma^{,rs}_{ij,j} + \varrho^{,rs} f^0_i + 2\varrho^{,r} f^{,s}_i + \varrho^0 f^{,rs}_i\right) S^{rs}_{\mathrm{b}}\delta u_i \,\mathrm{d}\Omega \\
&+ \int_{\partial\Omega_\sigma}\left(\sigma^{,rs}_{ij} n_j - \hat{t}^{,rs}_i\right) S^{rs}_{\mathrm{b}}\delta u_i \,\mathrm{d}(\partial\Omega) + \int_{\partial\Omega_u}\left(\sigma^{,rs}_{ij} n_j - t^{,rs}_i\right) S^{rs}_{\mathrm{b}}\delta u_i \,\mathrm{d}(\partial\Omega) \\
&+ \int_\Omega \left(C^{,rs}_{ijkl}\varepsilon^0_{kl} + 2C^{,r}_{ijkl}\varepsilon^{,s}_{kl} + C^0_{ijkl}\varepsilon^{,rs}_{kl} - \sigma^{,rs}_{ij}\right) S^{rs}_{\mathrm{b}}\delta\varepsilon_{ij}\,\mathrm{d}\Omega \qquad (3.27) \\
&+ \int_\Omega \left[\varepsilon^{,rs}_{ij} - \frac{1}{2}\left(u^{,rs}_{i,j} + u^{,rs}_{j,i}\right)\right] S^{rs}_{\mathrm{b}}\delta\sigma_{ij}\,\mathrm{d}\Omega + \int_{\partial\Omega_u}\left(u^{,rs}_i - \hat{u}^{,rs}_i\right) S^{rs}_{\mathrm{b}}\delta t_i \,\mathrm{d}(\partial\Omega) = 0
\end{aligned}
$$

As before, the second-order equation has been obtained by multiplying the R-variate PDF $p_R(b_1, b_2, \ldots, b_R)$ by the ϵ^2-terms and integrating over the domain of the random field variables $b_r(x_k)$.

The complete solution to this problem, i.e. evaluation of the second-order accurate mean value and the first-order accurate cross-covariance of the random field variables $u_i[b_r(x_k); x_k]$, $\varepsilon_{ij}[b_r(x_k); x_k]$ and $\sigma_{ij}[b_r(x_k); x_k]$, can be obtained by following the general procedure described in the preceding section, Eqs. (3.12)–(3.23). We note that in the context of the multi-field variational statements the probabilistic distributions for u_i, ε_{ij} and σ_{ij} are obtained as independent results (i.e. they are interrelated in the variational sense only), since they are presumed to be random field variables independently subjected to variation, cf. [82].

3.4 Stochastic Hamilton Principle

It will best serve our purpose of developing approximate solution techniques if we consider the Hamilton principle of Section 2.3.2, cf. Eq. (2.29), in its time-integrated form (2.35),(2.63), i.e. the d'Alembert principle. Since the latter principle can be formally viewed as an extension to the principle of virtual work, the perturbation approach to be adopted below will basically follow the procedure presented in Section 3.2 with due account for inertial and damping terms. For the sake of clarity of the derivation let us rewrite the d'Alembert equation (2.63) employing Eq. (2.1) to express $\delta\varepsilon_{ij}$ in terms of δu_i as

$$\begin{aligned}&\int_\Omega \varrho\, \ddot{u}_i\, \delta u_i\, \mathrm{d}\Omega + \int_\Omega \left(\alpha \varrho\, \dot{u}_i\, \delta u_i + \beta\, C_{ijkl} \dot{u}_{i,j}\, \delta u_{k,l}\right) \mathrm{d}\Omega \\ &\quad + \int_\Omega C_{ijkl} u_{i,j}\, \delta u_{k,l}\, \mathrm{d}\Omega = \int_\Omega \varrho f_i\, \delta u_i\, \mathrm{d}\Omega + \int_{\partial\Omega_\sigma} \hat{t}_i\, \delta u_i\, \mathrm{d}(\partial\Omega) \end{aligned} \tag{3.28}$$

Assuming again the set of (time-invariant) random field variables $\mathbf{b}(x_k) = \{b_r(x_k)\}$, $r = 1, 2, \ldots, R$, to represent randomness in any parameter entering effectively Eq. (3.28) (cross-sectional area, length, thickness, Young's modulus, mass density, damping parameter, etc.) we adopt the second-order expansions (3.5) about the spatial expectations $\mathbf{b}^0(x_k) = \{b_r^0(x_k)\}$. We note that in the dynamic analysis the last three functions on the left-hand side of Eqs. (3.5) are assumed to depend in general on the spatial variable x_k and time variable τ; the variables u_i^0, $u_i^{,r}$, $u_i^{,rs}$, etc. are consequently time-dependent as well. In addition, the proportionality factors α and β, assumed to be functions of $b_r(x_k)$, are expanded as

$$\begin{aligned}\alpha[\mathbf{b}(x_k); x_k] &= \alpha^0[\mathbf{b}^0(x_k); x_k] + \epsilon\, \alpha^{,r}[\mathbf{b}^0(x_k); x_k]\, \Delta b_r(x_k) \\ &\quad + \frac{1}{2}\epsilon^2 \alpha^{,rs}[\mathbf{b}^0(x_k); x_k]\, \Delta b_r(x_k) \Delta b_s(x_k) \\ \beta[\mathbf{b}(x_k); x_k] &= \beta^0[\mathbf{b}^0(x_k); x_k] + \epsilon\, \beta^{,r}[\mathbf{b}^0(x_k); x_k]\, \Delta b_r(x_k) \\ &\quad + \frac{1}{2}\epsilon^2 \beta^{,rs}[\mathbf{b}^0(x_k); x_k]\, \Delta b_r(x_k) \Delta b_s(x_k)\end{aligned} \tag{3.29}$$

Substituting the expansions (3.5) and (3.29) into Eq. (3.28), multiplying the second-order terms by the R-variate PDF $p_R(b_1, b_2, \ldots, b_R)$ and integrating them over the domain of the random field variables $b_r(x_k)$, and equating the coefficients of like power of ϵ yields the zeroth-, first- and second-order equations for the stochastic Hamilton (d'Alembert) principle as follows:

- Zeroth-order (ϵ^0 terms, one equation)

$$\begin{aligned}&\int_\Omega \varrho^0\, \ddot{u}_i^0\, \delta u_i\, \mathrm{d}\Omega + \int_\Omega \left(\alpha^0 \varrho^0 \dot{u}_i^0\, \delta u_i + \beta^0\, C_{ijkl}^0 \dot{u}_{i,j}^0\, \delta u_{k,l}\right) \mathrm{d}\Omega \\ &\quad + \int_\Omega C_{ijkl}^0 u_{i,j}^0\, \delta u_{k,l}\, \mathrm{d}\Omega = \int_\Omega \varrho^0 f_i^0\, \delta u_i\, \mathrm{d}\Omega + \int_{\partial\Omega_\sigma} \hat{t}_i^0\, \delta u_i\, \mathrm{d}(\partial\Omega)\end{aligned} \tag{3.30}$$

- First-order (ϵ^1 terms, R equations)

$$
\begin{aligned}
&\int_\Omega \varrho^0 \ddot{u}_i^{,r} \delta u_i \, \mathrm{d}\Omega + \int_\Omega \left(\alpha^0 \varrho^0 \dot{u}_i^{,r} \delta u_i + \beta^0 C_{ijkl}^0 \dot{u}_{i,j}^{,r} \delta u_{k,l}\right) \mathrm{d}\Omega \\
&\quad + \int_\Omega C_{ijkl}^0 u_{i,j}^{,r} \delta u_{k,l} \, \mathrm{d}\Omega = \int_\Omega \left(\varrho^{,r} f_i^0 + \varrho^0 f_i^{,r}\right) \delta u_i \, \mathrm{d}\Omega \\
&\quad + \int_{\partial\Omega_\sigma} \hat{t}_i^{,r} \delta u_i \, \mathrm{d}(\partial\Omega) - \int_\Omega \varrho^{,r} \ddot{u}_i^0 \delta u_i \, \mathrm{d}\Omega - \int_\Omega \Big[\left(\alpha^{,r} \varrho^0 + \alpha^0 \varrho^{,r}\right) \dot{u}_i^0 \delta u_i \\
&\quad + \left(\beta^{,r} C_{ijkl}^0 + \beta^0 C_{ijkl}^{,r}\right) \dot{u}_{i,j}^0 \delta u_{k,l}\Big] \mathrm{d}\Omega - \int_\Omega C_{ijkl}^{,r} u_{i,j}^0 \delta u_{k,l} \, \mathrm{d}\Omega
\end{aligned}
\tag{3.31}
$$

- Second-order (ϵ^2 terms, one equation)

$$
\begin{aligned}
&\int_\Omega \varrho^0 \ddot{u}_i^{,rs} S_{\mathrm{b}}^{rs} \delta u_i \, \mathrm{d}\Omega + \int_\Omega \left(\alpha^0 \varrho^0 \dot{u}_i^{,rs} S_{\mathrm{b}}^{rs} \delta u_i + \beta^0 C_{ijkl}^0 \dot{u}_{i,j}^{,rs} S_{\mathrm{b}}^{rs} \delta u_{k,l}\right) \mathrm{d}\Omega \\
&\quad + \int_\Omega C_{ijkl}^0 u_{i,j}^{,rs} S_{\mathrm{b}}^{rs} \delta u_{k,l} \, \mathrm{d}\Omega = \int_\Omega \left(\varrho^{,rs} f_i^0 + 2\varrho^{,r} f_i^{,s} + \varrho^0 f_i^{,rs}\right) S_{\mathrm{b}}^{rs} \delta u_i \, \mathrm{d}\Omega \\
&\quad + \int_{\partial\Omega_\sigma} \hat{t}_i^{,rs} S_{\mathrm{b}}^{rs} \delta u_i \, \mathrm{d}(\partial\Omega) - \int_\Omega \left(\varrho^{,rs} \ddot{u}_i^0 + 2\varrho^{,r} \ddot{u}_i^{,s}\right) S_{\mathrm{b}}^{rs} \delta u_i \, \mathrm{d}\Omega \\
&\quad - \int_\Omega \Big\{\Big[\left(\alpha^{,rs} \varrho^0 + 2\alpha^{,r} \varrho^{,s} + \alpha^0 \varrho^{,rs}\right) \dot{u}_i^0 + 2\left(\alpha^{,r} \varrho^0 + \alpha^0 \varrho^{,r}\right) \dot{u}_i^{,s}\Big] S_{\mathrm{b}}^{rs} \delta u_i \\
&\quad + \Big[\left(\beta^{,rs} C_{ijkl}^0 + 2\beta^{,r} C_{ijkl}^{,s} + \beta^0 C_{ijkl}^{,rs}\right) \dot{u}_{i,j}^0 \\
&\quad + 2\left(\beta^{,r} C_{ijkl}^0 + \beta^0 C_{ijkl}^{,r}\right) \dot{u}_{i,j}^{,s}\Big] S_{\mathrm{b}}^{rs} \delta u_{k,l}\Big\} \mathrm{d}\Omega \\
&\quad - \int_\Omega \left(C_{ijkl}^{,rs} u_{i,j}^0 + 2C_{ijkl}^{,r} u_{i,j}^{,s}\right) S_{\mathrm{b}}^{rs} \delta u_{k,l} \, \mathrm{d}\Omega
\end{aligned}
\tag{3.32}
$$

We note that the zeroth-order equation is identical to the deterministic version of the d'Alembert principle, cf. Eqs. (2.63) and (3.28). With the same operator acting on the left-hand side of Eqs. (3.30)–(3.32) and the system probabilistic characteristics translated entirely into the effective forces on the right-hand sides, the stochastic problem becomes 'deterministic' in the computational context; the formulation thus provides an effective variational basis to model and solve probabilistic dynamics problems through computational techniques used for the deterministic analysis.

To evaluate the first two statistical moment functions for the displacement, strain and stress fields it should be recalled that:

(i) The displacements field $u_i[b_r(x_k); x_k, \tau]$, strain field $\varepsilon_{ij}[b_r(x_k); x_k, \tau]$ and stress field $\sigma_{ij}[b_r(x_k); x_k, \tau]$, $r = 1, 2, \ldots, R$; $k = 1, 2, 3$, $\tau \in [0, \infty)$, are the three-dimensional and R-variate random fields, cf. Section 1.1.6.

(ii) The dynamical system considered is invariant to any time shift, i.e. a time shift in the load (continuous input time signal) will cause the same time shift in the system response (continuous output time signal).

(iii) The random field variables $b_r(x_k)$ are time-independent.

In view of these observations, the formulation introduced enables us to evaluate not only the second moment functions for the multi-dimensional and multi-variate random fields of displacement, strain and stress at any time $\tau = t$, $\tau \in [0, \infty)$, but also those of two arguments $\xi_1 = (x_k^{(1)}, t_1)$ and $\xi_2 = (x_k^{(2)}, t_2)$, cf. Eq. (1.91). By following the derivations similar to those of Section 3.2, Eqs. (3.12)–(3.23), we arrive at the following results:

- For three-dimensional and R-variate displacement random field u_i

  Second-order accurate expectation at $\tau = t$

$$E\Big[u_i[\mathbf{b}(x_k); x_k, t]\Big] = u_i^0(x_k, t) + \frac{1}{2} u_i^{,rs}(x_k, t)\, S_{\mathrm{b}}^{rs} \tag{3.33}$$

  First-order accurate cross-covariance at $\xi_1 = (x_k^{(1)}, t_1)$ and $\xi_2 = (x_k^{(2)}, t_2)$

$$\begin{aligned} &\mathrm{Cov}\Big(u_i[\mathbf{b}(x_k^{(1)}); x_k^{(1)}, t_1], u_j[\mathbf{b}(x_k^{(2)}); x_k^{(2)}, t_2]\Big) \\ &= S_{\mathbf{u}}^{ij}(x_k^{(1)}, t_1; x_k^{(2)}, t_2) = u_i^{,r}(x_k^{(1)}, t_1)\, u_j^{,s}(x_k^{(2)}, t_2)\, S_{\mathrm{b}}^{rs} \end{aligned} \tag{3.34}$$

- For three-dimensional and R-variate strain random field ε_{ij}

  Second-order accurate expectation at $\tau = t$

$$\begin{aligned} &E\Big[\varepsilon_{ij}[\mathbf{b}(x_k); x_k, t]\Big] \\ &= \frac{1}{2}\Big\{u_{i,j}^0(x_k, t) + u_{j,i}^0(x_k, t) + \frac{1}{2}\Big[u_{i,j}^{,rs}(x_k, t) + u_{j,i}^{,rs}(x_k, t)\Big] S_{\mathrm{b}}^{rs}\Big\} \\ &= \varepsilon_{ij}^0(x_k, t) + \frac{1}{2}\varepsilon_{ij}^{,rs}(x_k, t)\, S_{\mathrm{b}}^{rs} \end{aligned} \tag{3.35}$$

  First-order accurate cross-covariance at $\xi_1 = (x_k^{(1)}, t_1)$ and $\xi_2 = (x_k^{(2)}, t_2)$

$$\begin{aligned} &\mathrm{Cov}\Big(\varepsilon_{ij}[\mathbf{b}(x_k^{(1)}); x_k^{(1)}, t_1], \varepsilon_{kl}[\mathbf{b}(x_k^{(2)}); x_k^{(2)}, t_2]\Big) \\ &= S_{\varepsilon}^{ijkl}(x_k^{(1)}, t_1; x_k^{(2)}, t_2) \\ &= \frac{1}{4}\Big[u_{i,j}^{,r}(x_k^{(1)}, t_1) + u_{j,i}^{,r}(x_k^{(1)}, t_1)\Big]\Big[u_{k,l}^{,s}(x_k^{(2)}, t_2) + u_{l,k}^{,s}(x_k^{(2)}, t_2)\Big] S_{\mathrm{b}}^{rs} \\ &= \varepsilon_{ij}^{,r}(x_k^{(1)}, t_1)\, \varepsilon_{kl}^{,s}(x_k^{(2)}, t_2)\, S_{\mathrm{b}}^{rs} \end{aligned} \tag{3.36}$$

- For three-dimensional and R-variate strain random field σ_{ij}

  Second-order accurate expectation at $\tau = t$

$$\begin{aligned} &E\Big[\sigma_{ij}[\mathbf{b}(x_k); x_k, t]\Big] \\ &= C_{ijkl}^0(x_k)\, \varepsilon_{kl}^0(x_k, t) + \frac{1}{2}\Big[C_{ijkl}^{,rs}(x_k)\, \varepsilon_{kl}^0(x_k, t) \\ &\quad + 2C_{ijkl}^{,r}(x_k)\, \varepsilon_{kl}^{,s}(x_k, t) + C_{ijkl}^0(x_k)\, \varepsilon_{kl}^{,rs}(x_k, t)\Big] S_{\mathrm{b}}^{rs} \end{aligned} \tag{3.37}$$

First-order accurate cross-covariance at $\xi_1 = (x_k^{(1)}, t_1)$ and $\xi_2 = (x_k^{(2)}, t_2)$

$$\begin{aligned}
&\mathrm{Cov}\Big(\sigma_{ij}[\mathbf{b}(x_k^{(1)}); x_k^{(1)}, t_1], \sigma_{kl}[\mathbf{b}(x_k^{(2)}); x_k^{(2)}, t_2]\Big) \\
&\quad = S_\sigma^{ijkl}(x_k^{(1)}, t_1; x_k^{(2)}, t_2) \\
&\quad = \Big[C_{ijmn}^{,r}(x_k^{(1)})\, C_{kl\bar{m}\bar{n}}^{,s}(x_k^{(2)})\, \varepsilon_{mn}^{0}(x_k^{(1)}, t_1)\, \varepsilon_{\bar{m}\bar{n}}^{0}(x_k^{(2)}, t_2) \\
&\qquad + C_{ijmn}^{,r}(x_k^{(1)})\, C_{kl\bar{m}\bar{n}}^{0}(x_k^{(2)})\, \varepsilon_{mn}^{0}(x_k^{(1)}, t_1)\, \varepsilon_{\bar{m}\bar{n}}^{,s}(x_k^{(2)}, t_2) \\
&\qquad + C_{ijmn}^{0}(x_k^{(1)})\, C_{kl\bar{m}\bar{n}}^{,r}(x_k^{(2)})\, \varepsilon_{mn}^{,s}(x_k^{(1)}, t_1)\, \varepsilon_{\bar{m}\bar{n}}^{0}(x_k^{(2)}, t_2) \\
&\qquad + C_{ijmn}^{0}(x_k^{(1)})\, C_{kl\bar{m}\bar{n}}^{0}(x_k^{(2)})\, \varepsilon_{mn}^{,r}(x_k^{(1)}, t_1)\, \varepsilon_{\bar{m}\bar{n}}^{,s}(x_k^{(2)}, t_2)\Big]\, S_{\mathrm{b}}^{rs} \qquad (3.38)
\end{aligned}$$

Expressions (3.33)–(3.38) are quite similar to those for the static case derived in Section 3.2. The only difference is that the response functions in the form of multi-dimensional and multi-variate random fields are now time-dependent.

Chapter 4

Stochastic Finite Element Analysis

4.1 Finite Element Equations: Statics and Dynamics

In this chapter the basic equations for the stochastic finite element method will be derived followed by a discussion of algorithmic and implementation issues. We have already seen in Chapters 2 and 3 that the static finite element equations may be obtained directly as a specification of the dynamic finite element formulation. This observation will be taken advantage of in the present section in which we shall first present derivations leading to the spatially discretized equations of motion; at the end of Section 4.1 the equations will be specified to explicitly apply to the static case.

It should be clear by now that the sequence of variational statements given in Chapter 3 may serve as the basis for a spatially discretized formulation facilitating analysis on a digital computer. In order to apply the finite element technique let us first assume that the region of interest Ω has been discretized by a finite element mesh. The basic idea of the mean-based, second-order, second-moment analysis in SFEM is to expand, via Taylor series, all the vector and matrix stochastic field variables typical of deterministic FEM about the mean values of random variables $b_r(x_k)$, to retain only up to second-order terms and to use in the analysis only the first two statistical moments. In this way equations for the expectations and cross-covariances (autocovariances) of the nodal displacements can be obtained in terms of the nodal displacement derivatives with respect to the random variables.

In the framework of the FEM philosophy the fields $b_r(x_k)$ have to be represented by a set of basic random variables. Thus, it is necessary to discretize $b_r(x_k)$ by expressing them in terms of some point (say, nodal) values of the appropriate means and covariances. The following approximation is adopted:

$$b_r(x_k) = \varphi_{\bar{\alpha}}(x_k)\, b_{r\bar{\alpha}} \qquad r = 1, 2, \ldots, R\,; \;\; \bar{\alpha} = 1, 2, \ldots, \bar{N} \tag{4.1}$$

where $\varphi_{\bar{\alpha}}$ is the shape function for the $\bar{\alpha}$-th nodal point, $\bar{N}$ is the number of nodal points in the mesh and $b_{r\bar{\alpha}}$ is the matrix of random parameter nodal values; for a fixed r the vector $b_{r\bar{\alpha}}$, $\bar{\alpha} = 1, 2, \ldots, \bar{N}$, contains as its entries the successive nodal values of the random variables b_r. In order to maintain consistency with the genuine finite element approximation (2.42) the same shape functions are used in Eq. (4.1). To elaborate we rewrite Eq. (2.42) as

$$u_i(x_k) = \varphi_{\bar{\alpha}}(x_k)\, u_{i\bar{\alpha}} \tag{4.2}$$

where $u_{i\bar{\alpha}}$ is the matrix of nodal displacements, i.e. $u_{1\bar{\alpha}}$, $u_{2\bar{\alpha}}$ and $u_{3\bar{\alpha}}$ are three displacement components at node $\bar{\alpha}$, $\bar{\alpha} = 1, 2, \ldots, \bar{N}$. The matrix $u_{i\bar{\alpha}}$ can be related to the nodal displacement vector q_α of Eq. (2.42) by the Boolean transformation

$$u_{i\bar{\alpha}} = A_{i\bar{\alpha}\alpha}\, q_\alpha \qquad i = 1, 2, 3 \tag{4.3}$$

which, when substituted into Eq. (4.2), generates Eq. (2.42) provided we denote

$$\varphi_{i\alpha}(x_k) = \varphi_{\bar{\alpha}}(x_k)\, A_{i\bar{\alpha}\alpha} \tag{4.4}$$

For the sake of notational convenience we introduce now a vector of nodal random variables b_ρ, $\rho = 1, 2, \ldots, \tilde{N} = R \times \bar{N}$, related to the matrix $b_{r\bar{\alpha}}$ by an appropriate Boolean transformation

$$b_{r\bar{\alpha}} = A_{r\bar{\alpha}\rho}\, b_\rho \tag{4.5}$$

Eq. (4.1) then becomes

$$b_r(x_k) = \varphi_{\bar{\alpha}}(x_k)\, A_{r\bar{\alpha}\rho}\, b_\rho = \varphi_{r\rho}(x_k)\, b_\rho \tag{4.6}$$

which may be regarded as the random variable counterpart of the displacement expansion (2.42).

The approach employed here follows that suggested in [80] which can be called the *interpolation* method; it is based on representing the random field in terms of an interpolation rule involving a set of deterministic shape functions and the random nodal values of the field. The alternative approaches to the discretization of random fields are: (i) the *spatial averaging* method [62,118,119] in which the element random variable is defined as the spatial average of the random field over the element domain; (ii) the *midpoint* method [51,52,62] in which the element random variable is defined as the value of the random field at the centroid of the element; and (iii) the *series expansion* method [72,73] in which the random field is modelled as a series of shape functions with random coefficients and any discretization of the field domain is thus avoided.[1]

An important consideration in any (except the last) of the above approaches is the selection of the element size in the finite element mesh. Three factors can

[1]As noted in [63], with the exception of the midpoint method all the other approaches are strictly valid only in the case of Gaussian random fields as they involve sums of random variables, and it is only the Gaussian distribution that maintains its distribution form through such a process.

be identified to contribute to this problem in the context of stochastic analysis [63,77]. With a view to accuracy the element size is controlled by the correlation length of the field, i.e. a measure of the distance over which the correlation coefficient ϱ approaches a small value (a smaller random field element should be used if the correlation length is large). Investigations indicate that a size equal to one-half to one-quarter of the correlation length may be appropriate [53,62]. The second factor is the numerical stability of the transformation to the standard normal space. If the random field is excessively fine, the discretized element variables are highly correlated and their correlation matrix is nearly singular, thus possibly yielding the transformation numerically unstable. Hence, this factor provides a lower bound on the element size. Finally, in selecting random field meshes the problem of computational efficiency must be considered; large element sizes are clearly preferable from this viewpoint.

A general methodology addressing the above problem was suggested in [62]. It consists in using separate meshes for the finite element and for each of the random fields by adopting a basic finite element mesh based on both the standard (i.e. stress gradient-related) FEM requirements as well as the correlation length of all random fields, and choosing for each random field a mesh that is coincident with or coarser than the basic mesh at the same time making sure that the probability transformation remains stable.

It should be emphasized in passing that the issue of random field meshes remains largely unexplored and much work has to be done on this particular aspect of SFEM.

Returning to Eq. (4.1) we may write

$$E[b_r(x_k)] \;=\; b_r^0(x_k) \;=\; \varphi_{r\rho}(x_k)\, b_\rho^0 \tag{4.7}$$

$$\mathrm{Cov}\Big(b_r(x_k), b_s(x_k)\Big) \;=\; S_{\mathrm{b}}^{rs} \;=\; \varphi_{r\rho}(x_k)\, \varphi_{s\sigma}(x_k)\, S_{\mathrm{b}}^{\rho\sigma} \tag{4.8}$$

and

$$\Delta b_r(x_k) = \varphi_{r\rho}(x_k)\, \Delta b_\rho \tag{4.9}$$

where

$$\Delta b_\rho = b_\rho - b_\rho^0 \tag{4.10}$$

and b_ρ^0 and $S_{\mathrm{b}}^{\rho\sigma}$ stand for the mean value vector and the covariance matrix of the nodal random variable vector b_ρ, respectively. The remaining random field variables in the problem considered, i.e. elastic moduli $C_{ijkl}(x_k)$, mass density $\varrho(x_k)$, body forces $f_i(x_k, \tau)$, boundary tractions $\hat{t}_i(x_k, \tau)$ and displacements $u_i(x_k, \tau)$ are expanded using the same shape functions as:[1]

[1]According to the suggestions of Section 2.4.1 the 'element level' notation will be avoided wherever possible.

$$
\begin{aligned}
C_{ijkl}[b_r(x_k);x_k] &= C_{ijkl}[\varphi_{r\rho}(x_k)b_\rho;x_k] = \varphi_{\bar\alpha}(x_k)\,C_{ijkl\bar\alpha}(b_\rho)\\
\varrho[b_r(x_k);x_k] &= \varrho[\varphi_{r\rho}(x_k)b_\rho;x_k] = \varphi_{\bar\alpha}(x_k)\,\varrho_{\bar\alpha}(b_\rho)\\
f_i[b_r(x_k);x_k,\tau] &= f_i[\varphi_{r\rho}(x_k)b_\rho;x_k,\tau] = \varphi_{\bar\alpha}(x_k)\,f_{i\bar\alpha}(b_\rho;\tau)\\
\hat t_i[b_r(x_k);x_k,\tau] &= \hat t_i[\varphi_{r\rho}(x_k)b_\rho;x_k,\tau] = \varphi_{\bar\alpha}(x_k)\,\hat t_{i\bar\alpha}(b_\rho;\tau)\\
u_i[b_r(x_k);x_k,\tau] &= u_i[\varphi_{r\rho}(x_k)b_\rho;x_k,\tau] = \varphi_{\bar\alpha}(x_k)\,u_{i\bar\alpha}(b_\rho;\tau)\\
&= \varphi_{i\alpha}(x_k)\,q_\alpha(b_\rho;\tau)\\
\bar\alpha = 1,2,\ldots,\bar N\,;\quad &\alpha = 1,2,\ldots,N\,;\quad \rho = 1,2,\ldots,\tilde N\,;\quad i=1,2,3
\end{aligned}
\tag{4.11}
$$

where N is the total number of degrees of freedom in the discretized model. Employing the Taylor series expansions for the nodal random variables $C_{ijkl\bar\alpha}(b_\varrho)$, $\varrho_{\bar\alpha}(b_\varrho)$, $f_{i\bar\alpha}(b_\varrho;\tau)$, $\hat t_{i\bar\alpha}(b_\varrho;\tau)$ and $q_\alpha(b_\varrho;\tau)$ in the form

$$
\begin{aligned}
C_{ijkl\bar\alpha}(b_\varrho) &= C^0_{ijkl\bar\alpha}(b^0_\varrho) + \epsilon\,C^{,\rho}_{ijkl\bar\alpha}(b^0_\varrho)\,\Delta b_\rho + \frac{1}{2}\epsilon^2\,C^{,\rho\sigma}_{ijkl\bar\alpha}(b^0_\varrho)\,\Delta b_\rho\Delta b_\sigma\\
\varrho_{\bar\alpha}(b_\varrho) &= \varrho^0_{\bar\alpha}(b^0_\varrho) + \epsilon\,\varrho^{,\rho}_{\bar\alpha}(b^0_\varrho)\,\Delta b_\rho + \frac{1}{2}\epsilon^2\,\varrho^{,\rho\sigma}_{\bar\alpha}(b^0_\varrho)\,\Delta b_\rho\Delta b_\sigma\\
f_{i\bar\alpha}(b_\varrho;\tau) &= f^0_{i\bar\alpha}(b^0_\varrho;\tau) + \epsilon\,f^{,\rho}_{i\bar\alpha}(b^0_\varrho;\tau)\,\Delta b_\rho + \frac{1}{2}\epsilon^2\,f^{,\rho\sigma}_{i\bar\alpha}(b^0_\varrho;\tau)\,\Delta b_\rho\Delta b_\sigma\\
\hat t_{i\bar\alpha}(b_\varrho;\tau) &= \hat t^0_{i\bar\alpha}(b^0_\varrho;\tau) + \epsilon\,\hat t^{,\rho}_{i\bar\alpha}(b^0_\varrho;\tau)\,\Delta b_\rho + \frac{1}{2}\epsilon^2\,\hat t^{,\rho\sigma}_{i\bar\alpha}(b^0_\varrho;\tau)\,\Delta b_\rho\Delta b_\sigma\\
q_\alpha(b_\varrho;\tau) &= q^0_\alpha(b^0_\varrho;\tau) + \epsilon\,q^{,\rho}_\alpha(b^0_\varrho;\tau)\,\Delta b_\rho + \frac{1}{2}\epsilon^2\,q^{,\rho\sigma}_\alpha(b^0_\varrho;\tau)\,\Delta b_\rho\Delta b_\sigma
\end{aligned}
\tag{4.12}
$$

we have,[1] upon combining Eqs. (4.11) and (4.12)

$$
\begin{aligned}
C_{ijkl}[b_r(x_k);x_k] &= \varphi_{\bar\alpha}(x_k)\Big[C^0_{ijkl\bar\alpha}(b^0_\varrho) + \epsilon\,C^{,\rho}_{ijkl\bar\alpha}(b^0_\varrho)\,\Delta b_\rho\\
&\qquad + \frac{1}{2}\epsilon^2\,C^{,\rho\sigma}_{ijkl\bar\alpha}(b^0_\varrho)\,\Delta b_\rho\Delta b_\sigma\Big]\\
\varrho[b_r(x_k);x_k] &= \varphi_{\bar\alpha}(x_k)\Big[\varrho^0_{\bar\alpha}(b^0_\varrho) + \epsilon\,\varrho^{,\rho}_{\bar\alpha}(b^0_\varrho)\,\Delta b_\rho\\
&\qquad + \frac{1}{2}\epsilon^2\,\varrho^{,\rho\sigma}_{\bar\alpha}(b^0_\varrho)\,\Delta b_\rho\Delta b_\sigma\Big]\\
f_i[b_r(x_k);x_k,\tau] &= \varphi_{\bar\alpha}(x_k)\Big[f^0_{i\bar\alpha}(b^0_\varrho;\tau) + \epsilon\,f^{,\rho}_{i\bar\alpha}(b^0_\varrho;\tau)\,\Delta b_\rho\\
&\qquad + \frac{1}{2}\epsilon^2\,f^{,\rho\sigma}_{i\bar\alpha}(b^0_\varrho;\tau)\,\Delta b_\rho\Delta b_\sigma\Big]\\
\hat t_i[b_r(x_k);x_k,\tau] &= \varphi_{\bar\alpha}(x_k)\Big[\hat t^0_{i\bar\alpha}(b^0_\varrho;\tau) + \epsilon\,\hat t^{,\rho}_{i\bar\alpha}(b^0_\varrho;\tau)\,\Delta b_\rho\\
&\qquad + \frac{1}{2}\epsilon^2\,\hat t^{,\rho\sigma}_{i\bar\alpha}(b^0_\varrho;\tau)\,\Delta b_\rho\Delta b_\sigma\Big]\\
u_i[b_r(x_k);x_k,\tau] &= \varphi_{i\alpha}(x_k)\Big[q^0_\alpha(b^0_\varrho;\tau) + \epsilon\,q^{,\rho}_\alpha(b^0_\varrho;\tau)\,\Delta b_\rho\\
&\qquad + \frac{1}{2}\epsilon^2\,q^{,\rho\sigma}_\alpha(b^0_\varrho;\tau)\,\Delta b_\rho\Delta b_\sigma\Big]
\end{aligned}
\tag{4.13}
$$

[1]The index ϱ has been used instead of ρ to avoid confusion in the summation over the repeated index ρ.

Comparing Eqs. (3.5) (for the dynamic case) with Eqs. (4.13) and using Eq. (4.9) we end up with a set of relationships of the form

$$
\begin{aligned}
C^{0}_{ijkl}[b_r(x_k); x_k] &= \varphi_{\bar\alpha}(x_k)\, C^{0}_{ijkl\bar\alpha}(b^0_\varrho) \\
C^{,r}_{ijkl}[b_r(x_k); x_k]\, \varphi_{r\rho}(x_k) &= \varphi_{\bar\alpha}(x_k)\, C^{,\rho}_{ijkl\bar\alpha}(b^0_\varrho) \\
C^{,rs}_{ijkl}[b_r(x_k); x_k]\, \varphi_{r\rho}(x_k)\, \varphi_{s\sigma}(x_k) &= \varphi_{\bar\alpha}(x_k)\, C^{,\rho\sigma}_{ijkl\bar\alpha}(b^0_\varrho) \\
\varrho^{0}[b_r(x_k); x_k] &= \varphi_{\bar\alpha}(x_k)\, \varrho^{0}_{\bar\alpha}(b^0_\varrho) \\
\varrho^{,r}[b_r(x_k); x_k]\, \varphi_{r\rho}(x_k) &= \varphi_{\bar\alpha}(x_k)\, \varrho^{,\rho}_{\bar\alpha}(b^0_\varrho) \\
\varrho^{,rs}[b_r(x_k); x_k]\, \varphi_{r\rho}(x_k)\, \varphi_{s\sigma}(x_k) &= \varphi_{\bar\alpha}(x_k)\, \varrho^{,\rho\sigma}_{\bar\alpha}(b^0_\varrho) \\
f^{0}_{i}[b_r(x_k); x_k, \tau] &= \varphi_{\bar\alpha}(x_k)\, f^{0}_{i\bar\alpha}(b^0_\varrho; \tau) \\
f^{,r}_{i}[b_r(x_k); x_k, \tau]\, \varphi_{r\rho}(x_k) &= \varphi_{\bar\alpha}(x_k)\, f^{,\rho}_{i\bar\alpha}(b^0_\varrho; \tau) \\
f^{,rs}_{i}[b_r(x_k); x_k, \tau]\, \varphi_{r\rho}(x_k)\, \varphi_{s\sigma}(x_k) &= \varphi_{\bar\alpha}(x_k)\, f^{,\rho\sigma}_{i\bar\alpha}(b^0_\varrho; \tau) \\
\hat t^{0}_{i}[b_r(x_k); x_k, \tau] &= \varphi_{\bar\alpha}(x_k)\, \hat t^{0}_{i\bar\alpha}(b^0_\varrho; \tau) \\
\hat t^{,r}_{i}[b_r(x_k); x_k, \tau]\, \varphi_{r\rho}(x_k) &= \varphi_{\bar\alpha}(x_k)\, \hat t^{,\rho}_{i\bar\alpha}(b^0_\varrho; \tau) \\
\hat t^{,rs}_{i}[b_r(x_k); x_k, \tau]\, \varphi_{r\rho}(x_k)\, \varphi_{s\sigma}(x_k) &= \varphi_{\bar\alpha}(x_k)\, \hat t^{,\rho\sigma}_{i\bar\alpha}(b^0_\varrho; \tau) \\
u^{0}_{i}[b_r(x_k); x_k, \tau] &= \varphi_{i\alpha}(x_k)\, q^{0}_{\alpha}(b^0_\varrho; \tau) \\
u^{,r}_{i}[b_r(x_k); x_k, \tau]\, \varphi_{r\rho}(x_k) &= \varphi_{i\alpha}(x_k)\, q^{,\rho}_{\alpha}(b^0_\varrho; \tau) \\
u^{,rs}_{i}[b_r(x_k); x_k, \tau]\, \varphi_{r\rho}(x_k)\, \varphi_{s\sigma}(x_k) &= \varphi_{i\alpha}(x_k)\, q^{,\rho\sigma}_{\alpha}(b^0_\varrho; \tau)
\end{aligned}
\tag{4.14}
$$

The same sequence of derivations can be repeated for the damping factors α and β to obtain, cf. Eqs. (3.29)

$$
\begin{aligned}
\alpha[b_r(x_k); x_k] &= \varphi_{\bar\alpha}(x_k)\Big[\alpha^{0}_{\bar\alpha}(b^0_\varrho) + \epsilon\, \alpha^{,\rho}_{\bar\alpha}(b^0_\varrho)\, \Delta b_\rho + \tfrac{1}{2}\epsilon^2\, \alpha^{,\rho\sigma}_{\bar\alpha}(b^0_\varrho)\, \Delta b_\rho \Delta b_\sigma\Big] \\
\beta[b_r(x_k); x_k] &= \varphi_{\bar\alpha}(x_k)\Big[\beta^{0}_{\bar\alpha}(b^0_\varrho) + \epsilon\, \beta^{,\rho}_{\bar\alpha}(b^0_\varrho)\, \Delta b_\rho + \tfrac{1}{2}\epsilon^2\, \beta^{,\rho\sigma}_{\bar\alpha}(b^0_\varrho)\, \Delta b_\rho \Delta b_\sigma\Big]
\end{aligned}
\tag{4.15}
$$

and

$$
\begin{aligned}
\alpha^{0}[b_r(x_k); x_k] &= \varphi_{\bar\alpha}(x_k)\, \alpha^{0}_{\bar\alpha}(b^0_\varrho) \\
\alpha^{,r}[b_r(x_k); x_k]\, \varphi_{r\rho}(x_k) &= \varphi_{\bar\alpha}(x_k)\, \alpha^{,\rho}_{\bar\alpha}(b^0_\varrho) \\
\alpha^{,rs}[b_r(x_k); x_k]\, \varphi_{r\rho}(x_k)\, \varphi_{s\sigma}(x_k) &= \varphi_{\bar\alpha}(x_k)\, \alpha^{,\rho\sigma}_{\bar\alpha}(b^0_\varrho) \\
\beta^{0}[b_r(x_k); x_k] &= \varphi_{\bar\alpha}(x_k)\, \beta^{0}_{\bar\alpha}(b^0_\varrho) \\
\beta^{,r}[b_r(x_k); x_k]\, \varphi_{r\rho}(x_k) &= \varphi_{\bar\alpha}(x_k)\, \beta^{,\rho}_{\bar\alpha}(b^0_\varrho) \\
\beta^{,rs}[b_r(x_k); x_k]\, \varphi_{r\rho}(x_k)\, \varphi_{s\sigma}(x_k) &= \varphi_{\bar\alpha}(x_k)\, \beta^{,\rho\sigma}_{\bar\alpha}(b^0_\varrho)
\end{aligned}
\tag{4.16}
$$

Substituting the above finite element approximations into the zeroth-, first- and second-order variational statements (3.30)–(3.32), employing the standard perturbation procedure and using the arbitrariness of δq_α the following 'hierarchical' finite element equations of motion are obtained:

- Zeroth-order (ϵ^0 terms, one system of N linear simultaneous ordinary differential equations for $q^0_\alpha(b^0_\varrho;\tau)$, $\alpha = 1,2,\ldots,N$)

$$M^0_{\alpha\beta}(b^0_\varrho)\,\ddot{q}^0_\beta(b^0_\varrho;\tau) + C^0_{\alpha\beta}(b^0_\varrho)\,\dot{q}^0_\beta(b^0_\varrho;\tau) + K^0_{\alpha\beta}(b^0_\varrho)\,q^0_\beta(b^0_\varrho;\tau) = Q^0_\alpha(b^0_\varrho;\tau) \qquad (4.17)$$

- First-order (ϵ^1 terms, $\tilde{N}$ systems of N linear simultaneous ordinary differential equations for $q^{,\rho}_\alpha(b^0_\varrho;\tau)$, $\rho = 1,2,\ldots,\tilde{N}$, $\alpha = 1,2,\ldots,N$)

$$\begin{aligned} &M^0_{\alpha\beta}(b^0_\varrho)\,\ddot{q}^{,\rho}_\beta(b^0_\varrho;\tau) + C^0_{\alpha\beta}(b^0_\varrho)\,\dot{q}^{,\rho}_\beta(b^0_\varrho;\tau) + K^0_{\alpha\beta}(b^0_\varrho)\,q^{,\rho}_\beta(b^0_\varrho;\tau) = Q^{,\rho}_\alpha(b^0_\varrho;\tau) \\ &\quad - \left[M^{,\rho}_{\alpha\beta}(b^0_\varrho)\,\ddot{q}^0_\beta(b^0_\varrho;\tau) + C^{,\rho}_{\alpha\beta}(b^0_\varrho)\,\dot{q}^0_\beta(b^0_\varrho;\tau) + K^{,\rho}_{\alpha\beta}(b^0_\varrho)\,q^0_\beta(b^0_\varrho;\tau)\right] \qquad (4.18)\end{aligned}$$

- Second-order (ϵ^2 terms, one system of N linear simultaneous ordinary differential equations for $q^{(2)}_\alpha(b^0_\varrho;\tau)$, $\alpha = 1,2,\ldots,N$)

$$\begin{aligned} &M^0_{\alpha\beta}(b^0_\varrho)\ddot{q}^{(2)}_\beta(b^0_\varrho;\tau) + C^0_{\alpha\beta}(b^0_\varrho)\dot{q}^{(2)}_\beta(b^0_\varrho;\tau) + K^0_{\alpha\beta}(b^0_\varrho)q^{(2)}_\beta(b^0_\varrho;\tau) = \Big\{Q^{,\rho\sigma}_\alpha(b^0_\varrho;\tau) \\ &\quad - 2\left[M^{,\rho}_{\alpha\beta}(b^0_\varrho)\,\ddot{q}^{,\sigma}_\beta(b^0_\varrho;\tau) + C^{,\rho}_{\alpha\beta}(b^0_\varrho)\,\dot{q}^{,\sigma}_\beta(b^0_\varrho;\tau) + K^{,\rho}_{\alpha\beta}(b^0_\varrho)\,q^{,\sigma}_\beta(b^0_\varrho;\tau)\right] \\ &\quad - \left[M^{,\rho\sigma}_{\alpha\beta}(b^0_\varrho)\,\ddot{q}^0_\beta(b^0_\varrho;\tau) + C^{,\rho\sigma}_{\alpha\beta}(b^0_\varrho)\,\dot{q}^0_\beta(b^0_\varrho;\tau) + K^{,\rho\sigma}_{\alpha\beta}(b^0_\varrho)\,q^0_\beta(b^0_\varrho;\tau)\right]\Big\}\,S^{\rho\sigma}_{\mathrm{b}} \qquad (4.19)\end{aligned}$$

where

$$q^{(2)}_\alpha(b^0_\varrho;\tau) = q^{,\rho\sigma}_\alpha(b^0_\varrho;\tau)\,S^{\rho\sigma}_{\mathrm{b}} \qquad (4.20)$$

The right-hand side of Eq. (4.20) results from substituting Eq. (4.8) into the double sum $u^{,rs}_{i,j}\,S^{rs}_{\mathrm{b}}$ on the left-hand side of Eq. (3.32) and using the last equation of the approximations (4.14), i.e.

$$u^{,rs}_{i,j}\,S^{rs}_{\mathrm{b}} \;=\; u^{,rs}_{i,j}\,\varphi_{r\rho}\,\varphi_{s\sigma}\,S^{\rho\sigma}_{\mathrm{b}} \;=\; \varphi_{i\alpha,j}\,q^{,\rho\sigma}_\alpha\,S^{\rho\sigma}_{\mathrm{b}} \qquad (4.21)$$

In Eqs. (4.17)–(4.19) the zeroth-order mass, damping and stiffness matrices and load vector and their first and second mixed derivatives with respect to nodal random variables b_ρ are defined as follows, cf. Eqs. (2.44), (2.46), (2.66) and (2.67):

- Zeroth-order functions

$$M^0_{\alpha\beta}(b^0_\varrho) \;=\; \int_\Omega \varphi_{\bar{\alpha}}\,\varrho^0_{\bar{\alpha}}\,\varphi_{i\alpha}\,\varphi_{i\beta}\,\mathrm{d}\Omega \qquad (4.22)$$

$$C^0_{\alpha\beta}(b^0_\varrho) \;=\; \int_\Omega \varphi_{\bar{\alpha}}\,\varphi_{\bar{\beta}}\Big(\alpha^0_{\bar{\alpha}}\,\varrho^0_{\bar{\beta}}\,\varphi_{i\alpha}\,\varphi_{i\beta} + \beta^0_{\bar{\alpha}}\,C^0_{ijkl\bar{\beta}}\,B_{ij\alpha}B_{kl\beta}\Big)\,\mathrm{d}\Omega \qquad (4.23)$$

$$K^0_{\alpha\beta}(b^0_\varrho) \;=\; \int_\Omega \varphi_{\bar{\alpha}}\,C^0_{ijkl\bar{\alpha}}\,B_{ij\alpha}B_{kl\beta}\,\mathrm{d}\Omega \qquad (4.24)$$

$$Q^0_\alpha(b^0_\varrho;\tau) \;=\; \int_\Omega \varphi_{\bar{\alpha}}\,\varphi_{\bar{\beta}}\,\varrho^0_{\bar{\alpha}}\,f^0_{i\bar{\beta}}\,\varphi_{i\alpha}\,\mathrm{d}\Omega + \int_{\partial\Omega_\sigma} \varphi_{\bar{\alpha}}\,\hat{t}^0_{i\bar{\alpha}}\,\varphi_{i\alpha}\,\mathrm{d}(\partial\Omega) \qquad (4.25)$$

- First partial derivatives

$$M_{\alpha\beta}^{,\rho}(b_\varrho^0) = \int_\Omega \varphi_{\bar\alpha}\, \varrho_{\bar\alpha}^{,\rho}\, \varphi_{i\alpha}\, \varphi_{i\beta}\, \mathrm{d}\Omega \tag{4.26}$$

$$C_{\alpha\beta}^{,\rho}(b_\varrho^0) = \int_\Omega \varphi_{\bar\alpha}\, \varphi_{\bar\beta} \Big[\Big(\alpha_{\bar\alpha}^{,\rho}\, \varrho_{\bar\beta}^0 + \alpha_{\bar\alpha}^0\, \varrho_{\bar\beta}^{,\rho} \Big)\, \varphi_{i\alpha}\, \varphi_{i\beta} + \Big(\beta_{\bar\alpha}^{,\rho}\, C_{ijkl\bar\beta}^0 + \beta_{\bar\alpha}^0\, C_{ijkl\bar\beta}^{,\rho} \Big)\, B_{ij\alpha} B_{kl\beta} \Big]\, \mathrm{d}\Omega \tag{4.27}$$

$$K_{\alpha\beta}^{,\rho}(b_\varrho^0) = \int_\Omega \varphi_{\bar\alpha}\, C_{ijkl\bar\alpha}^{,\rho}\, B_{ij\alpha} B_{kl\beta}\, \mathrm{d}\Omega \tag{4.28}$$

$$Q_\alpha^{,\rho}(b_\varrho^0;\tau) = \int_\Omega \varphi_{\bar\alpha}\, \varphi_{\bar\beta} \Big(\varrho_{\bar\alpha}^{,\rho} f_{i\bar\beta}^0 + \varrho_{\bar\alpha}^0 f_{i\bar\beta}^{,\rho} \Big) \varphi_{i\alpha}\, \mathrm{d}\Omega + \int_{\partial\Omega_\sigma} \varphi_{\bar\alpha}\, \hat t_{i\bar\alpha}^{,\rho}\, \varphi_{i\alpha}\, \mathrm{d}(\partial\Omega) \tag{4.29}$$

- Second partial derivatives

$$M_{\alpha\beta}^{,\rho\sigma}(b_\varrho^0) = \int_\Omega \varphi_{\bar\alpha}\, \varrho_{\bar\alpha}^{,\rho\sigma}\, \varphi_{i\alpha}\, \varphi_{i\beta}\, \mathrm{d}\Omega \tag{4.30}$$

$$C_{\alpha\beta}^{,\rho\sigma}(b_\varrho^0) = \int_\Omega \varphi_{\bar\alpha}\, \varphi_{\bar\beta} \Big[\Big(\alpha_{\bar\alpha}^{,\rho\sigma}\, \varrho_{\bar\beta}^0 + \alpha_{\bar\alpha}^{,\rho}\, \varrho_{\bar\beta}^{,\sigma} + \alpha_{\bar\alpha}^{,\sigma}\, \varrho_{\bar\beta}^{,\rho} + \alpha_{\bar\alpha}^0\, \varrho_{\bar\beta}^{,\rho\sigma} \Big) \varphi_{i\alpha}\, \varphi_{i\beta} + \Big(\beta_{\bar\alpha}^{,\rho\sigma} C_{ijkl\bar\beta}^0 + \beta_{\bar\alpha}^{,\rho} C_{ijkl\bar\beta}^{,\sigma} + \beta_{\bar\alpha}^{,\sigma} C_{ijkl\bar\beta}^{,\rho} + \beta_{\bar\alpha}^0 C_{ijkl\bar\beta}^{,\rho\sigma} \Big) B_{ij\alpha} B_{kl\beta} \Big] \mathrm{d}\Omega \tag{4.31}$$

$$K_{\alpha\beta}^{,\rho\sigma}(b_\varrho^0) = \int_\Omega \varphi_{\bar\alpha}\, C_{ijkl\bar\alpha}^{,\rho\sigma}\, B_{ij\alpha} B_{kl\beta}\, \mathrm{d}\Omega \tag{4.32}$$

$$Q_\alpha^{,\rho\sigma}(b_\varrho^0;\tau) = \int_\Omega \varphi_{\bar\alpha}\, \varphi_{\bar\beta} \Big(\varrho_{\bar\alpha}^{,\rho\sigma} f_{i\bar\beta}^0 + \varrho_{\bar\alpha}^{,\rho} f_{i\bar\beta}^{,\sigma} + \varrho_{\bar\alpha}^{,\sigma} f_{i\bar\beta}^{,\rho} + \varrho_{\bar\alpha}^0 f_{i\bar\beta}^{,\rho\sigma} \Big) \varphi_{i\alpha}\, \mathrm{d}\Omega + \int_{\partial\Omega_\sigma} \varphi_{\bar\alpha}\, \hat t_{i\bar\alpha}^{,\rho\sigma}\, \varphi_{i\alpha}\, \mathrm{d}(\partial\Omega) \tag{4.33}$$

All the functions (4.22)–(4.33) are evaluated at the expectations b_ρ^0 of the nodal random variables b_ρ.

Quite similarly, from the variational statements (3.8)–(3.10) wherein the displacements are time-independent, the finite element equilibrium equations may be obtained as:

- Zeroth-order (ϵ^0 terms, one system of N linear simultaneous algebraic equations for $q_\alpha^0(b_\varrho^0)$, $\alpha = 1,2,\ldots,N$)

$$K_{\alpha\beta}^0(b_\varrho^0)\, q_\beta^0(b_\varrho^0) = Q_\alpha^0(b_\varrho^0) \tag{4.34}$$

- First-order (ϵ^1 terms, $\tilde N$ systems of N linear simultaneous algebraic equations for $q_\alpha^{,\rho}(b_\varrho^0)$, $\rho = 1,2,\ldots,\tilde N$, $\alpha = 1,2,\ldots,N$)

$$K_{\alpha\beta}^0(b_\varrho^0)\, q_\beta^{,\rho}(b_\varrho^0) = Q_\alpha^{,\rho}(b_\varrho^0) - K_{\alpha\beta}^{,\rho}(b_\varrho^0)\, q_\beta^0(b_\varrho^0) \tag{4.35}$$

- Second-order (ϵ^2 terms, one system of N linear simultaneous algebraic equations for $q_\alpha^{(2)}(b_\varrho^0)$, $\alpha = 1,2,\ldots,N$)

$$K_{\alpha\beta}^0(b_\varrho^0) q_\beta^{(2)}(b_\varrho^0) = \Big[Q_\alpha^{,\rho\sigma}(b_\varrho^0) - 2K_{\alpha\beta}^{,\rho}(b_\varrho^0)\, q_\beta^{,\sigma}(b_\varrho^0) - K_{\alpha\beta}^{,\rho\sigma}(b_\varrho^0)\, q_\beta^0(b_\varrho^0) \Big] S_{\mathrm{b}}^{\rho\sigma} \tag{4.36}$$

with $q_\alpha^{(2)}(b_\varrho^0) = q_\alpha^{,\rho\sigma}(b_\varrho^0) S_{\mathrm{b}}^{\rho\sigma}$.

Similarly as in Chapter 2 it is assumed here that the kinematic conditions have been fulfilled prior to establishing the finite element equations (4.17)–(4.19) (dynamic case) or (4.34)–(4.36) (static case).

4.2 Probabilistic Distribution Output

Having solved Eqs. (4.17)–(4.19) for $q^0_\alpha(b^0_\varrho;\tau)$, $q^{,\rho}_\alpha(b^0_\varrho;\tau)$ and $q^{(2)}_\alpha(b^0_\varrho;\tau)$ and their time derivatives, the probabilistic distributions for the nodal displacements, velocities and accelerations as well as for the element strains and stresses can be determined. As pointed out in Section 3.2, by the assumption that the coefficients of variation in the problem considered are not too large (variances of random variables are small when compared with their expected values) all the solutions can be obtained formally by setting $\epsilon = 1$ in the expansions. Thus, the random displacement field can now be expressed as, cf. Eq. $(4.12)_5$

$$q_\alpha(b_\varrho;\tau) = q^0_\alpha(b^0_\varrho;\tau) + q^{,\rho}_\alpha(b^0_\varrho;\tau)\,\Delta b_\rho + \frac{1}{2} q^{,\rho\sigma}_\alpha(b^0_\varrho;\tau)\,\Delta b_\rho \Delta b_\sigma \tag{4.37}$$

From now on, for the sake of brevity in further discussions we shall be frequently using a concise notation $q_\alpha(b^0_\varrho;\tau) = q_\alpha(\tau)$, etc.; the dependence of random fields on $b_r(x_m)$ or b^0_ϱ will not be explicitly indicated unless confusion is likely to arise.

By the definition of the nodal displacement expectations at any time instant $\tau = t$ and cross-covariances at $\xi_1 = (x^{(1)}_m, t_1)$ and $\xi_2 = (x^{(2)}_m, t_2)$, cf. Eqs. (3.12) and (3.16), we have, respectively

$$E[q_\alpha(t)] = \underbrace{\int_{-\infty}^{+\infty}\int_{-\infty}^{+\infty}\cdots\int_{-\infty}^{+\infty}}_{\tilde{N}\text{-fold}} q_\alpha(t)\, p_{\tilde{N}}(b_1, b_2, \ldots, b_{\tilde{N}})\, \mathrm{d}b_1 \mathrm{d}b_2 \ldots \mathrm{d}b_{\tilde{N}} \tag{4.38}$$

$$\begin{aligned}\mathrm{Cov}\Big(q_\alpha(t_1), q_\beta(t_2)\Big) &= S^{\alpha\beta}_{\mathbf{q}}(t_1, t_2) \\ &= \underbrace{\int_{-\infty}^{+\infty}\int_{-\infty}^{+\infty}\cdots\int_{-\infty}^{+\infty}}_{\tilde{N}\text{-fold}} \Big\{q_\alpha(t_1) - E[q_\alpha(t_1)]\Big\}\Big\{q_\beta(t_2) - E[q_\beta(t_2)]\Big\} \\ &\qquad \times\, p_{\tilde{N}}(b_1, b_2, \ldots, b_{\tilde{N}})\, \mathrm{d}b_1 \mathrm{d}b_2 \ldots \mathrm{d}b_{\tilde{N}}\end{aligned} \tag{4.39}$$

where $p_{\tilde{N}}(b_1, b_2, \ldots, b_{\tilde{N}})$ is the $\tilde{N}$-variate PDF. Similar relationships hold for the velocities and accelerations.

Substituting Eq. (4.37) into Eq. (4.38), using Eq. (4.20) and observing that the terms involving the first variation of Δb_ρ vanish, cf. Eq. (3.13), yields the second-order accurate expectations for the nodal displacements at any $\tau = t$ as

$$E[q_\alpha(t)] = q^0_\alpha(t) + \frac{1}{2} q^{,\rho\sigma}_\alpha(t)\, S^{\rho\sigma}_{\mathbf{b}} = q^0_\alpha(t) + \frac{1}{2} q^{(2)}_\alpha(t) \tag{4.40}$$

The second-order accurate cross-covariances for the nodal displacements at $\xi_1 = (x^{(1)}_m, t_1)$ and $\xi_2 = (x^{(2)}_m, t_2)$ are obtained by combining Eqs. (4.37) and (4.39)

$$\begin{aligned}S^{\alpha\beta}_{\mathbf{q}}(t_1, t_2) &= q^{,\rho}_\alpha(t_1)\, q^{,\sigma}_\beta(t_2)\, S^{\rho\sigma}_{\mathbf{b}} \\ &+ \frac{1}{2}\Big[q^{,\rho}_\alpha(t_1) q^{,\sigma\upsilon}_\beta(t_2) + q^{,\rho\sigma}_\alpha(t_1) q^{,\upsilon}_\beta(t_2)\Big] S^{\rho\sigma\upsilon}_{\mathbf{b}} + \frac{1}{4} q^{,\rho\sigma}_\alpha(t_1) q^{,\upsilon\omega}_\beta(t_2) S^{\rho\sigma\upsilon\omega}_{\mathbf{b}}\end{aligned} \tag{4.41}$$

where $S^{\rho\sigma\upsilon}_{\mathbf{b}}$ and $S^{\rho\sigma\upsilon\omega}_{\mathbf{b}}$ denote third- and fourth-order central moments of the nodal random variables b_ρ. If only the first-order accurate estimate of the nodal displacement cross-covariances is of interest, Eq. (4.41) reduces to

$$S^{\alpha\beta}_{\mathbf{q}}(t_1, t_2) = q^{,\rho}_\alpha(t_1)\, q^{,\sigma}_\beta(t_2)\, S^{\rho\sigma}_{\mathbf{b}} \tag{4.42}$$

which could be obtained directly by adopting the results presented in Section 1.1.5, cf. Eq. (1.67). Recall that the first-order accurate nodal displacement cross-covariances are consistent with the second-moment approach.

By using Eqs. $(2.44)_2$ and (4.37) the strain tensor (random field) at any point in the element assembly can be written as

$$\varepsilon_{ij}(x_m,\tau) \;=\; B_{ij\alpha}(x_m)\Big[q^0_\alpha(\tau) + q^{,\rho}_\alpha(\tau)\,\Delta b_\rho + \frac{1}{2}q^{,\rho\sigma}_\alpha(\tau)\,\Delta b_\rho \Delta b_\sigma\Big] \qquad (4.43)$$

Since the matrices $B_{ij\alpha}(x_m)$ are deterministic functions, the substitution of Eq. (4.43) into strain equations analogous to Eqs. (4.38) and (4.39) yields the second-order accurate expectations for the strain components at $\tau = t$ in the form

$$\begin{aligned} E[\varepsilon_{ij}(x_m,t)] \;&=\; B_{ij\alpha}(x_m)\Big[q^0_\alpha(t) + \frac{1}{2}q^{,\rho\sigma}_\alpha(t)\,S^{\rho\sigma}_{\mathrm{b}}\Big] \\ &=\; B_{ij\alpha}(x_m)\Big[q^0_\alpha(t) + \frac{1}{2}q^{(2)}_\alpha(t)\Big] \qquad x_m \in \Omega_e \end{aligned} \qquad (4.44)$$

and their first-order accurate cross-covariances at $\xi_1 = (x^{(1)}_m, t_1)$ and $\xi_2 = (x^{(2)}_m, t_2)$ as

$$\begin{aligned} \mathrm{Cov}\Big(\varepsilon_{ij}(x^{(1)}_m,t_1), \varepsilon_{kl}(x^{(2)}_m,t_2)\Big) \;&=\; S^{ijkl}_\varepsilon(x^{(1)}_m,t_1; x^{(2)}_m,t_2) \\ =\; B_{ij\alpha}(x^{(1)}_m)\,B_{kl\beta}(x^{(2)}_m)\,&q^{,\rho}_\alpha(t_1)\,q^{,\sigma}_\beta(t_2)\,S^{\rho\sigma}_{\mathrm{b}} \\ x^{(1)}_m \in \Omega_e\,;\quad & x^{(2)}_m \in \Omega_f \end{aligned} \qquad (4.45)$$

The time-dependent stress tensor (random field) results as

$$\begin{aligned} \sigma_{ij}(x_m,\tau) \;&=\; C_{ijkl}\,B_{kl\alpha}(x_m)\,q_\alpha(\tau) \\ &=\; \varphi_{\bar\alpha}(x_m)\,\Big[C^0_{ijkl\bar\alpha} + C^{,\rho}_{ijkl\bar\alpha}\,\Delta b_\rho + \frac{1}{2}C^{,\rho\sigma}_{ijkl\bar\alpha}\,\Delta b_\rho\Delta b_\sigma\Big] \\ &\quad\times B_{kl\alpha}(x_m)\Big[q^0_\alpha(\tau) + q^{,\varsigma}_\alpha(\tau)\Delta b_\varsigma + \frac{1}{2}q^{,\varsigma\omega}_\alpha(\tau)\Delta b_\varsigma\Delta b_\omega\Big] \end{aligned} \qquad (4.46)$$

Again, a similar procedure based upon stress equations of the type (4.38), (4.39) yields the second-order accurate expectations for the stress components at $\tau = t$ in the form

$$\begin{aligned} E[\sigma_{ij}(x_m,t)] \;&=\; \varphi_{\bar\alpha}(x_m)\,C^0_{ijkl\bar\alpha}B_{kl\alpha}(x_m)q^0_\alpha(t) + \frac{1}{2}\varphi_{\bar\alpha}(x_m)\Big[C^{,\rho\sigma}_{ijkl\bar\alpha}\,q^0_\alpha(t) \\ &\quad + 2\,C^{,\rho}_{ijkl\bar\alpha}\,q^{,\sigma}_\alpha(t) + C^0_{ijkl\bar\alpha}\,q^{,\rho\sigma}_\alpha(t)\Big]B_{kl\alpha}(x_m)\,S^{\rho\sigma}_{\mathrm{b}} \\ &=\; \varphi_{\bar\alpha}(x_m)\,C^0_{ijkl\bar\alpha}\,E[\varepsilon_{kl}(x_m,t)] \\ &\quad + \varphi_{\bar\alpha}(x_m)\Big[\frac{1}{2}C^{,\rho\sigma}_{ijkl\bar\alpha}\,q^0_\alpha(t) + C^{,\rho}_{ijkl\bar\alpha}\,q^{,\sigma}_\alpha(t)\Big]B_{kl\alpha}(x_m)\,S^{\rho\sigma}_{\mathrm{b}} \end{aligned} \qquad (4.47)$$

and their first-order accurate cross-covariances at $\xi_1 = (x^{(1)}_m, t_1)$ and $\xi_2 = (x^{(2)}_m, t_2)$ as

$$\begin{aligned} \mathrm{Cov}\Big(\sigma_{ij}(x^{(1)}_m,t_1), \sigma_{kl}(x^{(2)}_m,t_2)\Big) \;&=\; S^{ijkl}_\sigma(x^{(1)}_m,t_1; x^{(2)}_m,t_2) \\ =\; \varphi_{\bar\alpha}(x^{(1)}_m)\,\varphi_{\bar\beta}(x^{(2)}_m)\,\Big[&C^{,\rho}_{ijtu\bar\alpha}\,C^{,\sigma}_{klvw\bar\beta}\,q^0_\alpha(t_1)\,q^0_\beta(t_2) \\ +\; C^{,\rho}_{ijtu\bar\alpha}\,C^0_{klvw\bar\beta}\,q^{,\sigma}_\alpha(t_1)\,q^0_\beta(t_2) &+ C^0_{ijtu\bar\alpha}\,C^{,\rho}_{klvw\bar\beta}\,q^0_\alpha(t_1)\,q^{,\sigma}_\beta(t_2) \\ +\; C^0_{ijtu\bar\alpha}\,C^0_{klvw\bar\beta}\,q^{,\rho}_\alpha(t_1)\,q^{,\sigma}_\beta(t_2)\Big]\,&B_{tu\alpha}(x^{(1)}_m)\,B_{vw\beta}(x^{(2)}_m)\,S^{\rho\sigma}_{\mathrm{b}} \\ x^{(1)}_m \in \Omega_e\,;\quad & x^{(2)}_m \in \Omega_f \end{aligned} \qquad (4.48)$$

It goes without saying that all the above results can be directly employed in the stochastic static analysis as well.

4.3 Stochastic Free Vibrations

4.3.1 Finite Element Equations

As an alternative to the variational approach, the second-order perturbation method may be applied directly to equations of discretized structural problems with stochastic parameters. Without repeating the derivations for the problems discussed so far in the book, we shall now illustrate the alternative approach by considering the free vibration problem. The stochastic finite element equations describing the generalized eigenproblem of free vibrations have the form, cf. Eq. (2.86)

$$\Big[K_{\alpha\beta}(b_\rho) - \Omega_{(\alpha)}(b_\rho)\, M_{\alpha\beta}(b_\rho)\Big]\, \phi_{\beta\gamma}(b_\rho) = 0 \qquad \text{(no sum on } \alpha) \qquad (4.49)$$

$$\rho = 1, 2, \ldots, \tilde{N}\;;\quad \alpha, \beta, \gamma = 1, 2, \ldots, N$$

where $M_{\alpha\beta}(b_\rho)$, $K_{\alpha\beta}(b_\rho)$, $\Omega_{(\alpha)}(b_\rho)$ and $\phi_{\alpha\beta}(b_\rho)$ are the system mass, stiffness, eigenvalue and eigenvector matrices, respectively; b_ρ stands for the vector of nodal random variables, i.e. nodal values of the random variables $b_r(x_k)$, cf. Eq. (4.1); $\tilde{N}$ is the number of nodal random variables and N is the number of degrees of freedom in the system. We note that since $M_{\alpha\beta}$ and $K_{\alpha\beta}$ are functions of the nodal random variables b_ρ, $\Omega_{(\alpha)}$ and $\phi_{\alpha\beta}$ are implicit functions of b_ρ as well. The matrices $M_{\alpha\beta}$ and $K_{\alpha\beta}$ are assumed positive definite and differentiable with respect to the nodal random variables b_ρ; the eigenvalues are assumed to be distinct.[1] Given the stochastic properties of $M_{\alpha\beta}$ and $K_{\alpha\beta}$, the objective of the analysis is to evaluate probabilistic distributions (i.e. the first two moments) for the eigenvalues $\Omega_{(\alpha)}$ and eigenvectors $\phi_{\alpha\beta}$.

By employing the Taylor series we expand $M_{\alpha\beta}(b_\rho)$, $K_{\alpha\beta}(b_\rho)$, $\Omega_{(\alpha)}(b_\rho)$ and $\phi_{\alpha\beta}(b_\rho)$ about the spatial expectations b_ρ^0 of the nodal random variables b_ρ; retaining up to the second-order terms yields, cf. Eq. (4.12)

$$\begin{aligned}
M_{\alpha\beta}(b_\varrho) &= M^0_{\alpha\beta}(b^0_\varrho) + \epsilon\, M^{,\rho}_{\alpha\beta}(b^0_\varrho)\,\Delta b_\rho + \frac{1}{2}\epsilon^2\, M^{,\rho\sigma}_{\alpha\beta}(b^0_\varrho)\,\Delta b_\rho \Delta b_\sigma \\
K_{\alpha\beta}(b_\varrho) &= K^0_{\alpha\beta}(b^0_\varrho) + \epsilon\, K^{,\rho}_{\alpha\beta}(b^0_\varrho)\,\Delta b_\rho + \frac{1}{2}\epsilon^2\, K^{,\rho\sigma}_{\alpha\beta}(b^0_\varrho)\,\Delta b_\rho \Delta b_\sigma \\
\Omega_{(\alpha)}(b_\varrho) &= \Omega^0_{(\alpha)}(b^0_\varrho) + \epsilon\, \Omega^{,\rho}_{(\alpha)}(b^0_\varrho)\,\Delta b_\rho + \frac{1}{2}\epsilon^2\, \Omega^{,\rho\sigma}_{(\alpha)}(b^0_\varrho)\,\Delta b_\rho \Delta b_\sigma \\
\phi_{\alpha\beta}(b_\varrho) &= \phi^0_{\alpha\beta}(b^0_\varrho) + \epsilon\, \phi^{,\rho}_{\alpha\beta}(b^0_\varrho)\,\Delta b_\rho + \frac{1}{2}\epsilon^2\, \phi^{,\rho\sigma}_{\alpha\beta}(b^0_\varrho)\,\Delta b_\rho \Delta b_\sigma
\end{aligned} \qquad (4.50)$$

By introducing the expansions (4.50) into Eq. (4.49), multiplying the second-order terms by the $\tilde{N}$-variate PDF $p_{\tilde{N}}(b_1, b_2, \ldots, b_{\tilde{N}})$, integrating over the domain of b_ρ and collecting terms of equal orders with respect to the small parameter ϵ, we arrive at the zeroth-, first- and second-order equations for the stochastic eigenproblem in the following form:

---

[1] In contrast to structural optimization problems the occurrence of repeated eigenvalues in stochastic structural analysis is rare.

- Zeroth-order (ϵ^0 terms, one system of N equations)

$$\left[K^0_{\alpha\beta}(b^0_\varrho) - \Omega^0_{(\alpha)}(b^0_\varrho)\, M^0_{\alpha\beta}(b^0_\varrho)\right] \phi^0_{\beta\gamma}(b^0_\varrho) = 0 \qquad \text{(no sum on } \alpha) \qquad (4.51)$$

- First-order (ϵ^1 terms, $\tilde{N}$ systems of N equations, $\rho = 1, 2, \ldots, \tilde{N}$)

$$\begin{aligned}&\left[K^0_{\alpha\beta}(b^0_\varrho) - \Omega^0_{(\alpha)}(b^0_\varrho)\, M^0_{\alpha\beta}(b^0_\varrho)\right] \phi^{,\rho}_{\beta\gamma}(b^0_\varrho)\\ &\quad = -\left[K^{,\rho}_{\alpha\beta}(b^0_\varrho) - \Omega^{,\rho}_{(\alpha)}(b^0_\varrho)\, M^0_{\alpha\beta}(b^0_\varrho) - \Omega^0_{(\alpha)}(b^0_\varrho)\, M^{,\rho}_{\alpha\beta}(b^0_\varrho)\right] \phi^0_{\beta\gamma}(b^0_\varrho)\end{aligned}$$
$$\text{(no sum on } \alpha) \qquad (4.52)$$

- Second-order (ϵ^2 terms,one system of N equations)

$$\begin{aligned}&\left[K^0_{\alpha\beta}(b^0_\varrho) - \Omega^0_{(\alpha)}(b^0_\varrho)\, M^0_{\alpha\beta}(b^0_\varrho)\right] \phi^{(2)}_{\beta\gamma}(b^0_\varrho)\\ &\quad = -\Big\{\Big[K^{,\rho\sigma}_{\alpha\beta}(b^0_\varrho) - \Omega^{,\rho\sigma}_{(\alpha)}(b^0_\varrho)\, M^0_{\alpha\beta}(b^0_\varrho)\\ &\quad - 2\,\Omega^{,\rho}_{(\alpha)}(b^0_\varrho)\, M^{,\sigma}_{\alpha\beta}(b^0_\varrho) - \Omega^0_{(\alpha)}(b^0_\varrho)\, M^{,\rho\sigma}_{\alpha\beta}(b^0_\varrho)\Big] \phi^0_{\beta\gamma}(b^0_\varrho)\\ &\quad + 2\left[K^{,\rho}_{\alpha\beta}(b^0_\varrho) - \Omega^{,\rho}_{(\alpha)}(b^0_\varrho)\, M^0_{\alpha\beta}(b^0_\varrho) - \Omega^0_{(\alpha)}(b^0_\varrho)\, M^{,\rho}_{\alpha\beta}(b^0_\varrho)\right] \phi^{,\sigma}_{\beta\gamma}(b^0_\varrho)\Big\}\, S^{\rho\sigma}_{\mathrm{b}}\end{aligned}$$
$$\text{(no sum on } \alpha) \qquad (4.53)$$

where

$$\phi^{(2)}_{\alpha\beta}(b^0_\varrho) = \phi^{,\rho\sigma}_{\alpha\beta}(b^0_\varrho)\, S^{\rho\sigma}_{\mathrm{b}} \qquad (4.54)$$

The zeroth-order equation (4.51) is nothing other than the equation for the free vibration eigenproblem in the deterministic sense; it can thus be solved for the zeroth-order eigenpairs $\Omega^0_{(\alpha)}$ and $\phi^0_{\alpha\beta}$ by any conventional technique effective in dealing with the generalized eigenproblems, cf. [7,34,36,93], for instance. In contrast to the zeroth-order eigenproblem, the solution procedures for the first and second derivatives of eigenvalues and eigenvectors with respect to the random variables are not so straightforward. We may observe that: (i) the systems (4.52) and (4.53) are nonhomogeneous, (ii) the coefficient matrices on the left-hand side in Eqs. (4.52) and (4.53) are singular of rank $N-1$, i.e. they cannot be inverted. We shall show in the next section that the evaluation of the first partial derivatives $\Omega^{,\rho}_{(\alpha)}$ and the second mixed derivatives $\Omega^{,\rho\sigma}_{(\alpha)}$ (to be exact, $\Omega^{,\rho\sigma}_{(\alpha)} S^{\rho\sigma}_{\mathrm{b}}$) of the eigenvalues $\Omega_{(\alpha)}$ is quite simple, though. This will be followed in Section 4.3.3 by a discussion of the numerical techniques for finding first and second derivatives of the eigenvectors.

4.3.2 First and Second Derivatives of Eigenvalues

In terms of underlying mathematics the procedure for the determination of $\Omega^{,\rho}_{(\alpha)}$ is identical to that employed previously for finding the first-order structural design sensitivity of eigenvalues with respect to the variation of (deterministic) design parameters, Section 2.8.2.3. To be specific, let us pre-multiply both sides of the first-order equation (4.52) by the zeroth-order eigenvector matrix $\phi^0_{\alpha\beta}$ transposed to get

$$\phi^{0\mathrm{T}}_{\gamma\alpha}\left(K^0_{\alpha\beta}-\Omega^0_{(\alpha)}\,M^0_{\alpha\beta}\right)\phi^{,\rho}_{\beta\delta}-\phi^{0\mathrm{T}}_{\gamma\alpha}\,\Omega^{,\rho}_{(\alpha)}\,M^0_{\alpha\beta}\,\phi^0_{\beta\delta}$$
$$=-\,\phi^{0\mathrm{T}}_{\gamma\alpha}\left(K^{,\rho}_{\alpha\beta}-\Omega^0_{(\alpha)}\,M^{,\rho}_{\alpha\beta}\right)\phi^0_{\beta\delta} \tag{4.55}$$

Since $\Omega^0_{(\alpha)}$ is diagonal and $K^0_{\alpha\beta}$ and $M^0_{\alpha\beta}$ are symmetric, the first term on the left-hand side of Eq. (4.55) vanishes so that, cf. Eq. (4.51)

$$\left[\phi^{0\mathrm{T}}_{\gamma\alpha}\left(K^0_{\alpha\beta}-\Omega^0_{(\alpha)}\,M^0_{\alpha\beta}\right)\phi^{,\rho}_{\beta\delta}\right]^{\mathrm{T}}=\phi^{,\rho\mathrm{T}}_{\gamma\alpha}\left(K^0_{\alpha\beta}-\Omega^0_{(\alpha)}\,M^0_{\alpha\beta}\right)\phi^0_{\beta\delta}=0 \tag{4.56}$$

Also, noting that $\Omega^{,\rho}_{(\alpha)}$ is diagonal and using the orthonormality condition (2.87) the second term on the left-hand side of Eq. (4.55) reads

$$\phi^{0\mathrm{T}}_{\gamma\alpha}\,\Omega^{,\rho}_{(\alpha)}\,M^0_{\alpha\beta}\,\phi^0_{\beta\delta}=\Omega^{,\rho}_{(\alpha)}\,\phi^{0\mathrm{T}}_{\gamma\alpha}\,M^0_{\alpha\beta}\,\phi^0_{\beta\delta}=\Omega^{,\rho}_{(\alpha)}\,\delta_{\gamma\delta} \tag{4.57}$$

Substituting Eqs. (4.56) and (4.57) into Eq. (4.55) leads to the expression for the first partial derivative of eigenvalues (first-order eigenvalues in the context of SFEM equations (4.51)–(4.53)) as, cf. Eq. (2.155)

$$\Omega^{,\rho}_{(\alpha)}=\phi^{0\mathrm{T}}_{\gamma\alpha}\left(K^{,\rho}_{\alpha\beta}-\Omega^0_{(\alpha)}\,M^{,\rho}_{\alpha\beta}\right)\phi^0_{\beta\delta} \tag{4.58}$$

Following a similar procedure we may readily obtain the expression for the generalized second-order eigenvalues $\Omega^{(2)}_{(\alpha)}(b^0_\varrho)$, defined as the double sum of the second partial derivatives of eigenvalues $\Omega^{,\rho\sigma}_{(\alpha)}$ times the corresponding entries of the covariance matrix $S^{\rho\sigma}_{\mathrm{b}}$, i.e.

$$\Omega^{(2)}_{(\alpha)}(b^0_\varrho)=\Omega^{,\rho\sigma}_{(\alpha)}(b^0_\varrho)\,S^{\rho\sigma}_{\mathrm{b}} \tag{4.59}$$

Pre-multiplying both sides of the second-order equation (4.53) by the zeroth-order eigenvector matrix $\phi^0_{\alpha\beta}$ transposed yields

$$\phi^{0\mathrm{T}}_{\gamma\alpha}\left(K^0_{\alpha\beta}-\Omega^0_{(\alpha)}\,M^0_{\alpha\beta}\right)\phi^{(2)}_{\beta\delta}-\phi^{0\mathrm{T}}_{\gamma\alpha}\,\Omega^{(2)}_{(\alpha)}\,M^0_{\alpha\beta}\,\phi^0_{\beta\delta}$$
$$=-\left[\phi^{0\mathrm{T}}_{\gamma\alpha}\left(K^{,\rho\sigma}_{\alpha\beta}-2\,\Omega^{,\rho}_{(\alpha)}\,M^{,\sigma}_{\alpha\beta}-\Omega^0_{(\alpha)}\,M^{,\rho\sigma}_{\alpha\beta}\right)\phi^0_{\beta\delta}\right.$$
$$\left.+\,2\,\phi^{0\mathrm{T}}_{\gamma\alpha}\left(K^{,\rho}_{\alpha\beta}-\Omega^{,\rho}_{(\alpha)}\,M^0_{\alpha\beta}-\Omega^0_{(\alpha)}\,M^{,\rho}_{\alpha\beta}\right)\phi^{,\sigma}_{\beta\delta}\right]S^{\rho\sigma}_{\mathrm{b}} \tag{4.60}$$

This implies that, cf. Eq. (4.57)

$$\Omega^{(2)}_{(\alpha)}=\left[\phi^{0\mathrm{T}}_{\gamma\alpha}\left(K^{,\rho\sigma}_{\alpha\beta}-2\,\Omega^{,\rho}_{(\alpha)}\,M^{,\sigma}_{\alpha\beta}-\Omega^0_{(\alpha)}\,M^{,\rho\sigma}_{\alpha\beta}\right)\phi^0_{\beta\delta}\right.$$
$$\left.+\,2\,\phi^{0\mathrm{T}}_{\gamma\alpha}\left(K^{,\rho}_{\alpha\beta}-\Omega^{,\rho}_{(\alpha)}\,M^0_{\alpha\beta}-\Omega^0_{(\alpha)}\,M^{,\rho}_{\alpha\beta}\right)\phi^{,\sigma}_{\beta\delta}\right]S^{\rho\sigma}_{\mathrm{b}} \tag{4.61}$$

since, as before, the first term on the left-hand side of Eq. (4.60) equals zero, cf. Eq. (4.56). It is noted that the only quantities which remain to be evaluated on the right-hand side of Eq. (4.61) are the first-order eigenvectors $\phi^{,\rho}_{\alpha\beta}$. We shall elaborate on this in the next section.

4.3.3 First and Second Derivatives of Eigenvectors

Evaluation of the first- and second-order eigenvectors requires more sophisticated techniques than those employed for the eigenvalue derivatives. Two alternative approaches may be employed to compute the first partial derivatives of eigenvectors. The main idea behind the first approach [37] is differentiation of the M-orthonormality condition with respect to random variables to obtain an additional equation for each eigenpair. The advantage of the algorithm is that only the specified eigenpair needs to be known. However, the procedure requires multiplication of an $N \times (N+1)$ matrix by its transpose in order to establish a nonsingular system of N linear algebraic equations. In addition, the bandwidth structure of the original system is destroyed and the resulting system matrix is frequently ill-conditioned.

In the second approach [24,37] the eigenvector derivative is expressed as a linear combination of all the eigenvectors in the original system. Equations for the coefficients of the linear combination are formed by using the M-orthonormality and K-orthogonality conditions. The algorithm is simple in concept, does not require inversion of an additional matrix as in the first method, and the structure of the system matrix is not changed during the solution process. Moreover, the complete solution of the zeroth-order eigenproblem is usually not required, since the eigenvector derivatives may be adequately approximated by a few dominant zeroth-order eigenvectors.

By adopting the basic philosophy of the latter approach we shall now develop a procedure for establishing the first and second derivatives of eigenvectors with respect to random variables. To do this we rewrite the first-order equation (4.52) for the $\hat{\alpha}$-th eigenpair as

$$\left(K^0_{\alpha\beta} - \omega^0_{(\hat{\alpha})} M^0_{\alpha\beta}\right) \phi^{,\rho}_{\beta(\hat{\alpha})} \;=\; -\left(K^{,\rho}_{\alpha\beta} - \omega^{,\rho}_{(\hat{\alpha})} M^0_{\alpha\beta} - \omega^0_{(\hat{\alpha})} M^{,\rho}_{\alpha\beta}\right) \phi^0_{\beta(\hat{\alpha})} \tag{4.62}$$

where $(\omega^0_{(\hat{\alpha})}, \phi^0_{\alpha(\hat{\alpha})})$ and $(\omega^{,\rho}_{(\hat{\alpha})}, \phi^{,\rho}_{\alpha(\hat{\alpha})})$ are the $\hat{\alpha}$-th zero- and first-order eigenpair, respectively. Eq. (4.58) can be specified for the $\hat{\alpha}$-th first-order eigenvalue as

$$\omega^{,\rho}_{(\hat{\alpha})} \;=\; \phi^0_{\alpha(\hat{\alpha})} \left(K^{,\rho}_{\alpha\beta} - \omega^0_{(\hat{\alpha})} M^{,\rho}_{\alpha\beta}\right) \phi^0_{\beta(\hat{\alpha})} \tag{4.63}$$

Substituting Eq. (4.63) into Eq. (4.62) yields

$$\left(K^0_{\alpha\beta} - \omega^0_{(\hat{\alpha})} M^0_{\alpha\beta}\right) \phi^{,\rho}_{\beta(\hat{\alpha})} \;=\; R^\rho_{\alpha(\hat{\alpha})} \tag{4.64}$$

where

$$R^\rho_{\alpha(\hat{\alpha})} = -\Big[K^{,\rho}_{\alpha\beta} - \phi^0_{\gamma(\hat{\alpha})}\Big(K^{,\rho}_{\gamma\delta} - \omega^0_{(\hat{\alpha})} M^{,\rho}_{\gamma\delta}\Big)\phi^0_{\delta(\hat{\alpha})} M^0_{\alpha\beta} - \omega^0_{(\hat{\alpha})} M^{,\rho}_{\alpha\beta}\Big]\phi^0_{\beta(\hat{\alpha})} \tag{4.65}$$

Eq. (4.64) does not give a unique solution for $\phi^{,\rho}_{\alpha(\hat{\alpha})}$ since the system matrix is of rank $N-1$.

Let us proceed by assuming that the $\hat{\alpha}$-th first-order eigenvector $\phi^{,\rho}_{\alpha(\hat{\alpha})}$ can be expressed as a linear combination of all the zeroth-order eigenvectors, i.e.

$$\phi^{,\rho}_{\alpha(\hat{\alpha})} \;=\; \phi^0_{\alpha\beta}\, a^\rho_{\beta(\hat{\alpha})} \qquad\qquad \alpha, \beta, \hat{\alpha} = 1, 2, \ldots, N \tag{4.66}$$

Introducing Eq. (4.66) into Eq. (4.64) and pre-multiplying both sides of the resulting equation by $\phi^0_{\alpha\beta}$ transposed yields

$$\phi^{0\mathrm{T}}_{\gamma\alpha}\left(K^0_{\alpha\beta}-\omega^0_{(\hat\alpha)}\,M^0_{\alpha\beta}\right)\phi^0_{\beta\delta}\,a^\rho_{\delta(\hat\alpha)} \;=\; \phi^{0\mathrm{T}}_{\gamma\alpha}R^\rho_{\alpha(\hat\alpha)} \qquad \text{(no sum on } \hat\alpha) \tag{4.67}$$

or, denoting the unit matrix by I

$$\left(\Omega^0_{(\alpha)}-\omega^0_{(\hat\alpha)}\,I\right)a^\rho_{\beta(\hat\alpha)} \;=\; \phi^{0\mathrm{T}}_{\beta\alpha}\,R^\rho_{\alpha(\hat\alpha)} \qquad \text{(no sum on } \hat\alpha) \tag{4.68}$$

since

$$\begin{aligned}&\phi^{0\mathrm{T}}_{\gamma\alpha}\left(K^0_{\alpha\beta}-\omega^0_{(\hat\alpha)}\,M^0_{\alpha\beta}\right)\phi^0_{\beta\delta}\\ &\quad= \phi^{0\mathrm{T}}_{\gamma\alpha}K^0_{\alpha\beta}\phi^0_{\beta\delta}-\omega^0_{(\hat\alpha)}\phi^{0\mathrm{T}}_{\gamma\alpha}\,M^0_{\alpha\beta}\,\phi^0_{\beta\delta} \;=\; \Omega^0_{(\alpha)}-\omega^0_{(\hat\alpha)}\,I\end{aligned} \tag{4.69}$$

As $(\Omega^0_{(\alpha)}-\omega^0_{(\hat\alpha)}I)$ is a diagonal matrix with the entries $(\omega^0_{(\alpha)}-\omega^0_{(\hat\alpha)})$ at the α-th row we get

$$a^\rho_{\alpha(\hat\alpha)} \;=\; \frac{\phi^0_{\beta(\alpha)}\,R^\rho_{\beta(\hat\alpha)}}{\omega^0_{(\alpha)}-\omega^0_{(\hat\alpha)}} \qquad \text{for} \quad \alpha\neq\hat\alpha \tag{4.70}$$

with the value of the $\hat\alpha$-th component of $a^\rho_{\alpha(\hat\alpha)}$ undetermined. This means that the $\hat\alpha$-th first-order eigenvector can be expressed explicitly in terms of all the system zeroth-order eigenvectors, except for the $\hat\alpha$-th one. To determine $a^\rho_{\hat\alpha(\hat\alpha)}$ uniquely the M-orthonormality relationship

$$\phi_{\alpha(\hat\alpha)}\,M_{\alpha\beta}\,\phi_{\beta(\hat\alpha)} \;=\; 1 \qquad \text{(no sum on } \hat\alpha) \tag{4.71}$$

must be employed. Taking the first derivative of Eq. (4.71) with respect to random variables b_ρ gives

$$\phi^0_{\alpha(\hat\alpha)}\,M^{,\rho}_{\alpha\beta}\,\phi^0_{\beta(\hat\alpha)}+2\,\phi^0_{\alpha(\hat\alpha)}\,M^0_{\alpha\beta}\,\phi^{,\rho}_{\beta(\hat\alpha)} \;=\; 0 \qquad \text{(no sum on } \hat\alpha) \tag{4.72}$$

or, by Eq. (4.66)

$$\phi^0_{\alpha(\hat\alpha)}\,M^{,\rho}_{\alpha\beta}\,\phi^0_{\beta(\hat\alpha)}+2\,\phi^0_{\alpha(\hat\alpha)}\,M^0_{\alpha\beta}\,\phi^0_{\beta\delta}\,a^\rho_{\delta(\hat\alpha)} \;=\; 0 \qquad \text{(no sum on } \hat\alpha) \tag{4.73}$$

By imposing the eigenvector orthonormality condition the second term on the left-hand side of Eq. (4.73) reduces to

$$2\,\phi^0_{\alpha(\hat\alpha)}\,M^0_{\alpha\beta}\,\phi^0_{\beta\gamma}\,a^\rho_{\gamma(\hat\alpha)} \;=\; 2\,a^\rho_{\hat\alpha(\hat\alpha)} \qquad \text{(no sum on } \hat\alpha) \tag{4.74}$$

which, when substituted into Eq. (4.73), gives the expression for $a^\rho_{\hat\alpha(\hat\alpha)}$ as

$$a^\rho_{\hat\alpha(\hat\alpha)} \;=\; -\tfrac{1}{2}\,\phi^0_{\alpha(\hat\alpha)}\,M^{,\rho}_{\alpha\beta}\,\phi^0_{\beta(\hat\alpha)} \qquad \text{(no sum on } \hat\alpha) \tag{4.75}$$

Eqs. (4.70) and (4.75) form the complete solution for the first-order eigenvectors. We rewrite it compactly together as

$$a^\rho_{\alpha(\hat\alpha)} \;=\; \begin{cases}\dfrac{\phi^0_{\beta(\alpha)}\,R^\rho_{\beta(\hat\alpha)}}{\omega^0_{(\alpha)}-\omega^0_{(\hat\alpha)}} & \text{for} \quad \alpha\neq\hat\alpha\\[2ex] -\tfrac{1}{2}\,\phi^0_{\beta(\hat\alpha)}\,M^{,\rho}_{\beta\gamma}\,\phi^0_{\gamma(\hat\alpha)} & \text{for} \quad \alpha=\hat\alpha\end{cases} \tag{4.76}$$

It should be noted at this point that in engineering practice the first derivatives of eigenvectors are sometimes approximated by [37]

$$\phi^{,\rho}_{\alpha(\hat{\alpha})} = \phi^{0}_{\alpha\bar{\alpha}}\, a^{\rho}_{\bar{\alpha}(\hat{\alpha})} \qquad \bar{\alpha} = 1, 2, \ldots, \bar{N} \ll N \tag{4.77}$$

This bears resemblance to the basic idea of modal analysis, which states that a few most important modes of free vibrations need frequently be employed only to describe the dynamic response of the linear system with sufficient accuracy. The efficiency of the procedure applied to first-order eigenvector problems remains to be proved, though.

Another algorithm which can be treated as a variation of the one described above [91] enables one to solve the problem separately for each specified eigenvector thus making the complete zeroth-order solution redundant. However, the partitioning process in this procedure is as a rule problem-dependent which makes the computational implementation inefficient.

The procedure for evaluating the second derivatives of eigenvectors follows similar lines. We first rewrite the second-order equation (4.53) for the $\hat{\alpha}$-eigenpair as, cf. Eqs. (4.62)–(4.65)

$$\left(K^{0}_{\alpha\beta} - \omega^{0}_{(\hat{\alpha})}\, M^{0}_{\alpha\beta}\right) \phi^{(2)}_{\beta(\hat{\alpha})} = R^{(2)}_{\alpha(\hat{\alpha})} \qquad \text{(no sum on } \hat{\alpha}) \tag{4.78}$$

where

$$\begin{aligned} R^{(2)}_{\alpha(\hat{\alpha})} = & - \Big[\left(K^{,\rho\sigma}_{\alpha\beta} - \omega^{,\rho\sigma}_{(\hat{\alpha})}\, M^{0}_{\alpha\beta} - 2\omega^{,\rho}_{(\hat{\alpha})}\, M^{,\sigma}_{\alpha\beta} - \omega^{0}_{(\hat{\alpha})}\, M^{,\rho\sigma}_{\alpha\beta} \right) \phi^{0}_{\beta(\hat{\alpha})} \\ & + 2 \left(K^{,\rho}_{\alpha\beta} - \omega^{,\rho}_{(\hat{\alpha})}\, M^{0}_{\alpha\beta} - \omega^{0}_{(\hat{\alpha})}\, M^{,\rho}_{\alpha\beta} \right) \phi^{,\sigma}_{\beta(\hat{\alpha})} \Big]\, S^{\rho\sigma}_{\mathrm{b}} \end{aligned} \tag{4.79}$$

(no sum on $\hat{\alpha}$)

and, cf. Eq. (4.54)

$$\phi^{(2)}_{\alpha(\hat{\alpha})} = \phi^{,\rho\sigma}_{\alpha(\hat{\alpha})}\, S^{\rho\sigma}_{\mathrm{b}} \tag{4.80}$$

Similarly as before, the second-order eigenvector $\phi^{(2)}_{\alpha(\hat{\alpha})}$ is now approximated by a linear combination of all the zeroth-order system eigenvectors as

$$\phi^{(2)}_{\alpha(\hat{\alpha})} = \phi^{0}_{\alpha\beta}\, a^{(2)}_{\beta(\hat{\alpha})} \qquad \alpha, \beta, \hat{\alpha} = 1, 2, \ldots, N \tag{4.81}$$

Eq. (4.78) then becomes, cf. Eqs. (4.67)–(4.69)

$$\left(\Omega^{0}_{(\alpha)} - \omega^{0}_{(\hat{\alpha})}\, I\right) a^{(2)}_{\beta(\hat{\alpha})} = \phi^{0^{\mathrm{T}}}_{\beta\alpha}\, R^{(2)}_{\alpha(\hat{\alpha})} \qquad \text{(no sum on } \hat{\alpha}) \tag{4.82}$$

and we arrive at, cf. Eq. (4.70)

$$a^{(2)}_{\alpha(\hat{\alpha})} = \frac{\phi^{0}_{\beta(\alpha)}\, R^{(2)}_{\beta(\hat{\alpha})}}{\omega^{0}_{(\alpha)} - \omega^{0}_{(\hat{\alpha})}} \qquad \text{for} \quad \alpha \neq \hat{\alpha} \qquad \text{(no sum on } \hat{\alpha}) \tag{4.83}$$

with the value of $a^{(2)}_{\alpha(\hat{\alpha})}$ for $\alpha = \hat{\alpha}$ undetermined. To avoid the arbitrariness we take the second derivative of Eq. (4.71) with respect to random variables b_ρ to obtain

$$\begin{aligned}
&\phi^{0}_{\alpha(\hat{\alpha})}\, M^{,\rho\sigma}_{\alpha\beta}\, \phi^{0}_{\beta(\hat{\alpha})} + 4\, \phi^{0}_{\alpha(\hat{\alpha})}\, M^{,\rho}_{\alpha\beta}\, \phi^{,\sigma}_{\beta(\hat{\alpha})} \\
&\quad + 2\, \phi^{,\rho\mathrm{T}}_{\alpha(\hat{\alpha})}\, M^{0}_{\alpha\beta}\, \phi^{,\sigma}_{\beta(\hat{\alpha})} + 2\, \phi^{0}_{\alpha(\hat{\alpha})}\, M^{0}_{\alpha\beta}\, \phi^{,\rho\sigma}_{\beta(\hat{\alpha})} = 0 \qquad (4.84)
\end{aligned}$$

(no sum on $\hat{\alpha}$)

The third and fourth terms on the left-hand side of Eq. (4.84) are rewritten as, cf. Eq. (4.66) and Eqs. (4.54), (4.80), (4.81), respectively

$$\begin{aligned}
2\, \phi^{,\rho\mathrm{T}}_{\alpha(\hat{\alpha})}\, M^{0}_{\alpha\beta}\, \phi^{,\sigma}_{\beta(\hat{\alpha})} &= 2\left(\phi^{0}_{\alpha\gamma}\, a^{\rho}_{\gamma(\hat{\alpha})}\right)^{\mathrm{T}} M^{0}_{\alpha\beta} \left(\phi^{0}_{\beta\delta}\, a^{\sigma}_{\delta(\hat{\alpha})}\right) \\
&= 2\, a^{\rho}_{\gamma(\hat{\alpha})} \left(\phi^{0\mathrm{T}}_{\gamma\alpha}\, M^{0}_{\alpha\beta}\, \phi^{0}_{\beta\delta}\right) a^{\sigma}_{\delta(\hat{\alpha})} = 2\, a^{\rho}_{\alpha(\hat{\alpha})}\, a^{\sigma}_{\alpha(\hat{\alpha})} \qquad (4.85) \\
2\, \phi^{0}_{\alpha(\hat{\alpha})}\, M^{0}_{\alpha\beta}\, \phi^{,\rho\sigma}_{\beta(\hat{\alpha})}\, S^{\rho\sigma}_{\mathrm{b}} &= 2\, \phi^{0}_{\alpha(\hat{\alpha})}\, M^{0}_{\alpha\beta}\, \phi^{0}_{\beta\gamma}\, a^{(2)}_{\gamma(\hat{\alpha})} = 2\, a^{(2)}_{\hat{\alpha}(\hat{\alpha})} \qquad (4.86)
\end{aligned}$$

(no sum on $\hat{\alpha}$)

Pre-multiplying both sides of Eq. (4.84) by $S^{\rho\sigma}_{\mathrm{b}}$ and using Eqs. (4.85) and (4.86) yields the explicit expression for the $\hat{\alpha}$-th component of $a^{(2)}_{\alpha(\hat{\alpha})}$ as

$$a^{(2)}_{\hat{\alpha}(\hat{\alpha})} = -\left(\frac{1}{2}\phi^{0}_{\alpha(\hat{\alpha})}\, M^{,\rho\sigma}_{\alpha\beta}\, \phi^{0}_{\beta(\hat{\alpha})} + 2\phi^{0}_{\alpha(\hat{\alpha})}\, M^{,\rho}_{\alpha\beta}\, \phi^{,\sigma}_{\beta(\hat{\alpha})} + a^{\rho}_{\alpha(\hat{\alpha})}\, a^{\sigma}_{\alpha(\hat{\alpha})}\right) S^{\rho\sigma}_{\mathrm{b}} \qquad (4.87)$$

(no sum on $\hat{\alpha}$)

The last equation combined with Eq. (4.83) gives the unique value of $a^{(2)}_{\alpha(\hat{\alpha})}$ for any α, $\hat{\alpha} = 1, 2, \ldots, N$, in the form

$$a^{(2)}_{\alpha(\hat{\alpha})} = \begin{cases} \dfrac{\phi^{0}_{\beta(\alpha)}\, R^{(2)}_{\beta(\hat{\alpha})}}{\omega^{0}_{(\alpha)} - \omega^{0}_{(\hat{\alpha})}} & \text{for} \quad \alpha \neq \hat{\alpha} \\ -\left(\frac{1}{2}\phi^{0}_{\beta(\hat{\alpha})}\, M^{,\rho\sigma}_{\beta\gamma}\, \phi^{0}_{\gamma(\hat{\alpha})} + 2\phi^{0}_{\beta(\hat{\alpha})}\, M^{,\rho}_{\beta\gamma}\, \phi^{,\sigma}_{\gamma(\hat{\alpha})} + a^{\rho}_{\beta(\hat{\alpha})}\, a^{\sigma}_{\beta(\hat{\alpha})}\right) S^{\rho\sigma}_{\mathrm{b}} & \\ \qquad \text{for} \quad \alpha = \hat{\alpha} \qquad \text{(no sum on } \hat{\alpha}) & \end{cases} \qquad (4.88)$$

The last result completes the derivations necessary for the solution of the second-order eigenvector problem.

It should be emphasized that the formulations described above require only one second-order system to be solved instead of $\tilde{N}(\tilde{N}+1)/2$ systems, one for each second mixed derivative of the eigenvalues and corresponding eigenvectors with respect to nodal random variables (expansions (4.50) are symmetric with respect to ρ and σ). We also observe that no additional eigenproblem solution is required to calculate the first- and second-order eigenvalues and eigenvectors, provided the zero-order eigenpairs are known.

4.3.4 First Two Statistical Moments

Having explicit expression for the first- and second-order eigenvalues and eigenvectors their probabilistic distributions can be readily obtained. By the definition of the expectation and cross-covariance for eigenvalues we have

$$E[\omega_{(\hat{\alpha})}] = \underbrace{\int_{-\infty}^{+\infty}\int_{-\infty}^{+\infty}\cdots\int_{-\infty}^{+\infty}}_{\tilde{N}\text{-fold}} \omega_{(\hat{\alpha})}\, p_{\tilde{N}}(b_1, b_2, \ldots, b_{\tilde{N}})\, \mathrm{d}b_1 \mathrm{d}b_2 \ldots \mathrm{d}b_{\tilde{N}} \tag{4.89}$$

$$\begin{aligned}\mathrm{Cov}(\omega_{(\hat{\alpha})}, \omega_{(\hat{\beta})}) &= S_{\omega}^{\hat{\alpha}\hat{\beta}} \\ &= \underbrace{\int_{-\infty}^{+\infty}\int_{-\infty}^{+\infty}\cdots\int_{-\infty}^{+\infty}}_{\tilde{N}\text{-fold}} \left\{\omega_{(\hat{\alpha})} - E[\omega_{(\hat{\alpha})}]\right\}\left\{\omega_{(\hat{\beta})} - E[\omega_{(\hat{\beta})}]\right\} \\ &\qquad \times\, p_{\tilde{N}}(b_1, b_2, \ldots, b_{\tilde{N}})\, \mathrm{d}b_1 \mathrm{d}b_2 \ldots \mathrm{d}b_{\tilde{N}}\end{aligned} \tag{4.90}$$

where $p_{\tilde{N}}(b_1, b_2, \ldots, b_{\tilde{N}})$ is the $\tilde{N}$-variate PDF of the nodal random variables. The first two moments for the eigenvalue $\omega_{(\hat{\alpha})}$ may be explicitly expressed for a given value of the small parameter ϵ. Similarly as in the cases of statics and forced vibration analyses, with the assumption that the coefficients of variation of random variables defined by Eq. (1.20) are small, the formal solution can be obtained by selecting $\epsilon = 1$. The expansion $(4.50)_3$ can then be written for the $\hat{\alpha}$-th eigenvalue as

$$\omega_{(\hat{\alpha})}(b_\varrho) = \omega^0_{(\hat{\alpha})}(b^0_\varrho) + \omega^{,\rho}_{(\hat{\alpha})}(b^0_\varrho)\,\Delta b_\rho + \frac{1}{2}\,\omega^{,\rho\sigma}_{(\hat{\alpha})}(b^0_\varrho)\,\Delta b_\rho \Delta b_\sigma \tag{4.91}$$

Substituting Eq. (4.91) into Eq. (4.89), using Eq. (4.59) and noting that the terms including the first variation of Δb_ρ vanish, cf. Eq. (3.14), yields the second-order accurate expectation for the $\hat{\alpha}$-th eigenvalue as

$$E[\omega_{(\hat{\alpha})}] = \omega^0_{(\hat{\alpha})} + \frac{1}{2}\omega^{,\rho\sigma}_{(\hat{\alpha})}\, S^{\rho\sigma}_{\mathrm{b}} = \omega^0_{(\hat{\alpha})} + \frac{1}{2}\omega^{(2)}_{(\hat{\alpha})} \qquad \hat{\alpha} = 1, 2, \ldots, N \tag{4.92}$$

The first-order accurate cross-covariance for the $\hat{\alpha}$-th and $\hat{\beta}$-th eigenvalues can be determined by introducing Eq. (4.91) into Eq. (4.90) to obtain

$$\mathrm{Cov}(\omega_{(\hat{\alpha})}, \omega_{(\hat{\beta})}) = \omega^{,\rho}_{(\hat{\alpha})}\,\omega^{,\sigma}_{(\hat{\beta})}\, S^{\rho\sigma}_{\mathrm{b}} \qquad \hat{\alpha}, \hat{\beta} = 1, 2, \ldots, N \tag{4.93}$$

The first two moments for the eigenvector components are calculated in a similar way. By the definition of the eigenvector expectation and cross-covariance

$$E[\phi_{\alpha(\hat{\alpha})}] = \underbrace{\int_{-\infty}^{+\infty}\int_{-\infty}^{+\infty}\cdots\int_{-\infty}^{+\infty}}_{\tilde{N}\text{-fold}} \phi_{\alpha(\hat{\alpha})}\, p_{\tilde{N}}(b_1, b_2, \ldots, b_{\tilde{N}})\, \mathrm{d}b_1 \mathrm{d}b_2 \ldots \mathrm{d}b_{\tilde{N}} \tag{4.94}$$

$$\begin{aligned}\mathrm{Cov}(\phi_{\alpha(\hat{\alpha})}, \phi_{\beta(\hat{\beta})}) &= S_{\phi}^{\alpha\hat{\alpha}\beta\hat{\beta}} \\ &= \underbrace{\int_{-\infty}^{+\infty}\int_{-\infty}^{+\infty}\cdots\int_{-\infty}^{+\infty}}_{\tilde{N}\text{-fold}} \left\{\phi_{\alpha(\hat{\alpha})} - E[\phi_{\alpha(\hat{\alpha})}]\right\}\left\{\phi_{\beta(\hat{\beta})} - E[\phi_{\beta(\hat{\beta})}]\right\} \\ &\qquad \times\, p_{\tilde{N}}(b_1, b_2, \ldots, b_{\tilde{N}})\, \mathrm{d}b_1 \mathrm{d}b_2 \ldots \mathrm{d}b_{\tilde{N}}\end{aligned} \tag{4.95}$$

By using the expansion $(4.50)_4$ rewritten for the $\hat{\alpha}$-th eigenvector as

$$\phi_{\alpha(\hat{\alpha})}(b_\varrho) = \phi^0_{\alpha(\hat{\alpha})}(b^0_\varrho) + \phi^{,\rho}_{\alpha(\hat{\alpha})}(b^0_\varrho)\,\Delta b_\rho + \tfrac{1}{2}\,\phi^{,\rho\sigma}_{\alpha(\hat{\alpha})}(b^0_\varrho)\,\Delta b_\rho \Delta b_\sigma \qquad (4.96)$$

we arrive at the expressions for the second-order accurate expectation for the α-th component of the $\hat{\alpha}$-th eigenvector in the form

$$E[\phi_{\alpha(\hat{\alpha})}] = \phi^0_{\alpha(\hat{\alpha})} + \tfrac{1}{2}\,\phi^{,\rho\sigma}_{\alpha(\hat{\alpha})}\,S^{\rho\sigma}_{\mathrm{b}} = \phi^0_{\alpha(\hat{\alpha})} + \tfrac{1}{2}\,\phi^{(2)}_{\alpha(\hat{\alpha})} \qquad \alpha, \hat{\alpha} = 1, 2, \ldots, N \qquad (4.97)$$

and for the first-order accurate cross-covariance for the α-th component of the $\hat{\alpha}$-th eigenvector and the β-th component of the $\hat{\beta}$-th eigenvector in the form

$$\mathrm{Cov}(\phi_{\alpha(\hat{\alpha})}, \phi_{\beta(\hat{\beta})}) = \phi^{,\rho}_{\alpha(\hat{\alpha})}\,\phi^{,\sigma}_{\beta(\hat{\beta})}\,S^{\rho\sigma}_{\mathrm{b}} \qquad \alpha, \beta, \hat{\alpha}, \hat{\beta} = 1, 2, \ldots, N \qquad (4.98)$$

4.4 Stochastic Linearized Buckling Analysis

4.4.1 Finite Element Equations

As shown in Section 2.7 the essentially nonlinear phenomenon of structural buckling may be modelled as a linear problem by assuming that prebuckling deformations are small and applied loads are proportional to a single load parameter, and by using the property that the system geometric stiffness matrix is a linear and homogeneous function of stresses up to bifurcation. The buckling load as well as the corresponding buckling mode can then be computed by considering the condition for a bifurcation point to exit, i.e. by seeking a critical value of the load parameter such that the total stiffness matrix becomes singular. The fact that the linearized structural buckling analysis is a generalized eigenvalue problem makes it possible to use the results of the previous section to pose and solve the bifurcation problem for systems with stochastic parameters. However, before doing this some comments on algebraic differences between the buckling and free vibration problems seem to be in order. In contrast to the latter case, in which both the matrices entering the generalized eigenproblem may be assumed positive definite, the geometric stiffness does not in general exhibit this property. It is therefore useful to rewrite Eq. (2.120) for the $\hat{\alpha}$-th solution pair in the form

$$\left(K^{(\sigma)}_{\alpha\beta} - \tilde{\lambda}_{(\hat{\alpha})}\,K^{(e)}_{\alpha\beta}\right) v_{\beta(\hat{\alpha})} = 0 \qquad \text{(no sum on } \hat{\alpha}) \qquad (4.99)$$

where

$$\tilde{\lambda}_{(\hat{\alpha})} = \frac{1}{\lambda_{(\hat{\alpha})}} \qquad (4.100)$$

$v_{\alpha(\hat{\alpha})}$ being the $\hat{\alpha}$-th eigenvector (or eigenmode). The elastic constitutive small strain stiffness matrix $K^{(e)}_{\alpha\beta}$ is decomposed as

$$K^{(e)}_{\alpha\beta} = U^{\mathrm{T}}_{\alpha\gamma}\,U_{\gamma\beta} \qquad (4.101)$$

with $U_{\alpha\beta}$ being a nonsingular matrix. Substituting Eq. (4.101) into Eq. (4.99) and pre-multiplying both sides of the resulting equation by $U_{\alpha\beta}^{-T}$ the standard eigenproblem is obtained as

$$\left(\tilde{K}^{(\sigma)}_{\alpha\beta} - \tilde{\lambda}_{(\hat{\alpha})}\, I\right) \tilde{v}_{\beta(\hat{\alpha})} = 0 \tag{4.102}$$

where

$$\begin{aligned} \tilde{K}^{(\sigma)}_{\alpha\beta} &= U^{-T}_{\alpha\gamma}\, K^{(\sigma)}_{\gamma\delta}\, U^{-1}_{\delta\beta} \\ \tilde{v}_{\alpha(\hat{\alpha})} &= U_{\alpha\beta}\, v_{\beta(\hat{\alpha})} \end{aligned} \tag{4.103}$$

Since for symmetric matrices eigenvectors corresponding to different eigenvalues are orthogonal and the eigenvectors are defined up to a scalar multiplier only, we may assume

$$\tilde{v}_{\alpha(\hat{\alpha})}\, \tilde{v}_{\alpha(\hat{\beta})} = \delta_{\hat{\alpha}\hat{\beta}} \tag{4.104}$$

which implies

$$\tilde{v}_{\alpha(\hat{\alpha})}\, \tilde{K}^{(\sigma)}_{\alpha\beta}\, \tilde{v}_{\beta(\hat{\beta})} = \tilde{\lambda}_{(\hat{\alpha})}\, \delta_{\hat{\alpha}\hat{\beta}} \qquad \text{(no sum on } \hat{\alpha}) \tag{4.105}$$

By using Eqs. (4.101) and (4.103) we obtain

$$\begin{aligned} v_{\alpha(\hat{\alpha})}\, K^{(e)}_{\alpha\beta}\, v_{\beta(\hat{\beta})} &= \delta_{\hat{\alpha}\hat{\beta}} \\ v_{\alpha(\hat{\alpha})}\, K^{(\sigma)}_{\alpha\beta}\, v_{\beta(\hat{\beta})} &= \tilde{\lambda}_{(\hat{\alpha})}\, \delta_{\hat{\alpha}\hat{\beta}} \qquad \text{(no sum on } \hat{\alpha}) \end{aligned} \tag{4.106}$$

In other words, the eigenvectors in the buckling problem described by Eq. (4.99) are considered as $K^{(e)}_{\alpha\beta}$-orthonormal and $K^{(\sigma)}_{\alpha\beta}$-orthogonal.

We note that in contrast to the eigenvalue problem (2.120), wherein the smallest buckling load is associated with the *smallest* eigenvalue, by using the equivalent formulation (4.99) the *largest* eigenvalue defines the smallest buckling load.

We shall turn our attention now to the stochastic buckling problem which will be discussed with reference to the formulation (4.99). We assume $K^{(e)}_{\alpha\beta} = K^{(e)}_{\alpha\beta}(b_p)$ and $K^{(\sigma)}_{\alpha\beta} = K^{(\sigma)}_{\alpha\beta}[\sigma^*_{ij}(b_p)] = \hat{K}^{(\sigma)}_{\alpha\beta}(b_p)$, cf. Section 2.7 and, consequently, $\tilde{\lambda}_{(\hat{\alpha})} = \tilde{\lambda}_{(\hat{\alpha})}(b_p)$ and $v_{\beta(\hat{\alpha})} = v_{\beta(\hat{\alpha})}(b_p)$; b_p is, as before, the vector of nodal random variables while σ^*_{ij} are known stresses due to a test load, cf. Eq. (2.116), implicitly dependent on b_p. The functions $\tilde{\lambda}_{(\hat{\alpha})}$ and $v_{\beta(\hat{\alpha})}$ are also implicit in terms of b_p. Given the first two statistical moments for $K^{(e)}_{\alpha\beta}$ and $K^{(\sigma)}_{\alpha\beta}$, the problem is to calculate the first two moments for the buckling load factors (eigenvalues) $\lambda_{(\hat{\alpha})}$.

In order to see that the statistical moments for the initial stress matrix may be considered given we observe that

$$\begin{aligned} \hat{K}^{(\sigma)}_{\alpha\beta}(b_p) &= \int_\Omega \sigma^*_{ij}(b_p)\, \varphi_{k\alpha,i}\, \varphi_{k\beta,j}\, \mathrm{d}\Omega \\ \hat{K}^{(\sigma),p}_{\alpha\beta}(b_p) &= \int_\Omega \sigma^{*,p}_{ij}(b_p)\, \varphi_{k\alpha,i}\, \varphi_{k\beta,j}\, \mathrm{d}\Omega \\ \hat{K}^{(\sigma),p\varsigma}_{\alpha\beta}(b_p) &= \int_\Omega \sigma^{*,p\varsigma}_{ij}(b_p)\, \varphi_{k\alpha,i}\, \varphi_{k\beta,j}\, \mathrm{d}\Omega \end{aligned} \tag{4.107}$$

The partial derivatives of stresses may be computed if only we note that σ^*_{ij} is a solution to the linear elastic problem and

$$\begin{aligned}\sigma^*_{ij} &= C_{ijkl}\,B_{kl\alpha}\,q^*_\alpha \\ \sigma^{*,\rho}_{ij} &= C^{,\rho}_{ijkl}\,B_{kl\alpha}\,q^*_\alpha + C_{ijkl}\,B_{kl\alpha}\,q^{*,\rho}_\alpha \\ \sigma^{*,\rho\varsigma}_{ij} &= C^{,\rho\varsigma}_{ijkl}\,B_{kl\alpha}q^*_\alpha + C^{,\rho}_{ijkl}\,B_{kl\alpha}q^{*,\varsigma}_\alpha + C^{,\varsigma}_{ijkl}\,B_{kl\alpha}q^{*,\rho}_\alpha + C_{ijkl}\,B_{kl\alpha}q^{*,\rho\varsigma}_\alpha\end{aligned} \tag{4.108}$$

A more detailed discussion of this aspect is postponed to Chapter 9.

By employing the second-order perturbation method the Taylor expansions for $K^{(e)}_{\alpha\beta}$, $\hat{K}^{(\sigma)}_{\alpha\beta}$, $v_{\alpha(\hat{\alpha})}$ and $\tilde{\lambda}_{(\hat{\alpha})}$ about the spatial expectations b^0_ρ of the nodal random variables b_ρ read:

$$\begin{aligned}K^{(e)}_{\alpha\beta}(b_\varrho) &= K^{(e)0}_{\alpha\beta}(b^0_\varrho) + \epsilon\,K^{(e),\rho}_{\alpha\beta}(b^0_\varrho)\,\Delta b_\rho + \frac{1}{2}\epsilon^2\,K^{(e),\rho\varsigma}_{\alpha\beta}(b^0_\varrho)\,\Delta b_\rho\Delta b_\varsigma \\ \hat{K}^{(\sigma)}_{\alpha\beta}(b_\varrho) &= \hat{K}^{(\sigma)0}_{\alpha\beta}(b^0_\varrho) + \epsilon\,\hat{K}^{(\sigma),\rho}_{\alpha\beta}(b^0_\varrho)\,\Delta b_\rho + \frac{1}{2}\epsilon^2\,\hat{K}^{(\sigma),\rho\varsigma}_{\alpha\beta}(b^0_\varrho)\,\Delta b_\rho\Delta b_\varsigma \\ v_{\alpha(\hat{\alpha})}(b_\varrho) &= v^0_{\alpha(\hat{\alpha})}(b^0_\varrho) + \epsilon\,v^{,\rho}_{\alpha(\hat{\alpha})}(b^0_\varrho)\,\Delta b_\rho + \frac{1}{2}\epsilon^2\,v^{,\rho\varsigma}_{\alpha(\hat{\alpha})}(b^0_\varrho)\,\Delta b_\rho\Delta b_\varsigma \\ \tilde{\lambda}_{(\hat{\alpha})}(b_\varrho) &= \tilde{\lambda}^0_{(\hat{\alpha})}(b^0_\varrho) + \epsilon\,\tilde{\lambda}^{,\rho}_{(\hat{\alpha})}(b^0_\varrho)\,\Delta b_\rho + \frac{1}{2}\epsilon^2\,\tilde{\lambda}^{,\rho\varsigma}_{(\hat{\alpha})}(b^0_\varrho)\,\Delta b_\rho\Delta b_\varsigma\end{aligned} \tag{4.109}$$

As before, substituting Eqs. (4.109) into Eq. (4.99), pre-multiplying the second-order terms in the resulting equations by the $\tilde{N}$-variate PDF $p_{\tilde{N}}(b_1, b_2, \ldots, b_{\tilde{N}})$, integrating them over the domain of b_ρ and equating terms of like power of the small parameter ϵ, we obtain the zeroth-, first- and second-order equations for the stochastic structural buckling problem as follows:

- Zeroth-order (ϵ^0 terms, one system of N equations)

$$\left[\hat{K}^{(\sigma)0}_{\alpha\beta}(b^0_\varrho) - \tilde{\lambda}^0_{(\hat{\alpha})}(b^0_\varrho)\,K^{(e)0}_{\alpha\beta}(b^0_\varrho)\right] v^0_{\beta(\hat{\alpha})}(b^0_\varrho) = 0 \qquad \text{(no sum on } \hat{\alpha}) \tag{4.110}$$

- First-order (ϵ^1 terms, $\tilde{N}$ systems of N equations, $\rho = 1, 2, \ldots, \tilde{N}$)

$$\begin{aligned}&\left[\hat{K}^{(\sigma)0}_{\alpha\beta}(b^0_\varrho) - \tilde{\lambda}^0_{(\hat{\alpha})}(b^0_\varrho)\,K^{(e)0}_{\alpha\beta}(b^0_\varrho)\right] v^{,\rho}_{\beta(\hat{\alpha})}(b^0_\varrho) \\ &\quad = -\left[\hat{K}^{(\sigma),\rho}_{\alpha\beta}(b^0_\varrho) - \tilde{\lambda}^{,\rho}_{(\hat{\alpha})}(b^0_\varrho)K^{(e)0}_{\alpha\beta}(b^0_\varrho) - \tilde{\lambda}^0_{(\hat{\alpha})}(b^0_\varrho)K^{(e),\rho}_{\alpha\beta}(b^0_\varrho)\right] v^0_{\beta(\hat{\alpha})}(b^0_\varrho) \\ &\qquad\qquad \text{(no sum on } \hat{\alpha})\end{aligned} \tag{4.111}$$

- Second-order (ϵ^2 terms, one system of N equations)

$$\begin{aligned}&\left[\hat{K}^{(\sigma)0}_{\alpha\beta}(b^0_\varrho) - \tilde{\lambda}^0_{(\hat{\alpha})}(b^0_\varrho)\,K^{(e)0}_{\alpha\beta}(b^0_\varrho)\right] v^{(2)}_{\beta(\hat{\alpha})}(b^0_\varrho) \\ &= -\Big\{\Big[\hat{K}^{(\sigma),\rho\varsigma}_{\alpha\beta}(b^0_\varrho) - \tilde{\lambda}^{,\rho\varsigma}_{(\hat{\alpha})}(b^0_\varrho)\,K^{(e)0}_{\alpha\beta}(b^0_\varrho) \\ &\quad - 2\,\tilde{\lambda}^{,\rho}_{(\hat{\alpha})}(b^0_\varrho)\,K^{(e),\varsigma}_{\alpha\beta}(b^0_\varrho) - \tilde{\lambda}^0_{(\hat{\alpha})}(b^0_\varrho)\,K^{(e),\rho\varsigma}_{\alpha\beta}(b^0_\varrho)\Big]\,v^0_{\beta(\hat{\alpha})}(b^0_\varrho) \\ &\quad + 2\Big[\hat{K}^{(\sigma),\rho}_{\alpha\beta}(b^0_\varrho) - \tilde{\lambda}^{,\rho}_{(\hat{\alpha})}(b^0_\varrho)K^{(e)0}_{\alpha\beta}(b^0_\varrho) - \tilde{\lambda}^0_{(\hat{\alpha})}(b^0_\varrho)K^{(e),\rho}_{\alpha\beta}(b^0_\varrho)\Big]\,v^{,\varsigma}_{\beta(\hat{\alpha})}(b^0_\varrho)\Big\}\,S^{\rho\varsigma}_{\mathrm{b}} \\ &\qquad\qquad \text{(no sum on } \hat{\alpha})\end{aligned} \tag{4.112}$$

where $S^{\rho\varsigma}_{\mathrm{b}}$ is the covariance matrix of the nodal random variables b_ρ, and

$$v^{(2)}_{\alpha(\hat{\alpha})}(b^0_\varrho) = v^{,\rho\varsigma}_{\alpha(\hat{\alpha})}(b^0_\varrho)\,S^{\rho\varsigma}_{\mathrm{b}} \tag{4.113}$$

We note that the same operator acts on the left-hand side of all the zeroth-, first- and second-order equations (4.110)–(4.112). The 'deterministic' eigenvalue problem described by the zeroth-order equation (4.110) can be solved directly for the zeroth-order eigenvalues and eigenvectors by using any eigensolution procedure. An important point to emphasize here is selecting an optimum algorithm for this problem. Using a technique based on the forward iteration approach to solve Eq. (4.110) for an eigenpair is equivalent to carrying out an inverse iteration procedure on the original eigenproblem equation (2.120). Since the inverse iteration solution to Eq. (2.120) converges to the smallest eigenvalue and corresponding eigenvector, the forward iteration solution to Eq. (4.110) converges to the largest eigenvalue and corresponding eigenvector. The disadvantage of the forward iteration when compared with the inverse iteration is that the state-of-the-art software for the former formulation is rather scarce; in conventional finite element analyses the smallest eigenvalues have mainly been of the analysts' interest. Alternatively, algorithms typical of the inverse iteration procedures for large systems (such as the subspace iteration) may also be employed by imposing on the system a shift larger than the largest system eigenvalue prior to the solution [7].

4.4.2 First- and Second-Order Eigenvalues

Following the similar derivations as for the first and second derivatives of eigenvalues, cf. Section 4.3.2 above, the first- and second-order eigenvalues $\tilde{\lambda}^{,\rho}_{(\hat{\alpha})}$ and $\tilde{\lambda}^{,\rho\varsigma}_{(\hat{\alpha})}$ of the system (4.110)–(4.112) can be obtained with no additional (i.e. other than the zeroth-order) eigensolution required.

Pre-multiplying both sides of the first-order equation (4.111) by the $\hat{\alpha}$-th zeroth-order eigenvector $v^0_{\alpha(\hat{\alpha})}$ transposed we have

$$\begin{aligned} v^0_{\alpha(\hat{\alpha})} \left(\hat{K}^{(\sigma)0}_{\alpha\beta} - \tilde{\lambda}^0_{(\hat{\alpha})} K^{(e)0}_{\alpha\beta} \right) v^{,\rho}_{\beta(\hat{\alpha})} - \tilde{\lambda}^{,\rho}_{(\hat{\alpha})} v^0_{\alpha(\hat{\alpha})} K^{(e)0}_{\alpha\beta} v^0_{\beta(\hat{\alpha})} \\ = - v^0_{\alpha(\hat{\alpha})} \left(\hat{K}^{(\sigma),\rho}_{\alpha\beta} - \tilde{\lambda}^0_{(\hat{\alpha})} K^{(e),\rho}_{\alpha\beta} \right) v^0_{\beta(\hat{\alpha})} \end{aligned} \tag{4.114}$$

Since $K^{(e)0}_{\alpha\beta}$ and $\hat{K}^{(\sigma)0}_{\alpha\beta}$ are symmetric, the first term on the left-hand side of Eq. (4.114) equals zero by the definition of the buckling problem, cf. Eq. (4.110). By employing the orthonormality condition $(4.106)_1$ the second term on the left-hand side of this equation reduces to

$$\tilde{\lambda}^{,\rho}_{(\hat{\alpha})} v^0_{\alpha(\hat{\alpha})} K^{(e)0}_{\alpha\beta} v^0_{\beta(\hat{\alpha})} = \tilde{\lambda}^{,\rho}_{(\hat{\alpha})} \tag{4.115}$$

The expression for the $\hat{\alpha}$-th first-order eigenvalue then reads

$$\tilde{\lambda}^{,\rho}_{(\hat{\alpha})} = v^0_{\alpha(\hat{\alpha})} \left(\hat{K}^{(\sigma),\rho}_{\alpha\beta} - \tilde{\lambda}^0_{(\hat{\alpha})} K^{(e),\rho}_{\alpha\beta} \right) v^0_{\beta(\hat{\alpha})} \tag{4.116}$$

In accordance with the SFEM philosophy we define the $\hat{\alpha}$-th generalized second-order eigenvalue as

$$\tilde{\lambda}^{(2)}_{(\hat{\alpha})}(b^0_{\varrho}) = \tilde{\lambda}^{,\rho\varsigma}_{(\hat{\alpha})}(b^0_{\varrho})\, S^{\rho\varsigma}_{\mathrm{b}} \tag{4.117}$$

Pre-multiplying both sides of the second-order equation (4.112) by the $\hat{\alpha}$-th zeroth-order eigenvector $v^0_{\alpha(\hat{\alpha})}$ transposed yields

$$\begin{aligned} v^0_{\alpha(\hat{\alpha})} \left(\hat{K}^{(\sigma)0}_{\alpha\beta} - \tilde{\lambda}^0_{(\hat{\alpha})} K^{(e)0}_{\alpha\beta} \right) v^{(2)}_{\beta(\hat{\alpha})} - \tilde{\lambda}^{(2)}_{(\hat{\alpha})} v^0_{\alpha(\hat{\alpha})} K^{(e)0}_{\alpha\beta} v^0_{\beta(\hat{\alpha})} \\ = - \left[v^0_{\alpha(\hat{\alpha})} \left(\hat{K}^{(\sigma),\rho\varsigma}_{\alpha\beta} - 2\, \tilde{\lambda}^{,\rho}_{(\hat{\alpha})} K^{(e),\varsigma}_{\alpha\beta} - \tilde{\lambda}^0_{(\hat{\alpha})} K^{(e),\rho\varsigma}_{\alpha\beta} \right) v^0_{\beta(\hat{\alpha})} \right. \\ \left. + 2\, v^0_{\alpha(\hat{\alpha})} \left(\hat{K}^{(\sigma),\rho}_{\alpha\beta} - \tilde{\lambda}^{,\rho}_{(\hat{\alpha})} K^{(e)0}_{\alpha\beta} - \tilde{\lambda}^0_{(\hat{\alpha})} K^{(e),\rho}_{\alpha\beta} \right) v^{,\varsigma}_{\beta(\hat{\alpha})} \right] S^{\rho\varsigma}_{\mathrm{b}} \end{aligned} \tag{4.118}$$

Again, by using the definition of the buckling problem, cf. Eq. (4.110), the first term on the left-hand side of Eq. (4.118) vanishes. The second term on the left-hand side of this equation equals $\tilde{\lambda}^{(2)}_{(\hat{\alpha})}$ because of the orthonormality condition $(4.106)_1$. In other words

$$\begin{aligned} \tilde{\lambda}^{(2)}_{(\hat{\alpha})} = \left[v^0_{\alpha(\hat{\alpha})} \left(\hat{K}^{(\sigma),\rho\varsigma}_{\alpha\beta} - 2\, \tilde{\lambda}^{,\rho}_{(\hat{\alpha})} K^{(e),\varsigma}_{\alpha\beta} - \tilde{\lambda}^0_{(\hat{\alpha})} K^{(e),\rho\varsigma}_{\alpha\beta} \right) v^0_{\beta(\hat{\alpha})} \right. \\ \left. + 2\, v^0_{\alpha(\hat{\alpha})} \left(\hat{K}^{(\sigma),\rho}_{\alpha\beta} - \tilde{\lambda}^{,\rho}_{(\hat{\alpha})} K^{(e)0}_{\alpha\beta} - \tilde{\lambda}^0_{(\hat{\alpha})} K^{(e),\rho}_{\alpha\beta} \right) v^{,\varsigma}_{\beta(\hat{\alpha})} \right] S^{\rho\varsigma}_{\mathrm{b}} \end{aligned} \tag{4.119}$$

The only quantity which remains to be evaluated on the right-hand side of Eq. (4.119) is the $\hat{\alpha}$-th first-order eigenvector $v^{,\rho}_{\alpha(\hat{\alpha})}$. Since the system matrix involved on the left-hand side of Eq. (4.111) is singular of rank $N-1$, this nonhomogeneous equation does not give a unique solution for $\phi^{,\rho}_{\alpha(\hat{\alpha})}$. The computation of $v^{,\rho}_{\alpha(\hat{\alpha})}$ may proceed quite similarly as in Section 4.3.3. We approximate $v^{,\rho}_{\alpha(\hat{\alpha})}$ by a linear function of all the zeroth-order eigenvectors, i.e.

$$v^{,\rho}_{\alpha(\hat{\alpha})} = V^0_{\alpha\beta}\, a^{\rho}_{\beta(\hat{\alpha})} \qquad \alpha, \beta, \hat{\alpha} = 1, 2, \ldots, N \tag{4.120}$$

with $V^0_{\alpha\beta} = [v^0_{\alpha(1)} \;\; v^0_{\alpha(2)} \;\; \cdots \;\; v^0_{\alpha(N)}]$ being the zeroth-order eigenvector matrix. The question is how to determine the components of the coefficient vector $a^{\rho}_{\alpha(\hat{\alpha})}$. By substituting the linear combination (4.120) into Eq. (4.111) and pre-multiplying both sides of the resulting equation by $V^0_{\alpha\beta}$ transposed we get

$$V^{0\mathrm{T}}_{\gamma\alpha} \left(\hat{K}^{(\sigma)0}_{\alpha\beta} - \tilde{\lambda}^0_{(\hat{\alpha})} K^{(e)0}_{\alpha\beta} \right) V^0_{\beta\delta}\, a^{\rho}_{\delta(\hat{\alpha})} = V^{0\mathrm{T}}_{\gamma\alpha} R^{\rho}_{\alpha(\hat{\alpha})} \qquad \text{(no sum on } \hat{\alpha}) \tag{4.121}$$

where

$$R^{\rho}_{\alpha(\hat{\alpha})} = - \left(\hat{K}^{(\sigma),\rho}_{\alpha\beta} - \tilde{\lambda}^{,\rho}_{(\hat{\alpha})} K^{(e)0}_{\alpha\beta} - \tilde{\lambda}^0_{(\hat{\alpha})} K^{(e),\rho}_{\alpha\beta} \right) v^0_{\beta(\hat{\alpha})} \tag{4.122}$$

By employing the $K^{(e)}_{\alpha\beta}$-orthonormality and $\hat{K}^{(\sigma)}_{\alpha\beta}$-orthogonality conditions (4.106), Eq. (4.121) takes the form, cf. Eq. (4.69)

$$\left(\tilde{\Lambda}^0_{(\alpha)} - \tilde{\lambda}^0_{(\hat{\alpha})} I \right) a^{\rho}_{\beta(\hat{\alpha})} = V^{0\mathrm{T}}_{\beta\alpha} R^{\rho}_{\alpha(\hat{\alpha})} \qquad \text{(no sum on } \hat{\alpha}) \tag{4.123}$$

with $\tilde{\Lambda}^0_{(\alpha)}$ being the diagonal matrix of the zeroth-order eigenvalues. Because $(\tilde{\Lambda}^0_{(\alpha)} - \tilde{\lambda}^0_{(\hat{\alpha})} I)$ is a diagonal matrix with the entries $(\tilde{\lambda}^0_{(\alpha)} - \tilde{\lambda}^0_{(\hat{\alpha})})$ at the α-th row, the components of the coefficient vector $a^{\rho}_{\alpha(\hat{\alpha})}$ can be obtained as

$$a^{\rho}_{\alpha(\hat{\alpha})} = \frac{v^0_{\beta(\alpha)} R^{\rho}_{\beta(\hat{\alpha})}}{\tilde{\lambda}^0_{(\alpha)} - \tilde{\lambda}^0_{(\hat{\alpha})}} \qquad \text{for} \quad \alpha \neq \hat{\alpha} \tag{4.124}$$

with an arbitrary value assigned to the $\hat{\alpha}$-th component $a^{\rho}_{\hat{\alpha}(\hat{\alpha})}$.

To obtain the unique solution for $a^{\rho}_{\hat{\alpha}(\hat{\alpha})}$ the $K^{(e)}_{\alpha\beta}$-orthonormality condition $(4.106)_1$ has to be imposed. Differentiation of Eq. $(4.106)_1$ gives, cf. Eq. (4.120)

$$v^0_{\alpha(\hat{\alpha})}\, K^{(e),\rho}_{\alpha\beta}\, v^0_{\beta(\hat{\beta})} + 2\, v^0_{\alpha(\hat{\alpha})}\, K^{(e)0}_{\alpha\beta}\, V^0_{\beta\gamma}\, a^{\rho}_{\gamma(\hat{\alpha})} = 0 \qquad \text{(no sum on } \hat{\alpha}\text{)} \qquad (4.125)$$

By noting in turn that

$$v^0_{\alpha(\hat{\alpha})}\, K^{(e)0}_{\alpha\beta}\, V^0_{\beta\gamma}\, a^{\rho}_{\gamma(\hat{\alpha})} = a^{\rho}_{\hat{\alpha}(\hat{\alpha})} \qquad \text{(no sum on } \hat{\alpha}\text{)} \qquad (4.126)$$

Eq. (4.125) can be rewritten as

$$a^{\rho}_{\hat{\alpha}(\hat{\alpha})} = -\frac{1}{2}\, v^0_{\alpha(\hat{\alpha})}\, K^{(e),\rho}_{\alpha\beta}\, v^0_{\beta(\hat{\alpha})} \qquad \text{(no sum on } \hat{\alpha}\text{)} \qquad (4.127)$$

which, combined with Eq. (4.124), completes the solution for the $\hat{\alpha}$-th entry in the coefficient vector $a^{\rho}_{\beta(\hat{\alpha})}$. We compactly write

$$a^{\rho}_{\alpha(\hat{\alpha})} = \begin{cases} \dfrac{v^0_{\beta(\alpha)}\, R^{\rho}_{\beta(\hat{\alpha})}}{\tilde{\lambda}^0_{(\alpha)} - \tilde{\lambda}^0_{(\hat{\alpha})}} & \text{for} \quad \alpha \neq \hat{\alpha} \\ -\dfrac{1}{2}\, v^0_{\alpha(\hat{\alpha})}\, K^{(e),\rho}_{\alpha\beta}\, v^0_{\beta(\hat{\alpha})} & \text{for} \quad \alpha = \hat{\alpha} \end{cases} \qquad (4.128)$$

Substituting Eq. (4.128) into Eq. (4.120) gives the $\hat{\alpha}$-th first-order eigenvector $v^{,\rho}_{\alpha(\hat{\alpha})}$ which, when introduced in Eq. (4.119), yields the $\hat{\alpha}$-th generalized second-order eigenvalue for the buckling problem (4.99).

We recall that the index $\hat{\alpha}$ in the above equations will typically single out the eigenvalue $\tilde{\lambda}$ with the greatest mean (i.e. the one corresponding to the smallest buckling load mean); in some structural problems knowing other (higher) buckling loads may turn out useful as well.

4.4.3 First Two Moments for Buckling Load Factors

When $\tilde{\lambda}^0_{(\hat{\alpha})}$, $\tilde{\lambda}^{,\rho}_{(\hat{\alpha})}$ and $\tilde{\lambda}^{,\rho\varsigma}_{(\hat{\alpha})}$ are known, the probabilistic distributions for the buckling load factors $\lambda_{(\hat{\alpha})}$ (being an implicit function of the nodal random variables b_ρ) can be expressed explicitly with respect to the first two statistical moments of b_ρ. The formal solution is obtained by using the second-order expansions for $\lambda_{(\hat{\alpha})}(b_\rho)$ and $\tilde{\lambda}_{(\hat{\alpha})}(b_\rho)$ about the spatial expectations b^0_ρ in the form

$$\begin{aligned} \lambda_{(\hat{\alpha})}(b_\varrho) &= \lambda^0_{(\hat{\alpha})}(b^0_\varrho) + \lambda^{,\rho}_{(\hat{\alpha})}(b^0_\varrho)\, \Delta b_\rho + \frac{1}{2}\, \lambda^{,\rho\varsigma}_{(\hat{\alpha})}(b^0_\varrho)\, \Delta b_\rho \Delta b_\varsigma \\ \tilde{\lambda}_{(\hat{\alpha})}(b_\varrho) &= \tilde{\lambda}^0_{(\hat{\alpha})}(b^0_\varrho) + \tilde{\lambda}^{,\rho}_{(\hat{\alpha})}(b^0_\varrho)\, \Delta b_\rho + \frac{1}{2}\, \tilde{\lambda}^{,\rho\varsigma}_{(\hat{\alpha})}(b^0_\varrho)\, \Delta b_\rho \Delta b_\varsigma \end{aligned} \qquad (4.129)$$

As $\lambda_{(\hat{\alpha})} = 1/\tilde{\lambda}_{(\hat{\alpha})}$, Eq. (4.100), and all the functions involved on the right-hand side of Eqs. (4.129) are evaluated at the expectations b^0_ρ, we can express the zeroth, first and second partial derivatives of the buckling load factors, i.e. $\lambda^0_{(\hat{\alpha})}$,

$\lambda_{(\hat{\alpha})}^{,\rho}$ and $\lambda_{(\hat{\alpha})}^{,\rho\varsigma}$ in terms of the partial derivatives of $\tilde{\lambda}_{(\hat{\alpha})}$ in the system (4.99) as

$$\lambda_{(\hat{\alpha})}^0 = \frac{1}{\tilde{\lambda}_{(\hat{\alpha})}^0}$$

$$\lambda_{(\hat{\alpha})}^{,\rho} = -\frac{1}{(\tilde{\lambda}_{(\hat{\alpha})}^0)^2}\, \tilde{\lambda}_{(\hat{\alpha})}^{,\rho} \qquad\qquad (4.130)$$

$$\lambda_{(\hat{\alpha})}^{,\rho\varsigma} = -\frac{1}{(\tilde{\lambda}_{(\hat{\alpha})}^0)^3}\left(\tilde{\lambda}_{(\hat{\alpha})}^0\, \tilde{\lambda}_{(\hat{\alpha})}^{,\rho\varsigma} - 2\,\tilde{\lambda}_{(\hat{\alpha})}^{,\rho}\, \tilde{\lambda}_{(\hat{\alpha})}^{,\varsigma}\right)$$

Substituting Eqs. (4.130) into Eq. (4.129)$_1$ and using the definitions

$$E[\lambda_{(\hat{\alpha})}] = \underbrace{\int_{-\infty}^{+\infty}\int_{-\infty}^{+\infty}\cdots\int_{-\infty}^{+\infty}}_{\tilde{N}\text{-fold}} \lambda_{(\hat{\alpha})}\, p_{\tilde{N}}(b_1, b_2, \ldots, b_{\tilde{N}})\, db_1 db_2 \ldots db_{\tilde{N}} \qquad (4.131)$$

$$\begin{aligned}\text{Cov}(\lambda_{(\hat{\alpha})}, \lambda_{(\hat{\beta})}) &= S_\lambda^{\hat{\alpha}\hat{\beta}} \\ &= \underbrace{\int_{-\infty}^{+\infty}\int_{-\infty}^{+\infty}\cdots\int_{-\infty}^{+\infty}}_{\tilde{N}\text{-fold}}\left\{\lambda_{(\hat{\alpha})} - E[\lambda_{(\hat{\alpha})}]\right\}\left\{\lambda_{(\hat{\beta})} - E[\lambda_{(\hat{\beta})}]\right\} \\ &\qquad\qquad \times\, p_{\tilde{N}}(b_1, b_2, \ldots, b_{\tilde{N}})\, db_1 db_2 \ldots db_{\tilde{N}} \qquad (4.132)\end{aligned}$$

yields the second-order accurate expectation for the $\hat{\alpha}$-th buckling load factor as

$$E[\lambda_{(\hat{\alpha})}] = \frac{1}{\tilde{\lambda}_{(\hat{\alpha})}^0}\left[1 + \frac{\tilde{\lambda}_{(\hat{\alpha})}^{,\rho}\, \tilde{\lambda}_{(\hat{\alpha})}^{,\varsigma}\, S_b^{\rho\varsigma}}{(\tilde{\lambda}_{(\hat{\alpha})}^0)^2} - \frac{\tilde{\lambda}_{(\hat{\alpha})}^{(2)}}{2\,\tilde{\lambda}_{(\hat{\alpha})}^0}\right] \qquad (4.133)$$

and the first-order accurate cross-covariance for the $\hat{\alpha}$-th and $\hat{\beta}$-th buckling load factors as

$$\text{Cov}(\lambda_{(\hat{\alpha})}, \lambda_{(\hat{\beta})}) = \frac{\tilde{\lambda}_{(\hat{\alpha})}^{,\rho}\, \tilde{\lambda}_{(\hat{\beta})}^{,\varsigma}\, S_b^{\rho\varsigma}}{(\tilde{\lambda}_{(\hat{\alpha})}^0\, \tilde{\lambda}_{(\hat{\beta})}^0)^2} \qquad\qquad \hat{\alpha}, \hat{\beta} = 1, 2, \ldots, N \qquad (4.134)$$

4.4.4 First Two Moments for Buckling Loads

So far, only the probabilistic characteristics of the buckling load factors have been of interest and the reference load vector Q_α^* has not been considered (cf. Eq. (2.116) for the buckling problem of deterministic systems). We shall now describe procedures for calculation of bifurcation (buckling) loads assuming that the first two moments of the buckling load factors have already been found using the method presented in Section 4.4.3. We shall consider two separate cases of the applied load pattern: (i) a reference nodal load given deterministically, and (ii) a reference nodal load described stochastically as a function of the nodal random variables b_ρ, i.e. their components defined via the spatial expectations and covariances of b_ρ. Clearly, in both the cases the bifurcation loads obtained in the analysis are random field variables since the buckling load factors are inherently random.

For the one-parameter deterministic reference load, the $\hat{\alpha}$-th bifurcation load vector $Q_{\alpha(\hat{\alpha})}$ can be written as a linear function of the corresponding random variable $\lambda_{(\hat{\alpha})}$, i.e.

$$Q_{\alpha(\hat{\alpha})}(b_\varrho) = \lambda_{(\hat{\alpha})}(b_\varrho)\, Q^*_\alpha \tag{4.135}$$

Thus, the linear transformation of random variables introduced in Section 1.1.3 may be directly employed; we readily obtain the second-order accurate expectation for the $\hat{\alpha}$-th bifurcation load vector as

$$E[Q_{\alpha(\hat{\alpha})}] = E[\lambda_{(\hat{\alpha})}]\, Q^*_\alpha = \frac{1}{\tilde{\lambda}^0_{(\hat{\alpha})}} \left[1 + \frac{\tilde{\lambda}^{,\rho}_{(\hat{\alpha})}\, \tilde{\lambda}^{,\varsigma}_{(\hat{\alpha})}\, S^{\rho\varsigma}_{\mathrm{b}}}{(\tilde{\lambda}^0_{(\hat{\alpha})})^2} - \frac{\tilde{\lambda}^{(2)}_{(\hat{\alpha})}}{2\,\tilde{\lambda}^0_{(\hat{\alpha})}} \right] Q^*_\alpha \tag{4.136}$$

and the first-order accurate cross-covariance for the α-th component of the $\hat{\alpha}$-th bifurcation load vector and the β-th component of the $\hat{\beta}$-th one as

$$\begin{aligned} \mathrm{Cov}(Q_{\alpha(\hat{\alpha})}, Q_{\beta(\hat{\beta})}) &= \mathrm{Cov}(\lambda_{(\hat{\alpha})}, \lambda_{(\hat{\beta})})\, Q^*_\alpha\, Q^*_\beta \\ &= \frac{\tilde{\lambda}^{,\rho}_{(\hat{\alpha})}\, \tilde{\lambda}^{,\varsigma}_{(\hat{\beta})}\, S^{\rho\varsigma}_{\mathrm{b}}}{(\tilde{\lambda}^0_{(\hat{\alpha})}\, \tilde{\lambda}^0_{(\hat{\beta})})^2} Q^*_\alpha\, Q^*_\beta \\ &\qquad \alpha, \beta, \hat{\alpha}, \hat{\beta} = 1, 2, \ldots, N \end{aligned} \tag{4.137}$$

If we are interested in the smallest buckling load factor $\lambda_{(1)} = \lambda_{(\mathrm{cr})}$ and the corresponding buckling load $Q_{\alpha(1)} = Q_{\alpha(\mathrm{cr})}$ only, Eqs. (4.136) and (4.137) reduce to

$$E[Q_{\alpha(1)}] = \frac{1}{\tilde{\lambda}^0_{(1)}} \left[1 + \frac{\tilde{\lambda}^{,\rho}_{(1)}\, \tilde{\lambda}^{,\varsigma}_{(1)}\, S^{\rho\varsigma}_{\mathrm{b}}}{(\tilde{\lambda}^0_{(1)})^2} - \frac{\tilde{\lambda}^{(2)}_{(1)}}{2\,\tilde{\lambda}^0_{(1)}} \right] Q^*_\alpha \tag{4.138}$$

and

$$\mathrm{Cov}(Q_{\alpha(1)}, Q_{\beta(1)}) = \frac{\tilde{\lambda}^{,\rho}_{(1)}\, \tilde{\lambda}^{,\varsigma}_{(1)}\, S^{\rho\varsigma}_{\mathrm{b}}}{(\tilde{\lambda}^0_{(1)})^4}\, Q^*_\alpha\, Q^*_\beta \tag{4.139}$$

In the case of the stochastic reference load, the $\hat{\alpha}$-th bifurcation load vector $Q_{\alpha(\hat{\alpha})}$ can be expressed as[1]

$$Q_{\alpha(\hat{\alpha})}(b_\varrho) = \lambda_{(\hat{\alpha})}(b_\varrho)\, Q^*_\alpha(b_\varrho) \tag{4.140}$$

Using the second-order expansions for both $\lambda_{(\hat{\alpha})}$ (Eq. (4.129)$_1$) and Q^*_α

$$Q^*_\alpha(b_\varrho) = Q^{*0}_\alpha(b^0_\varrho) + Q^{*,\rho}_\alpha(b^0_\varrho)\, \Delta b_\rho + \tfrac{1}{2}\, Q^{*,\rho\varsigma}_\alpha(b^0_\varrho)\, \Delta b_\rho \Delta b_\varsigma \tag{4.141}$$

Eq. (4.140) becomes

$$\begin{aligned} Q_{\alpha(\hat{\alpha})} &= \lambda^0_{(\hat{\alpha})}\, Q^{*0}_\alpha + \left(\lambda^{,\rho}_{(\hat{\alpha})}\, Q^{*0}_\alpha + \lambda^0_{(\hat{\alpha})}\, Q^{*,\rho}_\alpha \right) \Delta b_\rho \\ &\quad + \tfrac{1}{2} \left(\lambda^{,\rho\varsigma}_{(\hat{\alpha})}\, Q^{*0}_\alpha + 2\, \lambda^{,\rho}_{(\hat{\alpha})}\, Q^{*,\varsigma}_\alpha + \lambda^0_{(\hat{\alpha})}\, Q^{*,\rho\varsigma}_\alpha \right) \Delta b_\rho \Delta b_\varsigma \end{aligned} \tag{4.142}$$

---

[1] The vector of nodal random variables $\mathbf{b}^{(\mathrm{system})} = \{b_\rho\}$ has to be extended prior to the solution to cover the load uncertainties, i.e. $\mathbf{b}^{(\mathrm{system})} = \{\mathbf{b}^{(\mathrm{structure})}, \mathbf{b}^{(\mathrm{reference\ load})}\}$.

Introducing Eq. (4.142) into the definition of the expectation and cross-covariance for the bifurcation load

$$E[Q_{\alpha(\hat{\alpha})}] = \underbrace{\int_{-\infty}^{+\infty}\int_{-\infty}^{+\infty}\cdots\int_{-\infty}^{+\infty}}_{\tilde{N}\text{-fold}} Q_{\alpha(\hat{\alpha})}\, p_{\tilde{N}}(b_1, b_2, \ldots, b_{\tilde{N}})\, \mathrm{d}b_1 \mathrm{d}b_2 \ldots \mathrm{d}b_{\tilde{N}} \tag{4.143}$$

$$\begin{aligned}\mathrm{Cov}(Q_{\alpha(\hat{\alpha})}, Q_{\beta(\hat{\beta})}) &= S_{\mathrm{Q}}^{\alpha\hat{\alpha}\beta\hat{\beta}} \\ &= \underbrace{\int_{-\infty}^{+\infty}\int_{-\infty}^{+\infty}\cdots\int_{-\infty}^{+\infty}}_{\tilde{N}\text{-fold}} \left\{Q_{\alpha(\hat{\alpha})} - E[Q_{\alpha(\hat{\alpha})}]\right\}\left\{Q_{\beta(\hat{\beta})} - E[Q_{\beta(\hat{\beta})}]\right\} \\ &\qquad \times\, p_{\tilde{N}}(b_1, b_2, \ldots, b_{\tilde{N}})\, \mathrm{d}b_1 \mathrm{d}b_2 \ldots \mathrm{d}b_{\tilde{N}}\end{aligned} \tag{4.144}$$

and using Eqs. (4.130) yields

$$\begin{aligned}E[Q_{\alpha(\hat{\alpha})}] &= \frac{1}{\tilde{\lambda}^0_{(\hat{\alpha})}} Q^{*0}_\alpha + \frac{1}{2\tilde{\lambda}^0_{(\hat{\alpha})}} \left[-\frac{\tilde{\lambda}^{(2)}_{(\hat{\alpha})}}{\tilde{\lambda}^0_{(\hat{\alpha})}} Q^{*0}_\alpha \right. \\ &\quad \left. + \frac{2\tilde{\lambda}^{,\rho}_{(\hat{\alpha})}}{\tilde{\lambda}^0_{(\hat{\alpha})}} \left(\frac{\tilde{\lambda}^{,\varsigma}_{(\hat{\alpha})}}{\tilde{\lambda}^0_{(\hat{\alpha})}} Q^{*0}_\alpha - Q^{*,\varsigma}_\alpha \right) S^{\rho\varsigma}_{\mathrm{b}} + Q^{*(2)}_\alpha \right]\end{aligned} \tag{4.145}$$

$$\begin{aligned}\mathrm{Cov}(Q_{\alpha(\hat{\alpha})}, Q_{\beta(\hat{\beta})}) &= \frac{1}{\tilde{\lambda}^0_{(\hat{\alpha})}\tilde{\lambda}^0_{(\hat{\beta})}} \left(\frac{\tilde{\lambda}^{,\rho}_{(\hat{\alpha})}\tilde{\lambda}^{,\varsigma}_{(\hat{\beta})}}{\tilde{\lambda}^0_{(\hat{\alpha})}\tilde{\lambda}^0_{(\hat{\beta})}} Q^{*0}_\alpha Q^{*0}_\beta + \frac{\tilde{\lambda}^{,\rho}_{(\hat{\alpha})}}{\tilde{\lambda}^0_{(\hat{\alpha})}} Q^{*0}_\alpha Q^{*,\varsigma}_\beta \right. \\ &\quad \left. + \frac{\tilde{\lambda}^{,\varsigma}_{(\hat{\beta})}}{\tilde{\lambda}^0_{(\hat{\beta})}} Q^{*,\rho}_\alpha Q^{*0}_\beta + Q^{*,\rho}_\alpha Q^{*,\varsigma}_\beta \right) S^{\rho\varsigma}_{\mathrm{b}}\end{aligned} \tag{4.146}$$

with $\alpha, \beta, \hat{\alpha}, \hat{\beta} = 1, 2, \ldots, N$, $\tilde{\lambda}^{(2)}_{(\hat{\alpha})}$ defined by Eq. (4.117) and $Q^{*(2)}_\alpha = Q^{*,\rho\varsigma}_\alpha S^{\rho\varsigma}_{\mathrm{b}}$.

Specifically, for the case of the smallest buckling load factor $\lambda_{(1)}$ and the associated buckling load $Q_{\alpha(1)}$, Eqs. (4.145) and (4.146) reduce to

$$\begin{aligned}E[Q_{\alpha(1)}] &= \frac{1}{\tilde{\lambda}^0_{(1)}} Q^{*0}_\alpha + \frac{1}{2\tilde{\lambda}^0_{(1)}} \left[-\frac{\tilde{\lambda}^{(2)}_{(1)}}{\tilde{\lambda}^0_{(1)}} Q^{*0}_\alpha \right. \\ &\quad \left. + \frac{2}{\tilde{\lambda}^0_{(1)}} \left(\frac{\tilde{\lambda}^{,\rho}_{(1)}\tilde{\lambda}^{,\varsigma}_{(1)}}{\tilde{\lambda}^0_{(1)}} Q^{*0}_\alpha - \tilde{\lambda}^{,\rho}_{(1)} Q^{*,\varsigma}_\alpha \right) S^{\rho\varsigma}_{\mathrm{b}} + Q^{*(2)}_\alpha \right]\end{aligned} \tag{4.147}$$

$$\begin{aligned}\mathrm{Cov}(Q_{\alpha(1)}, Q_{\beta(1)}) &= \frac{1}{(\tilde{\lambda}^0_{(1)})^2} \left[\frac{1}{\tilde{\lambda}^0_{(1)}} \left(\frac{\tilde{\lambda}^{,\rho}_{(1)}\tilde{\lambda}^{,\varsigma}_{(1)}}{\tilde{\lambda}^0_{(1)}} Q^{*0}_\alpha Q^{*0}_\beta + \tilde{\lambda}^{,\rho}_{(1)} Q^{*0}_\alpha Q^{*,\varsigma}_\beta \right. \right. \\ &\quad \left. \left. + \tilde{\lambda}^{,\varsigma}_{(1)} Q^{*,\rho}_\alpha Q^{*0}_\beta \right) + Q^{*,\rho}_\alpha Q^{*,\varsigma}_\beta \right] S^{\rho\varsigma}_{\mathrm{b}}\end{aligned} \tag{4.148}$$

It is pointed out in closing that statistical information on the eigenvectors $v_{\alpha(\hat{\alpha})}$ in the problem (4.99) may be required if one is interested in establishing the probabilistic distributions for postbuckling shapes of the structural system. In this case the first two moments for the components of the bifurcation modes can be found by following a procedure identical to that developed in Section 4.3.3 for the first and second mixed derivatives of eigenvectors in the stochastic generalized eigenproblem.

4.5 Computational Aspects

4.5.1 Notes on Solution Process

In the perturbational stochastic finite element analysis only the first two moments of random variables need to be known, whereas statistical techniques such as the Monte Carlo simulation generally require knowledge of probability density functions that are usually not available in practice. The Monte Carlo technique requires a large number of samples randomly generated from a suitable normal approximation of the probabilistic distributions. In SFEM, cf. Eqs. (4.17)–(4.19), we have to deal with one system of the zeroth-order equations, one system of the first-order equations for each of the random variables, and one system of the second-order equations. The number of linear system solutions required is thus equal to the number of nodal random variables plus two. Moreover, this approach does not restrict the analysis to some limits of random fields as in the statistical methods; it is applicable to both the homogeneous and inhomogeneous random fields and a normal approximation is not necessary. Although this nonstatistical technique is most effectively employed when the fluctuations of the random field variables are small, it performs quite well even when the structural response shows a coefficient of variation of up to, say, 0.15.

The zeroth-order mass, damping, stiffness matrices and load vector involved on the left-hand side of Eqs. (4.17)–(4.19), i.e. $M^0_{\alpha\beta}$, $C^0_{\alpha\beta}$, $K^0_{\alpha\beta}$ and Q^0_α, respectively, are those of the corresponding deterministic system and can be established by the conventional finite element methodology. The first and second mixed partial derivatives of $M_{\alpha\beta}$, $C_{\alpha\beta}$, $K_{\alpha\beta}$ and Q_α with respect to nodal random variables can be computed exactly if these matrices and vectors are expressible in a closed form in terms of the nodal random variables. In some FEM-based codes $M_{\alpha\beta}$, $C_{\alpha\beta}$, $K_{\alpha\beta}$ and Q_α may be implicitly generated; their derivatives can then be approximately calculated by using alternatively the perturbation technique, finite differences or the least squares fit method.

After the random field variables have been expressed in terms of their nodal values, the global mass, damping, stiffness matrices and load vectors of the structural system have been generated and their first and second partial derivatives with respect to nodal random variables have been evaluated, the formal solution to Eqs. (4.17)–(4.19) at each time step proceeds as follows:

(a) Equations (4.17) are solved for the zeroth-order generalized nodal accelerations $\ddot{q}^0_\alpha$, velocities $\dot{q}^0_\alpha$ and displacements q^0_α.

(b) The first-order load vector on the right-hand side of Eq. (4.18) is formed and the first-order equations are solved for $\ddot{q}^{,\rho}_\alpha$, $\dot{q}^{,\rho}_\alpha$ and $q^{,\rho}_\alpha$.

(c) The second-order load vector on the right-hand side of Eq. (4.19) is formed and the second-order equation is solved for $\ddot{q}^{(2)}_\alpha$, $\dot{q}^{(2)}_\alpha$ and $q^{(2)}_\alpha$.

(d) The expectations and cross-covariances (autocovariances) for the state variables are evaluated. (If the cross-covariances are required, the zeroth-, first- and second-order accelerations, velocities and displacements obtained at the corresponding time instants have to be saved.)

The above process of successive computations is typical of the perturbation technique used for solving systems of linear ordinary differential equations; the attractive feature of the algorithm in the FEM context is that it can be programmed 'in parallel'. To be specific, let us first observe that the same operator acts on the left-hand side of all the equations, which are to be integrated with respect to time by using a step-by-step algorithm. Assume that the equations of motion have been converted to the effective algebraic equations and the coefficient matrix was decomposed. At any time instant $\tau = t$ the zeroth-order load at t (input data), the first-order load at $t - \Delta t$ (computed) and the second-order load at $t - 2\Delta t$ (computed) are known. The forward reduction and backward substitution carried out simultaneously for all the three effective load vectors will give the corresponding solutions at three time instants differing by a backward shift Δt. Since the number of time steps required in stochastic dynamics is frequently large, the solution process being successive in nature can be computationally treated as a parallel process with shifting in phase for the higher-order solutions.

The computational effort may be considerably reduced if we note the zeroth-order mass, damping and stiffness matrices (and consequently, the effective coefficient matrix) have to be assembled (triangularized) only once prior to the solution. In addition, almost all operations required to calculate the right-hand side of the equation system and the first two moments can be performed by vector multiplications and at the element level.

Nonetheless, the computations required to solve equations and to calculate the first two moments for the state variables may be quite expensive owing to the appearance of the double sums of the type $(.)^{\rho\sigma} S_{\mathrm{b}}^{\rho\sigma}$. In particular, if the direct step-by-step integration schemes are used in the analysis of medium- or large-scale systems with a large number of time intervals required, the computational cost would be unacceptably expensive.

In order to solve the problem much more efficiently an approach based on the *random variable transformation* (or *standard normal transformation*) and the so-called *two-fold superposition technique* may be used. The main ideas behind the technique are:

- transformation of the system with correlated nodal random variables to a system with uncorrelated nodal random variables through a standard eigenproblem,
- transformation of the coupled system of equations of motion to an uncoupled system by a generalized eigenproblem,
- superposition of the resulting uncorrelated and modal responses to obtain complete solution.

The transformation of a set of correlated random variables to a set of uncorrelated ones seems to be closely related to dynamic problems, because of the additional eigensolution required. It turns out that this technique may be employed effectively for the stochastic analysis of structural statics as well.

Another aspect of the problem which should not be overlooked is that the application of the perturbation technique can produce invalid solutions owing to the appearance of *secular terms* which grow indefinitely with time. The secularity effects can be eliminated effectively by using the fast Fourier transform.

Furthermore, an alternative approach to the stochastic finite element analysis known as the *stochastic adjoint analysis* sometimes proves more effective than the direct approach.

All these aspects will be discussed below in more detail.

4.5.2 Random Variable Transformation

4.5.2.1 Transformed Uncorrelated Random Variables

As pointed out above, the formal solution to Eqs. (4.17)–(4.19) requires integration of $\tilde{N}+2$ systems of linear ordinary differential equations and evaluation of the first two moments of the nodal displacements, element strains and stresses. Since the second-order terms involved in the above expressions include the double sums of the form $(.)^{,\rho\sigma} S_{\mathrm{b}}^{\rho\sigma}$, the number of vector and matrix operations is proportional to $\tilde{N}(\tilde{N}+1)/2$. The question now is how to change the base of the system with correlated nodal random variables to an equivalent system in the corresponding standard normal space with uncorrelated (normalized) nodal random variables only. If this can be done the mixed derivatives will vanish and the full matrices of the second mixed derivatives with respect to nodal random variables will take the diagonal form, the entries being the second derivatives. The double sums will reduce then to single sums so that the number of matrix multiplications will be reduced to $\mathrm{O}(\tilde{N})$.

The random variable transformation can be performed through the standard eigenproblem [80]. We transform the set of correlated random variables b_ρ to a set of uncorrelated random variables c_ρ as, cf. Eq. (4.102)

$$\mathrm{Cov}(b_\rho, b_\sigma)\,\psi_{\sigma\varsigma} = S_{\mathrm{b}}^{\rho\sigma}\,\psi_{\sigma\varsigma} = \Theta_{(\rho)}\,\psi_{\rho\varsigma} \qquad \text{(no sum on } \rho) \tag{4.149}$$

where the covariance matrix $S_{\mathrm{b}}^{\rho\sigma}$ is assumed to be positive definite, $\Theta_{(\rho)}$ is an $\tilde{N}$-dimensional diagonal matrix of the variances for c_ρ and $\psi_{\rho\sigma}$ is an $\tilde{N}\times\tilde{N}$ orthonormal fundamental matrix, i.e.

$$\psi^{\mathrm{T}}_{\sigma\rho}\,\psi_{\rho\varsigma} = \delta_{\sigma\varsigma} \tag{4.150}$$

Having computed the solution to the eigenproblem (4.149) the uncorrelated nodal random variables c_ρ, their expectations and variances and derivatives of functions of c_ρ can be expressed in terms of b_ρ by using the superposition technique.

In practice, only a small number of the dominant normal modes, say $\check{N}$, $\check{N} \ll \tilde{N}$, is needed to simulate the major characteristics of many probabilistic distributions of the discretized random field b_ρ with an acceptable accuracy. By the properties of the correlation (covariance) matrix the *highest* standard normal modes (*largest* eigenvalues) are dominant,[1] in contrast to the modal structural analysis in which the dominant modes are the lowest eigenpairs. Thus, we can write the relationships between $c_{\check{\rho}}$ and b_ρ as

$$\left.\begin{aligned} c_{\check{\rho}} &= \psi_{\rho(\check{\rho})}\, b_\rho \\ E[c_{\check{\rho}}] &= \psi_{\rho(\check{\rho})}\, E[b_\rho] = \psi_{\rho(\check{\rho})}\, b_\rho^0 \\ \mathrm{Var}(c_{\check{\rho}}) &= \theta_{(\check{\rho})} \end{aligned}\right\} \quad \begin{aligned} &\sigma = 1, 2, \ldots, \tilde{N} \\ &\check{\rho} = 1, 2, \ldots, \check{N} \ll \tilde{N} \end{aligned} \tag{4.151}$$

Expressions for the first and second derivatives of any function (.) with respect to $c_{\check{\rho}}$ are written in terms of b_ρ as

$$\begin{aligned} \frac{\partial(.)}{\partial c_{\check{\rho}}} &= \psi_{\rho(\check{\rho})} \frac{\partial(.)}{\partial b_\rho} \\ \frac{\partial^2(.)}{\partial c_{\check{\rho}}^2} &= \psi_{\rho(\check{\rho})} \frac{\partial^2(.)}{\partial b_\rho\, \partial b_\sigma} \psi_{\sigma(\check{\rho})} \qquad \text{(no sum on } \check{\rho}) \end{aligned} \tag{4.152}$$

where

$$\begin{aligned} \psi_{\rho\sigma} &= \begin{bmatrix} \psi_{\rho(1)} & \psi_{\rho(2)} & \cdots & \psi_{\rho(\tilde{N})} \end{bmatrix} \\ \Theta_{(\rho)} &= \mathrm{diag} \begin{bmatrix} \theta_{(1)} & \theta_{(2)} & \cdots & \theta_{(\tilde{N})} \end{bmatrix} \end{aligned} \tag{4.153}$$

while

$$\frac{\partial^2(.)}{\partial b_\rho\, \partial b_\sigma} = \begin{bmatrix} \dfrac{\partial^2(.)}{\partial b_1^2} & \dfrac{\partial^2(.)}{\partial b_1\, \partial b_2} & \cdots & \dfrac{\partial^2(.)}{\partial b_1\, \partial b_{\tilde{N}}} \\ & \dfrac{\partial^2(.)}{\partial b_2^2} & \cdots & \dfrac{\partial^2(.)}{\partial b_2\, \partial b_{\tilde{N}}} \\ \text{symm.} & & \ddots & \vdots \\ & & & \dfrac{\partial^2(.)}{\partial b_{\tilde{N}}^2} \end{bmatrix} \tag{4.154}$$

4.5.2.2 Transformed Finite Element Equations

Stochastic Forced Vibrations and Statics. By substituting Eqs. (4.151) and (4.152) into Eqs. (4.17)–(4.19) we arrive at the following equations for the equivalent uncorrelated system:

[1] If the nodal random variable vector b_ρ is composed of three uncorrelated parts of random load, geometry and material, for instance, several highest modes for each of the three parts of $\mathrm{Cov}(b_\rho, b_\sigma)$ should be extracted and combined to obtain $c_{\check{\rho}}$.

- Zeroth-order (ϵ^0 terms, one system of N linear simultaneous ordinary differential equations for $q^0_\alpha(c^0_{\breve{\varrho}};\tau)$, $\alpha = 1,2,\ldots,N$)

$$M^0_{\alpha\beta}(c^0_{\breve{\varrho}})\,\ddot{q}^0_\beta(c^0_{\breve{\varrho}};\tau) + C^0_{\alpha\beta}(c^0_{\breve{\varrho}})\,\dot{q}^0_\beta(c^0_{\breve{\varrho}};\tau) + K^0_{\alpha\beta}(c^0_{\breve{\varrho}})\,q^0_\beta(c^0_{\breve{\varrho}};\tau) \;=\; Q^0_\alpha(c^0_{\breve{\varrho}};\tau) \tag{4.155}$$

- First-order (ϵ^1 terms, $\check{N}$ systems of N linear simultaneous ordinary differential equations for $q^{,\breve{\rho}}_\alpha(c^0_{\breve{\varrho}};\tau)$, $\breve{\rho} = 1,2,\ldots,\check{N}$, $\alpha = 1,2,\ldots,N$)

$$M^0_{\alpha\beta}(c^0_{\breve{\varrho}})\,\ddot{q}^{,\breve{\rho}}_\beta(c^0_{\breve{\varrho}};\tau) + C^0_{\alpha\beta}(c^0_{\breve{\varrho}})\,\dot{q}^{,\breve{\rho}}_\beta(c^0_{\breve{\varrho}};\tau) + K^0_{\alpha\beta}(c^0_{\breve{\varrho}})\,q^{,\breve{\rho}}_\beta(c^0_{\breve{\varrho}};\tau) \;=\; Q^{\breve{\rho}}_\alpha(c^0_{\breve{\varrho}};\tau) \tag{4.156}$$

- Second-order (ϵ^2 terms, one system of N linear simultaneous ordinary differential equations for $q^{(2)}_\alpha(c^0_{\breve{\varrho}};\tau)$, $\alpha = 1,2,\ldots,N$)

$$M^0_{\alpha\beta}(c^0_{\breve{\varrho}})\,\ddot{q}^{(2)}_\beta(c^0_{\breve{\varrho}};\tau) + C^0_{\alpha\beta}(c^0_{\breve{\varrho}})\,\dot{q}^{(2)}_\beta(c^0_{\breve{\varrho}};\tau) + K^0_{\alpha\beta}(c^0_{\breve{\varrho}})\,q^{(2)}_\beta(c^0_{\breve{\varrho}};\tau) \;=\; Q^{(2)}_\alpha(c^0_{\breve{\varrho}};\tau) \tag{4.157}$$

where the transformed nodal second-order displacements are defined as

$$q^{(2)}_\alpha(c^0_{\breve{\varrho}};\tau) \;=\; \sum_{\breve{\rho}=1}^{\check{N}} q^{,\breve{\rho}\breve{\rho}}_\alpha(c^0_{\breve{\varrho}};\tau)\,S^{\breve{\rho}}_{\mathrm{c}} \tag{4.158}$$

while the transformed first- and second-order load functions are

$$\begin{aligned}
Q^{\breve{\rho}}_\alpha(c^0_{\breve{\varrho}};\tau) \;&=\; Q^{,\breve{\rho}}_\alpha(c^0_{\breve{\varrho}};\tau) \\
&\quad - \Big[M^{,\breve{\rho}}_{\alpha\beta}(c^0_{\breve{\varrho}})\,\ddot{q}^0_\beta(c^0_{\breve{\varrho}};\tau) + C^{,\breve{\rho}}_{\alpha\beta}(c^0_{\breve{\varrho}})\,\dot{q}^0_\beta(c^0_{\breve{\varrho}};\tau) + K^{,\breve{\rho}}_{\alpha\beta}(c^0_{\breve{\varrho}})\,q^0_\beta(c^0_{\breve{\varrho}};\tau)\Big] \\
Q^{(2)}_\alpha(c^0_{\breve{\varrho}};\tau) \;&=\; \sum_{\breve{\rho}=1}^{\check{N}} \Big\{Q^{,\breve{\rho}\breve{\rho}}_\alpha(c^0_{\breve{\varrho}};\tau) \\
&\quad - 2\Big[M^{,\breve{\rho}}_{\alpha\beta}(c^0_{\breve{\varrho}})\,\ddot{q}^{,\breve{\rho}}_\beta(c^0_{\breve{\varrho}};\tau) + C^{,\breve{\rho}}_{\alpha\beta}(c^0_{\breve{\varrho}})\,\dot{q}^{,\breve{\rho}}_\beta(c^0_{\breve{\varrho}};\tau) + K^{,\breve{\rho}}_{\alpha\beta}(c^0_{\breve{\varrho}})\,q^{,\breve{\rho}}_\beta(c^0_{\breve{\varrho}};\tau)\Big] \\
&\quad - \Big[M^{,\breve{\rho}\breve{\rho}}_{\alpha\beta}(c^0_{\breve{\varrho}})\,\ddot{q}^0_\beta(c^0_{\breve{\varrho}};\tau) + C^{,\breve{\rho}\breve{\rho}}_{\alpha\beta}(c^0_{\breve{\varrho}})\,\dot{q}^0_\beta(c^0_{\breve{\varrho}};\tau) + K^{,\breve{\rho}\breve{\rho}}_{\alpha\beta}(c^0_{\breve{\varrho}})\,q^0_\beta(c^0_{\breve{\varrho}};\tau)\Big]\Big\}\,S^{\breve{\rho}}_{\mathrm{c}}
\end{aligned} \tag{4.159}$$

In Eqs. (4.156)–(4.159) $(.)^{,\breve{\rho}}$ and $(.)^{\breve{\rho}\breve{\rho}}$ denote the first and second derivatives with respect to the uncorrelated random variables $c_{\breve{\rho}}$. Thus, by employing the uncorrelated random variables $c^{\breve{\rho}}$ in Eqs. (4.17)–(4.19) the second mixed partial derivatives $(.)^{,\rho\sigma}$ with respect to the correlated random variables b_ρ reduce to the second derivatives $(.)^{\breve{\rho}\breve{\rho}}$ (no sum on $\breve{\rho}$) with respect to the uncorrelated random variables $c_{\breve{\rho}}$. It follows that the double sums over ρ and σ ($\rho,\sigma = 1,2,\ldots,\tilde{N}$) reduce to the single sums over $\breve{\rho}$ ($\breve{\rho} = 1,2,\ldots,\check{N}$).

It is also seen that the solution to Eqs. (4.155)–(4.157) requires only $\check{N} + 2$ integrations of the equations of motion. The value of $\check{N}$ is chosen depending on the probabilistic characteristics of the system considered. In the random variable transformation the eigenvalues (normalized variances) can be interpreted as weighting numbers for the corresponding mode shape necessary to approximate the covariance structure of the random field to a specified accuracy. For a one-dimensional and one-variate random field, for instance, as the correlation length increases from zero to infinity the number of largest eigenvalues needed to describe its uncertainties, i.e. $\check{N}$, decreases from $\tilde{N}$ to 1. In the case of the zero correlation length the random field variables are uncorrelated, all the $\tilde{N}$ eigenvalues are dominant and all the $\tilde{N}$ modes are needed, i.e. $\check{N} = \tilde{N}$. In the latter case the random field variables are perfectly correlated and there exists only one dominant eigenvalue; the only transformed random variable associated with the largest eigenvalue is then sufficient to capture the randomness of the field considered.

In Eqs. (4.155)–(4.157) the zeroth-order mass, damping and stiffness matrices, load vector and their first and second derivatives with respect to nodal uncorrelated random variables $c_{\check{p}}$ can be shown to be, cf. Eqs. (4.22)–(4.33):

- Zeroth-order functions

$$M^0_{\alpha\beta}(c^0_{\check{\varrho}}) = \int_\Omega \varphi_{\bar{\alpha}}\, \varrho^0_{\bar{\alpha}}\, \varphi_{i\alpha}\, \varphi_{i\beta}\, \mathrm{d}\Omega \tag{4.160}$$

$$C^0_{\alpha\beta}(c^0_{\check{\varrho}}) = \int_\Omega \varphi_{\bar{\alpha}}\, \varphi_{\bar{\beta}} \Big(\alpha^0_{\bar{\alpha}}\, \varrho^0_{\bar{\beta}}\, \varphi_{i\alpha}\, \varphi_{i\beta} + \beta^0_{\bar{\alpha}}\, C^0_{ijkl\bar{\beta}}\, B_{ij\alpha} B_{kl\beta} \Big)\, \mathrm{d}\Omega \tag{4.161}$$

$$K^0_{\alpha\beta}(c^0_{\check{\varrho}}) = \int_\Omega \varphi_{\bar{\alpha}}\, C^0_{ijkl\bar{\alpha}}\, B_{ij\alpha} B_{kl\beta}\, \mathrm{d}\Omega \tag{4.162}$$

$$Q^0_\alpha(c^0_{\check{\varrho}};\tau) = \int_\Omega \varphi_{\bar{\alpha}}\, \varphi_{\bar{\beta}}\, \varrho^0_{\bar{\alpha}}\, f^0_{i\bar{\beta}}\, \varphi_{i\alpha}\, \mathrm{d}\Omega + \int_{\partial\Omega_\sigma} \varphi_{\bar{\alpha}}\, \hat{t}^0_{i\bar{\alpha}}\, \varphi_{i\alpha}\, \mathrm{d}(\partial\Omega) \tag{4.163}$$

- First partial derivatives

$$M^{,\check{p}}_{\alpha\beta}(c^0_{\check{\varrho}}) = \int_\Omega \varphi_{\bar{\alpha}}\, \varrho^{,\check{p}}_{\bar{\alpha}}\, \varphi_{i\alpha}\, \varphi_{i\beta}\, \mathrm{d}\Omega \tag{4.164}$$

$$\begin{aligned} C^{,\check{p}}_{\alpha\beta}(c^0_{\check{\varrho}}) = \int_\Omega \varphi_{\bar{\alpha}}\, \varphi_{\bar{\beta}} \Big[\Big(\alpha^{,\check{p}}_{\bar{\alpha}}\, \varrho^0_{\bar{\beta}} + \alpha^0_{\bar{\alpha}}\, \varrho^{,\check{p}}_{\bar{\beta}} \Big) \varphi_{i\alpha}\, \varphi_{i\beta} \\ + \Big(\beta^{,\check{p}}_{\bar{\alpha}}\, C^0_{ijkl\bar{\beta}} + \beta^0_{\bar{\alpha}}\, C^{,\check{p}}_{ijkl\bar{\beta}} \Big) B_{ij\alpha} B_{kl\beta} \Big]\, \mathrm{d}\Omega \end{aligned} \tag{4.165}$$

$$K^{,\check{p}}_{\alpha\beta}(c^0_{\check{\varrho}}) = \int_\Omega \varphi_{\bar{\alpha}}\, C^{,\check{p}}_{ijkl\bar{\alpha}}\, B_{ij\alpha} B_{kl\beta}\, \mathrm{d}\Omega \tag{4.166}$$

$$Q^{,\check{p}}_\alpha(c^0_{\check{\varrho}};\tau) = \int_\Omega \varphi_{\bar{\alpha}}\, \varphi_{\bar{\beta}} \Big(\varrho^{,\check{p}}_{\bar{\alpha}} f^0_{i\bar{\beta}} + \varrho^0_{\bar{\alpha}} f^{,\check{p}}_{i\bar{\beta}} \Big) \varphi_{i\alpha}\, \mathrm{d}\Omega + \int_{\partial\Omega_\sigma} \varphi_{\bar{\alpha}}\, \hat{t}^{,\check{p}}_{i\bar{\alpha}}\, \varphi_{i\alpha}\, \mathrm{d}(\partial\Omega) \tag{4.167}$$

- Second partial derivatives (summation over $\check{p}$ is not implied)

$$M^{,\check{p}\check{p}}_{\alpha\beta}(c^0_{\check{\varrho}}) = \int_\Omega \varphi_{\bar{\alpha}}\, \varrho^{,\check{p}\check{p}}_{\bar{\alpha}}\, \varphi_{i\alpha}\, \varphi_{i\beta}\, \mathrm{d}\Omega \tag{4.168}$$

$$\begin{aligned} C^{,\check{p}\check{p}}_{\alpha\beta}(c^0_{\check{\varrho}}) = \int_\Omega \varphi_{\bar{\alpha}}\, \varphi_{\bar{\beta}} \Big[\Big(\alpha^{,\check{p}\check{p}}_{\bar{\alpha}}\, \varrho^0_{\bar{\beta}} + 2\, \alpha^{,\check{p}}_{\bar{\alpha}}\, \varrho^{,\check{p}}_{\bar{\beta}} + \alpha^0_{\bar{\alpha}}\, \varrho^{,\check{p}\check{p}}_{\bar{\beta}} \Big) \varphi_{i\alpha}\, \varphi_{i\beta} \\ + \Big(\beta^{,\check{p}\check{p}}_{\bar{\alpha}} C^0_{ijkl\bar{\beta}} + 2\beta^{,\check{p}}_{\bar{\alpha}}\, C^{,\check{p}}_{ijkl\bar{\beta}} + \beta^0_{\bar{\alpha}} C^{,\check{p}\check{p}}_{ijkl\bar{\beta}} \Big) B_{ij\alpha} B_{kl\beta} \Big] \mathrm{d}\Omega \end{aligned} \tag{4.169}$$

$$K^{,\check{p}\check{p}}_{\alpha\beta}(c^0_{\check{\varrho}}) = \int_\Omega \varphi_{\bar{\alpha}}\, C^{,\check{p}\check{p}}_{ijkl\bar{\alpha}}\, B_{ij\alpha} B_{kl\beta}\, \mathrm{d}\Omega \tag{4.170}$$

$$\begin{aligned} Q^{,\check{p}\check{p}}_\alpha(c^0_{\check{\varrho}};\tau) = \int_\Omega \varphi_{\bar{\alpha}}\, \varphi_{\bar{\beta}} \Big(\varrho^{,\check{p}\check{p}}_{\bar{\alpha}} f^0_{i\bar{\beta}} + 2\, \varrho^{,\check{p}}_{\bar{\alpha}} f^{,\check{p}}_{i\bar{\beta}} + \varrho^0_{\bar{\alpha}} f^{,\check{p}\check{p}}_{i\bar{\beta}} \Big) \varphi_{i\alpha}\, \mathrm{d}\Omega \\ + \int_{\partial\Omega_\sigma} \varphi_{\bar{\alpha}}\, \hat{t}^{,\check{p}\check{p}}_{i\bar{\alpha}}\, \varphi_{i\alpha}\, \mathrm{d}(\partial\Omega) \end{aligned} \tag{4.171}$$

For the case of structural statics the finite element model of the equilibrium equations for stochastic systems expressed in terms of the uncorrelated random variables $c_{\check{\rho}}$ reads, cf. Eqs. (4.34)–(4.36):

- Zeroth-order (ϵ^0 terms, one system of N linear simultaneous algebraic equations for $q^0_\alpha(c^0_{\check{\varrho}})$, $\alpha = 1, 2, \ldots, N$)

$$K^0_{\alpha\beta}(c^0_{\check{\varrho}})\, q^0_\beta(c^0_{\check{\varrho}}) \;=\; Q^0_\alpha(c^0_{\check{\varrho}}) \tag{4.172}$$

- First-order (ϵ^1 terms, $\check{N}$ systems of N linear simultaneous algebraic equations for $q^{,\check{\rho}}_\alpha(c^0_{\check{\varrho}})$, $\check{\rho} = 1, 2, \ldots, \check{N}$, $\alpha = 1, 2, \ldots, N$)

$$K^0_{\alpha\beta}(c^0_{\check{\varrho}})\, q^{,\check{\rho}}_\beta(c^0_{\check{\varrho}}) \;=\; Q^{,\check{\rho}}_\alpha(c^0_{\check{\varrho}}) - K^{,\check{\rho}}_{\alpha\beta}(c^0_{\check{\varrho}})\, q^0_\beta(c^0_{\check{\varrho}}) \tag{4.173}$$

- Second-order (ϵ^2 terms, one system of N linear simultaneous algebraic equations for $q^{(2)}_\alpha(c^0_{\check{\varrho}})$, $\alpha = 1, 2, \ldots, N$)

$$K^0_{\alpha\beta}(c^0_{\check{\varrho}}) q^{(2)}_\beta(c^0_{\check{\varrho}}) \;=\; Q^{(2)}_\alpha(c^0_{\check{\varrho}}) \tag{4.174}$$

where the vectors of transformed nodal second-order displacements and loads are represented respectively as

$$q^{(2)}_\alpha(c^0_{\check{\varrho}}) \;=\; \sum_{\check{\rho}=1}^{\check{N}} q^{,\check{\rho}\check{\rho}}_\alpha(c^0_{\check{\varrho}})\, S^{\check{\rho}}_{c} \tag{4.175}$$

$$Q^{(2)}_\alpha(c^0_{\check{\varrho}}) \;=\; \sum_{\check{\rho}=1}^{\check{N}} \left[Q^{,\check{\rho}\check{\rho}}_\alpha(c^0_{\check{\varrho}}) - 2K^{,\check{\rho}}_{\alpha\beta}(c^0_{\check{\varrho}})\, q^{,\check{\rho}}_\beta(c^0_{\check{\varrho}}) - K^{,\check{\rho}\check{\rho}}_{\alpha\beta}(c^0_{\check{\varrho}})\, q^0_\beta(c^0_{\check{\varrho}}) \right] S^{\check{\rho}}_{c} \tag{4.176}$$

Stochastic Free Vibrations. The standard normal transformation of random variables may be directly employed to reduce the computational effort in the stochastic analysis of free vibrations as developed in Section 4.3. The procedure for transforming Eqs. (4.51)–(4.53) expressed in terms of the correlated random variables b_ρ into the standard normal space follows similar lines as in the static and dynamic analysis. Thus, we arrive at the transformed zeroth-, first- and second-order equations for the stochastic eigenproblem as follows:

- Zeroth-order (ϵ^0 terms, one system of N equations)

$$\left[K^0_{\alpha\beta}(c^0_{\check{\varrho}}) - \Omega^0_{(\alpha)}(c^0_{\check{\varrho}})\, M^0_{\alpha\beta}(c^0_{\check{\varrho}}) \right] \phi^0_{\beta\gamma}(c^0_{\check{\varrho}}) \;=\; 0 \tag{4.177}$$

(no sum on α)

- First-order (ϵ^1 terms, $\check{N}$ systems of N equations, $\check{\rho} = 1, 2, \ldots, \check{N}$)

$$\left[K^0_{\alpha\beta}(c^0_{\check{\varrho}}) - \Omega^0_{(\alpha)}(c^0_{\check{\varrho}})\, M^0_{\alpha\beta}(c^0_{\check{\varrho}}) \right] \phi^{,\check{\rho}}_{\beta\gamma}(c^0_{\check{\varrho}}) \;=\; R^{\check{\rho}}_\alpha(c^0_{\check{\varrho}}) \tag{4.178}$$

(no sum on α)

- Second-order (ϵ^2 terms, one system of N equations)

$$\left[K^0_{\alpha\beta}(c^0_{\check{\varrho}}) - \Omega^0_{(\alpha)}(c^0_{\check{\varrho}})\, M^0_{\alpha\beta}(c^0_{\check{\varrho}}) \right] \phi^{(2)}_{\beta\gamma}(c^0_{\check{\varrho}}) \;=\; R^{(2)}_\alpha(c^0_{\check{\varrho}}) \tag{4.179}$$

(no sum on α)

where the generalized second derivatives of the eigenvectors with respect to the uncorrelated variables $c_{\check{p}}$ evaluated at the expectations $c^0_{\check{p}}$ are defined as

$$\phi^{(2)}_{\alpha\beta}(c^0_{\check{\varrho}}) = \sum_{\check{p}=1}^{\check{N}} \phi^{,\check{p}\check{p}}_{\alpha\beta}(c^0_{\check{\varrho}})\, S^{\check{p}}_{\rm c} \tag{4.180}$$

while the first- and second-order right-hand side vectors have the form

$$\begin{aligned} R^{\check{p}}_{\alpha}(c^0_{\check{\varrho}}) &= -\Big[K^{,\check{p}}_{\alpha\beta}(c^0_{\check{\varrho}}) \\ &\quad - \Omega^{,\check{p}}_{(\alpha)}(c^0_{\check{\varrho}})\, M^0_{\alpha\beta}(c^0_{\check{\varrho}}) - \Omega^0_{(\alpha)}(c^0_{\check{\varrho}})\, M^{,\check{p}}_{\alpha\beta}(c^0_{\check{\varrho}})\Big]\, \phi^0_{\beta\gamma}(c^0_{\check{\varrho}}) \\ R^{(2)}_{\alpha}(c^0_{\check{\varrho}}) &= -\sum_{\check{p}=1}^{\check{N}} \Big\{ \Big[K^{,\check{p}\check{p}}_{\alpha\beta}(c^0_{\check{\varrho}}) - \Omega^{,\check{p}\check{p}}_{(\alpha)}(c^0_{\check{\varrho}})\, M^0_{\alpha\beta}(c^0_{\check{\varrho}}) \\ &\quad - 2\,\Omega^{,\check{p}}_{(\alpha)}(c^0_{\check{\varrho}})\, M^{,\check{p}}_{\alpha\beta}(c^0_{\check{\varrho}}) - \Omega^0_{(\alpha)}(c^0_{\check{\varrho}})\, M^{,\check{p}\check{p}}_{\alpha\beta}(c^0_{\check{\varrho}})\Big]\, \phi^0_{\beta\gamma}(c^0_{\check{\varrho}}) \\ &\quad + 2\Big[K^{,\check{p}}_{\alpha\beta}(c^0_{\check{\varrho}}) - \Omega^{,\check{p}}_{(\alpha)}(c^0_{\check{\varrho}})\, M^0_{\alpha\beta}(c^0_{\check{\varrho}}) - \Omega^0_{(\alpha)}(c^0_{\check{\varrho}})\, M^{,\check{p}}_{\alpha\beta}(c^0_{\check{\varrho}})\Big]\phi^{,\check{p}}_{\beta\gamma}(c^0_{\check{\varrho}})\Big\} S^{\check{p}}_{\rm c} \end{aligned} \tag{4.181}$$

Stochastic Linearized Buckling. Similarly as above, we may readily obtain the transformed zeroth-, first- and second-order equations for the stochastic linearized buckling problem in the following form, cf. Eqs. (4.51)–(4.53):

- Zeroth-order (ϵ^0 terms, one system of N equations)

$$\Big[\hat{K}^{(\sigma)0}_{\alpha\beta}(c^0_{\check{\varrho}}) - \tilde{\lambda}^0_{(\hat{\alpha})}(c^0_{\check{\varrho}})\, K^{(e)0}_{\alpha\beta}(c^0_{\check{\varrho}})\Big]\, v^0_{\beta(\hat{\alpha})}(c^0_{\check{\varrho}}) = 0 \tag{4.182}$$

(no sum on $\hat{\alpha}$)

- First-order (ϵ^1 terms, $\check{N}$ systems of N equations, $\check{p} = 1, 2, \ldots, \check{N}$)

$$\Big[\hat{K}^{(\sigma)0}_{\alpha\beta}(c^0_{\check{\varrho}}) - \tilde{\lambda}^0_{(\hat{\alpha})}(c^0_{\check{\varrho}})\, K^{(e)0}_{\alpha\beta}(c^0_{\check{\varrho}})\Big]\, v^{,\check{p}}_{\beta(\hat{\alpha})}(c^0_{\check{\varrho}}) = R^{\check{p}}_{\alpha}(c^0_{\check{\varrho}}) \tag{4.183}$$

(no sum on $\hat{\alpha}$)

- Second-order (ϵ^2 terms, one system of N equations)

$$\Big[\hat{K}^{(\sigma)0}_{\alpha\beta}(c^0_{\check{\varrho}}) - \tilde{\lambda}^0_{(\hat{\alpha})}(c^0_{\check{\varrho}})\, K^{(e)0}_{\alpha\beta}(c^0_{\check{\varrho}})\Big]\, v^{(2)}_{\beta(\hat{\alpha})}(c^0_{\check{\varrho}}) = R^{(2)}_{\alpha}(c^0_{\check{\varrho}}) \tag{4.184}$$

(no sum on $\hat{\alpha}$)

where the following notation has been employed

$$v^{(2)}_{\alpha(\hat{\alpha})}(c^0_{\check{\varrho}}) = \sum_{\check{p}=1}^{\check{N}} v^{,\check{p}\check{\varsigma}}_{\alpha(\hat{\alpha})}(c^0_{\check{\varrho}})\, S^{\check{p}\check{\varsigma}}_{\rm c} \tag{4.185}$$

$$\begin{aligned} R^{\check{p}}_{\alpha}(c^0_{\check{\varrho}}) &= -\Big[\hat{K}^{(\sigma),\check{p}}_{\alpha\beta}(c^0_{\check{\varrho}}) \\ &\quad - \tilde{\lambda}^{,\check{p}}_{(\hat{\alpha})}(c^0_{\check{\varrho}})\, K^{(e)0}_{\alpha\beta}(c^0_{\check{\varrho}}) - \tilde{\lambda}^0_{(\hat{\alpha})}(c^0_{\check{\varrho}})\, K^{(e),\check{p}}_{\alpha\beta}(c^0_{\check{\varrho}})\Big]\, v^0_{\beta(\hat{\alpha})}(c^0_{\check{\varrho}}) \\ R^{(2)}_{\alpha}(c^0_{\check{\varrho}}) &= -\sum_{\check{p}=1}^{\check{N}} \Big\{ \Big[\hat{K}^{(\sigma),\check{p}\check{\varsigma}}_{\alpha\beta}(c^0_{\check{\varrho}}) - \tilde{\lambda}^{,\check{p}\check{\varsigma}}_{(\hat{\alpha})}(c^0_{\check{\varrho}})\, K^{(e)0}_{\alpha\beta}(c^0_{\check{\varrho}}) \\ &\quad - 2\,\tilde{\lambda}^{,\check{p}}_{(\hat{\alpha})}(c^0_{\check{\varrho}})\, K^{(e),\check{\varsigma}}_{\alpha\beta}(c^0_{\check{\varrho}}) - \tilde{\lambda}^0_{(\hat{\alpha})}(c^0_{\check{\varrho}})\, K^{(e),\check{p}\check{\varsigma}}_{\alpha\beta}(c^0_{\check{\varrho}})\Big]\, v^0_{\beta(\hat{\alpha})}(c^0_{\check{\varrho}}) \\ &\quad + 2\Big[\hat{K}^{(\sigma),\check{p}}_{\alpha\beta}(c^0_{\check{\varrho}}) - \tilde{\lambda}^{,\check{p}}_{(\hat{\alpha})}(c^0_{\check{\varrho}})\, K^{(e)0}_{\alpha\beta}(c^0_{\check{\varrho}}) - \tilde{\lambda}^0_{(\hat{\alpha})}(c^0_{\check{\varrho}})\, K^{(e),\check{p}}_{\alpha\beta}(c^0_{\check{\varrho}})\Big]\, v^{,\check{\varsigma}}_{\beta(\hat{\alpha})}(c^0_{\check{\varrho}})\Big\} S^{\check{p}\check{\varsigma}}_{\rm c} \end{aligned} \tag{4.186}$$

4.5.2.3 Response Probabilistic Distributions

General Remarks. Employing the standard normal transformation the first two moments of any state random variable can be immediately computed, in contrast to the use of mode superposition for forced-vibrations wherein the modal response expressed in the normal coordinates must be transformed into the original system of generalized coordinates. Here, the solution can be obtained through the second-order Taylor expansion for the state variable about the expectations $c^0_{\check{\varrho}}$ of $c_{\check{\rho}}$. For example, the random displacement field is expanded as, cf. Eq. (4.37)

$$q_\alpha(c_{\check{\varrho}};\tau) \;=\; q^0_\alpha(c^0_{\check{\varrho}};\tau) + q^{,\check{\rho}}_\alpha(c^0_{\check{\varrho}};\tau)\,\Delta c_{\check{\rho}} + \tfrac{1}{2}[q^{,\check{\rho}\check{\rho}}_\alpha(c^0_{\check{\varrho}};\tau)]^2\,(\Delta c_{\check{\rho}})^2 \qquad (4.187)$$

since the mixed derivatives $q^{\check{\rho}\check{\sigma}}_\alpha$ with respect to $c_{\check{\rho}}$ vanish when $\check{\rho} \neq \check{\sigma}$. The nodal displacement expectations at any time instant $\tau = t$ and cross-covariances at $\xi_1 = (x^{(1)}_k, t_1)$ and $\xi_2 = (x^{(2)}_k, t_2)$ become, cf. Eqs. (4.38) and (4.39)

$$E[q_\alpha(t)] \;=\; \underbrace{\int_{-\infty}^{+\infty}\int_{-\infty}^{+\infty}\cdots\int_{-\infty}^{+\infty}}_{\check{N}\text{-fold}} q_\alpha(t)\,p_{\check{N}}(c_1,c_2,\ldots,c_{\check{N}})\,\mathrm{d}c_1\mathrm{d}c_2\ldots\mathrm{d}c_{\check{N}} \qquad (4.188)$$

$$\begin{aligned}\mathrm{Cov}\Big(q_\alpha(t_1),q_\beta(t_2)\Big) \;&=\; S^{\alpha\beta}_{\mathbf{q}}(t_1,t_2)\\ &=\; \underbrace{\int_{-\infty}^{+\infty}\int_{-\infty}^{+\infty}\cdots\int_{-\infty}^{+\infty}}_{\check{N}\text{-fold}} \Big\{q_\alpha(t_1) - E[q_\alpha(t_1)]\Big\}\Big\{q_\beta(t_2) - E[q_\beta(t_2)]\Big\}\\ &\qquad\times\; p_{\check{N}}(c_1,c_2,\ldots,c_{\check{N}})\,\mathrm{d}c_1\mathrm{d}c_2\ldots\mathrm{d}c_{\check{N}}\end{aligned} \qquad (4.189)$$

where $p_{\check{N}}(c_1,c_2,\ldots,c_{\check{N}})$ is the $\check{N}$-variate PDF of the uncorrelated random variables $c_{\check{\rho}}$. We note that the lack of correlation for the variables $c_{\check{\rho}}$ is not a sufficient condition for their independence. We also note that the standard eigenvalue equation (4.149) gives the covariance matrix for $c_{\check{\rho}}$ in diagonal form, i.e.

$$\mathrm{Cov}(c_{\check{\rho}},c_{\check{\sigma}}) \;=\; \mathrm{Cov}(c_{\check{\rho}},c_{\check{\sigma}})\,\delta_{\check{\rho}\check{\sigma}} \;=\; \mathrm{Var}(c_{\check{\rho}}) \;=\; S^{\check{\rho}}_{\mathrm{c}} \qquad (4.190)$$

The aspects mentioned above are the only differences between the direct approach based on the original correlated random variables and that using the standard normal transformation; other derivations for both approaches are identical. For the sake of brevity of presentation we shall not discuss this procedure in detail. However, the specific formulae for expectations and covariances of the state variables in problems of forced- and free-vibrations and linearized buckling will be given below since they are useful in computer implementation. As before, all the expectations are evaluated up to second-order accuracy while the covariances are first-order accurate.

First Two Moments for Displacements, Strains and Stresses. In forced-vibration analysis, by using Eqs. (4.187)–(4.189) we arrive at the expectations for the nodal displacements at any time instant $\tau = t$ as, cf. Eq. (4.40)

$$E[q_\alpha(t)] \;=\; q^0_\alpha(t) + \tfrac{1}{2}\sum_{\check{\rho}=1}^{\check{N}} q^{,\check{\rho}\check{\rho}}_\alpha(t)\,S^{\check{\rho}}_{\mathrm{c}} \qquad (4.191)$$

and the displacement cross-covariances at $\xi_1 = (x_k^{(1)}, t_1)$ and $\xi_2 = (x_k^{(2)}, t_2)$ as, cf. Eq. (4.42)

$$\mathrm{Cov}\Big(q_\alpha(t_1), q_\beta(t_2)\Big) = \sum_{\check{p}=1}^{\check{N}} q_\alpha^{,\check{p}}(t_1)\, q_\beta^{,\check{p}}(t_2)\, S_{\mathrm{c}}^{\check{p}} \tag{4.192}$$

Consequently, the expectations for the strain tensor components at any point in the structure and at any time instant $\tau = t$ result in the form, cf. Eq. (4.44)

$$E[\varepsilon_{ij}(x_k, t)] = B_{ij\alpha}(x_k)\Big[q_\alpha^0(t) + \frac{1}{2}\sum_{\check{p}=1}^{\check{N}} q_\alpha^{,\check{p}\check{p}}(t)\, S_{\mathrm{c}}^{\check{p}}\Big] \qquad x_k \in \Omega_e \tag{4.193}$$

whereas the strain cross-covariances at $\xi_1 = (x_k^{(1)}, t_1)$ and $\xi_2 = (x_k^{(2)}, t_2)$ become, cf. Eq. (4.45)

$$\begin{aligned} &\mathrm{Cov}\Big(\varepsilon_{ij}(x_m^{(1)}, t_1), \varepsilon_{kl}(x_m^{(2)}, t_2)\Big) \\ &\quad = B_{ij\alpha}(x_m^{(1)})\, B_{kl\beta}(x_m^{(2)}) \sum_{\check{p}=1}^{\check{N}} q_\alpha^{,\check{p}}(t_1)\, q_\beta^{,\check{p}}(t_2)\, S_{\mathrm{c}}^{\check{p}} \\ &\quad x_m^{(1)} \in \Omega_e\,; \quad x_m^{(2)} \in \Omega_f \end{aligned} \tag{4.194}$$

The expression for the stress component expectations at any space point and any time $\tau = t$ takes the form, cf. Eq. (4.47)

$$\begin{aligned} E[\sigma_{ij}(x_k, t)] &= \varphi_{\bar{\alpha}}(x_m)\Big\{C_{ijkl\bar{\alpha}}^0\, B_{kl\alpha}(x_m)\, q_\alpha^0(t) + \frac{1}{2} B_{kl\alpha}(x_m) \\ &\quad \times \sum_{\check{p}=1}^{\check{N}} \Big[C_{ijkl\bar{\alpha}}^{,\check{p}\check{p}}\, q_\alpha^0(t) + 2C_{ijkl\bar{\alpha}}^{,\check{p}}\, q_\alpha^{,\check{p}}(t) + C_{ijkl\bar{\alpha}}^0\, q_\alpha^{,\check{p}\check{p}}(t)\Big]\, S_{\mathrm{c}}^{\check{p}}\Big\} \\ &= \varphi_{\bar{\alpha}}(x_m)\Big\{C_{ijkl\bar{\alpha}}^0\, E[\varepsilon_{kl}(x_m, t)] \\ &\quad + B_{kl\alpha}(x_m) \sum_{\check{p}=1}^{\check{N}} \Big[\frac{1}{2} C_{ijkl\bar{\alpha}}^{,\check{p}\check{p}}\, q_\alpha^0(t) + C_{ijkl\bar{\alpha}}^{,\check{p}}\, q_\alpha^{,\check{p}}(t)\Big]\, S_{\mathrm{c}}^{\check{p}}\Big\} \\ &\quad x_m \in \Omega_e \end{aligned} \tag{4.195}$$

whereas the stress cross-covariances at $\xi_1 = (x_k^{(1)}, t_1)$ and $\xi_2 = (x_k^{(2)}, t_2)$ become, cf. Eq. (4.48)

$$\begin{aligned} &\mathrm{Cov}\Big(\sigma_{ij}(x_m^{(1)}, t_1), \sigma_{kl}(x_m^{(2)}, t_2)\Big) \\ &\quad = \varphi_{\bar{\alpha}}(x_m^{(1)})\, \varphi_{\bar{\beta}}(x_m^{(2)})\, B_{tu\alpha}(x_m^{(1)})\, B_{vw\beta}(x_m^{(2)}) \\ &\quad \times \sum_{\check{p}=1}^{\check{N}} \Big[C_{ijtu\bar{\alpha}}^{,\check{p}}\, C_{klvw\bar{\beta}}^{,\check{p}}\, q_\alpha^0(t_1)\, q_\beta^0(t_2) + C_{ijtu\bar{\alpha}}^{,\check{p}}\, C_{klvw\bar{\beta}}^0\, q_\alpha^{,\check{p}}(t_1)\, q_\beta^0(t_2) \\ &\quad + C_{ijtu\bar{\alpha}}^0\, C_{klvw\bar{\beta}}^{,\check{p}}\, q_\alpha^0(t_1)\, q_\beta^{,\check{p}}(t_2) + C_{ijtu\bar{\alpha}}^0\, C_{klvw\bar{\beta}}^0\, q_\alpha^{,\check{p}}(t_1)\, q_\beta^{,\check{p}}(t_2)\Big]\, S_{\mathrm{c}}^{\check{p}} \\ &\quad x_m^{(1)} \in \Omega_e\,; \quad x_m^{(2)} \in \Omega_f \end{aligned} \tag{4.196}$$

First Two Moments for Eigenvalues and Eigenvectors. Similarly as in Section 4.3.1, we cannot arrive at the first and second derivatives of eigenpairs with respect to uncorrelated random variables by a direct solution to Eqs. (4.178)–(4.179) since these equations are nonhomogeneous and singular of rank $N-1$. We shall follow an indirect technique analogous to that used for solving this problem in terms of the correlated random variables b_ρ, Sections 4.3.2 and 4.3.3. Without tedious repetitions the first derivatives of eigenvalues with respect to the uncorrelated random variables $c_{\check{\rho}}$ can be shown to be, cf. Eq. (4.58)

$$\Omega^{,\check{\rho}}_{(\alpha)} = \phi^{0\mathrm{T}}_{\gamma\alpha}\left(K^{,\check{\rho}}_{\alpha\beta} - \Omega^{0}_{(\alpha)}\, M^{,\check{\rho}}_{\alpha\beta}\right)\phi^{0}_{\beta\delta} \tag{4.197}$$

while the (generalized) eigenvalue second derivatives are, cf. Eq. (4.61)

$$\begin{aligned}\Omega^{(2)}_{(\alpha)} &= \sum_{\check{\rho}=1}^{\check{N}} \Omega^{,\check{\rho}\check{\rho}}_{(\alpha)}(c^{0}_{\check{\varrho}})\, S^{\check{\rho}}_{\mathrm{c}} = \sum_{\check{\rho}=1}^{\check{N}} \Big[\phi^{0\mathrm{T}}_{\gamma\alpha}\Big(K^{,\check{\rho}\check{\rho}}_{\alpha\beta} - 2\,\Omega^{,\check{\rho}}_{(\alpha)}\, M^{,\check{\rho}}_{\alpha\beta} \\ &\quad - \Omega^{0}_{(\alpha)} M^{,\check{\rho}\check{\rho}}_{\alpha\beta}\Big)\phi^{0}_{\beta\delta} + 2\phi^{0\mathrm{T}}_{\gamma\alpha}\Big(K^{,\check{\rho}}_{\alpha\beta} - \Omega^{,\check{\rho}}_{(\alpha)} M^{0}_{\alpha\beta} - \Omega^{0}_{(\alpha)} M^{,\check{\rho}}_{\alpha\beta}\Big)\phi^{,\check{\rho}}_{\beta\delta}\Big] S^{\check{\rho}}_{\mathrm{c}}\end{aligned} \tag{4.198}$$

The $\hat{\alpha}$-th first derivative of eigenvectors with respect to $c_{\check{\rho}}$ can be approximated in terms of all the zeroth-order eigenvectors as, cf. Eq. (4.66)

$$\phi^{,\check{\rho}}_{\alpha(\hat{\alpha})} = \phi^{0}_{\alpha\beta}\, a^{\check{\rho}}_{\beta(\hat{\alpha})} \tag{4.199}$$

where, cf. Eqs. (4.76) and (4.65)

$$a^{\check{\rho}}_{\alpha(\hat{\alpha})} = \begin{cases} \dfrac{\phi^{0}_{\beta(\alpha)}\, R^{\check{\rho}}_{\beta(\hat{\alpha})}}{\omega^{0}_{(\alpha)} - \omega^{0}_{(\hat{\alpha})}} & \text{for} \quad \alpha \neq \hat{\alpha} \\ -\frac{1}{2}\, \phi^{0}_{\beta(\hat{\alpha})}\, M^{,\check{\rho}}_{\beta\gamma}\, \phi^{0}_{\gamma(\hat{\alpha})} & \text{for} \quad \alpha = \hat{\alpha} \end{cases} \tag{4.200}$$

$$R^{\check{\rho}}_{\alpha(\hat{\alpha})} = -\Big[K^{,\check{\rho}}_{\alpha\beta} - \phi^{0}_{\gamma(\hat{\alpha})}\Big(K^{,\check{\rho}}_{\gamma\delta} - \omega^{0}_{(\hat{\alpha})} M^{,\check{\rho}}_{\gamma\delta}\Big)\phi^{0}_{\delta(\hat{\alpha})} M^{0}_{\alpha\beta} - \omega^{0}_{(\hat{\alpha})} M^{,\check{\rho}}_{\alpha\beta}\Big]\phi^{0}_{\beta(\hat{\alpha})}$$

The $\hat{\alpha}$-th (generalized) second derivative of eigenvectors with respect to $c_{\check{\rho}}$ can be written as, cf. Eq. (4.81)

$$\phi^{(2)}_{\alpha(\hat{\alpha})} = \sum_{\check{\rho}=1}^{\check{N}} \phi^{,\check{\rho}\check{\rho}}_{\alpha(\hat{\alpha})}\, S^{\check{\rho}}_{\mathrm{c}} = \phi^{0}_{\alpha\beta}\, a^{(2)}_{\beta(\hat{\alpha})} \tag{4.201}$$

where, cf. Eqs. (4.88) and (4.79)

$$a^{(2)}_{\alpha(\hat{\alpha})} = \begin{cases} \dfrac{\phi^{0}_{\beta(\alpha)}\, R^{(2)}_{\beta(\hat{\alpha})}}{\omega^{0}_{(\alpha)} - \omega^{0}_{(\hat{\alpha})}} & \text{for} \quad \alpha \neq \hat{\alpha} \\ -\displaystyle\sum_{\check{\rho}=1}^{\check{N}}\Big(\tfrac{1}{2}\phi^{0}_{\beta(\hat{\alpha})}\, M^{,\check{\rho}\check{\rho}}_{\beta\gamma}\, \phi^{0}_{\gamma(\hat{\alpha})} + 2\phi^{0}_{\beta(\hat{\alpha})}\, M^{,\check{\rho}}_{\beta\gamma}\, \phi^{,\check{\rho}}_{\gamma(\hat{\alpha})} + a^{\check{\rho}}_{\beta(\hat{\alpha})}\, a^{\check{\rho}}_{\beta(\hat{\alpha})}\Big) S^{\check{\rho}}_{\mathrm{c}} & \\ \qquad\qquad \text{for} \quad \alpha = \hat{\alpha} & \end{cases} \tag{4.202}$$

$$\begin{aligned}R^{(2)}_{\alpha(\hat{\alpha})} &= -\sum_{\check{\rho}=1}^{\check{N}} \Big[\Big(K^{,\check{\rho}\check{\rho}}_{\alpha\beta} - \omega^{,\check{\rho}\check{\rho}}_{(\hat{\alpha})}\, M^{0}_{\alpha\beta} - 2\omega^{,\check{\rho}}_{(\hat{\alpha})}\, M^{,\check{\rho}}_{\alpha\beta} - \omega^{0}_{(\hat{\alpha})}\, M^{,\check{\rho}\check{\rho}}_{\alpha\beta}\Big)\phi^{0}_{\beta(\hat{\alpha})} \\ &\quad + 2\Big(K^{,\check{\rho}}_{\alpha\beta} - \omega^{,\check{\rho}}_{(\hat{\alpha})}\, M^{0}_{\alpha\beta} - \omega^{0}_{(\hat{\alpha})}\, M^{,\check{\rho}}_{\alpha\beta}\Big)\phi^{,\check{\rho}}_{\beta(\hat{\alpha})}\Big] S^{\check{\rho}}_{\mathrm{c}}\end{aligned}$$

First Two Moments for Buckling Factors and Loads. Quite similarly to the case of the generalized eigenproblem derived above, by pre-multiplying successively both sides of the first- and second-order equations in the system (4.182)–(4.184) by the $\hat{\alpha}$-th zeroth-order eigenvector $v^0_{\alpha(\hat{\alpha})}$ transposed and using the $K^{(e)}_{\alpha\beta}$-orthonormality condition $(4.106)_1$ we arrive at the $\hat{\alpha}$-th first-order eigenvalue as, cf. Eq. (4.116)

$$\tilde{\lambda}^{,\check{\rho}}_{(\hat{\alpha})} = v^0_{\alpha(\hat{\alpha})} \left(\hat{K}^{(\sigma),\check{\rho}}_{\alpha\beta} - \tilde{\lambda}^0_{(\hat{\alpha})} K^{(e),\check{\rho}}_{\alpha\beta} \right) v^0_{\beta(\hat{\alpha})} \tag{4.203}$$

and the $\hat{\alpha}$-th second-order eigenvalue as, cf. Eq. (4.119),

$$\begin{aligned} \tilde{\lambda}^{(2)}_{(\hat{\alpha})} &= \sum_{\check{\rho}=1}^{\check{N}} \tilde{\lambda}^{,\check{\rho}\check{\rho}}_{(\hat{\alpha})} S^{\check{\rho}}_{c} \\ &= \sum_{\check{\rho}=1}^{\check{N}} \Big[v^0_{\alpha(\hat{\alpha})} \left(\hat{K}^{(\sigma),\check{\rho}\check{\rho}}_{\alpha\beta} - 2\,\tilde{\lambda}^{,\check{\rho}}_{(\hat{\alpha})} K^{(e),\check{\rho}}_{\alpha\beta} - \tilde{\lambda}^0_{(\hat{\alpha})} K^{(e),\check{\rho}\check{\rho}}_{\alpha\beta} \right) v^0_{\beta(\hat{\alpha})} \\ &\quad + 2\, v^0_{\alpha(\hat{\alpha})} \left(\hat{K}^{(\sigma),\check{\rho}}_{\alpha\beta} - \tilde{\lambda}^{,\check{\rho}}_{(\hat{\alpha})} K^{(e)0}_{\alpha\beta} - \tilde{\lambda}^0_{(\hat{\alpha})} K^{(e),\check{\rho}}_{\alpha\beta} \right) v^{,\check{\rho}}_{\beta(\hat{\alpha})} \Big] S^{\check{\rho}}_{c} \end{aligned} \tag{4.204}$$

Furthermore, we assume that the $\hat{\alpha}$-th first-order eigenvector $v^{,\check{\rho}}_{\alpha(\hat{\alpha})}$, which is the only quantity to be computed in Eq. (4.204), can be represented by a linear function of all the zeroth-order eigenvectors, i.e.

$$v^{,\check{\rho}}_{\alpha(\hat{\alpha})} = V^0_{\alpha\beta}\, a^{\check{\rho}}_{\beta(\hat{\alpha})} \tag{4.205}$$

where $V^0_{\alpha\beta}$ is the zeroth-order eigenvector matrix. By following a procedure analogous to the process (4.121)–(4.127) we get the unique solution for the $\hat{\alpha}$-th entry of the coefficient vector $a^{\rho}_{\beta(\hat{\alpha})}$ in the form, cf. Eq. (4.128) and Eq. (4.122)

$$\begin{aligned} a^{\check{\rho}}_{\alpha(\hat{\alpha})} &= \begin{cases} \dfrac{v^0_{\beta(\alpha)} R^{\check{\rho}}_{\beta(\hat{\alpha})}}{\tilde{\lambda}^0_{(\alpha)} - \tilde{\lambda}^0_{(\hat{\alpha})}} & \text{for} \quad \alpha \neq \hat{\alpha} \\ -\dfrac{1}{2}\, v^0_{\alpha(\hat{\alpha})} K^{(e),\check{\rho}}_{\alpha\beta} v^0_{\beta(\hat{\alpha})} & \text{for} \quad \alpha = \hat{\alpha} \end{cases} \\ R^{\check{\rho}}_{\alpha(\hat{\alpha})} &= -\left(\hat{K}^{(\sigma),\check{\rho}}_{\alpha\beta} - \tilde{\lambda}^{,\check{\rho}}_{(\hat{\alpha})} K^{(e)0}_{\alpha\beta} - \tilde{\lambda}^0_{(\hat{\alpha})} K^{(e),\check{\rho}}_{\alpha\beta} \right) v^0_{\beta(\hat{\alpha})} \\ &\qquad\qquad (\text{no sum on } \hat{\alpha}) \end{aligned} \tag{4.206}$$

Now that we have $\tilde{\lambda}^0_{(\hat{\alpha})}$, $\tilde{\lambda}^{,\check{\rho}}_{(\hat{\alpha})}$ and $\tilde{\lambda}^{,\check{\rho}\check{\rho}}_{(\hat{\alpha})}$ the first two moments of the buckling load factor can be calculated. The first- and second-order buckling load factor are represented in terms of the derivatives of $\tilde{\lambda}_{(\hat{\alpha})}$ as, cf. Eqs. (4.129) and (4.130)

$$\begin{aligned} \lambda^0_{(\hat{\alpha})} &= \frac{1}{\tilde{\lambda}^0_{(\hat{\alpha})}} \\ \lambda^{,\check{\rho}}_{(\hat{\alpha})} &= -\frac{1}{(\tilde{\lambda}^0_{(\hat{\alpha})})^2}\, \tilde{\lambda}^{,\check{\rho}}_{(\hat{\alpha})} \\ \lambda^{,\check{\rho}\check{\rho}}_{(\hat{\alpha})} &= -\frac{1}{(\tilde{\lambda}^0_{(\hat{\alpha})})^3} \left[\tilde{\lambda}^0_{(\hat{\alpha})} \tilde{\lambda}^{,\check{\rho}\check{\rho}}_{(\hat{\alpha})} - 2(\tilde{\lambda}^{,\check{\rho}}_{(\hat{\alpha})})^2 \right] \qquad (\text{no sum on } \check{\rho}) \end{aligned} \tag{4.207}$$

Employing an expansion of the type (4.187) for $\lambda_{(\hat{\alpha})}$ about the uncorrelated random variables $c_{\check{\rho}}$ and using the definitions of the first two statistical moments leads to the expressions for the expectations of the $\hat{\alpha}$-th buckling load factor in the form, cf. Eq. (4.133)

$$E[\lambda_{(\hat{\alpha})}] = \frac{1}{\tilde{\lambda}^0_{(\hat{\alpha})}} \left[1 + \frac{1}{(\tilde{\lambda}^0_{(\hat{\alpha})})^2} \sum_{\check{\rho}=1}^{\check{N}} (\tilde{\lambda}^{,\check{\rho}}_{(\hat{\alpha})})^2 S^{\check{\rho}}_{\mathrm{c}} - \frac{\tilde{\lambda}^{(2)}_{(\hat{\alpha})}}{2\,\tilde{\lambda}^0_{(\hat{\alpha})}} \right] \tag{4.208}$$

and the cross-covariance for the $\hat{\alpha}$-th and $\hat{\beta}$-th buckling load factors as, cf. Eq. (4.134)

$$\mathrm{Cov}(\lambda_{(\hat{\alpha})}, \lambda_{(\hat{\beta})}) = \frac{1}{(\tilde{\lambda}^0_{(\hat{\alpha})}\,\tilde{\lambda}^0_{(\hat{\beta})})^2} \sum_{\check{\rho}=1}^{\check{N}} \tilde{\lambda}^{,\check{\rho}}_{(\hat{\alpha})}\,\tilde{\lambda}^{,\check{\rho}}_{(\hat{\beta})}\,S^{\check{\rho}}_{\mathrm{c}} \tag{4.209}$$

If the reference load pattern Q^*_α is deterministic, the probabilistic distributions of the buckling load are immediately obtained by using the linear transformation of random variables, Section 1.1.3. Thus, the expectation for the $\hat{\alpha}$-th buckling load vector is

$$E[Q_{\alpha(\hat{\alpha})}] = Q^*_\alpha\,E[\tilde{\lambda}^0_{(\hat{\alpha})}] \tag{4.210}$$

and the cross-covariance for the α-th component of the $\hat{\alpha}$-th buckling load vector and the β-th component of the $\hat{\beta}$-th one is, cf. Eq. (4.137)

$$\mathrm{Cov}(Q_{\alpha(\hat{\alpha})}, Q_{\beta(\hat{\beta})}) = Q^*_\alpha\,Q^*_\beta\,\mathrm{Cov}(\tilde{\lambda}_{(\hat{\alpha})}, \tilde{\lambda}_{(\hat{\beta})}) \tag{4.211}$$

When Q^*_α is described stochastically the derivations proceed analogously as in Eqs. (4.140)–(4.144) so that we get, cf. Eq. (4.145)

$$\begin{aligned} E[Q_{\alpha(\hat{\alpha})}] = {} & \frac{1}{\tilde{\lambda}^0_{(\hat{\alpha})}}\,Q^{*0}_\alpha + \frac{1}{2\,\tilde{\lambda}^0_{(\hat{\alpha})}} \Bigg\{ -\frac{\tilde{\lambda}^{(2)}_{(\hat{\alpha})}}{\tilde{\lambda}^0_{(\hat{\alpha})}}\,Q^{*0}_\alpha \\ & + \sum_{\check{\rho}=1}^{\check{N}} \left[\frac{2\tilde{\lambda}^{,\check{\rho}}_{(\hat{\alpha})}}{\tilde{\lambda}^0_{(\hat{\alpha})}} \left(\frac{\tilde{\lambda}^{,\check{\rho}}_{(\hat{\alpha})}}{\tilde{\lambda}^0_{(\hat{\alpha})}}\,Q^{*0}_\alpha - Q^{*,\check{\rho}}_\alpha \right) + Q^{*,\check{\rho}\check{\rho}}_\alpha \right] S^{\check{\rho}}_{\mathrm{c}} \Bigg\} \end{aligned} \tag{4.212}$$

and, cf. Eq. (4.146)

$$\begin{aligned} \mathrm{Cov}(Q_{\alpha(\hat{\alpha})}, Q_{\beta(\hat{\beta})}) = {} & \frac{1}{\tilde{\lambda}^0_{(\hat{\alpha})}\,\tilde{\lambda}^0_{(\hat{\beta})}} \sum_{\check{\rho}=1}^{\check{N}} \Bigg(\frac{\tilde{\lambda}^{,\check{\rho}}_{(\hat{\alpha})}\,\tilde{\lambda}^{,\check{\rho}}_{(\hat{\beta})}}{\tilde{\lambda}^0_{(\hat{\alpha})}\,\tilde{\lambda}^0_{(\hat{\beta})}}\,Q^{*0}_\alpha\,Q^{*0}_\beta \\ & + \frac{\tilde{\lambda}^{,\check{\rho}}_{(\hat{\alpha})}}{\tilde{\lambda}^0_{(\hat{\alpha})}}\,Q^{*0}_\alpha\,Q^{*,\check{\rho}}_\beta + \frac{\tilde{\lambda}^{,\check{\rho}}_{(\hat{\beta})}}{\tilde{\lambda}^0_{(\hat{\beta})}}\,Q^{*,\check{\rho}}_\alpha\,Q^{*0}_\beta + Q^{*,\check{\rho}}_\alpha\,Q^{*,\check{\rho}}_\beta \Bigg) S^{\check{\rho}}_{\mathrm{c}} \end{aligned} \tag{4.213}$$

It goes without saying that the smallest buckling load factor $\lambda_{(1)}$ and the corresponding bucking load $Q_{\alpha(1)}$ are readily obtained from the foregoing equations by setting $\hat{\alpha} = \hat{\beta} = 1$.

4.5.3 Two-Fold Superposition Technique

As shown in Section 4.5.2 the transformation of the set of correlated random variables to a set of uncorrelated variables enables one to significantly reduce the size of the problem to be solved. In a computational sense the stochastic finite element model of a large-scale problem becomes relatively independent of the complexity of random structural properties. It is so because: (i) the number of dominant uncorrelated variables necessary to simulate the correlation structure may be much smaller than the total number of correlated random variables in the problem; (ii) cumbersome algebraic operations such as double summations and multiplications of large matrices may be avoided; only the single sums and vector multiplications need to be performed to evaluate the partial derivatives, the right-hand side of the second-order equations and the first two statistical moments.

In terms of linear algebra, the solution process consisting of: (i) 'decoupling' $\tilde{N}+2$ correlated systems (4.17)–(4.19) to $\check{N}+2$ uncorrelated systems (4.155)–(4.157) and (ii) 'superposing' the dominant uncorrelated modes to obtain the desired result, can be treated as the *randomness superposition.*

Thus far we have been concerned with the computational aspects of a stochastic finite element model in the context of its probabilistic structure, rather than with the efficiency of a numerical technique known in deterministic finite element analysis and applied now to stochastic analysis. Such a discussion will be given in this section, in which we shall be concerned with the system transformation from the generalized finite-element coordinates to the normalized modal coordinates and vice versa. In the context of the so-called *two-fold superposition* procedure this will be referred to as the *modal superposition.*

Equations (4.155)–(4.157) represent $\check{N}+2$ systems of N linear simultaneous ordinary differential equations. (Recall that $\check{N}$ is the number of uncorrelated random variables used to represent the set of correlated random variables given in the original system and N is the overall number of degrees of freedom in the finite element model.) The differential equations with constant coefficients can be integrated over the time domain by using a direct scheme or an indirect algorithm such as those described in Section 2.5.2. As shown also in that section, the main disadvantages of the direct integration techniques are that they always deal with all N degrees of freedom in the system, and that the number of arithmetic operations required during the solution is proportional to the number of time steps used in the analysis. In the stochastic analysis of large-scale systems for which the global mass and damping matrices must sometimes be modelled in a consistent way and a long response duration may be required, the computational cost of the direct approach for solving $\check{N}+2$ systems of N equations would be unacceptably high and the mode superposition method turns out definitively better. Moreover, from the error assessment standpoint, the fact that only a few dominant uncorrelated degrees of freedom (variables) may appropriately represent the set of input nodal random variables via an eigenproblem suggests that the use of any integration procedure acting on all the FEM's degrees of freedom in the system can be inadequate and unnecessary.

The concept of the mode superposition technique and its performance in deterministic analysis were described in Section 2.5.2. In Eqs. (4.155)–(4.157) all the uncertain properties of the system are entirely translated to the right-hand sides; the matrices at the linear operator on the left-hand sides are the zeroth-order coefficients of the Taylor expansion and are evaluated at the spatial expectations $c^0_{\check{\rho}}$ of the uncorrelated variables $c_{\check{\rho}}$, i.e. they can be treated as deterministic quantities. In view of this the modal analysis can be employed straightforwardly for solving systems (4.155)–(4.157) provided the zeroth-order damping matrix $C^0_{\alpha\beta}$ is of the modal damping type. Since the same linear operator acts upon the left-hand side of all the equations, there exists only one generalized eigenproblem for all the zeroth-, first- and second-order systems.

To decouple Eqs. (4.155)–(4.157) let us change the basis of the system from the finite element coordinates q^0_α (and consequently, $q^{,\check{\rho}}_\alpha$ and $q^{(2)}_\alpha$) to the normalized coordinates $r^0_{\bar{\alpha}}$ ($r^{,\check{\rho}}_{\bar{\alpha}}$ and $r^{(2)}_{\bar{\alpha}}$), cf. Eq. (2.85)

$$\begin{aligned} q^0_\alpha(c^0_{\check{\varrho}};\tau) &= \phi^0_{\alpha\bar{\alpha}}(c^0_{\check{\varrho}})\, r^0_{\bar{\alpha}}(c^0_{\check{\varrho}};\tau) \\ q^{,\check{\rho}}_\alpha(c^0_{\check{\varrho}};\tau) &= \phi^0_{\alpha\bar{\alpha}}(c^0_{\check{\varrho}})\, r^{,\check{\rho}}_{\bar{\alpha}}(c^0_{\check{\varrho}};\tau) \\ q^{(2)}_\alpha(c^0_{\check{\varrho}};\tau) &= \phi^0_{\alpha\bar{\alpha}}(c^0_{\check{\varrho}})\, r^{(2)}_{\bar{\alpha}}(c^0_{\check{\varrho}};\tau) \end{aligned} \qquad \begin{aligned} &\alpha = 1,2,\ldots,N \\ &\bar{\alpha} = 1,2,\ldots,\bar{N} \ll N \\ &(\text{no sum on } \check{\rho}) \end{aligned} \qquad (4.214)$$

(Similar expressions for $\dot{q}^0_\alpha$, $\ddot{q}^0_\alpha$, $\dot{q}^{,\check{\rho}}_\alpha, \ldots$ are used.) The zeroth-order mode shape matrix

$$\phi^0_{\alpha\bar{\alpha}} = \left[\, \phi^0_{\alpha(1)}\ \phi^0_{\alpha(2)}\ \cdots\ \phi^0_{\alpha(\bar{N})} \,\right]$$

is established from the eigensolution of the (deterministic) undamped free vibration problem of the zeroth-order system (4.155) so that is satisfies the condition

$$\left[K^0_{\alpha\beta}(c^0_{\check{\varrho}}) - \Omega^0_{(\bar{\alpha})}(c^0_{\check{\varrho}})\, M^0_{\alpha\beta}(c^0_{\check{\varrho}})\right] \phi^0_{\beta\bar{\alpha}}(c^0_{\check{\varrho}}) = 0 \qquad (\text{no sum on } \bar{\alpha}) \qquad (4.215)$$

and its eigenvectors $\phi^0_{\alpha(\bar{\alpha})}$ are M^0-orthonormal, K^0- and C^0-orthogonal, i.e.

$$\begin{aligned} \phi^0_{\alpha(\bar{\alpha})}\, M^0_{\alpha\beta}\, \phi^0_{\beta(\bar{\beta})} &= \delta_{\bar{\alpha}\bar{\beta}} \\ \phi^0_{\alpha(\bar{\alpha})}\, K^0_{\alpha\beta}\, \phi^0_{\beta(\bar{\beta})} &= \omega^0_{(\bar{\alpha})}\, \delta_{\bar{\alpha}\bar{\beta}} \\ \phi^0_{\alpha(\bar{\alpha})}\, C^0_{\alpha\beta}\, \phi^0_{\beta(\bar{\beta})} &= 2\,\xi^0_{(\bar{\alpha})}\, \omega^0_{(\bar{\alpha})}\, \delta_{\bar{\alpha}\bar{\beta}} \end{aligned} \qquad (\text{no sum on } \bar{\alpha}) \qquad (4.216)$$

where $\omega^0_{(\bar{\alpha})}$ is the $\bar{\alpha}$-th zeroth-order natural frequency of the undamped system from Eq. (4.155) (square root of the $\bar{\alpha}$-th entry of the diagonal eigenvalue matrix $\Omega^0_{(\bar{\alpha})}$) and $\xi^0_{(\bar{\alpha})}$ is the zeroth-order damping factor associated with the $\bar{\alpha}$-th mode. As a result we obtain uncoupled differential equations which can be integrated independently for each modal coordinate. We have then from Eqs. (4.155)–(4.157), cf. Eq. (2.89):

- Zeroth-order (ϵ^0 terms, one system of $\bar{N}$ uncoupled linear ordinary differential equations for $r^0_{\bar{\alpha}}(c^0_{\check{\varrho}};\tau)$, $\bar{\alpha}=1,2,\ldots,\bar{N}$)

$$\ddot{r}^0_{\bar{\alpha}}(c^0_{\check{\varrho}};\tau) + 2\,\xi^0_{(\bar{\alpha})}(c^0_{\check{\varrho}})\,\omega^0_{(\bar{\alpha})}(c^0_{\check{\varrho}})\,\dot{r}^0_{\bar{\alpha}}(c^0_{\check{\varrho}};\tau) + [\omega^0_{(\bar{\alpha})}(c^0_{\check{\varrho}})]^2\,r^0_{\bar{\alpha}}(c^0_{\check{\varrho}};\tau) = R^0_{\bar{\alpha}}(c^0_{\check{\varrho}};\tau)$$
$$\text{(no sum on } \bar{\alpha}) \qquad (4.217)$$

- First-order (ϵ^1 terms, $\check{N}$ systems of $\bar{N}$ uncoupled linear ordinary differential equations for $r^{,\check{\rho}}_{\bar{\alpha}}(c^0_{\check{\varrho}};\tau)$, $\check{\rho}=1,2,\ldots,\check{N}$, $\bar{\alpha}=1,2,\ldots,\bar{N}$)

$$\ddot{r}^{,\check{\rho}}_{\bar{\alpha}}(c^0_{\check{\varrho}};\tau) + 2\,\xi^0_{(\bar{\alpha})}(c^0_{\check{\varrho}})\,\omega^0_{(\bar{\alpha})}(c^0_{\check{\varrho}})\,\dot{r}^{,\check{\rho}}_{\bar{\alpha}}(c^0_{\check{\varrho}};\tau) + [\omega^0_{(\bar{\alpha})}(c^0_{\check{\varrho}})]^2\,r^{,\check{\rho}}_{\bar{\alpha}}(c^0_{\check{\varrho}};\tau) = R^{\check{\rho}}_{\bar{\alpha}}(c^0_{\check{\varrho}};\tau)$$
$$\text{(no sum on } \bar{\alpha}) \qquad (4.218)$$

- Second-order (ϵ^2 terms, one system of $\bar{N}$ uncoupled linear ordinary differential equations for $r^{(2)}_{\bar{\alpha}}(c^0_{\check{\varrho}};\tau)$, $\bar{\alpha}=1,2,\ldots,\bar{N}$)

$$\ddot{r}^{(2)}_{\bar{\alpha}}(c^0_{\check{\varrho}};\tau) + 2\,\xi^0_{(\bar{\alpha})}(c^0_{\check{\varrho}})\,\omega^0_{(\bar{\alpha})}(c^0_{\check{\varrho}})\,\dot{r}^{(2)}_{\bar{\alpha}}(c^0_{\check{\varrho}};\tau) + [\omega^0_{(\bar{\alpha})}(c^0_{\check{\varrho}})]^2\,r^{(2)}_{\bar{\alpha}}(c^0_{\check{\varrho}};\tau) = R^{(2)}_{\bar{\alpha}}(c^0_{\check{\varrho}};\tau)$$
$$\text{(no sum on } \bar{\alpha}) \qquad (4.219)$$

The notation for the zeroth-, first- and second-order vectors of the modal displacements $r^0_{\bar{\alpha}}$, $r^{,\check{\rho}}_{\bar{\alpha}}$ and $r^{(2)}_{\bar{\alpha}}$ and modal loads $R^0_{\bar{\alpha}}$, $R^{\check{\rho}}_{\bar{\alpha}}$ and $R^{(2)}_{\bar{\alpha}}$ have been employed such that:

$$\begin{aligned} r^0_{\bar{\alpha}} &= \phi^{0\mathrm{T}}_{\bar{\alpha}\alpha}\,q^0_\alpha & \quad r^{,\check{\rho}}_{\bar{\alpha}} &= \phi^{0\mathrm{T}}_{\bar{\alpha}\alpha}\,q^{,\check{\rho}}_\alpha & \quad r^{(2)}_{\bar{\alpha}} &= \phi^{0\mathrm{T}}_{\bar{\alpha}\alpha}\,q^{(2)}_\alpha \\ R^0_{\bar{\alpha}} &= \phi^{0\mathrm{T}}_{\bar{\alpha}\alpha}\,Q^0_\alpha & \quad R^{\check{\rho}}_{\bar{\alpha}} &= \phi^{0\mathrm{T}}_{\bar{\alpha}\alpha}\,Q^{\check{\rho}}_\alpha & \quad R^{(2)}_{\bar{\alpha}} &= \phi^{0\mathrm{T}}_{\bar{\alpha}\alpha}\,Q^{(2)}_\alpha \end{aligned} \qquad (4.220)$$

Having solved Eqs. (4.217)–(4.219) for the modal response $r^0_{\bar{\alpha}}$, $r^{,\check{\rho}}_{\bar{\alpha}}$ and $r^{(2)}_{\bar{\alpha}}$ the displacement response q^0_α, $q^{,\check{\rho}}_\alpha$ and $q^{(2)}_\alpha$ of the uncorrelated systems expressed by Eqs. (4.155)–(4.157) can be recovered by modal superposition, cf. Eq. (4.214).

To sum up the findings of this section, the following steps have to be done in order to analyse the response of a stochastic system by using the two-fold superposition technique:

(a) The random field $b_r(x_k)$ is discretized using Eqs. (4.1)–(4.8), and the set of nodal random variables b_ρ is obtained.

(b) The zeroth-order matrices of mass $M^0_{\alpha\beta}(b^0_\varrho)$, damping $C^0_{\alpha\beta}(b^0_\varrho)$ and stiffness $K^0_{\alpha\beta}(b^0_\varrho)$ and load vector $Q^0_\alpha(b^0_\varrho;\tau)$ are assembled and the first- and second-order matrices $M^{,\rho}_{\alpha\beta}(b^0_\varrho)$, $C^{,\rho}_{\alpha\beta}(b^0_\varrho)$, $K^{,\rho}_{\alpha\beta}(b^0_\varrho)$ and $M^{,\rho\sigma}_{\alpha\beta}(b^0_\varrho)$, $C^{,\rho\sigma}_{\alpha\beta}(b^0_\varrho)$, $K^{,\rho\sigma}_{\alpha\beta}(b^0_\varrho)$ are determined; $\tilde{N}+2$ 'correlated' systems each of N linear simultaneous ordinary differential equations (4.17)–(4.19) may then be formed.

(c) The transformation of the correlated random variables b_ρ to the uncorrelated ones $c_{\check{\rho}}$ is carried out via the standard eigenproblem (4.149). By Eqs. (4.151)–(4.152), the zeroth-, first- and second-order functions involved in Eqs. (4.17)–(4.19) are transformed into the standard normal space of $c_{\check{\rho}}$, resulting in $M^0_{\alpha\beta}(c^0_{\check{\varrho}})$, $C^0_{\alpha\beta}(c^0_{\check{\varrho}})$, $K^0_{\alpha\beta}(c^0_{\check{\varrho}})$, $Q^0_\alpha(c^0_{\check{\varrho}};\tau)$; $M^{,\check{\rho}}_{\alpha\beta}(c^0_{\check{\varrho}})$, $C^{,\check{\rho}}_{\alpha\beta}(c^0_{\check{\varrho}})$, $K^{,\check{\rho}}_{\alpha\beta}(c^0_{\check{\varrho}})$ and $M^{,\check{\rho}\check{\rho}}_{\alpha\beta}(c^0_{\check{\varrho}})$, $C^{,\check{\rho}\check{\rho}}_{\alpha\beta}(c^0_{\check{\varrho}})$, $K^{,\check{\rho}\check{\rho}}_{\alpha\beta}(c^0_{\check{\varrho}})$. The 'uncorrelated' equations (4.155)–(4.157) so obtained consist of $\check{N}+2$ ($\check{N} \ll \tilde{N}$) systems of N simultaneous equations each.

(d) Equations (4.155)–(4.157) expressed in the generalized coordinates $q_\alpha^0(c_\varrho^0;\tau)$, $q_\alpha^{,\check{\rho}}(c_\varrho^0;\tau)$ and $q_\alpha^{(2)}(c_\varrho^0;\tau)$ are transformed into the normalized space of the modal coordinates $r_{\bar{\alpha}}^0(c_\varrho^0;\tau)$, $r_{\bar{\alpha}}^{,\check{\rho}}(c_\varrho^0;\tau)$ and $r_{\bar{\alpha}}^{(2)}(c_\varrho^0;\tau)$ through the generalized eigenproblem (4.215),(4.216) and Eqs. (4.220), leading to the system of Eqs. (4.217)–(4.219) involving $\check{N}+2$ uncoupled systems of $\bar{N}$ ($\bar{N} \ll N$) separate equations.

(e) Each of the $(\check{N}+2)\times\bar{N}$ linear ordinary differential equations (4.217)–(4.219) is solved independently for the zeroth-, first- and second-order modal responses, i.e. $r_{\bar{\alpha}}^0$, $r_{\bar{\alpha}}^{,\check{\rho}}$ and $r_{\bar{\alpha}}^{(2)}$, by any direct step-by-step integration scheme. The zeroth-, first- and second-order uncoupled equations are processed successively (in parallel with the backward shift Δt for the first-order solutions, $2\Delta t$ for the second-order solutions) and the right-hand side of the second-order equations is determined by randomness superposition, Eq. $(4.159)_2$.

(f) The zeroth-, first- and second-order 'uncorrelated' responses q_α^0, $q_\alpha^{,\check{\rho}}$ and $q_\alpha^{(2)}$ are evaluated from the modal responses by modal superposition, Eqs. (4.214).

(g) The first two moments for the time-dependent nodal displacements, element strains and stresses are evaluated by Eqs. (4.191)–(4.196), in which the cross-covariances are computed by randomness superposition.

4.5.4 Secularity Elimination

The formal solutions to dynamic systems obtained on the basis of the Taylor expansion do not generally have to converge. However, such sequential solutions are often more effective than those based on uniformly and absolutely convergent series, which is because a power expansion may give an adequate approximation by employing only a few terms. This is why they have been extensively used in many problems of applied mathematics and, particularly, in the field of computational engineering.

In the second-order perturbation method we accept a priori that the expansion is limited to the first two terms. This can produce invalid solutions owing to the appearance of the so-called *secular terms* which grow indefinitely with time. Such an unbounded solution may occur even for systems that are known to possess a bounded solution, such as conservative systems. The effect is much more significant in the stochastic sensitivity analysis, since the expectations of the sensitivity gradient coefficients are bilinear functions of the nodal displacements of the structural and adjoint systems, while their covariances are square functions of these quantities. It has been shown above that the natural frequencies (or linear operators on the left-hand sides) are the same for all the zeroth-, first- and second-order equations in the systems (4.17)–(4.19). This is also the case for the transformed equations (4.155)–(4.157) and the uncoupled systems (4.217)–(4.219). The zeroth-order equations are only excited by zeroth-order external forces, so generally no resonance occurs. However, the first- and second-order forcing sequences are functions of the zeroth-order response, which implies resonant excitation until the transient part of the zeroth-order solution is damped away. Hence, a rough application of the second-order perturbation approach whereby only the amplitude is altered may not always be satisfactory; and the resonant part involved on the right-hand side of the first- and second-order equations necessarily has to be removed or weighted to maintain the validity of the solution when a relatively long duration of the system response is required. In other words, a modification to prevent the occurrence of the secular terms must alter both the amplitude and the period of vibrations.

To deal with the secularity problem some theoretical methodologies have been developed in [18,90,96,105]. However, most of this work was concerned with single-degree-of-freedom systems treated analytically; little work has been done on the numerical elimination of secularities for multi-degree-of-freedom systems. An efficient algorithm based on the fast Fourier transform of the sine and cosine pair was developed in [81]. The numerical procedure presented below is an alternative form of it. Namely, the real-valued signals of the forcing functions are discretized and then converted into a complex-valued sequence of half length, the fast Fourier transform being carried out on the complex-valued sequence. The concept of the discrete Fourier transform, the philosophy of FFT and the advantages of this modern computational technique have been described in detail in Sections 1.2.2 and 1.2.3. We shall briefly discuss here how to use FFT to eliminate secularities in practical engineering computation.

Let us consider a time-varying forcing function $Q(\tau)$, $\tau \in [0, T]$, treated as an input signal of a discrete system, i.e. as a sequence of sample values $Q(t_n)$ in which $t_n = n\Delta t$, Δt being the sampling interval. Employing the concept of the zero-order hold we generate $Q(\tau)$ in the form of a staircase, a sampling interval apart, as

$$Q_n = Q(t_n) = Q(n\Delta t) \qquad \begin{array}{l} t_n \in \big[n\Delta t, (n+1)\Delta t\big) \\ n = 0, 1, \ldots, 2L \end{array} \tag{4.221}$$

where $2L = T/\Delta t$ is the sampling length. Next, the discrete real-valued signal is assembled and converted into a complex-valued sequence of the discrete time signal p_k of the form

$$p_k = p(t_k) = Q(t_{2k}) + \mathrm{i}\, Q(t_{2k+1}) \qquad k = 0, 1, \ldots, L \tag{4.222}$$

where the half sampling length L is known as the *Fourier transform degree* or the *complex sequence length.*

In what follows we shall use the angular frequency ω rather than the circular frequency f used for the derivations in Section 1.2.2. This is because with the symmetry of the factors outside the integrals the latter representation is more elegant in a mathematical sense, whereas in computational applications the former directly provides a physical interpretation for the solution of the system eigenvalue problem. In view of this the finite-range direct transform $P(\omega, T)$ of p_k can be discretized via a sequence of discrete signals in the form, cf. Eqs. (1.135) and (1.140)

$$\begin{aligned} P_j(\omega, T) \;=\; P(\omega_j, T) \;&=\; \frac{T}{L}\sum_{k=0}^{L-1} p(t_k)\,\exp(-\mathrm{i}\,\omega_j t_k) \\ &=\; \frac{T}{L}\sum_{k=0}^{L-1} p_k\,\exp\Big(-\mathrm{i}\frac{2\pi}{L}\,jk\Big) \end{aligned} \tag{4.223}$$

where $j = 0, 1, \ldots, L-1$

$$\omega_j = j\,\Delta\omega \qquad\qquad \Delta\omega = \frac{2\pi}{T} \qquad\qquad 0 \leq \omega_j \leq \Omega \tag{4.224}$$

and $\Omega = \pi/(\Delta t)$ is the angular Nyquist frequency, i.e. the highest frequency appearing in the data record corresponding to the sampling interval Δt. The choice of an appropriate value for the sampling interval and, if necessary, of a data processing technique for reducing the computational cost and simultaneously handling the aliasing problem were discussed in Section 1.2.2. The algorithm for the fast Fourier analysis on a complex-valued sequence, cf. Section 1.2.3, can be used to evaluate the discretized one-sided Fourier spectrum $P(\omega, T)$ of the time sequence of the input signal $p_k(\tau, T)$ in the form of a discretized frequency sequence.

It is observed from Eqs. (4.218) and (4.219) that since the first- and second-order modal forcing records are already decoupled, only the terms related to their corresponding natural frequencies will produce secularity. To remove the secular terms the coefficients $P_j(\omega)$ of the frequency sequence which lie within a specified range of the modal systems are eliminated or weighted adequately, thereby removing the resonant part from the Fourier spectra. In other words, the coefficients associated with the j-th frequencies adjacent to the natural frequency $\omega^0_{(\hat{\alpha})}$ are almost entirely eliminated (weighted), whereas the coefficients separated from this frequency domain are unaffected. Thus, in the direct Fourier transform the coefficients of a frequency sequence are assumed to cause secularities when they satisfy the following condition:

$$\omega^0_{(\hat{\alpha})} - \Delta\Omega \leq \omega_j \leq \omega^0_{(\hat{\alpha})} + \Delta\Omega \tag{4.225}$$

where $\Delta\Omega$ is a specified *frequency range.* There exist many data windows to filtrate secular frequencies; the frequency weighting functions mentioned below are frequently applied [96,105]:

$$\begin{aligned} &\text{triangular} && P_j := P_j\frac{|\,\omega^0_{(\hat{\alpha})} - \omega_j\,|}{\Delta\Omega} \\ &\text{cosine} && P_j := P_j\Big[1 - \cos\frac{\pi(\omega^0_{(\hat{\alpha})} - \omega_j)}{2\,\Delta\Omega}\Big] \\ &\text{cosine-squared} && P_j := \tfrac{1}{2}\,P_j\Big[1 - \cos\frac{\pi(\omega^0_{(\hat{\alpha})} - \omega_j)}{\Delta\Omega}\Big] \end{aligned} \tag{4.226}$$

Once the coefficients P_j of the frequency sequence are weighted (secular terms are eliminated) the discrete inverse Fourier transform of the form

$$\begin{aligned} p_k(\tau,\Omega) &= P(t_k,\Omega) = \frac{1}{T}\sum_{j=0}^{L-1} P(\omega_j)\,\exp(\mathrm{i}\,\omega_j t_k) \\ &= \frac{1}{T}\sum_{j=0}^{L-1} P_j\,\exp\Big(\mathrm{i}\frac{2\pi}{L}jk\Big) \qquad k=0,1,\dots,L-1 \end{aligned} \tag{4.227}$$

is carried out by the fast Fourier synthesis to obtain the complex-valued forcing record p_k with no resonant part included. Finally, the real-valued forcing sequence in which the secular terms are already eliminated can be recovered from the real and imaginary parts of the discrete output signal p_k as

$$\begin{aligned} Q_{2k}(\tau) &= \mathrm{Re}\,p_k(\tau) \\ Q_{2k+1}(\tau) &= \mathrm{Im}\,p_k(\tau) \end{aligned} \qquad k=0,1,\dots,L-1 \tag{4.228}$$

A drawback of stochastic finite element applications in structural dynamics is that the accuracy may deteriorate even with slight structural damping. Therefore, to ensure that all statistical results are valid for a long response duration, the secularity elimination cannot be neglected. On the other hand, to reduce computational cost there is a need to incorporate an effective algorithm to simulate the forcing function with a large sampling length in any stochastic finite element code. The use of the FFT is most suitable in this regard.

4.5.5 Adjoint Variable Technique

In some structural mechanics problems we are interested in the behaviour of only a few state variable components in the local domain of the whole region taken up by a structure, rather than in analysing all state variables in the entire region. Design sensitivity, strain localization, stress concentration and local stability are examples of such situations. The reliability analysis, wherein the failure probability frequently initiates in a small domain of interest may also be quoted. In such cases calculations following the general scheme presented in the preceding sections would be ineffective, since all the first-order nodal displacement vectors, with almost all entries not taking part in further computations, would have to be determined.

As shown in Chapter 3, the second-order perturbation approach results in stochastic finite element models of random systems which consist of three sets of equations. The zeroth-order equations describe the equilibrium of all generalized forces acting on the corresponding deterministic system, i.e. the system with parameters and state variables evaluated at the spatial expectations of random field variables. The first- and second-order equations have the physical interpretation of the first and second variations of the generalized forces with respect to the change of the random variables, respectively. The first- and second-order nodal displacements (state variables) can in turn be treated as the first- and second-order displacement sensitivities with respect to variations of design parameters which are described stochastically. In this sense, the deterministic structural design sensitivity problem and the stochastic system analysis based on the perturbation approach are quite similar.

In view of these observations the philosophy of the adjoint variable approach to structural design sensitivity analysis of deterministic systems, cf. Section 2.8, can be applied to the stochastic finite element analysis as well. It has been observed for the class of problems mentioned above that the second-order terms may contribute very little to the total value of the response expectations obtained. In other words, the first-order Taylor expansion can be employed resulting in the first-order accuracy of both the expectations and covariances of state variables. Without any loss of generality let us consider the application of the adjoint technique to the static problem described in terms of the uncorrelated random variables $c_{\check{\rho}}$, cf. (4.172)–(4.174). Because of the first-order accuracy required the first-order perturbations are used, so that Eqs. (4.172)–(4.174) become:

- Zeroth-order (ϵ^0 terms, one system of N linear simultaneous algebraic equations for $q^0_\alpha(c^0_\varrho)$, $\alpha = 1, 2, \ldots, N$)

$$K^0_{\alpha\beta}\, q^0_\beta \;=\; Q^0_\alpha \tag{4.229}$$

- First-order (ϵ^1 terms, $\check{N}$ systems of N linear simultaneous algebraic equations for $q^{,\check{\rho}}_\alpha(c^0_\varrho)$, $\check{\rho} = 1, 2, \ldots, \check{N}$, $\alpha = 1, 2, \ldots, N$)

$$K^0_{\alpha\beta}\, q^{,\check{\rho}}_\beta \;=\; Q^{,\check{\rho}}_\alpha - K^{,\check{\rho}}_{\alpha\beta}\, q^0_\beta \tag{4.230}$$

As an example, let us find the first-order accurate displacement expectations and covariance for the $\hat{\alpha}$-th and $\hat{\beta}$-th components. Employing the adjoint method we define an adjoint variable vector λ_α, with $\alpha = 1, 2, \ldots, N$, such that

$$K^0_{\alpha\beta}\, \lambda_{\beta(\hat{\alpha})} \;=\; \delta_{\alpha(\hat{\alpha})} \tag{4.231}$$

with $\delta_{\alpha(\hat{\alpha})}$ being the unit adjoint load vector with unit value at the $\hat{\alpha}$-th entry and zeros elsewhere

$$\delta_{\alpha(\hat{\alpha})} \;=\; \begin{cases} 1 & \text{for} \quad \alpha = \hat{\alpha} \\ 0 & \text{for} \quad \alpha \neq \hat{\alpha} \end{cases} \tag{4.232}$$

Since $K^0_{\alpha\beta}$ is symmetric the $\hat{\alpha}$-th components of the first-order displacement vectors $q^{,\check{\rho}}_\alpha$, $\check{\rho} = 1, 2, \ldots, \check{N}$, read

$$q^{,\check{\rho}}_{\hat{\alpha}} \;=\; \delta_{\alpha(\hat{\alpha})}\, \overset{-1}{K}{}^0_{\alpha\beta} \left(Q^{,\check{\rho}}_\beta - K^{,\check{\rho}}_{\beta\gamma}\, q^0_\gamma \right) \;=\; \lambda_{\alpha(\hat{\alpha})} \left(Q^{,\check{\rho}}_\alpha - K^{,\check{\rho}}_{\alpha\beta}\, q^0_\beta \right) \tag{4.233}$$

We note that: (i) the adjoint equation (4.231) describes the equilibrium of a *deterministic* system; (ii) the adjoint displacement vector $\lambda_{\alpha(\hat{\alpha})}$ is independent of random variables; it may be shown from Eq. (4.233) that all the $\hat{\alpha}$-th components of the displacement first derivatives with respect to $c_{\check{\rho}}$ over the entire region Ω are evaluated with one vector $\lambda_{\alpha(\hat{\alpha})}$, whose components are treated as influence coefficients (or weighting numbers) at the first-order loads; and (iii) the adjoint load vector $\delta_{\alpha(\hat{\alpha})}$ can be treated as an additional zeroth-order load vector so that q_α^0 and $\lambda_{\alpha(\hat{\alpha})}$ may be solved for simultaneously. These observations are important to reduce the amount of computations in the adjoint algorithm. Thus, the solution phase of the adjoint variable scheme proceeds as follows:

(a) The zeroth-order displacement vector and adjoint displacement vectors q_α^0 and $\lambda_{\alpha(\hat{\alpha})}$, $\lambda_{\alpha(\hat{\beta})}$ are solved for from the one zeroth-order system

$$\left[K_{\alpha\beta}^0\right]\left[q_\beta^0,\ \lambda_{\beta(\hat{\alpha})},\ \lambda_{\beta(\hat{\beta})}\right] = \left[Q_\alpha^0,\ \delta_{\alpha(\hat{\alpha})},\ \delta_{\alpha(\hat{\beta})}\right] \tag{4.234}$$

The first-order accurate displacement expectations for the $\hat{\alpha}$-th and $\hat{\beta}$-th components can be directly extracted from q_α^0 to get, cf. Eq. (3.15)

$$\begin{aligned} E[q_{\hat{\alpha}}] &= q_{\hat{\alpha}}^0 \\ E[q_{\hat{\beta}}] &= q_{\hat{\beta}}^0 \end{aligned} \tag{4.235}$$

(b) The $\hat{\alpha}$-th and $\hat{\beta}$-th displacement derivatives $q_{\hat{\alpha}}^{,\check{\rho}}$ and $q_{\hat{\beta}}^{,\check{\rho}}$, $\check{\rho} = 1, 2, \ldots, \check{N}$, are computed by Eq. (4.233), resulting in the first-order accurate displacement covariance for the $\hat{\alpha}$-th and $\hat{\beta}$-th components of the form (cf. Eq. (4.192) reduced to the static case)

$$\begin{aligned} &\operatorname{Cov}(q_{\hat{\alpha}}, q_{\hat{\beta}}) \\ &= \lambda_{\alpha(\hat{\alpha})}\left[\sum_{\check{\rho}=1}^{\check{N}}\left(Q_\alpha^{,\check{\rho}} - K_{\alpha\beta}^{,\check{\rho}}\, q_\beta^0\right) S_{\mathrm{c}}^{\check{\rho}} \left(Q_\gamma^{,\check{\rho}} - K_{\gamma\delta}^{,\check{\rho}}\, q_\delta^0\right)^{\mathrm{T}}\right]\lambda_{\gamma(\hat{\beta})} \end{aligned} \tag{4.236}$$

Clearly, the number of computations for the first two moments of the nodal displacements have been considerably reduced. By using the adjoint technique the number of solutions required is three, and the solutions are obtained simultaneously at the zeroth-order level, whereas the direct method requires $\check{N} + 1$ solutions (one zeroth-order and $\check{N}$ first-order solutions), and the zeroth-order and first-order equations must be solved separately.

The computational assessment of both the direct and adjoint methods employed in the design sensitivity analysis, as discussed in detail Section 2.8, remains valid in the stochastic case. Thus, the advantages and drawbacks of each method depend on the analysis type, number of load cases, number of constraints, etc. In other words, the choice of one or other method is strongly problem-dependent.

It is repeated in passing that the above computational aspects of the adjoint method have been discussed in the framework of the first-order accurate description of stochastic systems.

4.5.6 Concluding Remarks on SFEM Implementation

The change of basis for random variables in problems of statics, free and forced vibrations and structural buckling enables us to effectively analyse these problems for medium- and large-scale structures. Taking into account the dominance of the highest normal modes in the stochastic response, a large number of random variables representing a broad class of various uncertainties of arbitrary probabilistic distributions can be included simultaneously in one large system to be transformed. A solution of the transformed system, which becomes relatively independent of the size of the initial random variable vector, can be obtained on a small computer at a reasonable computational cost. In this sense the stochastic finite element approach exceeds by far other statistical techniques when dealing with large-scale problems.

The fact that only a few highest modes (random variable transformation) are needed to approximate randomness in the system has the same significance as the fact that several lowest modes (generalized coordinate transformation) are necessary to describe the dynamic response of the system. A common feature of both procedures is that they are based on the change of basis of coordinates through an orthogonal transformation. Clearly, if all the uncorrelated variables in the random variable transformation and all the natural frequencies in the coordinate transformation are used in the analysis, the exact solution should be obtained, i.e. the one corresponding to the original system with correlated random variables solved by a direct step-by-step integration scheme. The stochastic finite element algorithm based on the two-fold superposition procedure seems to offer the most efficient tool for the stochastic analysis of structures.

The advantages of the adjoint method over the direct method make themselves visible in some cases of 'local' analyses if only the first-order expansion is believed to suffice. Effective applications of the adjoint technique combined with the second-order expansion approach are still a problem to be investigated, though. We note that by the definition of the covariance functions the generalized second-order displacements in SFEM analysis are obtained from a single equation, i.e. there are $\check{N}+2$ solutions required instead of $D(D+1)/2$ equations required in the second-order design sensitivity analysis (D being the number of design variables). As shown in [66], to calculate second-order sensitivities with respect to design variables the smallest number of solutions may be achieved by a mixed direct-adjoint method and is equal to $D+2$. In this sense the applicability of the adjoint and direct approaches in SFEM seems to be equivalent. On the other hand, to establish the right-hand side of the second-order equation the adjoint problem must be solved for each component of all the first-order displacement vectors resulting in additional vector multiplications that are not needed in the direct approach. Thus, the adjoint method may be more efficient only if the number of degrees of freedom is smaller than the number of uncorrelated random variables used in the analysis. This will rarely occur in computational practice, though.

It should be emphasized that the accuracy of the solution may turn out to be sensitive to variations in the size of the stochastic finite elements. To ensure an acceptable accuracy the SFEM mesh should be fine enough so that fluctuations of the random field variables can be well approximated. The fluctuations are measured by the correlation length which, in turn, controls the element size. Thus, a fine SFEM mesh must be employed when the correlation length is large. This criterion provides an upper bound on the element size. On the other hand, if the SFEM mesh is excessively fine the nodal random variables may be highly correlated, in which case their correlation (covariance) matrix may become ill-conditioned and the random variable transformation numerically unstable. This criterion provides a lower bound on the element size.

When a combination of the random field discretization and a displacement-type finite element discretization is employed, we observe that: (i) the shape functions adopted for the random variables need not be the same as those used for displacements; (ii) the nodal points used to define the nodal random variables need not coincide with those used to describe the nodal displacements; and (iii) the requirements for the displacement gradient and fluctuation rate (correlation length) and for the stability of the random variable transformation may be different in various domains of the structure considered. Thus, it sometimes proves advantageous to choose different meshes for the displacement field and each of the random fields [62,63]. One may set a finite element mesh satisfying the requirement for the displacement gradient and correlation length, and then select for each random field a mesh coincident with (or coarser than) the chosen finite element mesh such that the corresponding transformations are numerically stable. However, this problem-dependent approach requires complicated procedures for effectively generating various grids in the finite element model. Moreover, the overall influences of the element size on the accuracy and stability of analysis do not seem to have been fully investigated in the literature and require future work to increase the applicability and effectiveness of the approach.

4.6 Numerical Illustrations

Before presenting more advanced computer solutions we shall start this section by discussing some very elementary examples without making reference to computer programs at all.

Example 4.1 Let us begin with probably the simplest structural mechanics problem possible. Fig. 4.1 shows a cantilever bar of length l, cross-sectional

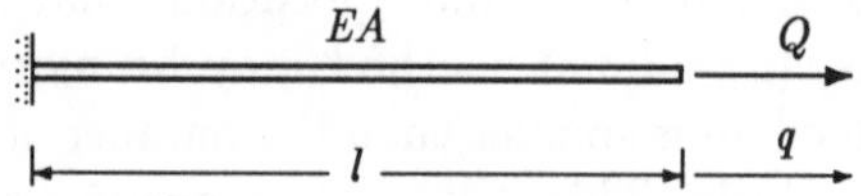

Figure 4.1 Cantilever bar subjected to axial force.

area A and Young's modulus E subjected to axial load Q. Assuming that some of the parameters A, E, l and Q are random, the problem is to evaluate the first two statistical moments for the displacement q at the free end of the bar.

By incorporating randomness in one of the material, geometry or load parameters available into the finite element equilibrium equation $kq = Q$ with $k = EA/l$, the stochastic finite element equations describing static response of the single-random-variable system are obtained as, cf. Eqs. (4.34)–(4.36)

$$\begin{aligned} k^0 q^0 &= Q^0 \\ k^0 q^{,b} &= Q^{,b} - k^{,b} q^0 \\ k^0 q^{(2)} &= \left(Q^{,bb} - 2k^{,b} q^{,b} - k^{,bb} q^0\right) \mathrm{Var}(b) \qquad \text{with} \qquad q^{(2)} = q^{,bb}\, \mathrm{Var}(b) \end{aligned}$$

where $(.)^0$, $(.)^{,b}$ and $(.)^{,bb}$ denote the zeroth, first and second derivatives with respect to the random variable b; these functions are evaluated at the spatial expectation of b. The second-order accurate expectation and first-order accurate variance for the displacement can be expressed as, cf. Eqs. (4.40) and (4.42)

$$E[q] = q^0 + \frac{1}{2} q^{(2)} ; \qquad \mathrm{Var}(q) = (q^{,b})^2\, \mathrm{Var}(b)$$

We now consider all the four possible cases of the single-random-variable problem at hand.

• *Case 1:* The cross-sectional area A is assumed to be a random variable, i.e. $b \equiv A$, while E, l and Q are deterministic. Derivatives of the stiffness and load with respect to A are written as

$$\begin{aligned} k^0 &= \frac{EA^0}{l} ; \qquad & k^{,A} &= \frac{E}{l} ; \qquad & k^{,AA} &= 0 \\ Q^0 &= Q ; & Q^{,A} &= 0 ; & Q^{,AA} &= 0 \end{aligned}$$

and, consequently, the zeroth-, first- and second-order SFEM equations become

$$\begin{aligned} \frac{EA^0}{l} q^0 &= Q \\ \frac{EA^0}{l} q^{,A} &= -\frac{E}{l} q^0 \\ \frac{EA^0}{l} q^{(2)} &= -\frac{2E}{l} q^{,A}\, \mathrm{Var}(A) \qquad \text{with} \qquad q^{(2)} = q^{,AA}\, \mathrm{Var}(A) \end{aligned}$$

Having solved for the zeroth-, first- and second-order displacements as

$$q^0 = \frac{Ql}{EA^0} ; \qquad q^{,A} = -\frac{Ql}{E(A^0)^2} ; \qquad q^{(2)} = \frac{2Ql}{E(A^0)^3} \mathrm{Var}(A)$$

we arrive at the first two statistical moments of the displacement q in the form

$$E[q] = \frac{Ql}{EA^0} \left[1 + \frac{1}{(A^0)^2} \mathrm{Var}(A)\right] ; \qquad \mathrm{Var}(q) = \frac{Q^2\, l^2}{E^2 (A^0)^4} \mathrm{Var}(A)$$

• *Case 2:* Young's modulus E is assumed to be a random variable, i.e. $b \equiv E$, while A, l and Q are deterministic. Quite similarly as in the previous case with A being replaced by E we readily obtain

$$E[q] = \frac{Ql}{E^0 A}\left[1 + \frac{1}{(E^0)^2}\mathrm{Var}(E)\right] \; ; \qquad \mathrm{Var}(q) = \frac{Q^2\, l^2}{(E^0)^4 A^2}\mathrm{Var}(E)$$

• *Case 3:* The length l is assumed to be a random variable, i.e. $b \equiv l$, while E, A and Q are deterministic. We obtain in turn:

derivatives of the stiffness and load with respect to l[1]

$$k^0 = \frac{EA}{l^0} \; ; \qquad k^{,l} = -\frac{EA}{(l^0)^2} \; ; \qquad k^{,ll} = \frac{2EA}{(l^0)^3} \; ;$$
$$Q^0 = Q \; ; \qquad Q^{,l} = 0 \; ; \qquad Q^{,ll} = 0$$

the zeroth-, first- and second-order SFEM equations

$$\frac{EA}{l^0}\, q^0 = Q$$
$$\frac{EA}{l^0}\, q^{,l} = \frac{EA}{(l^0)^2}\, q^0$$
$$\frac{EA}{l^0}\, q^{(2)} = -\frac{2EA}{(l^0)^2}\overbrace{\left(\frac{1}{l^0}\, q^0 - q^{,l}\right)}^{=0}\mathrm{Var}(l) \qquad \text{with} \qquad q^{(2)} = q^{,ll}\,\mathrm{Var}(l)$$

the displacement expectation and variance

$$E[q] = q^0 = \frac{Ql^0}{EA} \; ; \qquad \mathrm{Var}(q) = \frac{Q^2}{(EA)^2}\mathrm{Var}(l)$$

• *Case 4:* The load Q is assumed to be a random variable, i.e. $b \equiv Q$, while E, A and l are deterministic. We obtain in turn:

derivatives of the stiffness and load with respect to Q

$$k^0 = \frac{EA}{l} \; ; \qquad k^{,Q} = 0 \; ; \qquad k^{,QQ} = 0$$
$$Q^{,Q} = 1 \; ; \qquad Q^{,QQ} = 0$$

the zeroth-, first- and second-order SFEM equations

$$\frac{EA}{l}\, q^0 = Q^0$$
$$\frac{EA}{l}\, q^{,Q} = 1$$
$$\frac{EA}{l}\, q^{(2)} = 0$$

---

[1]To avoid confusion it is emphasized that 'l' is not a running index here; we simply use the notation $(.)^{,l} = \partial(.)/\partial l$, l being the random length of the bar.

the displacement expectation and variance

$$E[q] = q^0 = \frac{Q^0 l}{EA}\,; \qquad \mathrm{Var}(q) = \frac{l^2}{(EA)^2}\mathrm{Var}(Q)$$

It is seen from the last two cases in which length l and load Q were consecutively random that only the first-order expansion was required to describe the problem. We may also note that for such a single-random-variable problem as considered above: (i) both the finite element and analytical formulations are identical and (ii) the expressions for the first two displacement moments may be easily verified using the derivations of Section 1.1.3 devoted to the linear transformation of random variables.

Let us now consider the same one-degree-of-freedom problem in the case of a multi-random-variable system. By assuming, for instance, that the variables l and Q are now random and stochastically dependent while E and A are deterministic, we obtain the following SFEM equations:

one equation for the zeroth-order generalized displacement q^0

$$k^0\, q^0 = Q^0$$

two equations for the first-order generalized displacements $q^{,l}$ and $q^{,Q}$

$$\begin{aligned} k^0\, q^{,l} &= Q^{,l} - k^{,l}\, q^0 \\ k^0\, q^{,Q} &= Q^{,Q} - k^{,Q}\, q^0 \end{aligned}$$

one equation for the second-order generalized displacement $q^{(2)}$

$$\begin{aligned} k^0\, q^{(2)} &= Q^{(2)} \\ &= \left(Q^{,\rho\sigma} - 2k^{,\rho}\, q^{,\sigma} - k^{,\rho\sigma} q^0\right) S_{lQ}^{\rho\sigma} \end{aligned}$$

with the indices ρ and σ running over the sequence l and Q and

$$q^{(2)} = q^{,\rho\sigma} S_{lQ}^{\rho\sigma}$$

$$S_{lQ}^{\rho\sigma} = \mathbf{S}_{lQ} = \begin{bmatrix} \mathrm{Var}(l) & \mathrm{Cov}(l,Q) \\ \text{symm.} & \mathrm{Var}(Q) \end{bmatrix}$$

Using the following notation:

$$\mathbf{q}^{(1)} = \{\, q^{,l} \quad q^{,Q} \,\}$$

$$\mathbf{k}^{(1)} = \{\, k^{,l} \quad k^{,Q} \,\} = \left\{ -\frac{EA}{(l^0)^2} \quad 0 \right\}$$

$$\mathbf{K}^{(2)} = \begin{bmatrix} k^{,ll} & k^{,lQ} \\ \text{symm.} & k^{,QQ} \end{bmatrix} = \begin{bmatrix} \dfrac{2EA}{(l^0)^3} & 0 \\ 0 & 0 \end{bmatrix}$$

and noting that the terms $Q^{,\rho\sigma}$ on the right-hand side of the second-order equation for $q^{(2)}$ form a matrix with zero entries, the second-order generalized load $Q^{(2)}$ can be expressed in matrix form as

$$Q^{(2)} = -\left(2\mathbf{k}^{(1)\mathrm{T}}\,\mathbf{S}_{lQ}\,\mathbf{q}^{(1)} + \mathbf{K}^{(2)}\colon\mathbf{S}_{lQ}\,q^{0}\right)$$

where the double dot product of two two-dimensional matrices implies

$$\mathbf{K}^{(2)}\colon\mathbf{S}_{lQ} = k^{,\rho\sigma}\,S_{lQ}^{\rho\sigma}$$

With this notation the zeroth-, first- and second-order equilibrium equations generate a SFEM equation system which can be explicitly written as

$$\frac{EA}{l^0}\,q^0 = Q^0$$

$$\begin{bmatrix} \dfrac{EA}{l^0} & 0 \\ 0 & \dfrac{EA}{l^0} \end{bmatrix}\begin{bmatrix} q^{,l} \\ q^{,Q} \end{bmatrix} = \begin{bmatrix} \dfrac{EA}{(l^0)^2}\,q^0 \\ 1 \end{bmatrix}$$

$$\frac{EA}{l^0}\,q^{(2)} = \frac{2EA}{(l^0)^2}\left[\underbrace{\left(\frac{1}{l^0}\,q^0 - q^{,l}\right)}_{=0}\mathrm{Var}(l) - q^{,Q}\,\mathrm{Cov}(l,Q)\right]$$

Solving for the zeroth-, first- and second-order generalized displacements yields, respectively,

$$q^0 = \frac{Q^0 l^0}{EA}$$

$$q^{,l} = \frac{Q^0}{EA}\,; \qquad q^{,Q} = \frac{l^0}{EA}$$

$$q^{(2)} = -\frac{2}{EA}\,\mathrm{Cov}(l,Q)$$

The second-order accurate expectation and first-order accurate covariance for the displacement can then be obtained as

$$E[q] = q^0 + \tfrac{1}{2}q^{(2)} = \frac{Q^0 l^0}{EA} - \frac{1}{EA}\,\mathrm{Cov}(l,Q)$$

$$\begin{aligned}\mathrm{Var}(q) &= \mathbf{q}^{(1)\mathrm{T}}\mathbf{S}_{lQ}\,\mathbf{q}^{(1)} \\ &= \frac{1}{(EA)^2}\left[(Q^0)^2\,\mathrm{Var}(l) + 2Q^0 l^0\,\mathrm{Cov}(l,Q) + (l^0)^2\,\mathrm{Var}(Q)\right]\end{aligned}$$

By examining the perturbation solution to the two-random-variable cantilever bar problem it is observed that for stochastically independent variables l and Q: (i) the second-order perturbations are useless and (ii) the displacement variance may be evaluated as the algebraic sum of the corresponding solutions to two stochastic systems each with a separate uncorrelated random variable.

Example 4.2 A single-degree-of-freedom dynamical system with the spring stiffness k, dashpot viscous damping coefficient c and discrete mass m, cf. Fig. 4.2, is considered. Assume that: (i) the system spring is linear; (ii) the excitation is

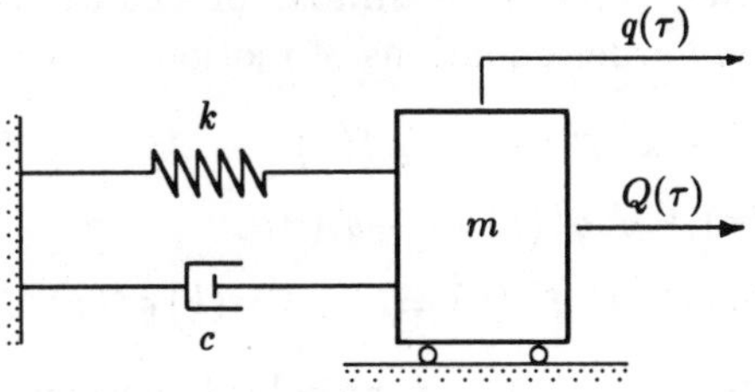

Figure 4.2 Single-degree-of-freedom dynamic system.

given in the form of an externally applied force as $Q(\tau) = \hat{Q} f(\tau)$ where $f(\tau)$ is a deterministic function of time while the magnitude $\hat{Q}$ is time-independent; and (iii) the initial conditions are homogeneous, i.e. $q(0) = 0$ and $\dot{q}(0) = 0$. Assuming that some of the structural and load parameters are random the problem is to evaluate the time distribution of the first two statistical moments for the displacements $q(\tau)$.

There exist various approaches to derive the time response of the deterministic system equivalent to that of Fig. 4.2 under an arbitrary excitation $f(\tau)$, cf. [22,86], for instance. One approach, based on the Fourier transformation, cf. Section 1.2.2, is to represent the excitation through a Fourier series so that the excitation can be considered harmonic; letting the period T of the expanded forcing sequence tend to infinity leads to the response to nonperiodical excitation via a Fourier integral. Another method, based on the concept of the Dirac delta distribution, cf. Section 1.2.1, expresses the forcing function in terms of an infinite sequence of adjacent impulses as their sampling intervals approach zero; the system response $q(\tau)$ at any time instant $\tau = t$ is obtained by superposition of the unit impulse responses by means of the convolution integral

$$q(t) = \int_0^t Q(\tau)\, u(t-\tau)\, \mathrm{d}\tau = \hat{Q} \int_0^t f(\tau)\, u(t-\tau)\, \mathrm{d}\tau \qquad \tau \in [0, T]$$

where $u(\tau)$ is the response to the Dirac-type excitation $\delta(\tau)$ which may be expressed as

$$u(\tau) = u(m, c, k; \tau) = \frac{1}{m\omega_c} \exp(-\xi\omega\tau) \sin \omega_c \tau \qquad \tau \in [0, T]$$

with the damped free vibration frequency ω_c, viscous damping factor ξ and natural frequency ω defined as

$$\omega_c = \omega \left(1 - \xi^2\right)^{\frac{1}{2}}; \qquad \xi = \frac{c}{2m\omega}; \qquad \omega = \left(\frac{k}{m}\right)^{\frac{1}{2}}$$

We consider now two cases of system randomness: (1) k being a random variable alone and (2) k and Q being two randomly dependent variables.

• *Case 1:* The spring stiffness k is assumed to be a random variable while m, c and $\hat{Q}$ are deterministic. Following a similar procedure as in Example 4.1 the zeroth-, first- and second-order equations of motion are written as follows:[1]

$$\begin{aligned} m\,\ddot{q}^0(\tau) + c\,\dot{q}^0(\tau) + k^0\,q^0(\tau) &= \hat{Q}\,f(\tau) \\ m\,\ddot{q}^{,k}(\tau) + c\,\dot{q}^{,k}(\tau) + k^0\,q^{,k}(\tau) &= -q^0(\tau) \\ m\,\ddot{q}^{(2)}(\tau) + c\,\dot{q}^{(2)}(\tau) + k^0\,q^{(2)}(\tau) &= -2\,\mathrm{Var}(k)\,q^{,k}(\tau) \end{aligned}$$

The three 'deterministic' equations can be solved successively for q^0, $q^{,k}$ and $q^{(2)}$ at any time $\tau = t$, $t \in [0, T]$, as

$$\begin{aligned} q^0(t) &= \hat{Q}\int_0^t f(\tau)\,u(m, c, k^0; t-\tau)\,\mathrm{d}\tau = \hat{Q}\int_0^t f(\tau)\,u^0(t-\tau)\,\mathrm{d}\tau \\ q^{,k}(t) &= -\int_0^t q^0(\tau)\,u^0(t-\tau)\,\mathrm{d}\tau \\ q^{(2)}(t) &= -\,\mathrm{Var}(k)\int_0^t q^{,k}(\tau)\,u^0(t-\tau)\,\mathrm{d}\tau \end{aligned}$$

Having solved for the zeroth-, first- and second-order responses the displacement expectation at any time instant $\tau = t$ and cross-covariance at any (t_1, t_2) are readily obtained as

$$\begin{aligned} E[q(t)] &= q^0(t) + \tfrac{1}{2}q^{(2)}(t) \\ &= \int_0^t \left[\hat{Q}\,f(\tau) - \mathrm{Var}(k)q^{,k}(\tau)\right] u^0(t-\tau)\,\mathrm{d}\tau \\ \mathrm{Cov}\big(q(t_1), q(t_2)\big) &= q^{,k}(t_1)\,q^{,k}(t_2)\,\mathrm{Var}(k) \\ &= \left[\int_0^{t_1} q^0(\tau)\,u^0(t_1-\tau)\mathrm{d}\tau\right]\left[\int_0^{t_2} q^0(\nu)\,u^0(t_2-\nu)\mathrm{d}\nu\right]\mathrm{Var}(k) \end{aligned}$$

• *Case 2:* The spring stiffness k and load magnitude $\hat{Q}$ are assumed to be dependent random variables while m and c are deterministic. Since all the second and mixed derivatives with respect to the random variables are zero we get the SFEM equation system in the form:

one equation for the zeroth-order generalized displacement q^0

$$m\,\ddot{q}^0(\tau) + c\,\dot{q}^0(\tau) + k^0\,q^0(\tau) = \hat{Q}^0\,f(\tau)$$

two equations for the first-order generalized displacements $q^{,k}$ and $q^{,\hat{Q}}$

$$\begin{aligned} m\,\ddot{q}^{,k}(\tau) + c\,\dot{q}^{,k}(\tau) + k^0\,q^{,k}(\tau) &= -q^0(\tau) \\ m\,\ddot{q}^{,\hat{Q}}(\tau) + c\,\dot{q}^{,\hat{Q}}(\tau) + k^0\,q^{,\hat{Q}}(\tau) &= f(\tau) \end{aligned}$$

[1] Similarly as before in Example 4.1, 'k' is clearly not a running index here but stands for the random stiffness variable k.

one equation for the second-order generalized displacement $q^{(2)}$

$$m\,\ddot{q}^{(2)}(\tau) + c\,\dot{q}^{(2)}(\tau) + k^0\,q^{(2)}(\tau) = -2\left[\mathrm{Var}(k)\,q^{,k}(\tau) + \mathrm{Cov}(k,\hat{Q})\,q^{,\hat{Q}}(\tau)\right]$$

These equations may be solved sequentially for the zeroth-, first- and second-order generalized displacements at any time instant $\tau = t$ to get, respectively

$$q^0(t) = \hat{Q}^0 \int_0^t f(\tau)\,u^0(t-\tau)\,\mathrm{d}\tau$$

$$\left.\begin{aligned} q^{,k}(t) &= -\int_0^t q^0(\tau)\,u^0(t-\tau)\,\mathrm{d}\tau \\ q^{,\hat{Q}}(t) &= \int_0^t f(\tau)\,u^0(t-\tau)\,\mathrm{d}\tau \end{aligned}\right\}$$

$$q^{(2)}(t) = -2\int_0^t \left[\mathrm{Var}(k)\,q^{,k}(\tau) + \mathrm{Cov}(k,\hat{Q})\,q^{,Q}(\tau)\right] u^0(t-\tau)\,\mathrm{d}\tau$$

This implies that

$$\begin{aligned} E[q(t)] &= q^0(t) + \tfrac{1}{2}q^{(2)}(t) \\ &= \int_0^t \left[\hat{Q}^0 f(\tau) - \mathrm{Var}(k)q^{,k}(\tau) - \mathrm{Cov}(k,\hat{Q})\,q^{,Q}(\tau)\right] u^0(t-\tau)\,\mathrm{d}\tau \end{aligned}$$

$$\begin{aligned} \mathrm{Cov}\big(q(t_1),q(t_2)\big) &= \begin{bmatrix} q^{,k}(t_1) \\ q^{,\hat{Q}}(t_1) \end{bmatrix}^{\mathrm{T}} \begin{bmatrix} \mathrm{Var}(k) & \mathrm{Cov}(k,\hat{Q}) \\ \mathrm{Cov}(k,\hat{Q}) & \mathrm{Var}(\hat{Q}) \end{bmatrix} \begin{bmatrix} q^{,k}(t_2) \\ q^{,\hat{Q}}(t_2) \end{bmatrix} \\ &= \left[\int_0^{t_1} q^0(\tau)\,u^0(t_1-\tau)\,\mathrm{d}\tau\right]\left[\int_0^{t_2} q^0(\nu)\,u^0(t_2-\nu)\,\mathrm{d}\nu\right]\mathrm{Var}(k) \\ &\quad - \left[\int_0^{t_1} q^0(\tau)\,u^0(t_1-\tau)\,\mathrm{d}\tau\right]\left[\int_0^{t_2} f(\nu)\,u^0(t_2-\nu)\,\mathrm{d}\nu\right]\mathrm{Cov}(k,\hat{Q}) \\ &\quad - \left[\int_0^{t_1} f(\tau)\,u^0(t_1-\tau)\,\mathrm{d}\tau\right]\left[\int_0^{t_2} q^0(\nu)\,u^0(t_2-\nu)\,\mathrm{d}\nu\right]\mathrm{Cov}(k,\hat{Q}) \\ &\quad + \left[\int_0^{t_1} f(\tau)\,u^0(t_1-\tau)\,\mathrm{d}\tau y\right]\left[\int_0^{t_2} f(\nu)\,u^0(t_2-\nu)\,\mathrm{d}\nu\right]\mathrm{Var}(\hat{Q}) \end{aligned}$$

Example 4.3 The static response of a two-element cantilever bar subjected to axial loads, Fig. 4.3, is considered in this example. Deterministic input data for

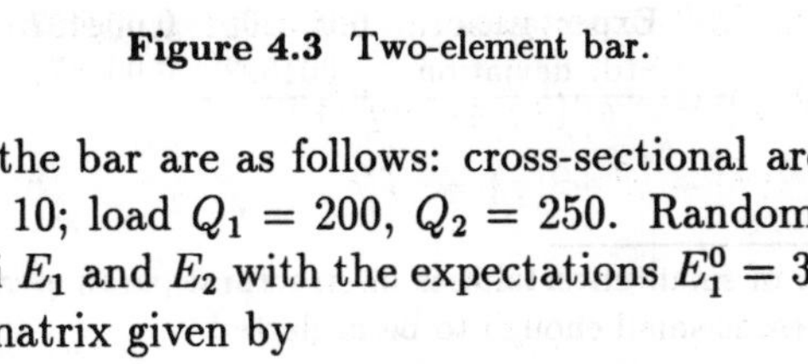

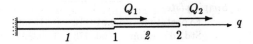

Figure 4.3 Two-element bar.

the two elements of the bar are as follows: cross-sectional area $A_1 = 4$, $A_2 = 2$; length $l_1 = 15$, $l_2 = 10$; load $Q_1 = 200$, $Q_2 = 250$. Randomness is assumed in both Young's moduli E_1 and E_2 with the expectations $E_1^0 = 3000$ and $E_2^0 = 2500$ and the covariance matrix given by

$$\mathrm{Cov}(E_\rho, E_\sigma) = \begin{bmatrix} 2.5 \times 10^5 & 2.0 \times 10^5 \\ \text{symm.} & 1.6 \times 10^5 \end{bmatrix}$$

The SFEM model has two degrees of freedom, two bar finite elements and two random variables. The displacement expectations and covariances computed at nodes 1 and 2 are obtained as

$$E[q_1] = 0.578125\ ; \qquad E[q_2] = 1.090925$$

$$\mathrm{Cov}(q_\alpha, q_\beta) = \begin{bmatrix} 8.789062 \times 10^{-3} & 1.628906 \times 10^{-2} \\ \text{symm.} & 3.018906 \times 10^{-2} \end{bmatrix}$$

This result may easily be verified by performing simple hand calculations.

Example 4.4 In this example the static deflections of a simply supported square plate and a clamped square plate with random thickness subjected to a concentrated centre load are examined. Deterministic input data are: plate dimensions $l_x = l_y = 100$, Young's modulus $E = 1.0 \times 10^7$, Poisson's ratio $\nu = 0.3$ and load $Q = 100$. The expectation of the thickness $E[t(x, y)]$ is equal to 1.0 and the homogeneous covariance function for the thickness is assumed to be

$$\mathrm{Cov}\Big(t(x_\rho, y_\rho), t(x_\sigma, y_\sigma)\Big) = \sigma_t^2 \exp\left(\frac{x_\rho - x_\sigma}{d_x}\right)^2 \exp\left(\frac{y_\rho - y_\sigma}{d_y}\right)^2$$

with standard deviation $\sigma_t = 0.1$ and d_x and d_y the decay factors in the x and y directions determined from the end-point correlations $\mu(t[x(0), y], t[x(l), y])$ and $\mu(t[x, y(0)], t[x, y(l)])$ (see [73] for further details). In each problem the thickness end-point correlation coefficients are set to 1 and the deflections at the centre of the plates are calculated. Because of symmetry[1] only a quarter of the plate is considered; it is discretized by an 8×8-element mesh. The SFEM solutions are compared with those based on the basic random variable (BRV) technique [73], cf. Table 4.1. Since with the stochastic finite element model the thickness of each element is modelled as a random variable the finite element mesh controls the

Table 4.1 Square plates. Stochastic deflection.

	SFEM	BRV
Simple plate		
Expectation	0.013112	0.013386
Std. deviation	0.003473	0.003894
Fixed plate		
Expectation	0.006205	0.006467
Std. deviation	0.001592	0.001877

[1]By the assumption of small uncertainties in structural parameters the asymmetry due to the thickness randomness is small enough to be neglected.

accuracy of the thickness approximation and the accuracy of the displacement interpolation. This can be less efficient than the BRV because the displacement field is usually more complex than the covariances of the thickness. On the other hand, in the BRV more difficult integrals must be evaluated owing to the higher order polynomials. Moreover, the accuracy of the covariances in the latter approach depends on the number of Legendre functions used in the approximation and the adaptation of the procedures to existing 'deterministic' finite element codes is not straightforward, particularly if various types of random variables are simultaneously required in a problem.

Example 4.5 The statement of the problem is depicted in Fig. 4.4. The spring–

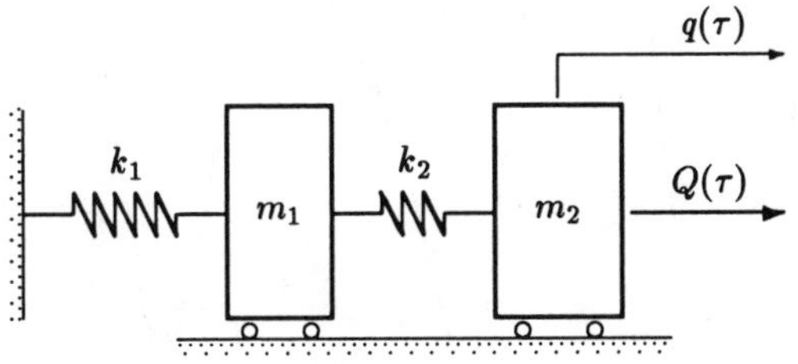

Figure 4.4 Two-degree-of-freedom spring–mass system.

mass system with two degrees of freedom is modelled through a two-element cantilever bar with concentrated masses. Deterministic input data are: structural masses $m_1 = 0.372$ and $m_2 = 0.248$; proportional damping factor $\xi = 0.03$; and sinusoidal forcing function $Q(\tau) = 25.0 \times 10^6 \sin(2000\tau)$. The random spring constants are normally distributed with expectations $k_1 = 24.0 \times 10^6$, and $k_2 = 12.0 \times 10^6$; the coefficient of variation α is equal to 0.1. The system of eight equations is solved by a combination of the two-fold superposition method and the Wilson θ-algorithm ($\theta = 1.4$) for each of the uncoupled modal equations. The number of sampling intervals is equal to 1024 with the time increment $\Delta t = 1.2 \times 10^{-5}$. The elimination of secularities is performed on sequences of 1024 Fourier terms and the frequency range $\Delta\Omega$ is equal to 0.15 of the first natural frequency. The expectation and standard deviation for the displacement at the free end of the system are plotted in Fig. 4.5 and compared with the results obtained by Monte Carlo simulation (MCS). It is seen that the expectation computed by the SFEM model compares well with the MCS [79], whereas differences in the variance are more significant, which is probably due to the different frequency ranges used for eliminating secularities.

Example 4.6 In this example we consider the effects of secularity elimination on the SFEM solution in the problem of wave propagation in a cantilever bar. Deterministic input data are: cross-sectional area $A=6$, length $l=1000$, mass density $\varrho=0.00776$, and Heaviside's forcing function $Q(\tau)=25.0\times10^4$. The expectation, correlation function and coefficient of variation of the Young's modulus E_ρ along

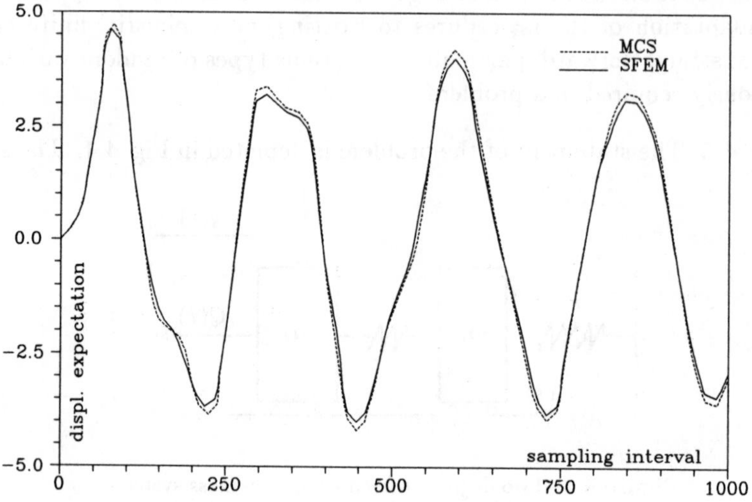

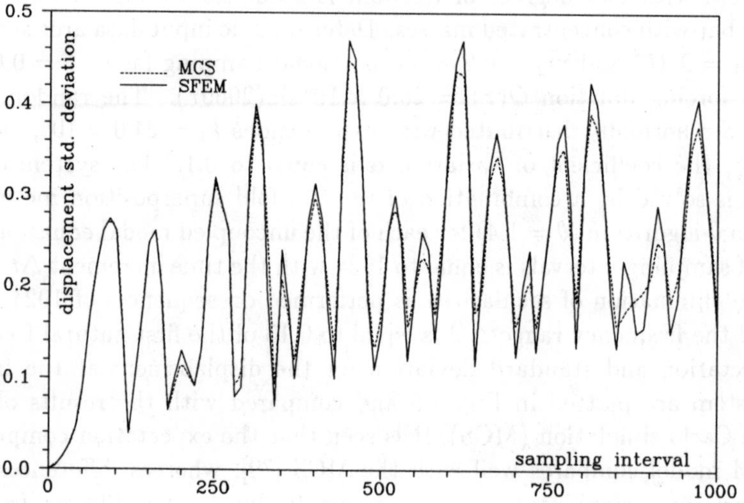

Figure 4.5 Spring–mass system. Stochastic displacement response.

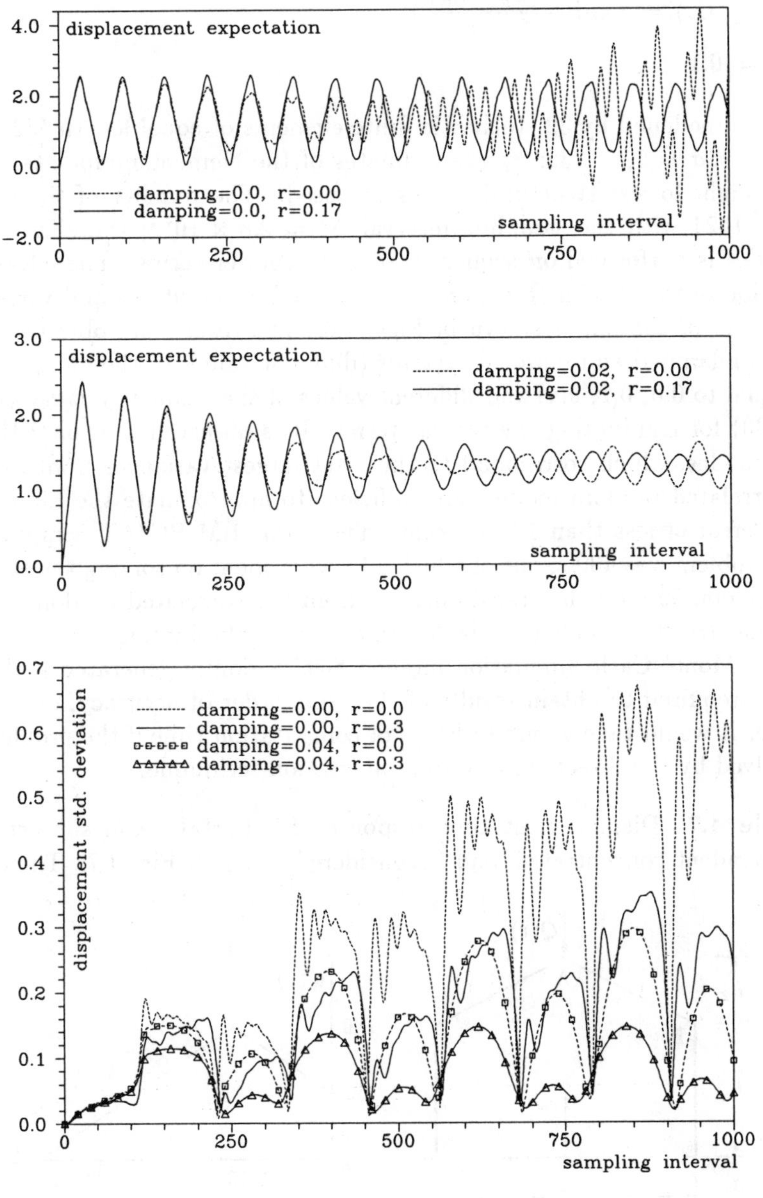

Figure 4.6 32-element bar. Secularity elimination effects.

the axis of the bar are assumed as follows:

$$E[E_\rho] = E^0\left(1 + \frac{x_\rho}{\vartheta l}\right) ; \qquad E^0 = 3.0 \times 10^7 ; \qquad \vartheta = 10$$

$$\mu(E_\rho, E_\sigma) = \exp\left(-\frac{|x_\rho - x_\sigma|}{\vartheta l}\right)$$

$$\alpha = 0.1$$

The bar is modelled by 32 stochastic finite elements of equal length (32 random variables, $\rho, \sigma = 1, \ldots, 32$; x_ρ are ordinates of the element midpoints). In the analysis eight lowest structural modes are used. The number of time steps is equal to 1024 with the sampling interval $\Delta t = 2.8 \times 10^{-4}$; the elimination of secularities is performed on sequences of 1024 Fourier terms. The effect of the elimination of the secular terms on the computed expectations and variances at the free end of the bar is shown in Figs. 4.6. The results are obtained for the cases of undamped and damped systems (different values of the damping factor ξ are equal to 0.0, 0.2, 0.4) and different values of the frequency range ($r = 0.0$, 0.17, 0.30) for eliminating the secular terms. It is observed that only the eight highest modes, which correspond to the eight largest variances from the set of 32 uncorrelated random modes, are sufficient to approximate the random field with an error of less than 1.5 per cent. The total IBM PC/AT computer time for the problem was 430 s, out of which 43 s were spent on solving the structural eigenproblem, 50 s on the transformation from the correlated random variables to the uncorrelated random variables and 56 s on eliminating secularities. By using the Monte Carlo simulation method 800 randomly generated realizations would be required to obtain results of the same order of accuracy; on the same computer it would take about 12 hours of computation time if the problem were to be solved by the direct step-by-step integration technique.

Example 4.7 Displacement time response of a portal beam subjected to a time-dependent concentrated load is considered next, cf. Fig. 4.7. The element

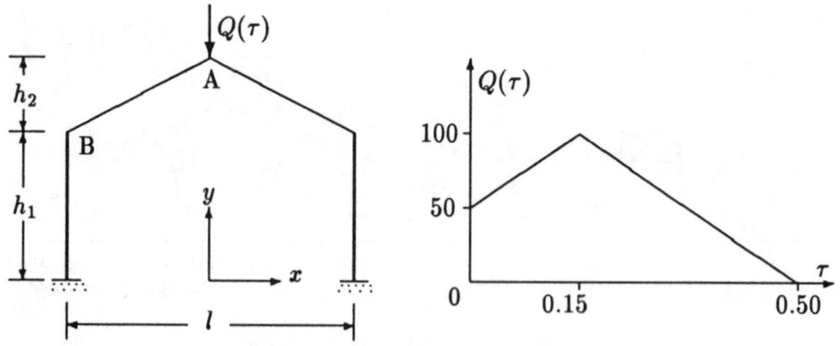

Figure 4.7 100-element portal frame.

cross-sectional areas A_ρ, $\rho = 1, \ldots, 100$, are assumed as random variables. The vertical and the oblique beams are discretized by 25 equal-length elements each.

The respective expectation, correlation function and coefficient of variation for the cross-sectional areas are assumed as follows:

$$E[A_\rho] = A^0 = 10.0$$
$$\mu(A_\rho, A_\sigma) = \exp\Big(-\frac{|x_\rho - x_\sigma|}{\lambda}\Big)$$
$$\alpha = 0.05$$

with $\lambda = 0.5$ while $x_1 = 0.0$, $x_2 = 0.01$, $\ldots$, $x_{100} = 1.0$. The following deterministic data are assumed: base length $l = 200$, height $h_1 = 100$, height $h_2 = 50$, Young's modulus $E = 2.0 \times 10^7$, Poisson's ratio $\nu = 0.2$, mass density $\varrho = 0.001$ and damping factor $\xi = 0.05$. To solve the initial-value problem the two-fold superposition technique is used with 10 lowest eigenpairs; the set of 100 correlated random variables is transformed to a set of uncorrelated variables, out of which the 10 highest modes are used in the calculations. The equations are integrated with respect to time using 512 sampling intervals (with a time step length of $\Delta t = 0.001$). The secular terms are eliminated using the frequency range factor $r = 0.15$ and 1024 Fourier terms. Fig. 4.8 shows the time behaviour of the expectations, variances and autocovariances for the vertical displacement $q_y(\tau)$ at the apex A and the horizontal displacement $q_x(\tau)$ at the point B; the deterministic solutions are also given for comparison.

Example 4.8 This example is concerned with the time response of the displacements of a cylindrical shell clamped at the boundaries and subjected to a concentrated time-varying load at the midpoint A, cf. Fig. 4.9. The element thicknesses $t_\rho(x, y)$ are assumed to be random variables; their expectation, correlation function and coefficient of variation are given, respectively, as follows:

$$E[t_\rho] = t^0 = 0.05$$
$$\mu(t_\rho, t_\sigma) = \exp\Big(-\frac{|x_\rho - x_\sigma|}{\lambda_x}\Big)\exp\Big(-\frac{|y_\rho - y_\sigma|}{\lambda_y}\Big)$$
$$\alpha = 0.05$$

with $\lambda_x = 0.2d$ and $\lambda_y = 0.1d$. The following deterministic data are employed: length $l = 9.6$, diameter $d = 8.0$, Young's modulus $E = 1.0 \times 10^5$, Poisson's ratio $\nu = 0.3$, mass density $\varrho = 0.02$ and damping factor $\xi = 0.05$. Because of symmetry (the coefficient of variation α is small) only one-quarter of the shell is considered. The finite element mesh includes 60 constant-thickness rectangular elements (60 random variables, $\rho = 1, 2, \ldots, 60$), and the total number of degrees of freedom is 313. The initial-value problem is solved by using the mode superposition technique with the 10 lowest eigenpairs. The set of 60 correlated random variables is transformed to a set of uncorrelated variables, out of which the 10 highest variables are used in computations. The equations are integrated with respect to time using 512 time steps (with a the step length of $\Delta t = 0.001$). The secular terms are eliminated using the secularity elimination factor $r = 0.15$ and 1024 Fourier terms. The time behaviour of the expectations and standard deviation for the z-displacement at the midpoint A (compared with the deterministic solution) are displayed in Fig. 4.9.

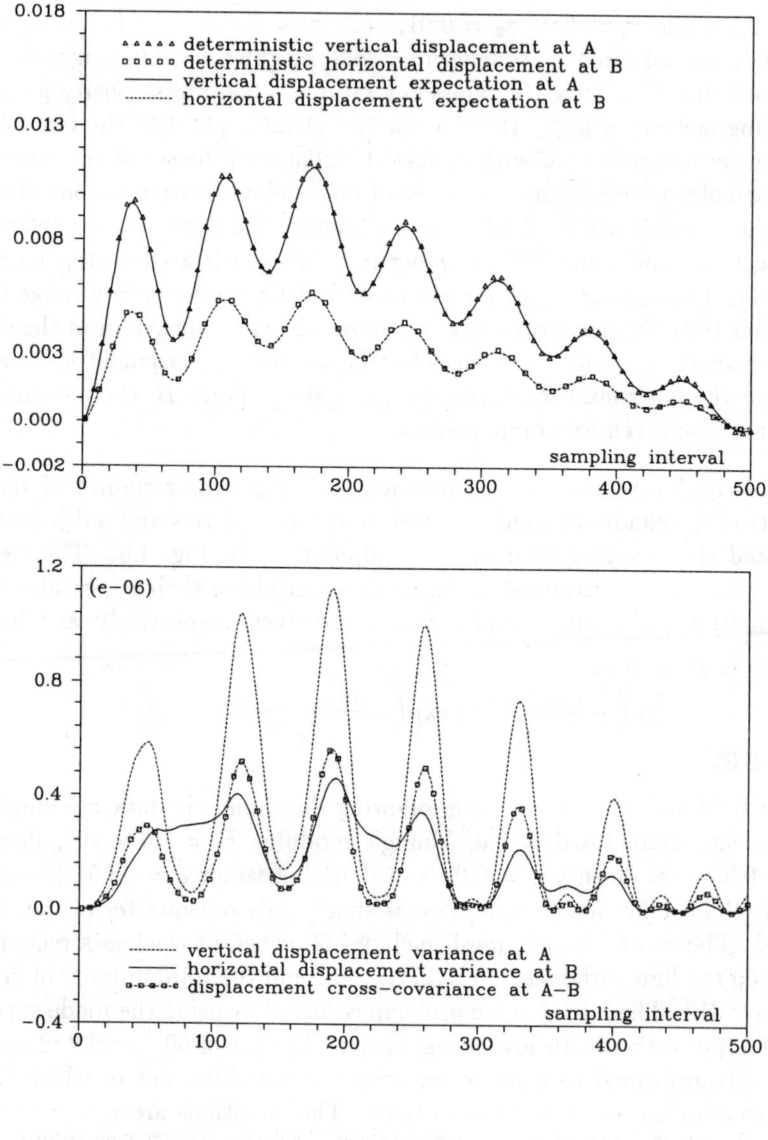

Figure 4.8 100-element frame. Displacement time response.

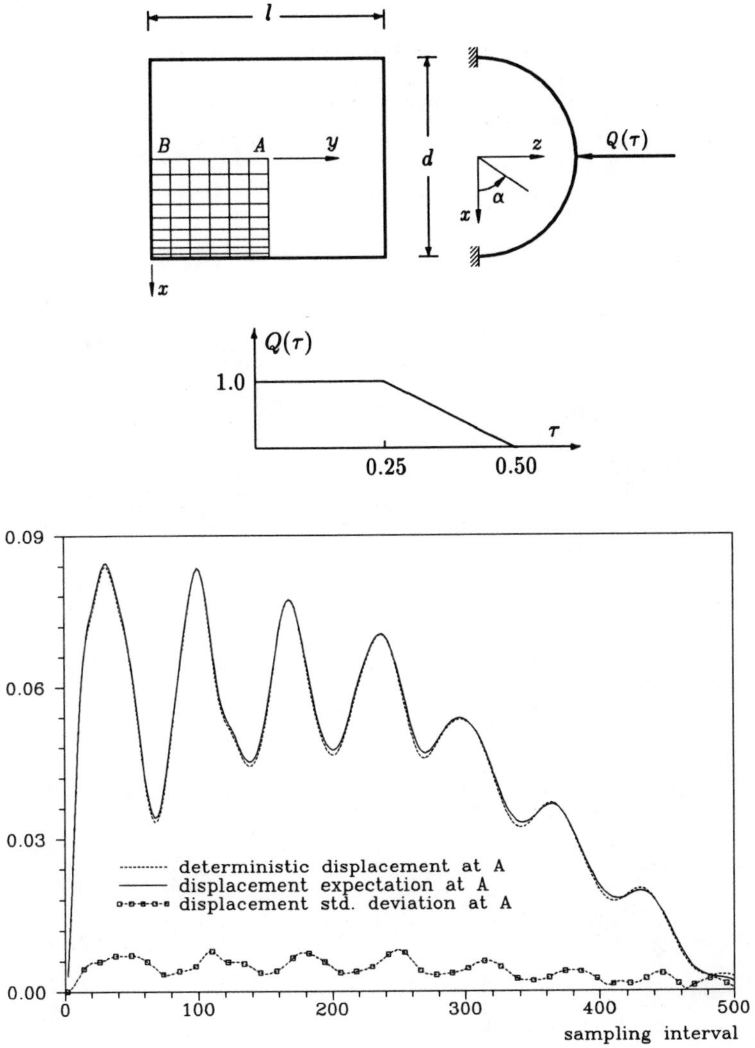

Figure 4.9 60-element cylindrical shell. Displacement time response.

Chapter 5

Stochastic Sensitivity: Static Problems

5.1 Finite Element Formulation

As we repeatedly stressed in Section 2.8 of this book, structural design sensitivity analysis provides information that is useful for any complete structural analysis algorithm. This information is also important in its own right since it represents trends for the structural performance functions and can therefore be used for optimization purposes. In view of the stochastic finite element techniques described in Chapter 4 it seems highly interesting, and important in practice, to consider structural response sensitivity with respect to design parameters subject to probabilistic variations. As we shall see below, design sensitivity and stochastic analyses are formally strikingly similar in terms of the methodology and computer implementation, which greatly facilitate the combined analysis.

The techniques of stochastic structural design sensitivity (SSDS) analysis are concerned with the change of stochastic structural response due to variations in stochastically and/or deterministically described design parameters, i.e. with the evaluation of probabilistic distributions of the sensitivity gradients. In this chapter we consider the SSDS problem for systems under static loads; Chapter 7 deals with the dynamic case. For the sake of brevity only the finite element formulation of the SSDS problems will be considered; in other words, we shall base our discussion in this chapter on the FEM static equations although considering sensitivity for nondiscretized structures followed by the application of FEM may be considered formally more appropriate. A number of test examples will be discussed at the end of the appropriate chapters; ready-to-run computer programs will be described in Chapters 6 and 8, respectively.

Let us consider the static structural response of a linear elastic system with N degrees of freedom defined by the functional

$$\mathcal{G}(h^d, b_\rho) = G\left[q_\alpha(h^d, b_\rho), h^d\right] \qquad d = 1,2,\ldots,D\ ;\ \ \rho = 1,2,\ldots,\tilde{N}\ ;\ \ \alpha = 1,2,\ldots,N \tag{5.1}$$

where G is a given function of its arguments, h^d is a D-dimensional vector of design variables, b_p is an $\tilde{N}$-dimensional vector of random variables defined at selected points in the structure and q_α is an N-dimensional vector of nodal displacement-type parameters. Some or all components of the vectors h^d and b_p may coincide. The nodal displacements satisfy the matrix equilibrium equation

$$K_{\alpha\beta}(h^d, b_p)\, q_\beta(h^d, b_p) \;=\; Q_\alpha(h^d, b_p) \tag{5.2}$$

Since the stiffness matrix $K_{\alpha\beta}$ and the load vector Q_α are functions of the design parameter vector h^d and the random variable vector b_p, the nodal displacement vector q_α is assumed to be an implicit function of these variables as well. The functions $K_{\alpha\beta}(h^d, b_p)$ and $Q_\alpha(h^d, b_p)$ are explicitly given; we assume these functions to be continuously differentiable with respect to h^d and b_p which implies that $q_\beta(h^d, b_p)$ have the same properties.

Given the equilibrium equation (5.2) our objective is to evaluate the probabilistic distributions of the gradient of the functional (5.1) with respect to the design parameters h^d. Since the design variables and random field variables appear in the coefficients of the linear operator, the system equations are generally nonlinear random functions with respect to these variables. Differentiation of Eq. (5.1) with respect to the design variables h^d by the Leibnitz rule leads to

$$\mathcal{G}^{,d} \;=\; G^{,d} + G_{,\alpha}\, q_\alpha^{,d} \tag{5.3}$$

where the symbols $(.)^{,d}$ and $(.)_{,\alpha}$ denote, as before, the first partial derivatives with respect to the d-th design variable and the α-th nodal displacement, respectively. In order to rewrite Eq. (5.3) explicitly in terms of variations in the design parameters the adjoint variable technique will be employed here since it offers certain distinct advantages over the direct differentiation technique when used for the stochastic structural design studies. Defining an adjoint variable vector $\lambda = \{\lambda_\alpha(h^d, b_p)\}$, $\alpha = 1, 2, \ldots, N$, and noting that the stiffness matrix is positive definite and symmetric, the adjoint system equations takes the form, cf. Eq. (2.129)

$$K_{\alpha\beta}(h^d, b_p)\, \lambda_\beta(h^d, b_p) \;=\; G_{,\alpha}(h^d, b_p) \tag{5.4}$$

It is obvious from Eq. (5.4) that the adjoint variables λ_α are implicit functions of the random variables b_p. The expression for the coefficients of the structural response sensitivity gradient becomes, cf. Eq. (2.130)

$$\mathcal{G}^{,d} \;=\; G^{,d} + \lambda_\alpha \left(Q_\alpha^{,d} - K_{\alpha\beta}^{,d}\, q_\beta \right) \tag{5.5}$$

So far the derivations for the stochastic sensitivity problem have been identical to the ones presented in Section 2.8.1 for the deterministic sensitivity problem, the only difference being that all the quantities involved now are random functions.

In order to more explicitly incorporate into the analysis uncertainties in structural material, element geometry and external load the stochastic finite element technique as presented in the preceding chapter will now be employed. To this aim we first approximate the random adjoint variable vector λ_α by expanding it about the expectations b^0_ϱ of the random variables b_ρ as, cf. Eq. (4.12)

$$\begin{aligned}\lambda_\alpha(h^d, b_\varrho) &= \lambda^0_\alpha(h^d, b^0_\varrho) + \epsilon\, \lambda^{,\rho}_\alpha(h^d, b^0_\varrho)\, \Delta b_\rho \\ &+ \frac{1}{2}\, \epsilon^2\, \lambda^{,\rho\sigma}_\alpha(h^d, b^0_\varrho)\, \Delta b_\rho\, \Delta b_\sigma\end{aligned} \tag{5.6}$$

Furthermore, the random functions $G_{.\alpha}$, $G^{.d}$, $K^{.d}_{\alpha\beta}$ and $Q^{.d}_\alpha$, are similarly expanded to yield

$$\begin{aligned}G_{.\alpha}(h^d, b_\varrho) &= G^0_{.\alpha}(h^d, b^0_\varrho) + \epsilon\, G^{,\rho}_{.\alpha}(h^d, b^0_\varrho)\, \Delta b_\rho \\ &+ \frac{1}{2}\, \epsilon^2\, G^{,\rho\sigma}_{.\alpha}(h^d, b^0_\varrho)\, \Delta b_\rho\, \Delta b_\sigma \\ G^{.d}(h^d, b_\varrho) &= G^{.d0}(h^d, b^0_\varrho) + \epsilon\, G^{.d,\rho}(h^d, b^0_\varrho)\, \Delta b_\rho \\ &+ \frac{1}{2}\, \epsilon^2\, G^{.d,\rho\sigma}(h^d, b^0_\varrho)\, \Delta b_\rho\, \Delta b_\sigma \\ K^{.d}_{\alpha\beta}(h^d, b_\varrho) &= K^{.d0}_{\alpha\beta}(h^d, b^0_\varrho) + \epsilon\, K^{.d,\rho}_{\alpha\beta}(h^d, b^0_\varrho)\, \Delta b_\rho \\ &+ \frac{1}{2}\, \epsilon^2\, K^{.d,\rho\sigma}_{\alpha\beta}(h^d, b^0_\varrho)\, \Delta b_\rho\, \Delta b_\sigma \\ Q^{.d}_\alpha(h^d, b_\varrho) &= Q^{.d0}_\alpha(h^d, b^0_\varrho) + \epsilon\, Q^{.d,\rho}_\alpha(h^d, b^0_\varrho)\, \Delta b_\rho \\ &+ \frac{1}{2}\, \epsilon^2\, Q^{.d,\rho\sigma}_\alpha(h^d, b^0_\varrho)\, \Delta b_\rho\, \Delta b_\sigma\end{aligned} \tag{5.7}$$

where the notation explained in the preceding chapter is employed.

The functions involving the partial derivatives of stiffness and loading with respect to the random variables can be calculated exactly by partial differentiation or, if this turns out to be awkward, by the finite difference technique or by the least squares fit method as mentioned in Section 2.8.1 and 4.5.1. The partial derivatives $G^{,\rho}_{.\alpha}$, $G^{.d,\rho}$ and $G^{,\rho\sigma}_{.\alpha}$, $G^{.d,\rho\sigma}$ can generally be determined by one of the latter methods only as they are implicit functions of random variables.

Substituting Eqs. (5.6) and (5.7) into Eqs. (5.2) and (5.4) and equating the coefficients of like power of ϵ the following sequence of equation sets is generated for both the primary and adjoint systems, cf. Eqs. (4.34)–(4.36):

- Zeroth-order (ϵ^0 terms, one pair of systems of N linear simultaneous algebraic equations for $q^0_\alpha(h^d, b^0_\varrho)$ and $\lambda^0_\alpha(h^d, b^0_\varrho)$, respectively)

$$\begin{aligned}K^0_{\alpha\beta}(h^d, b^0_\varrho)\, q^0_\beta(h^d, b^0_\varrho) &= Q^0_\alpha(h^d, b^0_\varrho) \\ K^0_{\alpha\beta}(h^d, b^0_\varrho)\, \lambda^0_\beta(h^d, b^0_\varrho) &= G^0_{.\alpha}(h^d, b^0_\varrho)\end{aligned} \tag{5.8}$$

- First-order (ϵ^1 terms, $\tilde{N}$ pairs of systems of N linear simultaneous algebraic equations for $q^{,\rho}_\alpha(h^d, b^0_\varrho)$ and $\lambda^{,\rho}_\alpha(h^d, b^0_\varrho)$, respectively)

$$\begin{aligned}K^0_{\alpha\beta}(h^d, b^0_\varrho)\, q^{,\rho}_\beta(h^d, b^0_\varrho) &= Q^{,\rho}_\alpha(h^d, b^0_\varrho) - K^{,\rho}_{\alpha\beta}(h^d, b^0_\varrho)\, q^0_\beta(h^d, b^0_\varrho) \\ K^0_{\alpha\beta}(h^d, b^0_\varrho)\, \lambda^{,\rho}_\beta(h^d, b^0_\varrho) &= G^{,\rho}_{.\alpha}(h^d, b^0_\varrho) - K^{,\rho}_{\alpha\beta}(h^d, b^0_\varrho)\, \lambda^0_\beta(h^d, b^0_\varrho)\end{aligned} \tag{5.9}$$

- Second-order (ϵ^2 terms, one pair of systems of N linear simultaneous algebraic equations for $q_\alpha^{(2)}(h^d, b_\varrho^0)$ and $\lambda_\alpha^{(2)}(h^d, b_\varrho^0)$, respectively)

$$\begin{aligned} K_{\alpha\beta}^0(h^d, b_\varrho^0)\, q_\beta^{(2)}(h^d, b_\varrho^0) &= \Big[Q_\alpha^{,\rho\sigma}(h^d, b_\varrho^0) \\ &\quad - 2\, K_{\alpha\beta}^{,\rho}(h^d, b_\varrho^0)\, q_\beta^{,\sigma}(h^d, b_\varrho^0) - K_{\alpha\beta}^{,\rho\sigma}(h^d, b_\varrho^0)\, q_\beta^0(h^d, b_\varrho^0) \Big]\, S_{\mathrm{b}}^{\rho\sigma} \\ K_{\alpha\beta}^0(h^d, b_\varrho^0)\, \lambda_\beta^{(2)}(h^d, b_\varrho^0) &= \Big[G_{\cdot\alpha}^{,\rho\sigma}(h^d, b_\varrho^0) \\ &\quad - 2\, K_{\alpha\beta}^{,\rho}(h^d, b_\varrho^0)\, \lambda_\beta^{,\sigma}(h^d, b_\varrho^0) - K_{\alpha\beta}^{,\rho\sigma}(h^d, b_\varrho^0)\, \lambda_\beta^0(h^d, b_\varrho^0) \Big]\, S_{\mathrm{b}}^{\rho\sigma} \end{aligned} \tag{5.10}$$

where $d = 1, 2, \ldots, D$; $\alpha, \beta = 1, 2, \ldots, N$; $\rho, \sigma = 1, 2, \ldots, \tilde{N}$ and

$$\begin{aligned} q_\alpha^{(2)}(h^d, b_\varrho^0) &= q_\alpha^{,\rho\sigma}(h^d, b_\varrho^0)\, S_{\mathrm{b}}^{\rho\sigma} \\ \lambda_\alpha^{(2)}(h^d, b_\varrho^0) &= \lambda_\alpha^{,\rho\sigma}(h^d, b_\varrho^0)\, S_{\mathrm{b}}^{\rho\sigma} \end{aligned} \tag{5.11}$$

Note again that the pair of second-order equations has been obtained by multiplying the $\tilde{N}$-variate probability density function $p_{\tilde{N}}(b_1, b_2, \ldots, b_{\tilde{N}})$ by the ϵ^2-terms and integrating over the domain of the random field variables b_ρ, cf. Eq. (3.11).

Having solved systems of Eqs. (5.8)–(5.10) the probabilistic distributions for the sensitivity gradient coefficients of structural response can be evaluated. The first two moments for the sensitivity gradient coefficients are defined as

$$E[\mathcal{G}^{\cdot d}] = \underbrace{\int_{-\infty}^{+\infty}\int_{-\infty}^{+\infty}\cdots\int_{-\infty}^{+\infty}}_{\tilde{N}\text{-fold}} \mathcal{G}^{\cdot d}\, p_{\tilde{N}}(b_1, b_2, \ldots, b_{\tilde{N}})\, \mathrm{d}b_1 \mathrm{d}b_2 \ldots \mathrm{d}b_{\tilde{N}} \tag{5.12}$$

$$\begin{aligned} \mathrm{Cov}(\mathcal{G}^{\cdot d}, \mathcal{G}^{\cdot e}) &= S_{\mathcal{G}\cdot}^{de} \\ &= \underbrace{\int_{-\infty}^{+\infty}\int_{-\infty}^{+\infty}\cdots\int_{-\infty}^{+\infty}}_{\tilde{N}\text{-fold}} \big\{ \mathcal{G}^{\cdot d} - E[\mathcal{G}^{\cdot d}] \big\} \big\{ \mathcal{G}^{\cdot e} - E[\mathcal{G}^{\cdot e}] \big\} \\ &\qquad \times\, p_{\tilde{N}}(b_1, b_2, \ldots, b_{\tilde{N}})\, \mathrm{d}b_1 \mathrm{d}b_2 \ldots \mathrm{d}b_{\tilde{N}} \end{aligned} \tag{5.13}$$

which, after using Eq. (5.5) leads to

$$E[\mathcal{G}^{\cdot d}] = E[G^{\cdot d}] + E[\lambda_\alpha\, Q_\alpha^{\cdot d}] - E[\lambda_\alpha\, K_{\alpha\beta}^{\cdot d}\, q_\beta] \tag{5.14}$$

$$\begin{aligned} S_{\mathcal{G}\cdot}^{de} &= \underbrace{\int_{-\infty}^{+\infty}\int_{-\infty}^{+\infty}\cdots\int_{-\infty}^{+\infty}}_{\tilde{N}\text{-fold}} \Big\{ G^{\cdot d} + \lambda_\alpha \left(Q_\alpha^{\cdot d} - K_{\alpha\beta}^{\cdot d}\, q_\beta \right) - E[\mathcal{G}^{\cdot d}] \Big\} \\ &\qquad \times \Big\{ G^{\cdot e} + \lambda_\gamma \left(Q_\gamma^{\cdot e} - K_{\gamma\delta}^{\cdot e}\, q_\delta \right) - E[\mathcal{G}^{\cdot e}] \Big\} \\ &\qquad \times\, p_{\tilde{N}}(b_1, b_2, \ldots, b_{\tilde{N}})\, \mathrm{d}b_1 \mathrm{d}b_2 \ldots \mathrm{d}b_{\tilde{N}} \end{aligned} \tag{5.15}$$

Employing the expansion equations for q_α, λ_α, $K_{\alpha\beta}^{\cdot d}$, $Q_\alpha^{\cdot d}$ and $G^{\cdot d}$ in Eqs. (5.14) and (5.15) yields the second-order accurate expectations and the first-order accurate covariances of the sensitivity gradient coefficients as

$$\begin{aligned} E[\mathcal{G}^{\cdot d}] &= G^{0.d} - \tfrac{1}{2} \lambda_\alpha^0\, K_{\alpha\beta}^{0.d}\, q_\beta^{(2)} + \left(\lambda_\alpha^0 + \tfrac{1}{2} \lambda_\alpha^{(2)} \right) \mathcal{A}_\alpha^d \\ &\quad + \tfrac{1}{2} \left(G^{\cdot d,\rho\sigma} + 2\, \lambda_\alpha^{,\rho}\, \mathcal{B}_\alpha^{d\sigma} + \lambda_\alpha^0\, \mathcal{C}_\alpha^{d\rho\sigma} \right) S_{\mathrm{b}}^{\rho\sigma} \end{aligned} \tag{5.16}$$

$$\begin{aligned} S_{\mathcal{G}\cdot}^{de} &= \Big[G^{\cdot d,\rho}\, G^{\cdot e,\sigma} + \lambda_\alpha^{,\rho} \left(G^{\cdot d,\sigma} \mathcal{A}_\alpha^e + G^{\cdot e,\sigma} \mathcal{A}_\alpha^d \right) \\ &\quad + \lambda_\alpha^0 \left(G^{\cdot d,\rho}\, \mathcal{B}_\alpha^{e\sigma} + G^{\cdot e,\rho}\, \mathcal{B}_\alpha^{d\sigma} \right) + \lambda_\alpha^{,\rho}\, \lambda_\beta^{,\sigma}\, \mathcal{A}_\alpha^d\, \mathcal{A}_\beta^e \\ &\quad + \lambda_\alpha^{,\rho}\, \lambda_\beta^0 \left(\mathcal{A}_\alpha^d\, \mathcal{B}_\beta^{e\sigma} + \mathcal{A}_\beta^e\, \mathcal{B}_\alpha^{d\sigma} \right) + \lambda_\alpha^0\, \lambda_\beta^0\, \mathcal{B}_\beta^{d\rho}\, \mathcal{B}_\alpha^{e\sigma} \Big]\, S_{\mathrm{b}}^{\rho\sigma} \end{aligned} \tag{5.17}$$

where $d, e = 1, 2, \ldots, D$; $\rho, \sigma = 1, 2, \ldots, \tilde{N}$; $\alpha, \beta = 1, 2, \ldots, N$ and the following notation has been employed:

$$\begin{aligned}
\mathcal{A}^d_\alpha &= Q^{0.d}_\alpha - K^{0.d}_{\alpha\beta}\, q^0_\beta \\
\mathcal{B}^{d\rho}_\alpha &= Q^{.d,\rho}_\alpha - K^{.d,\rho}_{\alpha\beta}\, q^0_\beta - K^{0.d}_{\alpha\beta}\, q^{,\rho}_\beta \\
\mathcal{C}^{d\rho\sigma}_\alpha &= Q^{.d,\rho\sigma}_\alpha - K^{.d,\rho}_{\alpha\beta}\, q^{,\sigma}_\beta - K^{.d,\sigma}_{\alpha\beta}\, q^{,\rho}_\beta - K^{.d,\rho\sigma}_{\alpha\beta}\, q^0_\beta
\end{aligned} \tag{5.18}$$

Note that Eq. (5.17) has been obtained by noting that: (i) the terms involving the first variation vanish by the definition of the first two moments; (ii) the covariance matrix $S^{\rho\sigma}_{\mathrm{b}}$ is symmetric; and (iii) the repeated indices are dummy. The system of Eqs. (5.8)–(5.10) and (5.16), (5.17) describes in closed form the stochastic structural sensitivity problem in the finite element framework. The formulation requires only $2(\tilde{N}+2)$ solutions of algebraic equation systems while the number of sample systems needed to be solved in a statistical approach would be much larger if we wanted to obtain results of similar accuracy.

Knowing that the sensitivity gradient coefficients are random variables with the expectations and covariances given by Eqs. (5.16) and (5.17), respectively, we may attempt to determine the first-order change in the response functional (5.1) due to the variations in random (and/or deterministic) design parameters. In other words, we may seek the expectation and covariance of the random variable $\Delta\mathcal{G}$ defined as

$$\Delta\mathcal{G} = \mathcal{G}^{.d}\,\Delta h^d \qquad \text{(summation over } d = 1, 2, \ldots, D) \tag{5.19}$$

given the probabilistic characteristics of the design variables Δh^d. The explicit expressions for $E[\Delta\mathcal{G}]$ and $\mathrm{Var}(\Delta\mathcal{G})$ can be directly inferred from the derivations of Section 1.1.5.

Let us assume we have computed $E[\mathcal{G}^{.d}]$ and $S^{de}_{\mathcal{G}^.}$ and are given some values of $E[\Delta h^d]$ and $S^{de}_{\Delta h}$; let $\mathcal{G}^{.d}$ and Δh^d be uncorrelated (a more general case implies just a slightly more complex notation). We have by Eq. (1.61)

$$E[\Delta\mathcal{G}] = E[\Delta\mathcal{G}^{.d}]\,E[\Delta h^d] \qquad \text{(summation over } d = 1, 2, \ldots, D) \tag{5.20}$$

and, by Eq. (1.63)

$$\mathrm{Var}(\Delta\mathcal{G}) = E[\Delta h^d]\,E[\Delta h^e]\,S^{de}_{\mathcal{G}^.} + E[\Delta\mathcal{G}^{.d}]\,E[\Delta\mathcal{G}^{.e}]\,S^{de}_{\Delta h} \qquad \text{(summation over } d, e = 1, 2, \ldots, D) \tag{5.21}$$

because

$$\begin{aligned}
\frac{\partial\Delta\mathcal{G}}{\partial\mathcal{G}^{.d}} &= \Delta h^d \\
\frac{\partial\Delta\mathcal{G}}{\partial\Delta h^d} &= \mathcal{G}^{.d}
\end{aligned} \tag{5.22}$$

and the entries of the covariance matrix for the random $2D$-dimensional vector $\{\, \mathcal{G}^{\cdot 1} \;\; \mathcal{G}^{\cdot 2} \; \ldots \; \mathcal{G}^{\cdot D} \;\; \Delta h^{\cdot 1} \;\; \Delta h^{\cdot 2} \; \ldots \; \Delta h^{\cdot D} \}$ are

$$S^{\bar{d}\bar{e}}_{(\mathcal{G}\cdot)\Delta h} = \begin{bmatrix} \left[S^{de}_{\mathcal{G}\cdot} \right] & 0 \\ 0 & \left[S^{de}_{\Delta h} \right] \end{bmatrix} \qquad \begin{aligned} d, e &= 1, 2, \ldots, D \\ \bar{d}, \bar{e} &= 1, 2, \ldots, 2D \end{aligned} \tag{5.23}$$

By specifying the probabilistic characteristics of the variations in the design variables Δh^d we may thus compute by Eqs. (5.20) and (5.21) changes experienced by the response functional $\mathcal{G}$. In particular, the above procedure makes it possible to assess the change in $\mathcal{G}$ due to: (a) changes in design variable expectations alone (i.e. keeping their covariance matrix fixed) or (b) changes in the covariance matrix of the design variables alone (i.e. keeping their expectations fixed).

5.2 Computational Aspects

In the preceding section the stochastic sensitivity problem of the structural statics has been formulated. The computational standpoint suggests that: (i) for a single (or few) sensitivity functionals (5.1) the adjoint variable approach is more effective than the direct variable approach; (ii) the SFEM technique turns out to be numerically much more efficient than the statistical techniques; and (iii) problems of the structural sensitivity and stochastic finite elements are similar both in terms of the basic formulation and computer implementation which greatly facilitates the combined analysis. We shall show below that the computations become much less costly still when the standard normal transformation from the correlated space to an uncorrelated space of random variables is employed.

Equations (5.8)–(5.10) suggest that once both the zeroth-order nodal displacement vectors q^0_α of the primary structure and λ^0_α of the adjoint structure are known, the first- and second-order primary and adjoint displacement vectors $q^{;\rho}_\alpha$, $\lambda^{;\rho}_\alpha$ and $q^{(2)}_\alpha$, $\lambda^{(2)}_\alpha$, respectively, can be found in succession. Since the primary and adjoint systems have the same coefficient matrix on the left-hand sides, the system matrix needs to be decomposed only once prior to the solution phase by using any factorization scheme, cf. Eqs. (2.55), (2.59) and (2.62). Thus, the solution phase involves only $2(\tilde{N}+2)$ forward reductions (2.57) and backward substitutions (2.58) including: (i) two for q^0_α and λ^0_α, (ii) $2\tilde{N}$ for $q^{;\rho}_\alpha$ and $\lambda^{;\rho}_\alpha$, and (iii) two for $q^{(2)}_\alpha$ and $\lambda^{(2)}_\alpha$. In particular, if the response functional $\mathcal{G}$ is defined as a linear function of the nodal displacements the procedure may proceed in parallel, i.e. for each equal-order primary–adjoint system of equations the forward–backward process may be performed simultaneously.

Although almost all operations required to calculate the right-hand side of the first- and second-order equations and to evaluate probabilistic distributions for the sensitivity gradient can be carried out by vector multiplications and at the element level, the computational cost would remain unacceptably high owing to the double sums of the type $(.)^{\rho\sigma} S^{\rho\sigma}_{\mathrm{b}}$. To reduce the double summation to a single summation so that the number of algebraic operations is reduced from $O(\tilde{N}^2)$ to $O(\tilde{N})$, the transformation (4.149) from a set of correlated random variables to an uncorrelated set can be employed. Recall that when there exist various groups of correlated random variables involved in the system vector of random variables, uncorrelation of these groups must be assumed; and each dominant part taken from every group is then used to assemble the system vector of uncorrelated variables. By employing this transformation in Eqs. (5.8)–(5.10) the second mixed derivatives reduce to second derivatives with respect to single variables only, whereas the double sums from 1 to $\tilde{N}$ reduce to the single sums from 1 to $\check{N}$, leading to the transformed equations for the primary and adjoint systems in the form:

- Zeroth-order (ϵ^0 terms, one pair of systems of N linear simultaneous algebraic equations for $q^0_\alpha(h^d, c^0_{\breve{\varrho}})$ and $\lambda^0_\alpha(h^d, c^0_{\breve{\varrho}})$, respectively)

$$
\begin{aligned}
K^0_{\alpha\beta}(h^d, c^0_{\breve{\varrho}})\, q^0_\beta(h^d, c^0_{\breve{\varrho}}) &= Q^0_\alpha(h^d, c^0_{\breve{\varrho}}) \\
K^0_{\alpha\beta}(h^d, c^0_{\breve{\varrho}})\, \lambda^0_\beta(h^d, c^0_{\breve{\varrho}}) &= G^0_{.\alpha}(h^d, c^0_{\breve{\varrho}})
\end{aligned}
\tag{5.24}
$$

- First-order (ϵ^1 terms, $\check{N}$ pairs of systems of N linear simultaneous algebraic equations for $q^{,\breve{p}}_\alpha(h^d, c^0_{\breve{\varrho}})$ and $\lambda^{,\breve{p}}_\alpha(h^d, c^0_{\breve{\varrho}})$, respectively)

$$
\begin{aligned}
K^0_{\alpha\beta}(h^d, c^0_{\breve{\varrho}})\, q^{,\breve{p}}_\beta(h^d, c^0_{\breve{\varrho}}) &= Q^{,\breve{p}}_\alpha(h^d, c^0_{\breve{\varrho}}) - K^{,\breve{p}}_{\alpha\beta}(h^d, c^0_{\breve{\varrho}})\, q^0_\beta(h^d, c^0_{\breve{\varrho}}) \\
K^0_{\alpha\beta}(h^d, c^0_{\breve{\varrho}})\, \lambda^{,\breve{p}}_\beta(h^d, c^0_{\breve{\varrho}}) &= G^{,\breve{p}}_{.\alpha}(h^d, c^0_{\breve{\varrho}}) - K^{,\breve{p}}_{\alpha\beta}(h^d, c^0_{\breve{\varrho}})\, \lambda^0_\beta(h^d, c^0_{\breve{\varrho}})
\end{aligned}
\tag{5.25}
$$

- Second-order (ϵ^2 terms, one pair of systems of N linear simultaneous algebraic equations for $q^{(2)}_\alpha(h^d, c^0_{\breve{\varrho}})$ and $\lambda^{(2)}_\alpha(h^d, c^0_{\breve{\varrho}})$, respectively)

$$
\begin{aligned}
K^0_{\alpha\beta}(h^d, c^0_{\breve{\varrho}})\, q^{(2)}_\beta(h^d, c^0_{\breve{\varrho}}) &= Q^{(2)}_\alpha(h^d, c^0_{\breve{\varrho}}) \\
K^0_{\alpha\beta}(h^d, c^0_{\breve{\varrho}})\, \lambda^{(2)}_\beta(h^d, c^0_{\breve{\varrho}}) &= R^{(2)}_\alpha(h^d, c^0_{\breve{\varrho}})
\end{aligned}
\tag{5.26}
$$

with $d = 1, 2, \ldots, D$; $\alpha, \beta = 1, 2, \ldots, N$; $\breve{p} = 1, 2, \ldots, \tilde{N}$ and

$$
\begin{aligned}
q^{(2)}_\alpha(h^d, c^0_{\breve{\varrho}}) &= \sum_{\breve{p}=1}^{\check{N}} q^{,\breve{p}\breve{p}}_\alpha(h^d, c^0_{\breve{\varrho}})\, S^{\breve{p}}_{\mathrm{c}} \\
\lambda^{(2)}_\alpha(h^d, c^0_{\breve{\varrho}}) &= \sum_{\breve{p}=1}^{\check{N}} \lambda^{,\breve{p}\breve{p}}_\alpha(h^d, c^0_{\breve{\varrho}})\, S^{\breve{p}}_{\mathrm{c}} \\
Q^{(2)}_\alpha(h^d, c^0_{\breve{\varrho}}) &= \sum_{\breve{p}=1}^{\check{N}} \Big[Q^{,\breve{p}\breve{p}}_\alpha(h^d, c^0_{\breve{\varrho}}) \\
&\quad - 2K^{,\breve{p}}_{\alpha\beta}(h^d, c^0_{\breve{\varrho}})\, q^{,\breve{p}}_\beta(h^d, c^0_{\breve{\varrho}}) - K^{,\breve{p}\breve{p}}_{\alpha\beta}(h^d, c^0_{\breve{\varrho}})\, q^0_\beta(h^d, c^0_{\breve{\varrho}}) \Big] S^{\breve{p}}_{\mathrm{c}} \\
R^{(2)}_\alpha(h^d, c^0_{\breve{\varrho}}) &= \sum_{\breve{p}=1}^{\check{N}} \Big[G^{,\breve{p}\breve{p}}_{.\alpha}(h^d, c^0_{\breve{\varrho}}) \\
&\quad - 2K^{,\breve{p}}_{\alpha\beta}(h^d, c^0_{\breve{\varrho}})\, \lambda^{,\breve{p}}_\beta(h^d, c^0_{\breve{\varrho}}) - K^{,\breve{p}\breve{p}}_{\alpha\beta}(h^d, c^0_{\breve{\varrho}})\, \lambda^0_\beta(h^d, c^0_{\breve{\varrho}}) \Big] S^{\breve{p}}_{\mathrm{c}}
\end{aligned}
\tag{5.27}
$$

In Eqs. (5.24)–(5.27) the symbols $(.)^{,\check{\rho}}$ and $(.)^{,\check{\rho}\check{\rho}}$ denote the first and second derivatives with respect to the uncorrelated random variables $c^{\check{\rho}}$. Similarly to the solution to Eqs. (5.8)–(5.10) only one stiffness matrix is assembled and factorized for Eqs. (5.24)–(5.26) and a parallel computation scheme may also be used. The primary–adjoint system (5.24)–(5.26) requires only $2(\check{N}+2)$ solutions of linear algebraic equations, though.

By using the definition of the sensitivity expectation and covariance (which is similar to the one given by Eqs. (5.12) and (5.13)) written in terms of the uncorrelated random variables $c_{\check{\rho}}$ we obtain expressions for the second-order accurate expectations and the first-order accurate covariances of the sensitivity gradient coefficients in the form, cf. Eqs. (5.16) and (5.17)

$$\begin{aligned} E[\mathcal{G}^{\cdot e}] &= G^{0.d} - \tfrac{1}{2}\lambda^0_\alpha K^{0.d}_{\alpha\beta}\, q^{(2)}_\beta + \left(\lambda^0_\alpha + \tfrac{1}{2}\lambda^{(2)}_\alpha\right)\mathcal{A}^d_\alpha \\ &\quad + \tfrac{1}{2}\sum_{\check{\rho}=1}^{\check{N}}\left(G^{\cdot d,\check{\rho}\check{\rho}} + 2\,\lambda^{,\check{\rho}}_\alpha \mathcal{B}^{d\check{\rho}}_{.\alpha} + \lambda^0_\alpha \mathcal{C}^{d\check{\rho}\check{\rho}}_{.\alpha}\right) S^{\rho\sigma}_{\mathrm{b}} \end{aligned} \tag{5.28}$$

$$\begin{aligned} S^{de}_{\mathcal{G}\cdot} &= \sum_{\check{\rho}=1}^{\check{N}}\Big[G^{\cdot d,\check{\rho}}\, G^{\cdot e,\check{\rho}} + \lambda^{,\check{\rho}}_\alpha\left(G^{\cdot d,\check{\rho}}\mathcal{A}^e_\alpha + G^{\cdot e,\check{\rho}}\mathcal{A}^d_\alpha\right) \\ &\quad + \lambda^0_\alpha\left(G^{\cdot d,\check{\rho}}\,\mathcal{B}^{e\check{\rho}}_\alpha + G^{\cdot e,\check{\rho}}\,\mathcal{B}^{d\check{\rho}}_\alpha\right) + \lambda^{,\check{\rho}}_\alpha\,\lambda^{,\check{\rho}}_\beta\,\mathcal{A}^d_\alpha\,\mathcal{A}^e_\beta \\ &\quad + \lambda^{,\check{\rho}}_\alpha\,\lambda^0_\beta\left(\mathcal{A}^d_\alpha\,\mathcal{B}^{e\check{\rho}}_\beta + \mathcal{A}^e_\beta\,\mathcal{B}^{d\check{\rho}}_\alpha\right) + \lambda^0_\alpha\,\lambda^0_\beta\,\mathcal{B}^{d\check{\rho}}_\beta\,\mathcal{B}^{e\check{\rho}}_\alpha\Big]\, S^{\check{\rho}}_{\mathrm{c}} \end{aligned} \tag{5.29}$$

where $\mathcal{A}^d_\alpha$ has already been given in Eq. $(5.18)_1$ as is defined in the corresponding deterministic problem (i.e. it is independent of random variables), while

$$\begin{aligned} \mathcal{B}^{d\check{\rho}}_\alpha &= Q^{\cdot d,\check{\rho}}_\alpha - K^{\cdot d,\check{\rho}}_{\alpha\beta}\, q^0_\beta - K^{0.d}_{\alpha\beta}\, q^{,\check{\rho}}_\beta \\ \mathcal{C}^{d\check{\rho}\check{\rho}}_\alpha &= Q^{\cdot d,\check{\rho}\check{\rho}}_\alpha - 2\,K^{\cdot d,\check{\rho}}_{\alpha\beta}\, q^{,\check{\rho}}_\beta - K^{\cdot d,\check{\rho}\check{\rho}}_{\alpha\beta}\, q^0_\beta \qquad \text{(no sum on } \check{\rho}) \end{aligned} \tag{5.30}$$

It is seen that since only a few highest modes of $S^{\check{\rho}}_{\mathrm{c}}$, $\check{\rho} = 1, 2, \ldots, \check{N} \ll \tilde{N}$, are required to accurately approximate the main properties of many random quantities, the computation effort is significantly reduced. In other words, by using only dominant parts of the uncorrelated set of random variables the approach is well suited for stochastic sensitivity analysis of large systems at low computational cost. Since all the operations can be carried out by deterministic analysis procedures the algorithms worked out can be immediately adapted to fit into existing deterministic finite element codes.

A disadvantage of the SFEM modeling of SSDS problems is that with local averages of design variables across each element the finite element mesh controls the accuracy of the design parameter approximations and, consequently, the accuracy of the displacement field. This can be inefficient, since the discretized displacement field is usually more complex than the discretized design parameter fields; SSDS requires as a rule a finer finite element mesh than a typical displacement-stress problem does. In the framework of a finite element adaptive formulation, cf. [126] for instance, it is essential and interesting to investigate the sensitivity of the solution accuracy to the density of finite element meshes.

5.3 Numerical Illustrations

Example 5.1 Consider the static behaviour of a 100-element beam of unit length $l = 1.0$ clamped at both ends, Fig. 5.1. As design variables we take element cross-sectional areas, wherein randomness is assumed. We then have a

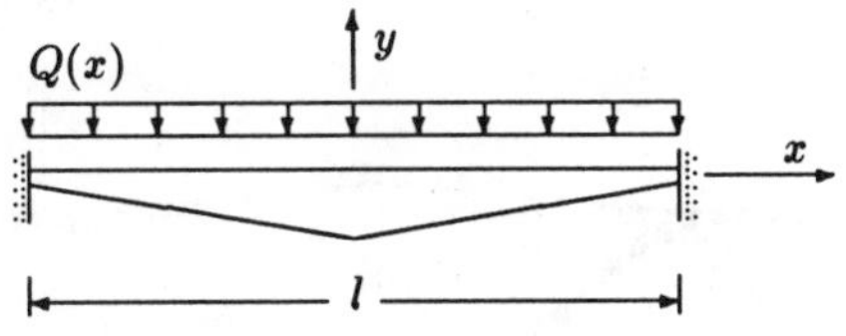

Figure 5.1 100-element clamped–clamped beam subjected to distributed load.

set of 100 random design variables A_ρ, $\rho = 1, \ldots, 100$. The expectation, correlation function and coefficient of variation of the cross-sectional areas are given, respectively, as follows:

$$E[A_\rho] = A^0 \left(1.0 + \frac{\vartheta x_\rho}{l}\right) ; \qquad E[A_{100-\rho+1}] = E[A_\rho]$$

$$A^0 = 5.0 \times 10^{-3} ; \qquad \vartheta = 0.3 ; \qquad \rho = 1, 2, \ldots, 50$$

$$\mu(A_\rho, A_\sigma) = \exp\left(- \frac{|x_\rho - x_\sigma|}{\lambda}\right) ; \qquad \lambda = 0.1 , \qquad \rho, \sigma = 1, 2, \ldots, 100$$

$$\alpha = 0.07$$

where the ordinate x_ρ is referred to the midpoint of the ρ-th element. The following deterministic data are adopted: distributed load $Q(x) = f + \gamma A^0$, with uniformly distributed load $f = 49.61$ per unit length, weight density $\gamma = 7.7126$; Young's modulus $E = 2.0 \times 10^5$; Poisson's ratio $\nu = 0.3$; and moments of inertia $I_y = I_z = \beta(A_\rho)^2$, $I_x = I_y + I_z$, $\beta = 1/6$. The structural response functional is defined as

$$\mathcal{G} = \frac{|q(x)|}{q^{(A)}} - 1 \leq 0$$

With the unit numerical values of the given allowable displacement and slope at the midpoint of the beam ($x = 0.5$), the application of AVM comes down to considering an 'adjoint' beam subjected to a unit concentrated force at the beam midpoint in the case of the displacement analysis and to a unit moment at this point in the case of the slope analysis. With the expectations of the vertical displacement and the slope at the midpoint of the beam of -1.411097×10^{-2} and 2.076255×10^{-3} (compared with values of -1.390371×10^{-2} and 2.044906×10^{-3} for the deterministic case) the expectations and standard deviation of the sensitivity coefficients for both cases are given in Fig. 5.2, in which they are compared with deterministic results for varying and constant cross-sectional areas

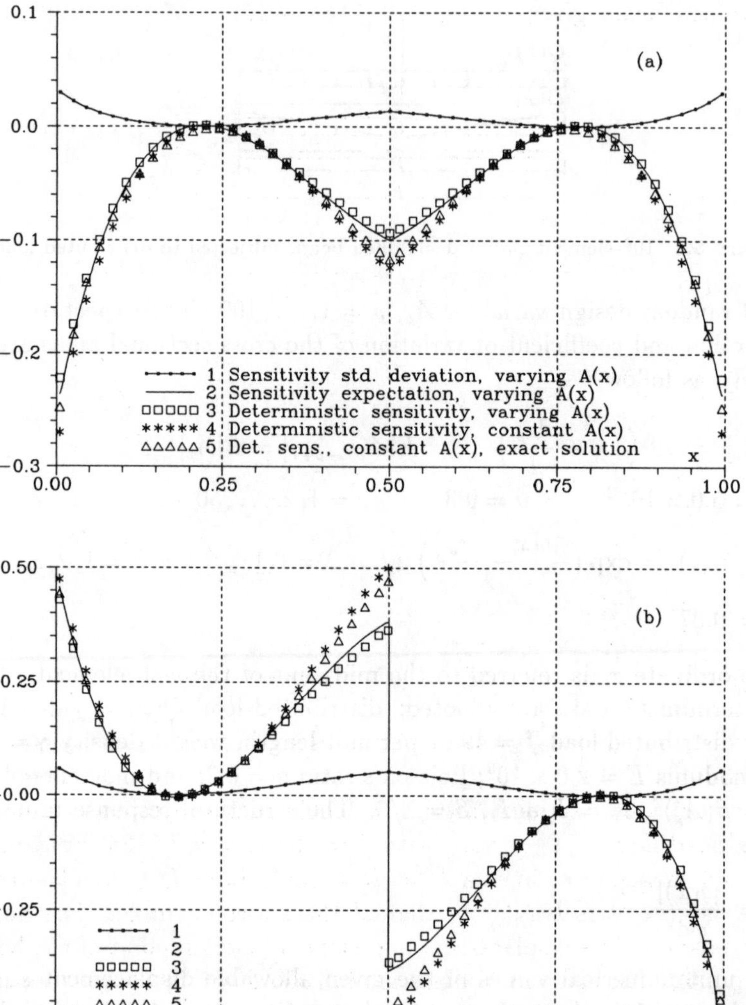

Figure 5.2 100-element beam sensitivity: (a) displacement analysis, (b) slope analysis.

and with the exact solution [44]. It is noted that relative differences in the expected values of the sensitivities are several times larger than the differences in the displacements. The effects of system asymmetry due to random variations in cross-sectional areas are very small.

Example 5.2 This example concerns the displacement and slope responses of the clamped beam given in Fig. 5.3. The beam is of unit length $l = 1.0$

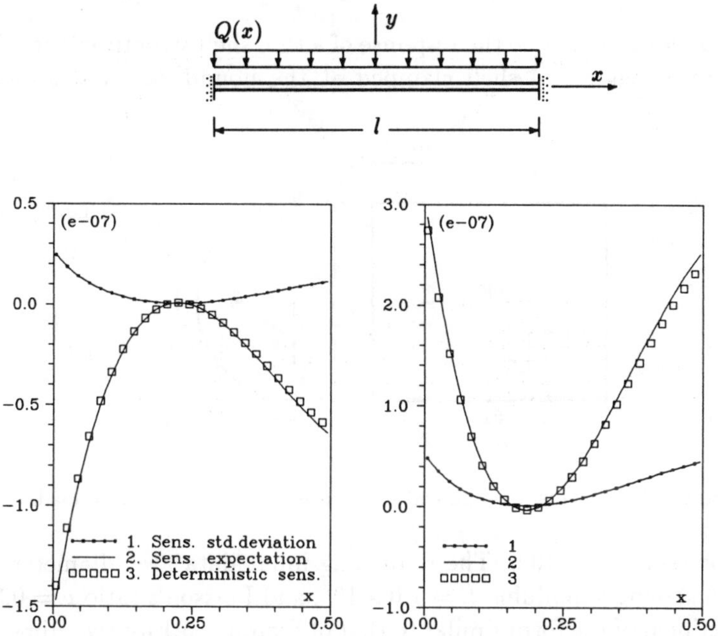

Figure 5.3 100-element beam sensitivity. Young's modulus as random design variable.

and of constant cross-sectional area $A = 0.005$; it is subjected to a uniformly distributed load $Q(x) = (f + \gamma A)$, with $f = 49.61$, $\gamma = 7.7126$; Poisson's ratio $\nu = 0.3$; and moments of inertia $I_y = I_z = \beta A^2$, $I_x = I_y + I_z$. The response functional has the form similar to that of the above example. The numerical values of the admissible displacement and slope at the midpoint of the beam are both set to 0.02. The element Young's moduli E_ρ, $\rho = 1, 2, \ldots, 100$, are taken as random design variables. The expectation, correlation function and coefficient of variation of the Young's moduli are given, respectively, as follows:

$$E[E_\rho] = E^0 \left(1.0 + \frac{\vartheta x_\rho}{l}\right) ; \qquad E[E_{100-\rho+1}] = E[E_\rho]$$

$$E^0 = 2.0 \times 10^5 ; \qquad \vartheta = 0.3 ; \qquad \rho = 1, 2, \ldots, 50$$

$$\mu(E_\rho, E_\sigma) = \exp\Big(- \frac{|x_\rho - x_\sigma|}{\lambda}\Big) ; \qquad \lambda = 0.1 , \qquad \rho, \sigma = 1, 2, \ldots, 100$$

$$\alpha = 0.09$$

It can be observed from the numerical results in both Examples 5.1 and 5.2 that some perturbations in probabilistic distributions appear in region where $0.20 < x < 0.26$ and $0.74 < x < 0.80$: the sensitivity gradients decrease to zero, whereas the values of the coefficient of variation considerably increase. The stochastic finite element technique employed is apparently less efficient in this case, which is due to the fact that the approximate displacement field is modelled in a more sophisticated way than the discretized random field (the random design variables are modelled by taking local averages over each element).

Example 5.3 In this example the response of a thin shell structure is considered. Fig. 5.4 shows a cylindrical shell clamped at the boundaries under uniformly

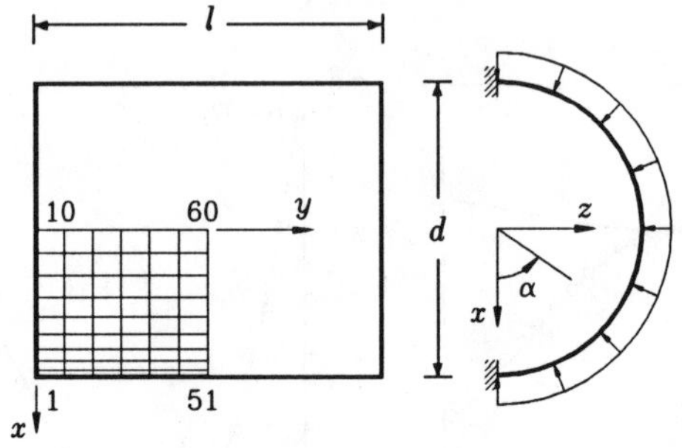

Figure 5.4 60-element cylindrical shell subjected to a distributed load.

distributed pressure $Q = 100$. The remaining input data are: diameter $d = 8$, length $l = 9.6$, Young's modulus $E = 1.0 \times 10^7$, and Poisson's ratio $\nu = 0.3$. The response functional of the form similar to that of Example 5.1 for two-dimensional case is defined for the midpoint A of the shell, with an allowable translation in the z-direction $q_z^{(A)} = 0.01$. Uncertainty in the shell thickness $t_\rho(x, y)$ is assumed as follows:

$$E[t_\rho] = t^0 = 0.1$$

$$\mu(t_\rho, t_\sigma) = \vartheta \exp\left(-\frac{|x_\rho - x_\sigma|}{\lambda}\right) \exp\left(-\frac{|y_\rho - y_\sigma|}{\lambda}\right)$$

$$\vartheta = \frac{3}{dl}\,; \qquad \lambda = 1.25dl$$

$$\alpha = 0.1$$

Owing to the symmetry in geometry, load and boundary conditions (see also the footnote on the page 154) only one-quarter of the shell is considered. The finite element mesh includes 60 rectangular elements of constant thickness (60 random design variables), and the total number of degrees of freedom is 313. The computed values of the expectations and standard deviations of the sensitivity coefficients are plotted in Fig. 5.5.

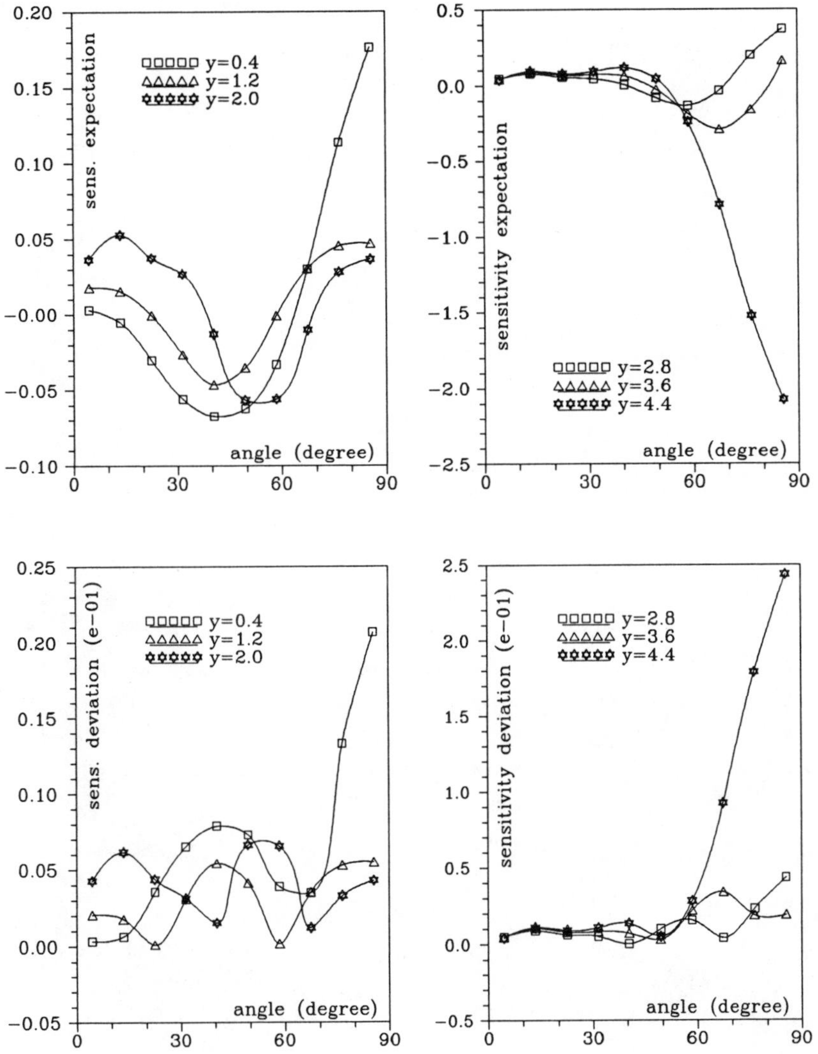

Figure 5.5 60-element shell. Variations of sensitivity versus coordinates.

Chapter 6

Program SFESTA

A CODE FOR THE DETERMINISTIC AND STOCHASTIC ANALYSIS OF STATICS AND STATIC SENSITIVITY OF 3D TRUSSES

6.1 Data Preparation

On the basis of the developments in Chapters 2, 4 and 5 a computer program called SFESTA has been written for the static analysis of truss-type structures with parameters described deterministically and/or stochastically. The code has been worked out for educational and research environments and aims at dealing with static deterministic and stochastic problems of small and medium-scale three-dimensional trusses. The system has been written for IBM compatible personal computers in Fortran 77 extended (it includes *do while, do enddo, block if, on-line comments*, etc.; the MS Fortran Compiler 5.0 and its compatible versions is recommended). Minimum hardware requirements are 640KB of RAM and a hard disk drive of 10MB.

SFESTA can be used for the deterministic structural analysis of displacement-stress problems[1] and static sensitivity problems. The class of structural sensitivity functionals included is assumed to be linear in terms of the displacements; functionals of more complicated form may be incorporated after only small modifications to the code. Design sensitivity variables available are cross-sectional area, length and Young's modulus in any of the structural elements.

For stochastic systems randomness in the geometry and the material of structural elements is incorporated into SFESTA by using a combination of the finite element technique and second-order perturbation method; the adjoint variable technique is employed in the sensitivity part of the analysis. In the version given below, each random field variable representing the field value at a specific finite element is approximated by its spatial average over the element. The second-order accurate expectations and first-order accurate covariances for the displacement response can be evaluated. In the sensitivity analysis the response functionals and design variables are defined similarly as in the deterministic case; the design

[1]This analysis option is based on a considerably modified version of the SAP-IV code [11].

variables assumed to be possibly random are cross-sectional area, length and Young's modulus in any of the structural elements. The results of the conventional displacement-stress analysis as well as the probabilistic distributions for the design sensitivity gradient coefficients, i.e. their second-order accurate expectations and first-order accurate standard deviations, are obtained on output.

To better appreciate the material being presented in this chapter the reader should be familiar with the basic aspects of using the Fortran language. The authors' programming experience suggests that the fastest way to initially familiarize oneself with a well-documented program is by reading format statements and their associated `read`, `write` commands. Thus, to introduce the main features of SFESTA we start with a description of some representative `format` statements; in the next section a representative set of SFESTA's routines will be provided and described. The complete source code of the program can be obtained from the Publisher; using it, together with the thorough discussion of the selected parts of the program presented below, it should be relatively easy for the reader to familiarize himself with the overall structure of and the data flow in the whole program.

The current version of SFESTA includes a main module and 30 routines. Modules developed in the code are grouped so that they are functionally independent and may be efficiently compiled, linked and run in the nonoverlay as well as overlay mode. The main procedure inputs the problem control informations, sets up the low- and high-speed storage required by the solution process, selects the appropriate solution mode (single- or multi-block scheme depending on the operating memory and the problem size) and monitors the process of setting up and solving the equations. It is seen from the specification of the analysis type parameter `nsta` in the following output `format` statement

```
2000 format (1h ,20a4                                               ///
    * ' P R O B L E M   C O N T R O L   I N F O R M A T I O N '  //4x,
    * ' NUMBER OF NODAL POINTS (numnp)..........................',i4/4x,
    * ' NUMBER OF FINITE ELEMENT GROUPS (neltyp)..............',i4/4x,
    * ' NUMBER OF LOAD CASES (ll)..............................',i4/4x,
    * ' ANALYSIS TYPE (nsta)...................................',i4/4x,
    * '   nsta=1 - DETERMINISTIC STATICS                        '   /4x,
    * '        2 - DETERMINISTIC STATIC SENSITIVITY              '   /4x,
    * '        3 - STOCHASTIC STATICS                            '   /4x,
    * '        4 - STOCHASTIC STATIC SENSITIVITY                 '   /4x,
    * ' OPERATING MODE (modex)..................................',i4/4x,
    * '  modex=0 - PROBLEM EXECUTION                             '   /4x,
    * '        1 - DATA CHECK ONLY                               '   /4x,
    * ' NUMBER OF EQUATIONS REQUIRED PER BLOCK (keqb).........',i4/4x,
    * ' NUMBER OF DISPLACEMENT CONSTRAINTS (nc)...............',i4  )
```

that four analysis options are available in SFESTA: two for the displacement-stress and sensitivity analysis of deterministic systems and the remaining two for those of stochastic systems. Besides the analysis type parameter `nsta` and the character field (20a4) reserved for the alphanumeric information (title) which identifies the problem considered, other formal parameters (lower-case letter expressions in parentheses) are defined as follows:

- numnp – total number of nodal points in the finite element model considered, each of the numnp nodes having three translational degrees of freedom,

- neltyp – number of different element groups in the model; neltyp may be used to specify various types of random and design variables for a single SFEM model and for element output-data processing (in contrast to frontal solvers, the efficiency of the Gauss-elimination solver implemented in SFESTA does not depend on the element numbering),

- ll – number of external load cases considered; for stochastic problems ll is automatically set equal to one (1) irrespective of any (nonzero) value entered for ll,

- modex – if modex=1 SFESTA runs in the data-check mode, verifying and printing all the input data as generated by the input routine bypassing high-cost computations in the execution mode invoked when modex=0 (or default),

- keqb – number of degrees of freedom per memory block; for large-scale problems or small computers, by setting keqb=0 (or default) SFESTA automatically divides the stiffness matrices and load vectors into submatrices and solves the equation system in the multi-block mode; a nonzero value of keqb may also be used to control the number of blocks or to test an extended version of SFESTA,

- nc – number of displacement-type functionals (constraints) in the sensitivity analysis; it is automatically set equal to one (1) in stochastic sensitivity problems regardless of a nonzero value typed on the problem control line.

Labels of the input and output format statements in the entire SFESTA are specified by four-digit numbers; the first digit of an input format label is the number one (1) while an output format label begins with the number two (2).

Random variables are defined in the program as the mean values of the random fields over each element domain by using the spatial averaging approach. Different random and design variable types available in SFESTA can be identified by examining an output format statement in the routine eltruss (described below). This output format statement provides control parameters and element data for each from neltyp element groups involved in the finite element model considered (the maximum value of neltyp is assumed equal to 30) and has the following form:

```
2000 format (///          ' 3 / D   T R U S S   E L E M E N T S '   /4x,
    *  ' NUMBER OF ELEMENTS (ntruss)...........................',i5/4x,
    *  ' NUMBER OF MATERIAL SETS (nummat)......................',i5/4x,
    *  ' FLAG FOR SENSITIVITY OR STOCHASTIC ANALYSIS (isens)...',i5/4x,
    *  '   isens=0 - NONE OF THESE                             '   /4x,
    *  '         1 - ONE OR TWO OF THESE                       '   /4x,
    *  ' RANDOM OR DESIGN VARIABLE TYPE (istype)...............',i5/4x,
    *  '  istype=0 - IF isens=0                                '   /4x,
    *  '         1 - CROSS-SECTIONAL AREA                      '   /4x,
    *  '         2 - YOUNG'S MODULUS                           '   /4x,
    *  '         4 - LENGTH                                    '     )
```

By assembling finite elements in separate groups with different geometry and/or material properties one may define different types of random and design variables in a single FEM model. To be specific, the zero value of `isens` ascribed to an element group means that all the elements in this group are described deterministically and no design variables are defined. For `isens=1` random variables or random design variables are defined for all the elements in this group by setting `istype` equal to one of the integer values 1,2,4; the analysis is then defined in accordance with the analysis type parameter `nsta`. Thus, deterministic design variables, random variables and random design variables may be specified such that they do not necessarily coincide with the finite element mesh. From the implementation standpoint the calculation of the first and second derivatives of the global stiffness matrix with respect to random and design variables can be carried out at the element level.

6.2 Fortran Routines

This section is devoted to some representative routines of SFESTA. The program is believed to be typical of the methodology for computer implementation of stochastic finite elements in a deterministic code; the procedures described below which deal with sensitivity as well as stochastic problems can readily be adapted to fit into any existing FEM-based program.

For the sake of clarity of presentation, routines are written so that they are as self-explanatory as possible, in particular:

- specifications of input and output variables are given at the beginning of each subroutine,
- many off-line and on-line comments are provided,
- variables and parameters are named suggestively; locally inactive parameters involved in the `common` statements are specified with the suffix `nul`,
- more-than-six-character variable names should not cause ambiguity even when using earlier versions of Fortran compilers, since the first six characters of any longer name are always uniquely specific.

6.2.1 Main Routine sfesta

The main procedure reads in the control parameters, opens scratch files and sets up the low- and high-speed storage required in the problem considered, declares the operating memory and determines the size of the model, checks if the system equations are to be solved in the multi-block mode, and monitors the macro process of model set-up and solution. All modules developed in SFESTA interact with the global database by means of a database control system included in this procedure.

```
*  ...........................................................
7                                                       format( '
    .S F E S T A   -   DETERMINISTIC AND STOCHASTIC ANALYSIS       '/ '
    .OF STATICS AND SENSITIVITY RESPONSE OF 3D-TRUSS SYSTEMS       '///)
*  ...........................................................

*       isens     Flag for random or design variable identification
*                 (0, no; 1, yes)
*       istype    Flag for random and/or design variable type
*                 (inactive if isens.eq.0; 1, cross-sectional area;
*                 2, Young's modulus; 4, el. length)
*       ll        Number of load cases
*       mband     System equation (half) bandwidth
*       modex     Operating mode
*                 (0, execution; 1, data check)
*       mtot      RAM-checking parameter ('Mother' array length)
*       nblock    Number of core blocks used to save system matrices
*       nc        Number of constraints (response functionals)
*       neltyp    Number of finite element groups
*       neq       Number of system degrees of freedom (d.o.f.'s)
*       neqb      Number of d.o.f.'s per block
*       npmax     Max. number of design variables in a group
*       nsta      Analysis type (0,1,2,3,4, see format 2000)
*       numnp     Number of system nodal points
*       mpar(.)   Numbers of design variables in each group
*       npar(.)   Element control parameters

*       a(.)      'Mother' array
*       t(.)      Time log array
*       title(.)  Problem heading line

        common /elpar/ npar(5),numnp,mband,neltyp,neq,mtot,n1,n2,n3,n4,n5
        common /acode/ modex,nsta
        common /sstiv/ isens,istype,nc,npmax,mpar(30)
        common /equat/ nblock,neqb,ll

        integer         year,month,day,yea,mont,dayss
        logical         sensitivity
        character*4     title(20)
        character*1     tmpname(64)
        character*64    filename
        equivalence     (tmpname,filename)
```

```
      real*8         t(7)

      real*8         a(44501)                          ! 'Mother' array
      mtot =          44500                    ! RAM-checking parameter

*     Enter specifications of input and output data files

      write (0,'(a\)') ' Input file specification: '
    3 if(ichar(tmpname(1)).eq.27) STOP 'Terminated by the user'
      read (0,'(a64)') filename
      open (5,file=filename,status='old',iostat=ierror)
      if(ierror.ne.0) then
        write (0,'(a\)') ' E R R O R !   Type again: '
        goto 3
      endif

      write (0,'(a\)') ' Output file specification: '
    4 if(ichar(tmpname(1)).eq.27) STOP 'Terminated by the user'
      read (0,'(a64)') filename
      open (6,file=filename,status='unknown',err=5)
      goto 9
    5 write (0,'(a\)') ' E R R O R !    Type again: '
      goto 4

    9 write (0,'(/)')
      write (6,7)
      call second(t(1))                              ! Start time counter

*     Read in problem control data

      read (5,1000) title,numnp,neltyp,ll,nsta,modex,keqb,nc
      if(numnp.eq.0)              stop 'E R R O R !  Zero nodal points'
      if(ll.lt.1)                 stop 'E R R O R !  Zero load cases'
      if(nsta.lt.1.or.nsta.gt.4) stop 'E R R O R !  Wrong analysis type'
      if(nsta.ge.3) then
        ll=1
        if(nsta.eq.3) then
          nc=0
        else
          nc=1
        endif
      endif
      if(modex.ne.0) modex=1
      write (6,2000) title,numnp,neltyp,ll,nsta,modex,keqb,nc

      call tmpfiles (nsta)                          ! Open scratch files
      rewind 8
                                                   ! For data check mode
      if(modex.eq.1) write (8) title,numnp,neltyp,ll,nsta
      rewind 10

      n1=1
```

```
*     Input and generate nodal point data

      n2=n1+(3*numnp+1)/2                 ! Operating memory preparation
      n3=n2+numnp                               ! for each solution phase
      n4=n3+numnp
      n5=n4+numnp
      n6=n5+numnp
      if(n6.gt.mtot) call error (n6-mtot)     ! Check if memory exceeded
      call nodeinp (a(n1),a(n2),a(n3),a(n4),a(n5),numnp,neq)
      call second (t(2))

*     Input and generate element data. Form element stiffness matrices
*     and their derivatives with respect to random and design variables

      write (0,'('' * Input and generate element data'')')
      rewind 1
      rewind 2
      if(nsta.ge.2) rewind 14
      mband=0
      numel=0
      npmax=0

      do ngroup=1,neltyp                         ! Loop over element groups
        read (5,1001) npar
        npar(1)=1
        if(modex.eq.1) write (8) npar
        write (1) npar
        numel =numel+npar(2)            ! Total number of finite elements
        isens =npar(4)            ! Flag for random/design variable groups
        istype=npar(5)             ! Random/design variable type in group
        if(isens.eq.1) then
          npar2=npar(2)     ! Number of random/design variables in group
          mpar(ngroup)=npar2
          if(npar2.gt.npmax) npmax=npar2
        else
          mpar(ngroup)=0
        endif
        call elmgroups (a,ngroup,1)             ! Element data processing
      enddo
      call second (t(3))

*     Determine block size for system equations

      sensitivity=nsta.eq.2.or.nsta.eq.4
      if(sensitivity) ll=ll+nc
      neqb=(mtot-4*ll)/(mband+ll+1)/2
      if(keqb.lt.2) keqb=99999
      if(keqb.lt.neqb) neqb=keqb
      ntmp=(mtot-mband)/(2*(mband+ll)+1)
      if(ntmp.lt.neqb) neqb=ntmp
      ntmp=(mtot-mband-ll*(mband-2))/(3*ll+mband+1)
      if(ntmp.lt.neqb) neqb=ntmp
      if(neqb.gt.neq) neqb=neq
```

```
      nblock=(neq-1)/neqb +1

      write (6,2001) neq,mband,neqb,nblock

*     Input (zeroth-order) external loads

      if(sensitivity) ll=ll-nc
      n3=n2+neqb*ll
      n4=n3+3*ll
      call loadinp (a(n1),a(n2),a(n3),numnp,neqb,ll)

*     Input displacement constraints for sensitivity analysis

      if(sensitivity) then
        n3=n2+neqb*nc
        n4=n3+3*nc
        call displinp (a(n1),a(n2),a(n3),numnp,neqb,nc)
      endif
      call second (t(4))

*     Form (zeroth-order) global stiffness matrix

      ne2b=neqb+neqb
      n2=n1+neqb*mband
      n3=n2+neqb*ll
      if(sensitivity) n3=n3+neqb*nc
      n4=n3+4*ll
      nn2=n1+ne2b*mband
      nn3=nn2+ne2b*ll
      if(sensitivity) nn3=nn3+ne2b*nc
      nn4=nn3+4*ll
      call sysstiff (a(n1),a(nn2),a(nn3),
     *               numel,nblock,ne2b,ll,mband,anorm,nvv)
      call second (t(5))

*     Monitor solution process

      if(modex.eq.1) then                             ! Data-check mode
        t(6)=t(5)
        if(nsta.ge.3) goto 10
        go to 20
      endif
                                                      ! Solution mode:
      call detersta (a)             ! deterministic statics or sensitivity
      call second (t(6))                              ! (nsta=1 and 2)
   10 if(nsta.ge.3) then
        call randsta (a)               ! stochastic statics or sensitivity
        call second (t(7))                            ! (nsta=3 and 4)
      else
        t(7)=t(6)
      endif
```

```
*     Log computational time

   20 tt=0.0d0
      do i=1,6
        t(i)=t(i+1)-t(i)
        tt=tt+t(i)
      enddo
      write (6,2003) (t(k),k=1,6),tt

      year=yea (0)
      month=mont (0)
      day=dayss (0)
      write (6,'(//'' date: '',i4,'' -'',i3,'' -'',i3/)') year,month,day

 1000 format (20a4/8i5)
 1001 format (5i5)
 2000 format (1h ,20a4                                          ///
     * ' P R O B L E M   C O N T R O L   I N F O R M A T I O N ' //4x,
     * ' NUMBER OF NODAL POINTS (numnp)........................',i4/4x,
     * ' NUMBER OF FINITE ELEMENT GROUPS (neltyp)..............',i4/4x,
     * ' NUMBER OF LOAD CASES (ll).............................',i4/4x,
     * ' ANALYSIS TYPE (nsta)..................................',i4/4x,
     * '   nsta=1 - DETERMINISTIC STATICS                      '   /4x,
     * '        2 - DETERMINISTIC STATIC SENSITIVITY           '   /4x,
     * '        3 - STOCHASTIC STATICS                         '   /4x,
     * '        4 - STOCHASTIC STATIC SENSITIVITY              '   /4x,
     * ' OPERATING MODE (modex)................................',i4/4x,
     * '  modex=0 - PROBLEM EXECUTION                          '   /4x,
     * '        1 - DATA CHECK ONLY                            '   /4x,
     * ' NUMBER OF EQUATIONS REQUIRED PER BLOCK (keqb).........',i4/4x,
     * ' NUMBER OF DISPLACEMENT CONSTRAINTS (nc)...............',i4   )
 2001 format (///     ' E Q U A T I O N   P A R A M E T E R S ' //4x,
     * ' TOTAL NUMBER OF EQUATIONS (neq).......................',i5/4x,
     * ' BANDWIDTH (mband).....................................',i5/4x,
     * ' NUMBER OF EQUATIONS PER BLOCK (neqb)..................',i5/4x,
     * ' NUMBER OF BLOCKS (nblock).............................',i5   )
 2003 format (///      ' O V E R A L L   T I M E   L O G '    ///4x,
     * ' NODAL POINT DATA PROCESSING.....................',f6.0 /4x,
     * ' ELEMENT STIFFNESS (DERIVATIVE) FORMATION.........',f6.0 /4x,
     * ' NODAL LOAD (SENSITIVITY CONSTRAINT) INPUT........',f6.0 /4x,
     * ' GLOBAL STIFFNESS FORMATION.......................',f6.0 /4x,
     * ' DETERMINISTIC STATICS (SENSITIVITY) SOLUTION.....',f6.0 /4x,
     * ' STOCHASTIC STATICS (SENSITIVITY) SOLUTION........',f6.0//4x,
     * ' OVERALL SOLUTION TIME............................',f6.0,4h sec)

      end
```

The flow of data in the main module of SFESTA is believed to be clear enough so that no additional comments on the particular computation phases will be given. Some routines which characterize computational aspects of the stochastic finite element implementation will be presented below. It is pointed out in passing that:

- at the beginning of each solution step the overall length of all the actual (formal) array variables, i.e. the current size of the 'mother' array `a(.)`, is determined anew,

- the memory-check routine `error` compares the required length of `a(.)` with the nominal (maximum) size declared and terminates the computations if the nominal length has been exceeded,

- at the end of each solution step the routine `second` monitors the time which elapsed during the step computations and saves it for the total time log assessment.

6.2.2 Subroutines `constrinp` and `detersens`

The option with the analysis type parameter `nsta=2` specified on the problem control data line, cf. Section 6.1, monitors the sensitivity solution for deterministic structural systems. The finite element formulation and computational aspects of the deterministic static sensitivity problem were discussed in Section 2.8.1.

It is assumed in SFESTA that the structural response functional $\mathcal{G}$ is defined implicitly with respect to the design variables h^d and that the loads are independent of h^d. As an example let us take a typical functional used to control the displacement response q_α at any point in a structural system subjected to static loads. The functional can be defined as

$$\mathcal{G}[q_\alpha(h^d)] = \frac{|q_\alpha(h^d)|}{q^{(A)}} - 1 \leq 0 \tag{6.1}$$

with $q^{(A)}$ being a prescribed allowable value.

The routine `constrinp` serves the purposes of inputting displacement constraints of the type (6.1) and forming the load vectors for the adjoint system equations. In SFESTA the number of such functions is specified by the parameter `nc` on the problem control data line; `nc` should not exceed 100. The adjoint load vectors computed are saved in blocks associated with the stiffness matrix blocks.

```
      subroutine constrinp (id,dgdq,temp,numnp,neqb,nc)

*     To read in displacement constraints and form adjoint load vectors

*     ON INPUT:
*       modex       Operating mode
*                   (0, execution; 1, data check)
*       nc          Number of constraints (response functionals)
*       neqb        Number of d.o.f.'s per block
*       numnp       Number of system nodal points
*       id(.,.)     System equation numbers (d.o.f. numbers)
*       qallow(.)   Admissible values of nodal displacements
```

```
*      ON OUTPUT:
*        idofnc(.)  Adjoint load component numbers
*        dgdq(.,.)  Adjoint load block

*        temp(.,.)  Working array

      implicit       real*8 (a-h,o-z)

      common /acode/ modex,nnul
      common /idonc/ idofnc(100)
      dimension      id(numnp,3),dgdq(neqb,nc),temp(3,nc),qallow(3)

      write (0,
     *  '('' * Input displacement constraints. Form adjoint loads'')')

      write (6,2002)                                   ! Printing title

      if(modex.eq.0) then                ! Zeroing first adjoint load block
        do k=1,nc
          do i=1,neqb
            dgdq(i,k)=0.0d0
          enddo
        enddo
      endif
      rewind 15
      kshf=0

      do 100 nn=1,numnp                     ! Loop over system nodal points

      do i=1,3
        do j=1,nc
          temp(i,j)=0.0d0
        enddo
      enddo

      if(nn.eq.1) goto 30
   15 if(n.ne.nn) goto 40
      do i=1,3
        temp(i,l)=qallow(i)
      enddo
   30 read (5,1001) n,l,qallow                             ! n, nodal point
      if(n.eq.0) goto 15                             ! l, constraint number
      write(6,2001) n,l,qallow                   ! qallow, allowable values
      goto 15

   40 if(modex.eq.0) then                                ! If solution mode
        do j=1,3                                  ! Loop over nodal d.o.f.'s
          ii=id(nn,j)-kshf                         ! Relative index in block
          if(ii.gt.0) then
            do k=1,nc                                  ! Loop over constraints
              tempjk=temp(j,k)
              if(tempjk.ne.0.0d0) then
                dgdq(ii,k)=1.0d0/tempjk            ! Adjoint load comp. module
```

```
              idofnc(k)=id(nn,j)              ! Adj. load component index
            endif                           ! ... in global equation system
          enddo
          if(ii.eq.neqb) then
            write (15) dgdq         ! Saving current adjoint load block
            kshf=kshf+neqb                 ! Index shift for next block
            do k=1,nc
              do i=1,neqb
                dgdq(i,k)=0.0d0      ! Zeroing next adjoint load block
              enddo
            enddo
          endif
        endif
      enddo
    endif
 100 continue

    if(modex.eq.0) write (15) dgdq     ! Saving last adjoint load block

1001 format (2i5,3d10.0)
2001 format (2(3x,i4),1p3d19.9)
2002 format (///            'D E S I G N   S E N S I T I V I T Y',
   *                                      ' C O N S T R A I N T S'//
   * 3x,4hNODE,3x,4hD.C.,2(13x,6hX-AXIS,13x,6hY-AXIS,13x,6hZ-AXIS)/
   * 7h NUMBER,7h NUMBER,3(10x,12hALLOWABLE)/)

    end
```

We note that: (i) for the response functional (6.1) each adjoint load vector has only one nonzero component; and (ii) `constrinp` computes the module of the adjoint load vector only, its direction (algebraic sign) being identified in accordance with the associated displacement component in the last phase of the sensitivity evaluation. In terms of computations, data input and processing for the displacement constraints and adjoint load vectors are similar to reading in and generating external nodal load vectors.

By employing the adjoint variable technique the expression for static sensitivity gradient $\mathcal{G}^{\cdot d}$ of functionals of the form (6.1) with respect to the design variables h^d can be written as, cf. Eq. (2.130)

$$\mathcal{G}^{\cdot d} = -\lambda_\alpha K^{\cdot d}_{\alpha\beta} q_\beta \qquad \alpha = 1, 2, \ldots, N \tag{6.2}$$

where $K^{\cdot d}_{\alpha\beta}$ is the derivative matrix of the system stiffness matrix with respect to h^d, N is the number of degrees of freedom in the system while λ_α is the adjoint variable vector. Since the response functional $\mathcal{G}$ is assumed linear in terms of the nodal displacements, the adjoint load vectors involved in the adjoint equations (2.129) are independent of the primary displacement vectors; they are dependent on the vectors' direction, though. In other words, both the primary and adjoint equation systems may be solved simultaneously, while the direction of the adjoint displacement vectors and, consequently, the algebraic sign of the sensitivity gradient coefficients can be fixed at the end of computations.

The subroutine **detersens** evaluates the sensitivity coefficients for deterministic systems under static loads. In this routine the number of load cases and displacement constraints may be given arbitrarily.

```
      subroutine detersens (qlambda,dgdh,ll6,nc6,dkdh,lm,
     *                      nblock,neltyp,neqb,neqblk,llnc,lc,ll,npmax)

*     To compute deterministic sensitivity gradient coefficients

*     ON INPUT:
*       isens          Flag for design variable identification
*                      (0, no; 1, yes)
*       istype         Design variable type
*                      (inactive if isens.eq.0; 1, cross-sectional area;
*                      2, Young's modulus; 4, el. length)
*       lc             Number equal to ll+nc
*       ll             Number of primary load cases
*       llnc           Number equal to ll*nc
*       nblock         Number of core blocks used to save system matrices
*       nc             Number of constraints (response functionals)
*       neltyp         Number of design variable groups
*       neqb           Number of equations per block
*       neqblk         Number equal to nblock*neqb
*       npmax          Max. number of design variables in a group
*       numnp          Number of system nodal points
*       idofnc(.)      Adjoint load component numbers
*       lm(.)          Degrees of freedom associated with design var.
*       mpar(.)        Numbers of design variables in each group
*       dkdh(.,.)      Stiffness derivatives with respect to design var.
*       qlambda(.,.)   Primary (q) and adjoint (lambda) displacements

*     ON OUTPUT:
*       dgdh(.,.)      Sensitivity gradient coefficients

*       ll6(.),nc6(.) Working arrays

      implicit       real*8 (a-h,o-z)

      common /sstiv/ isens,istype,nc,npnul,mpar(30)
      common /idonc/ idofnc(100)
      dimension      qlambda(neqblk,lc),dgdh(llnc,npmax),dkdh(6,6),
     *               ll6(llnc),nc6(llnc),lm(6)

*     Get primary and adjoint displacements into a global block

      rewind 2
      ne=neqblk
      ns=ne+1-neqb
      do nb=1,nblock                         ! Loop over displacement blocks
        read (2) ((qlambda(i,l),i=ns,ne),l=1,lc)
        ns=ns-neqb                           ! First d.o.f. number in block
        ne=ne-neqb                           ! Last d.o.f. number in block
      enddo
```

```
*     Determine load-case and constraint pointers for printing head

      j=0
      do l=1,ll                                   ! Loop over load cases
        do i=1,nc                                 ! Loop over constraints
          j=j+1
          ll6(j)=l                                ! Load case pointers
          nc6(j)=i                                ! Constraints pointers
        enddo
      enddo

*     Evaluate and print sensitivity gradient coefficients

      i6=(llnc+5)/6                   ! Number of 6-column printing pages
      rewind 14

      do 10 m=1,neltyp                ! Loop over design variables groups

      write (6,2001) m                                       ! Heading line
      nes=mpar(m)              ! Number of design variables in m-th group

      do n=1,nes                    ! Loop over design variables in group

        read (14) lm,dkdh              ! Reading in stiffness derivatives
                                                 ! and associated d.o.f.'s
        do l=1,ll                                   ! Loop over load cases
          lll=l*nc-nc
          do i=1,nc                                 ! Loop over constraints
            ii=ll+i
            ln=lll+i
            temp1=0.0d0
            do k=1,6                ! Outer loop over associated d.o.f.'s
              kk=lm(k)
              if(kk.gt.0) then
                temp2=0.0d0
                do j=1,6            ! Inner loop over associated d.o.f.'s
                  jj=lm(j)
                  if(jj.gt.0) temp2=temp2+dkdh(k,j)*qlambda(jj,l)
                enddo
                temp1=temp1+qlambda(kk,ii)*temp2
              endif
            enddo
            j=idofnc(i)           ! d.o.f. number of nonzero adjoint load
            if(qlambda(j,l).gt.0.0d0) then
              dgdh(ln,n) = -temp1                  ! Updating algebraic sign
            else                                                   ! ... of
              dgdh(ln,n) =  temp1    ! sensitivity gradient coefficients
            endif
          enddo
        enddo
      enddo
```

```
      do in=1,i6                                  ! Print 6-column pages
        ns=6*in-5
        ne=ns+5
        if(ne.gt.llnc) ne=llnc
        write (6,2010) (ll6(i),i=ns,ne)      ! Printing load case numbers
        write (6,2011) (nc6(i),i=ns,ne)     ! Printing constraint numbers
        write (6,'('' '')')
        do n=1,nes                           ! Loop over design variables
          write (6,2012) n,(dgdh(i,n),i=ns,ne)    ! Printing final result
        enddo
      enddo

   10 continue

 2001 format (///       ' D E S I G N   S E N S I T I V I T Y ',
     * ' G R A D I E N T S   -   V A R I A B L E   G R O U P ', i5//)
 2010 format (1x,7hELEMENT,6(:8x,9hLOAD CASE,i3))
 2011 format (2x, 6hNUMBER,6(:11x,  6hDESIGN,i3))
 2012 format (i8,1p6d20.10)

      end
```

By examining the above routine it is seen that the first step in the sensitivity computations is to get into an in-core block all the primary and adjoint displacements which have been computed and saved in the low-speed memory blocks. Since the solver implemented in SFESTA acts in the multi-block mode, once the backsubstitution process is performed the nodal displacements are saved block-wise in reverse order. To be specific, let us take an illustrative example of an algebraic system with 7 degrees of freedoms (i.e. `neq=7`) and `lc` right-hand sides. The system is solved for the unknown array `q(1:7,1:lc)` in the multi-block mode; the number of equations per block is preset to 3 (`neqb=3`). Thus, the number of solution blocks is `nblock=3` and the three Fortran arrays are written to a scratch file in the following order:

record 1: solution block no. 3

```
q(7,1)  q(7,2)  ...  q(7,lc)
   0       0    ...     0
   0       0    ...     0
```

record 2: solution block no. 2

```
q(4,1)  q(4,2)  ...  q(4,lc)
q(5,1)  q(5,2)  ...  q(5,lc)
q(6,1)  q(6,2)  ...  q(6,lc)
```

record 3: solution block no. 1

```
q(1,1)  q(1,2)  ...  q(1,lc)
q(2,1)  q(2,2)  ...  q(2,lc)
q(3,1)  q(3,2)  ...  q(3,lc)
```

(the zero entries written in the first record have a symbolic character only; these entries with indefinite values will not be dealt with later). If the single-block mode was employed, i.e. `nblock=1` and `neqb=neq`, the single record for the unknowns would be saved in the natural order of a Fortran array as

```
q(1,1)  q(1,2)  ...  q(1,lc)
q(2,1)  q(2,2)  ...  q(2,lc)
 ...     ...          ...
q(7,1)  q(7,2)  ...  q(7,lc)
```

6.2.3 Subroutines displcovar and disp1covar

With the analysis type `nsta=3` or `nsta=4` entered on the problem control data line, cf. Section 6.1, SFESTA carries out the solution process for structural static response and sensitivity of stochastic systems under static loads, respectively.

Random variables are defined by their spatial mean values and covariance matrix; mean values of the random variables are specified identically as in the deterministic analysis while input data for the covariance matrix can be handled in two different ways: (i) nonzero entries of the upper triangular part of the matrix (including the diagonal ones) and the corresponding row and column indices for these coefficients are entered in the common input data file, and (ii) the entire upper triangular part of the covariance matrix is set up by the user prior to the solution and is saved in the compact form of a binary unformatted file. Data processing for the input covariance matrix in SFESTA is similar to that employed in the SFEM-based code for dynamical systems SFEDYN, cf. Chapter 8; for a more complete treatment the reader is referred to Section 8.2.3.

Owing to the length limitation of the book, in this section we provide and discuss just two routines concerned with the evaluation of covariances for nodal displacements; in the next section a routine for the calculation of expectations and standard deviations for sensitivity gradient coefficients is additionally discussed.

The subroutine `displcovar` with its kernel part `disp1covar` computes the first-order accurate covariance for the α-th and β-th displacement components by using the expression, cf. Eq. (4.42)

$$\mathrm{Cov}(q_\alpha, q_\beta) = q_\alpha^{,\rho}\, q_\beta^{,\sigma}\, S_{\mathrm{b}}^{\rho\sigma} \tag{6.3}$$

The number of random variables specified by the parameter `nrand` is determined on the basis of the element data processing extending over all the element groups. Since the generalized nodal displacements are saved in `nblock` blocks of the size[1] `neqb`×`1` the number `neqb`×`neqb` of printing blocks for displacement covariances equals `nblock(nblock+1)/2` due to the symmetry. In the two routines the blocks including variance components are called the diagonal (or 'master') blocks while the term 'trailing' is used to specify off-diagonal blocks.

[1]For stochastic systems only one load case is allowed in the analysis.

```
      subroutine displcovar (covinp,discov,qmaster,qtrail,
     *                       nblock,neq,neqb,nrand,nt,i12)

*     To evaluate covariances for nodal displacements in blocks

*     ON INPUT:
*       i12           Flag for stochastic analysis type
*                     (1, displacement-stress only; 2, sensitivity)
*       nblock        Number of displacement blocks
*       neq           Number of system equations (system d.o.f.'s)
*       neqb          Number of equations per block
*       neqblk        Number equal to nblock*neqb
*       npmax         Max. number of design variables in group
*       nt            Scratch file unit (used to save final result)
*       qmaster(.,.)  Master block of first-order displacements
*       qtrail(.,.)   Trailing block of first-order displacements

*     ON OUTPUT:
*       discov(.,.)   Displacement covariance block on printing

      implicit  real*8 (a-h,o-z)

      dimension covinp(nrand),discov(neqb,neqb),
     *          qmaster(neqb,nrand),qtrail(neqb,nrand)

      iblock=0
                          ! Loops over nblock*(nblock+1)/2 output blocks
      do 10 m=1,nblock            ! Outer loop: over nblock master blocks

      mback1=701+nblock-m                   ! Block number (reverse order)
      call fileopen (mback1,17,1)                    ! Open scratch file
      read (17) qmaster                ! Read in first-order displacements
      if(i12.eq.2) then
         close (17,status='keep')         ! Save for sensitivity analysis
      else
         close (17,status='delete')
      endif
      ieqrow=neqb*m-neqb        ! First d.o.f. number of block (row index)
      ieqcol=ieqrow
                                     ! Covariance master-master calculation
      call displcovar (discov,qmaster,qmaster,covinp,
     *                 neq,neqb,nrand,ieqrow,ieqcol,iblock,nt)
                                 ! Master-trailing process for covariances
      m1=m+1
      if(m1.gt.nblock) return
                         ! Inner loop: over (nblock-m-1) trailing blocks
                      ! bypassing lower-triangular covariance components
      do 10 n=m1,nblock

      mback2=701+nblock-n
      call fileopen (mback2,17,1)
      read  (17) qtrail                ! Read in first-order displacements
      close (17,status='keep')
```

```
      ieqcol=neqb*n-neqb   ! First d.o.f. number of block (column index)
                          ! Covariance master-trailing calculation
      call displcovar (discov,qmaster,qtrail,covinp,
     *                 neq,neqb,nrand,ieqrow,ieqcol,iblock,nt)
   10 continue
      end
```

The routine displcovar, which is the main part of displcovar, evaluates the upper-triangular blocks of the displacement covariance matrix. The synopsis for this routine is not given here since all the formal variables and parameters have already been specified in displcovar.

```
      subroutine displcovar (discov,qmaster,qtrail,covinp,
     *                       neq,neqb,nrand,ieqrow,ieqcol,iblock,nt)

*     To compute and save displacement covariances in each block

      implicit  real*8 (a-h,o-z)
      dimension covinp(nrand),discov(neqb,neqb),
     *          qmaster(neqb,nrand),qtrail(neqb,nrand)

      iblock=iblock+1
      write (0,'(1h+,15x,''block nr'',i6)') iblock
      rewind 13                                       ! Initialization
      do l=1,neqb
        do i=1,neqb
          discov(i,l)=0.0d0
        enddo
      enddo

      do 10 i=1,nrand                         ! Loop over random variables
      read (13) covinp                      ! Read in input covariance row
      do 5 j=1,neqb                        ! Outer loop over block d.o.f.'s
      ieqrj=ieqrow+j                    ! Row index of covariance component
      if(ieqrj.gt.neq) goto 10                ! Last system d.o.f. is over
      tmp=0.0d0
      do k=1,nrand                        ! Summing up [dq/db](j,k)*Cov(k,i)
        tmp=tmp+qmaster(j,k)*covinp(k)
      enddo
      do l=1,neqb                          ! Inner loop over block d.o.f.'s
        ieqcl=ieqcol+l                ! Column index of covariance component
        if(ieqcl.gt.neq) goto 5                        ! Last system d.o.f.
                     ! Summing up [dq/db](j,k)*Cov(k,i)*[dq/db](l,i)
        discov(j,l)=discov(j,l)+tmp*qtrail(l,i)
      enddo
    5 continue
   10 continue
                 ! Save row-column indices and value of upper-triangular
                                         ! covariance entries for printing
      write (nt) ieqrow,ieqcol,discov

      end
```

A short routine `fileopen` is used to open temporary files using a single common file unit. It is simply a useful trick employed to handle the limitation in the overall number (20 for a typical PC/AT) of file units opened simultaneously during the program's execution. Scratch files opened by `fileopen` have the specification `block.???` with the extension being used to define the block number. In the case of `nblock=21`, for instance, by using `fileopen` the global stiffness matrix is saved in 21 files named `block.001`, `block.002`, ..., `block.021`, while the first-order displacement-variable vectors are stored on `block.701`, `block.702`, ..., `block.721`, etc. An alternative technique for handling scratch files based on the direct access mode adopted in the code SFEDYN is presented in Chapter 8.

6.2.4 Subroutine randsens

The option for static sensitivity analysis of stochastic systems is monitored in SFESTA by the analysis type parameter `nsta=4`. Similarly as in the deterministic sensitivity analysis (`nsta=2`), cf. Section 6.2.2, we assume that: (i) the structural response functional $\mathcal{G}$ is expressed implicitly in terms of the design variables h^d, and (ii) load functions are independent of h^d and specified deterministically. Thus, the functional of the form (6.1) can be used; the difference is that the design variables h^d are assumed to be random and defined by their expectations and covariances. The expressions for the second-order accurate expectations and first-order accurate variances for the sensitivity gradient coefficients can be written in simplified form as, cf. Eqs. (5.16)–(5.18)

$$E[\mathcal{G}^{\cdot d}] = -\left[\lambda^0_\alpha\, K^{0.d}_{\alpha\beta}\left(q^0_\beta + \frac{1}{2}\,q^{(2)}_\beta\right) + \frac{1}{2}\,\lambda^{(2)}_\alpha\, K^{0.d}_{\alpha\beta}\, q^0_\beta + \lambda^{,\rho}_\alpha\, K^{0.d}_{\alpha\beta}\, q^{,\rho}_\beta\, S^{\rho\sigma}_{\rm b}\right] \tag{6.4}$$

$$\mathrm{Var}(\mathcal{G}^{\cdot d}) = \left(\lambda^{,\rho}_\alpha\, K^{0.d}_{\alpha\beta}\, q^0_\beta + \lambda^0_\alpha\, K^{0.d}_{\alpha\beta}\, q^{,\rho}_\beta\right)\left(\lambda^{,\sigma}_\alpha\, K^{0.d}_{\alpha\beta}\, q^0_\beta + \lambda^0_\alpha\, K^{0.d}_{\alpha\beta}\, q^{,\sigma}_\beta\right) S^{\rho\sigma}_{\rm b} \tag{6.5}$$

the expanded form of which is given here for the sake of convenience in computer implementation. In Eqs. (6.4) and (6.5) $\alpha, \beta = 1, 2, \ldots, N$ while $d, \rho, \sigma = 1, 2, \ldots, \tilde{N} = D$ which is because in the present version of SFESTA the spatial averaging strategy is employed and all the design variables are assumed to be random. In `randsens` the numbers D and N are specified by `nrand` and `neq`, respectively.

```
      subroutine randsens (q1,q2,dkq1,q1dkq0,covinp,
     *                     average,sdeviat,neqblk,neqb,nblock,nrand)

*     To evaluate sensitivity expectations and deviations
*     of structural systems subjected to static loads

*     ON INPUT:
*       nblock      Number of core blocks used to save system matrices
*       neqb        Number of equations per block
*       neqblk      Number equal to nblock*neqb
*       nrand       Total number of random variables
*       lm(.)       Degrees of freedom associated with design variable
```

```
*        covinp(.)    Input covariance row
*        dkq0(.,1)    Stiff. derivatives times 0th primary displacements
*        dkq0(.,2)    Stiff. derivatives times 0th adjoint displacements
*        dkq1(.,.)    Stiff. derivatives times 1st primary displacements
*        q1(.,.,1)    First-order primary displacements (file block.7??)
*        q1(.,.,2)    First-order adjoint displacements
*        q2(.,1)      Second-order primary displacements (file block.8??)
*        q2(.,2)      Second-order adjoint displacements

*      ON OUTPUT:
*        average(.)   Expectations for sensitivity gradient coefficients
*        sdeviat(.)   Deviations for sensitivity gradient coefficients

*        q1dkq0(.)    Working array
*        dknul1(.), dknul2(.)  Inactive arrays

      implicit        real*8 (a-h,o-z)

      common /elems/ lm(6),dknul1(36),dkq0(6,2),dknul2(80)
      dimension       q1(neqblk,nrand,2),q2(neqblk,2),
     *                dkq1(6,nrand),q1dkq0(nrand),covinp(nrand),
     *                average(nrand),sdeviat(nrand)

*     Assemble primary and adjoint displacement blocks in core

      ne=neqblk                          ! Last d.o.f. number of last block
      ns=ne+1-neqb                      ! First d.o.f. number of last block

      do m=1,nblock                          ! Loop over displacement blocks
        m700=700+m
        call fileopen (m700,17,1)
                ! Read in first-order primary and adjoint displacements
        read  (17) (((q1(j,n,i),j=ns,ne),n=1,nrand),i=1,2)
        close (17,status='delete')
        m800=800+m
        call fileopen (m800,17,1)
               ! Read in second-order primary and adjoint displacements
        read  (17) ((q2(j,i),j=ns,ne),i=1,2)
        close (17,status='delete')
        ns=ns-neqb                      ! First d.o.f. number of next block
        ne=ne-neqb                       ! Last d.o.f. number of next block
      enddo

*     Process computation of sensitivity expectations and deviations

      rewind 4
      rewind 15
      write(0,'('''')')

      do 10 n=1,nrand                   ! Loop over random design variables

      write (0,'(1h+,15x,''design variable nr'',i5)') n
      read (15) lm,dkq0
```

```
      do m=1,nrand
        q1dkq0(m)=0.0d0
      enddo
      smean2=0.0d0

      do k=1,6                      ! Loop over element (variable) d.o.f.'s
        lmk=lm(k)
        if(lmk.gt.0) then                  ! For unconstrained d.o.f.
          tmp1=dkq0(k,1)
          tmp2=dkq0(k,2)
                        ! Second-order terms of sensitivity expectations
          smean2=smean2+tmp1*q2(lmk,2)+tmp2*q2(lmk,1)
          do m=1,nrand                     ! Loop over random variables
            q1dkq0(m)=q1dkq0(m)+tmp1*q1(lmk,m,2)+tmp2*q1(lmk,m,1)
          enddo
        endif
      enddo

      read (4) dkq1
      rewind 13
      variance=0.0d0
      smean1=0.0d0

      do m=1,nrand                       ! Outer loop over random variables
        read (13) covinp                     ! Read in input covariance row
        do k=1,6                    ! Loop over element (variable) d.o.f.'s
          lmk=lm(k)
          if(lmk.gt.0) then
            tmp1=0.0d0
            do j=1,nrand                 ! Inner loop over random variables
              tmp1=tmp1+dkq1(k,j)*covinp(j)
            enddo
                         ! First-order terms of sensitivity expectations
            smean1=smean1+tmp1*q1(lmk,m,2)
          endif
        enddo
        tmp1=0.0d0
        do j=1,nrand
          tmp1=tmp1+q1dkq0(j)*covinp(j)
        enddo
        variance=variance+tmp1*q1dkq0(m)         ! Sensitivity variances
      enddo

      average(n)=smean1+smean2                ! Sensitivity expectations
      if(variance.gt.0.0d0) then       ! Sensitivity standard deviations
        sdeviat(n)=dsqrt(variance)
      else
        sdeviat(n)=0.0d0
      endif

   10 continue

      write (0,'(1h+,40(1h ))')
```

```
*     Print sensitivity expectations and deviations

      n4=(nrand+3)/4                  ! Number of 4-column printing pages
      write (6,2001)                              ! Printing heading line
      do m=1,n4                                          ! Loop over pages
        ns=4*m-3                                ! First variable on page
        ne=ns+3                                  ! Last variable on page
        if(ne.gt.nrand) ne=nrand
        write (6,2002) (n,average(n),n=ns,ne)
        write (6,2003) (n,sdeviat(n),n=ns,ne)
        write (6,'('''')')
      enddo

 2001 format(///       ' E X P E C T A T I O N S   E [ N ]   A N D ',
     *       ' S T A N D A R D   D E V I A T I O N S   V a r ( N ) ',
     *                       ' O F   S E N S I T I V I T I E S '//
     *                  ' (N = 1,..., NUMBER OF DESIGN VARIABLES)'//)
 2002 format(4(:1x,2hE(,i3,4h) = ,1pd19.12,4x))
 2003 format(4(:1x,2hD(,i3,4h) = ,1pd19.12,4x))

      end
```

It should be pointed out in closing that for any stochastic sensitivity analysis using SFESTA only one load case and one response functional (displacement constraint) can be considered in a single program run. For problems in which various (up to 30) groups of random design variables are defined, the variables are numbered uniformly in all groups in the order corresponding to the group numbering. For instance, a printout of the sensitivity distributions for the system with three 50-variable groups will list the design variables in the first, second and third group numbered 1–50, 51–100 and 101–150, respectively. A different scheme (believed to be more convenient) in which random design variables are numbered separately for each group has been adopted in the program SFEDYN, cf. Chapter 8.

For the sake of simplicity and clarity of presentation the full printing mode has been employed on SFESTA output, i.e. the first two moments for all the nodal displacements and all the sensitivity gradient coefficients are always printed. This may clearly turn out inefficient in the analysis of medium- or large-scale systems wherein only the probabilistic distributions for displacement response at a few selected points may need to be considered. Similarly, to preserve the illustrative character of SFESTA the standard normal transformation from a correlated set of random variables to an uncorrelated set has not been employed. Nevertheless, the random variable transformation as well as concise output printing at selected points may be incorporated in the program with no difficulty following the discussion of Chapter 8, in which a complete treatment of these computational aspects is presented through a description of the appropriate Fortran routines.

Chapter 7

Stochastic Sensitivity: Dynamic Problems

7.1 Finite Element Formulation

Structural response sensitivity of multi-degree-of-freedom dynamic systems is considered in this chapter. From a purely formal viewpoint the analysis is a combination of the deterministic dynamic sensitivity analysis (Section 2.8.2) and the stochastic structural analysis (Chapter 4); thus it is the most general case considered in this book. Putting it another way, we develop numerical tools for the assessment of spatial probabilistic distributions describing time response sensitivity with respect to variations in spatial probabilistic distributions characterizing structural parameters. Both the time interval and time instant response sensitivities discussed in Section 2.8.2 for the deterministic systems are considered here in the context of stochastic behaviour. The formal presentation is restricted to the case of the terminal time condition given explicitly in terms of the terminal time; the extension towards the inclusion of the terminal time condition given implicitly is straightforward, cf. Section 2.8.2.

In the present section we formulate equations describing the problem considered. Again, similarly as in Chapter 5, the analysis will be carried out entirely in the framework of the FEM description. This means, in particular, that the structural probabilistic characteristics are represented in terms of the nodal random variables b_ρ, $\rho = 1, 2, \ldots, \tilde{N}$.

The derivations in this section will be followed by a detailed discussion of some computational aspects in Section 7.2 and a presentation of numerous test examples in Section 7.3.

7.1.1 Time Interval Sensitivity

Given a linear elastic system with N degrees of freedom, consider the system response over the time interval $[0, T]$ described by the integral functional

$$\mathcal{G}(h^d, b_\rho) = \int_0^T G\,[q_\alpha(h^d, b_\rho; \tau), h^d]\,\mathrm{d}\tau$$
$$d = 1, 2, \ldots, D\,;\ \rho = 1, 2, \ldots, \tilde{N}\,;\ \alpha = 1, 2, \ldots, N \tag{7.1}$$

where G is a given function of its arguments, h^d is a D-dimensional design variable vector, b_ρ is an $\tilde{N}$-dimensional random variable vector and q_α is an N-dimensional vector of nodal displacement-type variables. Some or all entries in the vectors h^d and b_ρ may coincide.

The general form of the structural response functional (7.1) can be specified to represent most of the cost and constraint functions that are typically taken to quantify the structural response over a time interval. If all the quantities involved in Eq. (7.1) are evaluated at the spatial expectations of the random variables b_ρ, the functional becomes equivalent to the functional (2.132) used in the deterministic sensitivity analysis.

The nodal displacement-type variables $q_\alpha(h^d, b_\rho; \tau)$ are implicit functions of the random and design variables and satisfy the spatially discretized equation of motion of the form

$$M_{\alpha\beta}(h^d, b_\rho)\,\ddot{q}_\beta(h^d, b_\rho; \tau) + C_{\alpha\beta}(h^d, b_\rho)\,\dot{q}_\beta(h^d, b_\rho; \tau)$$
$$+\ K_{\alpha\beta}(h^d, b_\rho)\,q_\beta(h^d, b_\rho; \tau) = Q_\alpha(h^d, b_\rho; \tau) \tag{7.2}$$

together with the homogeneous initial conditions

$$q_\alpha(h^d, b_\rho; 0) = 0 \qquad \dot{q}_\alpha(h^d, b_\rho; 0) = 0 \tag{7.3}$$

where the mass, damping and stiffness matrices $M_{\alpha\beta}(h^d, b_\rho)$, $C_{\alpha\beta}(h^d, b_\rho)$ and $K_{\alpha\beta}(h^d, b_\rho)$ as well as the load vector $Q_\alpha(h^d, b_\rho; \tau)$ are explicitly given functions of the random and design variables. As shown in Section 2.8.2.3 the assumption of homogeneity of the initial conditions (7.3) is not restrictive. The objective of the stochastic analysis of dynamic structural sensitivity is to determine the probabilistic distributions of the change in the structural response functionals given by Eq. (7.1) with the constraint given by Eqs. (7.2) and (7.3) under the variations in the design parameters, i.e. to evaluate the time behaviour of the spatial expectations and covariances of the sensitivity gradient coefficients.

Assume that the function $G\,[q_\alpha(h^d, b_\rho; \tau), h^d]$ is continuous in the entire time interval $[0, T]$ considered. Furthermore, suppose that the functions $M_{\alpha\beta}(h^d, b_\rho)$, $C_{\alpha\beta}(h^d, b_\rho)$, $K_{\alpha\beta}(h^d, b_\rho)$ and $Q_\alpha(h^d, b_\rho; \tau)$, and consequently the solution vector $q_\alpha(h^d, b_\rho; \tau)$, are continuously differentiable with respect to the design variables h^d and the random variables b_ρ. Since the terminal time T is explicitly given, differentiation of Eq. (7.1) with respect to h^d leads to, cf. Eq. (2.136)

$$\mathcal{G}^{,d}(h^d, b_\rho) = \int_0^T \left[G^{,d}(h^d, b_\rho; \tau) + G_{,\alpha}(h^d, b_\rho; \tau)\, q_\alpha^{,d}(h^d, b_\rho; \tau) \right] \mathrm{d}\tau \tag{7.4}$$

where the symbols $(.)^{\cdot d}$ and $(.)_{.\alpha}$ denote, as always in this book, the first partial derivatives with respect to the d-th design variable and the α-th nodal displacement, respectively.

In order to express the sensitivity gradient coefficients $\mathcal{G}^{\cdot d}$ in terms of the variation in h^d the displacement sensitivity matrix $q_\alpha^{\cdot d}$ involved in Eq. (7.4) has to be written explicitly with respect to h^d. To do this the adjoint variable approach introduced in Section 2.8.2.1 may be employed. Let us partially differentiate Eq. (7.2) with respect to the design variables h^d to obtain[1]

$$M_{\alpha\beta}\,\ddot{q}_\beta^{\cdot d}(\tau) + C_{\alpha\beta}\,\dot{q}_\beta^{\cdot d}(\tau) + K_{\alpha\beta}\,q_\beta^{\cdot d}(\tau) - R_\alpha^d(\tau) \;=\; 0 \tag{7.5}$$

where

$$R_\alpha^d(\tau) \;=\; Q_\alpha^{\cdot d}(\tau) - \left[M_{\alpha\beta}^{\cdot d}\,\ddot{q}_\beta(\tau) + C_{\alpha\beta}^{\cdot d}\,\dot{q}_\beta(\tau) + K_{\alpha\beta}^{\cdot d}\,q_\beta(\tau)\right] \tag{7.6}$$

Pre-multiplying all terms involved in Eq. (7.5) by the transpose of an adjoint variable vector defined as $\lambda(\tau) = \{\lambda_\alpha(\tau)\}$, $\alpha = 1, 2, \ldots, N$, integrating the result over the time interval $[0, T]$ and noting that the integrand equals zero at any time τ, $\tau \in [0, T]$, yields

$$\int_0^T \lambda_\alpha(\tau)\left[M_{\alpha\beta}\,\ddot{q}_\beta^{\cdot d}(\tau) + C_{\alpha\beta}\,\dot{q}_\beta^{\cdot d}(\tau) + K_{\alpha\beta}\,q_\beta^{\cdot d}(\tau) - R_\alpha^d(\tau)\right] \mathrm{d}\tau \;=\; 0 \tag{7.7}$$

By using the homogeneous initial conditions (7.3) and integrating over time τ by parts the first two terms under the integral (7.7) become

$$\begin{aligned}\int_0^T \lambda_\alpha(\tau) M_{\alpha\beta}\,\ddot{q}_\beta^{\cdot d}(\tau)\,\mathrm{d}\tau &= \lambda_\alpha(T) M_{\alpha\beta}\,\dot{q}_\beta^{\cdot d}(T) - \int_0^T \dot{\lambda}_\alpha(\tau) M_{\alpha\beta}\,\dot{q}_\beta^{\cdot d}(\tau)\,\mathrm{d}\tau \\ &= \lambda_\alpha(T) M_{\alpha\beta}\,\dot{q}_\beta^{\cdot d}(T) - \dot{\lambda}_\alpha(T) M_{\alpha\beta}\,q_\beta^{\cdot d}(T) + \int_0^T \ddot{\lambda}_\alpha(\tau) M_{\alpha\beta}\,q_\beta^{\cdot d}(\tau)\,\mathrm{d}\tau\end{aligned} \tag{7.8}$$

and

$$\int_0^T \lambda_\alpha(\tau)\,C_{\alpha\beta}\,\dot{q}_\beta^{\cdot d}(\tau)\,\mathrm{d}\tau = \lambda_\alpha(T)\,C_{\alpha\beta}\,q_\beta^{\cdot d}(T) - \int_0^T \dot{\lambda}_\alpha(\tau)\,C_{\alpha\beta}\,q_\beta^{\cdot d}(\tau)\,\mathrm{d}\tau \tag{7.9}$$

Substituting Eqs. (7.7) and (7.9) into Eq. (7.7) leads to

$$\begin{aligned}&\lambda_\alpha(T)\,M_{\alpha\beta}\,\dot{q}_\beta^{\cdot d}(T) - \left[\dot{\lambda}_\alpha(T)\,M_{\alpha\beta} - \lambda_\alpha(T)\,C_{\alpha\beta}\right] q_\beta^{\cdot d}(T) \\ &+\int_0^T\left\{\left[\ddot{\lambda}_\alpha(\tau)M_{\alpha\beta} - \dot{\lambda}_\alpha(\tau)C_{\alpha\beta} + \lambda_\alpha(\tau)K_{\alpha\beta}\right]q_\beta^{\cdot d}(\tau) - \lambda_\alpha(\tau)R_\alpha^d(\tau)\right\}\mathrm{d}\tau = 0\end{aligned} \tag{7.10}$$

Since Eq. (7.10) holds for any arbitrary value of the adjoint variable vector λ_α (cf. Eq. (7.7)), the functions λ_α may be found so that the coefficients at $q_\alpha^{\cdot d}$ and $\dot{q}_\alpha^{\cdot d}$ in Eqs. (7.4) and (7.10) are equal. If such a solution for λ_α exists the components of the displacement sensitivity matrix $q_\alpha^{\cdot d}$ in Eq. (7.4) can be expressed through λ_α. To be specific, by equating the coefficients at $q_\alpha^{\cdot d}$ under

[1] As before, the arguments h^d and b_ρ will be suppressed in places to make the presentation more compact.

the integrals in Eqs. (7.4) and (7.10) we arrive at the differential equations of motion for the adjoint system in the form, cf. Eq. (2.142)

$$M_{\alpha\beta}(h^d, b_p)\,\ddot{\lambda}_\beta(h^d, b_p; \tau) - C_{\alpha\beta}(h^d, b_p)\,\dot{\lambda}_\beta(h^d, b_p; \tau) + K_{\alpha\beta}(h^d, b_p)\,\lambda_\beta(h^d, b_p; \tau) = G_{.\alpha}(h^d, b_p; \tau) \tag{7.11}$$

whereas the homogeneous terminal conditions for the adjoint displacement and adjoint velocity vectors at the terminal time T can be obtained from the first two terms in Eq. (7.10) as, cf. Eqs. (2.143) and (2.144)

$$\lambda_\alpha(h^d, b_p; T) = 0 \qquad \dot{\lambda}_\alpha(h^d, b_p; T) = 0 \tag{7.12}$$

(The corresponding terms in Eq. (7.4) are nonexistent since the terminal time is assumed independent of the design variables.) It is noted that the adjoint variable vector selected to satisfy the stochastic terminal problem (7.11),(7.12) stands for the function of random and design variables, i.e. $\lambda_\alpha = \lambda_\alpha(h^d, b_p; \tau)$.

The problem described by the adjoint equations (7.11) and (7.12) is referred to as the *stochastic terminal problem* and the equation systems (7.2),(7.3) and (7.11),(7.12) form the *stochastic initial-terminal problem* for the time interval SSDS analysis.

Having solved Eqs. (7.2),(7.3) and (7.11),(7.12) for the displacement response of the primary and adjoint systems, respectively, the expressions for the sensitivity gradient coefficients $\mathcal{G}^{\cdot d}$ can be written in terms of the adjoint variables λ_α, i.e. the displacement sensitivity components q^d_α in Eq. (7.4) can be eliminated. Eq. (7.10), after using Eqs. (7.11) and (7.12), becomes

$$\int_0^T G_{.\alpha}(h^d, b_p; \tau)\, q^{;d}_\alpha(h^d, b_p; \tau)\, \mathrm{d}\tau = \int_0^T \lambda_\alpha(h^d, b_p; \tau)\, R^d_\alpha(h^d, b_p; \tau)\, \mathrm{d}\tau \tag{7.13}$$

which, when introduced into Eq. (7.4), leads to the expressions for $\mathcal{G}^{\cdot d}$ as

$$\mathcal{G}^{\cdot d}(h^d, b_p) = \int_0^T \left[G^{\cdot d}(h^d, b_p; \tau) + \lambda_\alpha(h^d, b_p; \tau)\, R^d_\alpha(h^d, b_p; \tau) \right] \mathrm{d}\tau \tag{7.14}$$

To incorporate randomness in structural geometry, material and loading parameters into the formulation, the Taylor expansion truncated at the second-order terms can be used. Similarly to the derivations presented in Chapter 4 for the linearized stochastic differential equations of motion, from the primary equation (7.2) we get systems of zeroth-, first- and second-order equations. The resulting equations have formally similar forms to Eqs. (4-12)–(4-14) since the same linear operator evaluated at the expectations of random variables appears on the left-hand sides and all the uncertainty properties of the system are moved to the right-hand sides as the zeroth-, first- and second-order nodal loads. The difference is that in the present equations the mass, damping and stiffness matrices and the nodal accelerations, velocities and displacements and their first and second mixed partial derivatives with respect to random variables are (explicit or implicit) functions of the design variables.

To be consistent with the random-parameter space b_ρ the finite element coordinates q_α of the primary system (7.2) and λ_α of the adjoint system (7.11) are expanded about their expectations with the small parameter ϵ. Retaining up to the second-order perturbation yields, respectively

$$\begin{aligned} q_\alpha(h^d, b_\varrho; \tau) &= q^0_\alpha(h^d, b^0_\varrho; \tau) + \epsilon\, q^{,\rho}_\alpha(h^d, b^0_\varrho; \tau)\, \Delta b_\rho \\ &\quad + \frac{1}{2}\, \epsilon^2\, q^{,\rho\sigma}_\alpha(h^d, b^0_\varrho; \tau)\, \Delta b_\rho\, \Delta b_\sigma \\ \lambda_\alpha(h^d, b_\varrho; \tau) &= \lambda^0_\alpha(h^d, b^0_\varrho; \tau) + \epsilon\, \lambda^{,\rho}_\alpha(h^d, b^0_\varrho; \tau)\, \Delta b_\rho \\ &\quad + \frac{1}{2}\, \epsilon^2\, \lambda^{,\rho\sigma}_\alpha(h^d, b^0_\varrho; \tau)\, \Delta b_\rho\, \Delta b_\sigma \end{aligned} \tag{7.15}$$

Also, the second-order expansion is made correspondingly for the functions $\mathcal{G}$ and Q_α, $M_{\alpha\beta}$, $C_{\alpha\beta}$, $K_{\alpha\beta}$ as

$$\begin{aligned} G(h^d, b_\varrho; \tau) &= G^0(h^d, b^0_\varrho; \tau) + \epsilon\, G^{,\rho}(h^d, b^0_\varrho; \tau)\, \Delta b_\rho \\ &\quad + \frac{1}{2}\, \epsilon^2\, G^{,\rho\sigma}(h^d, b^0_\varrho; \tau)\, \Delta b_\rho\, \Delta b_\sigma \\ Q_\alpha(h^d, b_\varrho; \tau) &= Q^0_\alpha(h^d, b^0_\varrho; \tau) + \epsilon\, Q^{,\rho}_\alpha(h^d, b^0_\varrho; \tau)\, \Delta b_\rho \\ &\quad + \frac{1}{2}\, \epsilon^2\, Q^{,\rho\sigma}_\alpha(h^d, b^0_\varrho; \tau)\, \Delta b_\rho\, \Delta b_\sigma \\ M_{\alpha\beta}(h^d, b_\varrho) &= M^0_{\alpha\beta}(h^d, b^0_\varrho) + \epsilon\, M^{,\rho}_{\alpha\beta}(h^d, b^0_\varrho)\, \Delta b_\rho \\ &\quad + \frac{1}{2}\, \epsilon^2\, M^{,\rho\sigma}_{\alpha\beta}(h^d, b^0_\varrho)\, \Delta b_\rho\, \Delta b_\sigma \\ C_{\alpha\beta}(h^d, b_\varrho) &= C^0_{\alpha\beta}(h^d, b^0_\varrho) + \epsilon\, C^{,\rho}_{\alpha\beta}(h^d, b^0_\varrho)\, \Delta b_\rho \\ &\quad + \frac{1}{2}\, \epsilon^2\, C^{,\rho\sigma}_{\alpha\beta}(h^d, b^0_\varrho)\, \Delta b_\rho\, \Delta b_\sigma \\ K_{\alpha\beta}(h^d, b_\varrho) &= K^0_{\alpha\beta}(h^d, b^0_\varrho) + \epsilon\, K^{,\rho}_{\alpha\beta}(h^d, b^0_\varrho)\, \Delta b_\rho \\ &\quad + \frac{1}{2}\, \epsilon^2\, K^{,\rho\sigma}_{\alpha\beta}(h^d, b^0_\varrho)\, \Delta b_\rho\, \Delta b_\sigma \end{aligned} \tag{7.16}$$

Substituting the above expansions into Eqs. (7.2) and (7.3) and Eqs. (7.11) and (7.12) and equating the terms of the zeroth-, first- and second-power of ϵ, three equation sets are generated for both initial and terminal problems as follows:

- Zeroth-order (ϵ^0 terms, one pair of systems of N linear simultaneous ordinary differential equations for $q^0_\alpha(h^d, b^0_\varrho; \tau)$ and $\lambda^0_\alpha(h^d, b^0_\varrho; \tau)$, $\tau \in [0, T]$, respectively)

$$\begin{aligned} & M^0_{\alpha\beta}(h^d, b^0_\varrho)\, \ddot{q}^0_\beta(h^d, b^0_\varrho; \tau) + C^0_{\alpha\beta}(h^d, b^0_\varrho)\, \dot{q}^0_\beta(h^d, b^0_\varrho; \tau) \\ &\qquad + K^0_{\alpha\beta}(h^d, b^0_\varrho)\, q^0_\beta(h^d, b^0_\varrho; \tau) = Q^0_\alpha(h^d, b^0_\varrho; \tau) \\ & q^0_\alpha(h^d, b^0_\varrho; 0) = 0\, ; \quad \dot{q}^0_\alpha(h^d, b^0_\varrho; 0) = 0 \\ & M^0_{\alpha\beta}(h^d, b^0_\varrho)\, \ddot{\lambda}^0_\beta(h^d, b^0_\varrho; \tau) - C^0_{\alpha\beta}(h^d, b^0_\varrho)\, \dot{\lambda}^0_\beta(h^d, b^0_\varrho; \tau) \\ &\qquad + K^0_{\alpha\beta}(h^d, b^0_\varrho)\, \lambda^0_\beta(h^d, b^0_\varrho; \tau) = G^0_{.\alpha}(h^d, b^0_\varrho; \tau) \\ & \lambda^0_\alpha(h^d, b^0_\varrho; T) = 0\, ; \quad \dot{\lambda}^0_\alpha(h^d, b^0_\varrho; T) = 0 \end{aligned} \tag{7.17}$$

- First-order (ϵ^1 terms, $\tilde{N}$ pairs of systems of N linear simultaneous ordinary differential equations for $q_\alpha^{,\rho}(h^d, b_\varrho^0; \tau)$ and $\lambda_\alpha^{,\rho}(h^d, b_\varrho^0; \tau)$, $\tau \in [0, T]$, respectively)

$$\begin{aligned} &M_{\alpha\beta}^0(h^d, b_\varrho^0)\, \ddot{q}_\beta^{,\rho}(h^d, b_\varrho^0; \tau) + C_{\alpha\beta}^0(h^d, b_\varrho^0)\, \dot{q}_\beta^{,\rho}(h^d, b_\varrho^0; \tau) \\ &\qquad + K_{\alpha\beta}^0(h^d, b_\varrho^0)\, q_\beta^{,\rho}(h^d, b_\varrho^0; \tau) = Q_\alpha^\rho(h^d, b_\varrho^0; \tau) \\ &q_\alpha^{,\rho}(h^d, b_\varrho^0; 0) = 0\,; \quad \dot{q}_\alpha^{,\rho}(h^d, b_\varrho^0; 0) = 0 \\ &M_{\alpha\beta}^0(h^d, b_\varrho^0)\, \ddot{\lambda}_\beta^{,\rho}(h^d, b_\varrho^0; \tau) - C_{\alpha\beta}^0(h^d, b_\varrho^0)\, \dot{\lambda}_\beta^{,\rho}(h^d, b_\varrho^0; \tau) \\ &\qquad + K_{\alpha\beta}^0(h^d, b_\varrho^0)\, \lambda_\beta^{,\rho}(h^d, b_\varrho^0; \tau) = G_\alpha^\rho(h^d, b_\varrho^0; \tau) \\ &\lambda_\alpha^{,\rho}(h^d, b_\varrho^0; T) = 0\,; \quad \dot{\lambda}_\alpha^{,\rho}(h^d, b_\varrho^0; T) = 0 \\ &\qquad\qquad \text{with} \quad \rho = 1, 2, \ldots, \tilde{N} \end{aligned} \tag{7.18}$$

- Second-order (ϵ^2 terms, one pair of systems of N linear simultaneous ordinary differential equations for $q_\alpha^{(2)}(h^d, b_\varrho^0; \tau)$ and $\lambda_\alpha^{(2)}(h^d, b_\varrho^0; \tau)$, $\tau \in [0, T]$, respectively)

$$\begin{aligned} &M_{\alpha\beta}^0(h^d, b_\varrho^0)\, \ddot{q}_\beta^{(2)}(h^d, b_\varrho^0; \tau) + C_{\alpha\beta}^0(h^d, b_\varrho^0)\, \dot{q}_\beta^{(2)}(h^d, b_\varrho^0; \tau) \\ &\qquad + K_{\alpha\beta}^0(h^d, b_\varrho^0)\, q_\beta^{(2)}(h^d, b_\varrho^0; \tau) = Q_\alpha^{(2)}(h^d, b_\varrho^0; \tau) \\ &q_\alpha^{(2)}(h^d, b_\varrho^0; 0) = 0\,; \quad \dot{q}_\alpha^{(2)}(h^d, b_\varrho^0; 0) = 0 \\ &M_{\alpha\beta}^0(h^d, b_\varrho^0)\, \ddot{\lambda}_\beta^{(2)}(h^d, b_\varrho^0; \tau) - C_{\alpha\beta}^0(h^d, b_\varrho^0)\, \dot{\lambda}_\beta^{(2)}(h^d, b_\varrho^0; \tau) \\ &\qquad + K_{\alpha\beta}^0(h^d, b_\varrho^0)\, \lambda_\beta^{(2)}(h^d, b_\varrho^0; \tau) = G_\alpha^{(2)}(h^d, b_\varrho^0; \tau) \\ &\lambda_\alpha^{(2)}(h^d, b_\varrho^0; T) = 0\,; \quad \dot{\lambda}_\alpha^{(2)}(h^d, b_\varrho^0; T) = 0 \end{aligned} \tag{7.19}$$

where $d = 1, 2, \ldots, D$; $\alpha, \beta = 1, 2, \ldots, N$. The second-order primary and adjoint displacement vectors are defined as, cf. Eqs. (4.20) and (5.11)

$$\begin{aligned} q_\alpha^{(2)}(h^d, b_\varrho^0; \tau) &= q_\alpha^{,\rho\sigma}(h^d, b_\varrho^0; \tau)\, S_{\mathrm{b}}^{\rho\sigma} \\ \lambda_\alpha^{(2)}(h^d, b_\varrho^0; \tau) &= \lambda_\alpha^{,\rho\sigma}(h^d, b_\varrho^0; \tau)\, S_{\mathrm{b}}^{\rho\sigma} \end{aligned} \qquad \rho, \sigma = 1, 2, \ldots, \tilde{N} \tag{7.20}$$

while the first- and second-order primary and adjoint generalized load vectors are denoted by

$$\begin{aligned} Q_\alpha^\rho(h^d, b_\varrho^0; \tau) &= Q_\alpha^{,\rho}(h^d, b_\varrho^0; \tau) - \Big[M_{\alpha\beta}^{,\rho}(h^d, b_\varrho^0)\, \ddot{q}_\beta^0(h^d, b_\varrho^0; \tau) \\ &\quad + C_{\alpha\beta}^{,\rho}(h^d, b_\varrho^0)\, \dot{q}_\beta^0(h^d, b_\varrho^0; \tau) + K_{\alpha\beta}^{,\rho}(h^d, b_\varrho^0)\, q_\beta^0(h^d, b_\varrho^0; \tau) \Big] \\ G_\alpha^\rho(h^d, b_\varrho^0; \tau) &= G_{.\alpha}^{,\rho}(h^d, b_\varrho^0; \tau) - \Big[M_{\alpha\beta}^{,\rho}(h^d, b_\varrho^0)\, \ddot{\lambda}_\beta^0(h^d, b_\varrho^0; \tau) \\ &\quad - C_{\alpha\beta}^{,\rho}(h^d, b_\varrho^0)\, \dot{\lambda}_\beta^0(h^d, b_\varrho^0; \tau) + K_{\alpha\beta}^{,\rho}(h^d, b_\varrho^0)\, \lambda_\beta^0(h^d, b_\varrho^0; \tau) \Big] \end{aligned} \tag{7.21}$$

$$
\begin{aligned}
Q^{(2)}_{\alpha}(h^d, b^0_{\varrho}; \tau) = \Big\{ & Q^{,\rho\sigma}_{\alpha}(h^d, b^0_{\varrho}; \tau) - 2\Big[M^{,\rho}_{\alpha\beta}(h^d, b^0_{\varrho})\, \ddot{q}^{,\sigma}_{\beta}(h^d, b^0_{\varrho}; \tau) \\
& + C^{,\rho}_{\alpha\beta}(h^d, b^0_{\varrho})\, \dot{q}^{,\sigma}_{\beta}(h^d, b^0_{\varrho}; \tau) + K^{,\rho}_{\alpha\beta}(h^d, b^0_{\varrho})\, q^{,\sigma}_{\beta}(h^d, b^0_{\varrho}; \tau)\Big] \\
& - \Big[M^{,\rho\sigma}_{\alpha\beta}(h^d, b^0_{\varrho})\, \ddot{q}^{0}_{\beta}(h^d, b^0_{\varrho}; \tau) + C^{,\rho\sigma}_{\alpha\beta}(h^d, b^0_{\varrho})\, \dot{q}^{0}_{\beta}(h^d, b^0_{\varrho}; \tau) \\
& + K^{,\rho\sigma}_{\alpha\beta}(h^d, b^0_{\varrho})\, q^{0}_{\beta}(h^d, b^0_{\varrho}; \tau)\Big]\Big\}\, S^{\rho\sigma}_{\mathrm{b}} \\
G^{(2)}_{\alpha}(h^d, b^0_{\varrho}; \tau) = \Big\{ & G^{,\rho\sigma}_{.\alpha}(h^d, b^0_{\varrho}; \tau) - 2\Big[M^{,\rho}_{\alpha\beta}(h^d, b^0_{\varrho})\, \ddot{\lambda}^{,\sigma}_{\beta}(h^d, b^0_{\varrho}; \tau) \\
& - C^{,\rho}_{\alpha\beta}(h^d, b^0_{\varrho})\, \dot{\lambda}^{,\sigma}_{\beta}(h^d, b^0_{\varrho}; \tau) + K^{,\rho}_{\alpha\beta}(h^d, b^0_{\varrho})\, \lambda^{,\sigma}_{\beta}(h^d, b^0_{\varrho}; \tau)\Big] \\
& - \Big[M^{,\rho\sigma}_{\alpha\beta}(h^d, b^0_{\varrho})\, \ddot{\lambda}^{0}_{\beta}(h^d, b^0_{\varrho}; \tau) - C^{,\rho\sigma}_{\alpha\beta}(h^d, b^0_{\varrho})\, \dot{\lambda}^{0}_{\beta}(h^d, b^0_{\varrho}; \tau) \\
& + K^{,\rho\sigma}_{\alpha\beta}(h^d, b^0_{\varrho})\, \lambda^{0}_{\beta}(h^d, b^0_{\varrho}; \tau)\Big]\Big\}\, S^{\rho\sigma}_{\mathrm{b}}
\end{aligned}
\tag{7.22}
$$

It is seen from Eqs. (7.17)–(7.19) that: (i) there exist two linear operators acting on the left-hand sides, the only difference between the operators being the algebraic sign at the damping terms, and (ii) all the zero-, first- and second-order initial conditions for the primary equations and terminal conditions for the adjoint equations are homogeneous.

Once the initial-terminal systems (7.17)–(7.19) are solved for the zero-, first- and second-order primary and adjoint accelerations, velocities and displacements, the formal solution to the time interval SSDS problem can be obtained by setting $\epsilon = 1$ in the second-order expansions (7.15)–(7.16) (and consequently in the expansions for the first derivatives of these functions with respect to design variables). Apart from the first two statistical moments for the time response of the nodal displacements and element strains and stresses calculated from Eqs. (4.40)–(4.48), by using the definition (5.12)–(5.13) extended to the dynamic case the second-order accurate expectations and first-order accurate cross-covariances for the time interval sensitivity gradient coefficients can be expressed respectively as

$$
\begin{aligned}
E[\mathcal{G}^{\cdot d}] = \int_0^T \Big\{ & G^{0.d}(\tau) \\
& - \tfrac{1}{2}\lambda^0_{\alpha}(\tau)\,\mathcal{F}^{d(2)}_{\alpha}(\tau) + \Big[\lambda^0_{\alpha}(\tau) + \tfrac{1}{2}\lambda^{(2)}_{\alpha}(\tau)\Big]\mathcal{A}^d_{\alpha}(\tau) \\
& + \tfrac{1}{2}\Big[G^{\cdot d,\rho\sigma}(\tau) + 2\lambda^{,\rho}_{\alpha}(\tau)\mathcal{B}^{d\sigma}_{\alpha}(\tau) + \lambda^0_{\alpha}(\tau)\,\mathcal{C}^{d\rho\sigma}_{\alpha}(\tau)\Big] S^{\rho\sigma}_{\mathrm{b}}\Big\}\,\mathrm{d}\tau
\end{aligned}
\tag{7.23}
$$

$$
\begin{aligned}
\mathrm{Cov}(\mathcal{G}^{\cdot d}, \mathcal{G}^{\cdot e}) = {} & S^{de}_{\mathcal{G}^{\cdot}} \\
= {} & \Big(\int_0^T\!\!\int_0^T \Big\{ G^{\cdot d,\rho}(\tau)\Big[G^{\cdot e,\sigma}(v) + \lambda^{,\sigma}_{\alpha}(v)\mathcal{A}^e_{\alpha}(v) + \lambda^0_{\alpha}(v)\,\mathcal{B}^{e\sigma}_{\alpha}(v)\Big] \\
& + G^{\cdot e,\rho}(\tau)\Big[\lambda^{,\sigma}_{\alpha}(v)\,\mathcal{A}^d_{\alpha}(v) + \lambda^0_{\alpha}(v)\,\mathcal{B}^{d\sigma}_{\alpha}(v)\Big] \\
& + \lambda^{,\rho}_{\alpha}(\tau)\lambda^{,\sigma}_{\beta}(v)\mathcal{A}^d_{\alpha}(\tau)\mathcal{A}^e_{\beta}(v) + \lambda^0_{\alpha}(\tau)\lambda^0_{\beta}(v)\,\mathcal{B}^{d\rho}_{\alpha}(\tau)\,\mathcal{B}^{e\sigma}_{\beta}(v) \\
& + \lambda^{,\rho}_{\alpha}(\tau)\lambda^0_{\beta}(v)\Big[\mathcal{A}^d_{\alpha}(\tau)\mathcal{B}^{e\sigma}_{\beta}(v) + \mathcal{A}^e_{\beta}(\tau)\mathcal{B}^{d\sigma}_{\alpha}(v)\Big]\Big\}\mathrm{d}\tau\,\mathrm{d}v\Big) S^{\rho\sigma}_{\mathrm{b}}
\end{aligned}
\tag{7.24}
$$

where τ and v are dummy (time) variables of integration, and the following notation is employed:

$$
\begin{aligned}
\mathcal{A}^d_\alpha(\tau) &= Q^{0.d}_\alpha(\tau) - \mathcal{D}^d_\alpha(\tau) \\
\mathcal{B}^{d\rho}_\alpha(\tau) &= Q^{.d,\rho}_\alpha(\tau) - \mathcal{E}^{d\rho}_\alpha(\tau) - \mathcal{H}^{d\rho}_\alpha(\tau) \\
\mathcal{C}^{d\rho\sigma}_\alpha(\tau) &= Q^{.d,\rho\sigma}_\alpha(\tau) - \mathcal{K}^{d\rho\sigma}_\alpha(\tau) - \mathcal{K}^{d\sigma\rho}_\alpha(\tau) - \mathcal{L}^{d\rho\sigma}_\alpha(\tau) \\
\mathcal{D}^d_\alpha(\tau) &= M^{0.d}_{\alpha\beta}\,\ddot{q}^0_\beta(\tau) + C^{0.d}_{\alpha\beta}\,\dot{q}^0_\beta(\tau) + K^{0.d}_{\alpha\beta}\,q^0_\beta(\tau) \\
\mathcal{E}^{d\rho}_\alpha(\tau) &= M^{0.d}_{\alpha\beta}\,\ddot{q}^{,\rho}_\beta(\tau) + C^{0.d}_{\alpha\beta}\,\dot{q}^{,\rho}_\beta(\tau) + K^{0.d}_{\alpha\beta}\,q^{,\rho}_\beta(\tau) \\
\mathcal{F}^{d(2)}_\alpha(\tau) &= M^{0.d}_{\alpha\beta}\,\ddot{q}^{(2)}_\beta(\tau) + C^{0.d}_{\alpha\beta}\,\dot{q}^{(2)}_\beta(\tau) + K^{0.d}_{\alpha\beta}\,q^{(2)}_\beta(\tau) \\
\mathcal{H}^{d\rho}_\alpha(\tau) &= M^{.d,\rho}_{\alpha\beta}\,\ddot{q}^0_\beta(\tau) + C^{.d,\rho}_{\alpha\beta}\,\dot{q}^0_\beta(\tau) + K^{.d,\rho}_{\alpha\beta}\,q^0_\beta(\tau) \\
\mathcal{K}^{d\rho\sigma}_\alpha(\tau) &= M^{.d,\rho}_{\alpha\beta}\,\ddot{q}^{,\sigma}_\beta(\tau) + C^{.d,\rho}_{\alpha\beta}\,\dot{q}^{,\sigma}_\beta(\tau) + K^{.d,\rho}_{\alpha\beta}\,q^{,\sigma}_\beta(\tau) \\
\mathcal{L}^{d\rho\sigma}_\alpha(\tau) &= M^{.d,\rho\sigma}_{\alpha\beta}\,\ddot{q}^0_\beta(\tau) + C^{.d,\rho\sigma}_{\alpha\beta}\,\dot{q}^0_\beta(\tau) + K^{.d,\rho\sigma}_{\alpha\beta}\,q^0_\beta(\tau)
\end{aligned}
\qquad (7.25)
$$

with $\tau \in [0,T]$; $d,e = 1,2,\ldots,D$; $\rho,\sigma = 1,2,\ldots,\tilde{N}$; $\alpha,\beta = 1,2,\ldots,N$.

7.1.2 Time Instant Sensitivity

To consider the time instant SSDS problem the integral functional (2.146) employed for deterministic systems can be extended to the stochastic case. The system response at the time instant $\tau = t$ can now be expressed in terms of the time series convolution of shifted impulses as

$$
\begin{aligned}
\mathcal{G}(h^d, b_\rho; t) = \int_0^t G\left[q_\alpha(h^d, b_\rho; \tau), h^d\right] \delta(t-\tau)\,\mathrm{d}\tau \qquad & t \in [0,T] \\
\alpha = 1,2,\ldots,N\ ;\ \ \rho = 1,2,\ldots,\tilde{N}\ ;\ \ d = 1,2,\ldots,D &
\end{aligned}
\qquad (7.26)
$$

As before, the symbol t denotes the running terminal time while $\delta(t-\tau)$ is the Dirac delta distribution. The function $G[q_\alpha(h^d, b_\rho; \tau), h^d]$ is assumed to be continuous in the entire time interval [0,T] and continuously differentiable with respect to nodal displacement-type variables as well as with respect to random and design variables. The difference of the sampling time signal (7.26) from that given in Eq. (2.146) is that its weighting number $G(h^d, b_\rho; \tau)\mathrm{d}\tau$ of the unit impulse $\delta(t-\tau)$ is a function of random field variables b_ρ. In general, the function is expressed in terms of b_ρ in either an explicit or implicit way via the nodal displacement-type finite element coordinates $q_\alpha(h^d, b_\rho; \tau)$. The state variables are the solution to the initial problem described by the SFEM equation of motion

$$
\begin{aligned}
M_{\alpha\beta}(h^d, b_\rho)\,\ddot{q}_\beta(h^d, b_\rho; \tau) + C_{\alpha\beta}(h^d, b_\rho)\,\dot{q}_\beta(h^d, b_\rho; \tau) & \\
+ K_{\alpha\beta}(h^d, b_\rho)\,q_\beta(h^d, b_\rho; \tau) &= Q_\alpha(h^d, b_\rho; \tau) \qquad (7.27) \\
q_\alpha(h^d, b_\rho; 0) = 0\ ;\ \ \dot{q}_\alpha(h^d, b_\rho; 0) &= 0
\end{aligned}
$$

Since the running terminal time t takes on some a priori selected value in the time interval $[0,T]$, noting that

$$\int_0^t G^{,d}(h^d, b_\rho; \tau)\,\delta(t-\tau)\,\mathrm{d}\tau \;=\; G^{,d}(h^d, b_\rho; t) \tag{7.28}$$

yields, by using the chain rule of differentiation, an expression for the derivative of $\mathcal{G}$ with respect to h^d in the form, cf. Eq. (2.147)

$$\mathcal{G}^{,d}(t) \;=\; G^{,d}(t) + \int_0^t G_{,\alpha}(\tau)\, q_\alpha^{,d}(\tau)\,\delta(t-\tau)\,\mathrm{d}\tau \qquad t \in [0,T] \tag{7.29}$$

To express the displacement sensitivity matrix $q_\alpha^{,d}(\tau)$ explicitly with respect to the variations in h^d, a similar procedure to that used for the time interval SSDS can be employed, cf. Eqs. (7.5)–(7.12). This includes the following steps:

(i) defining an adjoint vector $\lambda = \{\lambda_\alpha(\tau)\}$, which is initially assumed independent of random and design variables,

(ii) taking differentiation of Eq. (7.27) with respect to design variables h^d, premultiplying the result by $\lambda_\alpha(\tau)$ transposed and integrating by parts with respect to τ over the running time interval $[0,t]$,

(iii) equating the coefficients at q_α and $q_\alpha^{,d}$ in the resulting equation and Eq. (7.29) to get an adjoint equation system for $\lambda = \lambda_\alpha(h^d, b_\rho; \tau)$ together with the running terminal conditions at $\tau = t$ for $\lambda_\alpha(h^d, b_\rho; \tau)$ and $\dot{\lambda}_\alpha(h^d, b_\rho; \tau)$.

As a result we arrive at the time instant terminal problem expressed as

$$\begin{aligned} &M_{\alpha\beta}(h^d, b_\rho)\,\ddot{\lambda}_\beta(h^d, b_\rho; \tau) - C_{\alpha\beta}(h^d, b_\rho)\,\dot{\lambda}_\beta(h^d, b_\rho; \tau) \\ &\qquad + K_{\alpha\beta}(h^d, b_\rho)\,\lambda_\beta(h^d, b_\rho; \tau) \;=\; G_{,\alpha}(h^d, b_\rho; t)\,\delta(t-\tau) \\ &\lambda_\alpha(h^d, b_\rho; t) = 0\,; \quad \dot{\lambda}_\alpha(h^d, b_\rho; t) = 0\,; \qquad \tau \in [0,t]\,;\; t \in [0,T] \end{aligned} \tag{7.30}$$

which differs from Eq. (7.11) by the impulse forcing function of magnitude equal the derivative of G with respect to q_α evaluated at the running terminal time t. (Recall that in contrast to the direct differentiation sensitivity analysis the derivative is taken with respect to the displacement components of interest only.)

The dynamic response sensitivity gradient coefficients can be obtained at any time instant $\tau = t$ as, cf. Eq. (7.14)

$$\mathcal{G}^{,d}(h^d, b_\rho; t) \;=\; G^{,d}(h^d, b_\rho; t) + \int_0^t \lambda_\alpha(h^d, b_\rho; \tau)\, R_\alpha^d(h^d, b_\rho; \tau)\,\mathrm{d}\tau \tag{7.31}$$

with $t \in [0,T]$ and $R_\alpha^d(\tau)$ defined by Eq. (7.6).

Employing the perturbation procedure the expansions (7.15) and (7.16) are introduced into the stochastic initial-terminal equations (7.27) and (7.30) and the terms of like order of ϵ in the resulting expressions are collected; we thus obtain the following set of governing equations generated for the time instant SSDS problem, cf. Eqs. (7.17)–(7.19):

- Zeroth-order (ϵ^0 terms, one pair of systems of N linear simultaneous ordinary differential equations for $q^0_\alpha(h^d, b^0_\varrho; \tau)$ and $\lambda^0_\alpha(h^d, b^0_\varrho; \tau)$, respectively)

$$\begin{aligned}
&M^0_{\alpha\beta}(h^d, b^0_\varrho)\, \ddot{q}^0_\beta(h^d, b^0_\varrho; \tau) + C^0_{\alpha\beta}(h^d, b^0_\varrho)\, \dot{q}^0_\beta(h^d, b^0_\varrho; \tau) \\
&\qquad + K^0_{\alpha\beta}(h^d, b^0_\varrho)\, q^0_\beta(h^d, b^0_\varrho; \tau) \;=\; Q^0_\alpha(h^d, b^0_\varrho; \tau) \\
&q^0_\alpha(h^d, b^0_\varrho; 0) = 0\,; \quad \dot{q}^0_\alpha(h^d, b^0_\varrho; 0) = 0\,; \quad \tau \in [0, t]\,; \quad t \in [0, T] \\
&M^0_{\alpha\beta}(h^d, b^0_\varrho)\, \ddot{\lambda}^0_\beta(h^d, b^0_\varrho; \tau) - C^0_{\alpha\beta}(h^d, b^0_\varrho)\, \dot{\lambda}^0_\beta(h^d, b^0_\varrho; \tau) \\
&\qquad + K^0_{\alpha\beta}(h^d, b^0_\varrho)\, \lambda^0_\beta(h^d, b^0_\varrho; \tau) \;=\; G^0_{.\alpha}(h^d, b^0_\varrho; t)\, \delta(t-\tau) \\
&\lambda^0_\alpha(h^d, b^0_\varrho; t) = 0\,; \quad \dot{\lambda}^0_\alpha(h^d, b^0_\varrho; t) = 0\,; \quad \tau \in [0, t]\,; \quad t \in [0, T]
\end{aligned} \tag{7.32}$$

- First-order (ϵ^1 terms, $\tilde{N}$ pairs of systems of N linear simultaneous ordinary differential equations for $q^{,\rho}_\alpha(h^d, b^0_\varrho; \tau)$ and $\lambda^{,\rho}_\alpha(h^d, b^0_\varrho; \tau)$, respectively)

$$\begin{aligned}
&M^0_{\alpha\beta}(h^d, b^0_\varrho)\, \ddot{q}^{,\rho}_\beta(h^d, b^0_\varrho; \tau) + C^0_{\alpha\beta}(h^d, b^0_\varrho)\, \dot{q}^{,\rho}_\beta(h^d, b^0_\varrho; \tau) \\
&\qquad + K^0_{\alpha\beta}(h^d, b^0_\varrho)\, q^{,\rho}_\beta(h^d, b^0_\varrho; \tau) \;=\; Q^\rho_\alpha(h^d, b^0_\varrho; \tau) \\
&q^{,\rho}_\alpha(h^d, b^0_\varrho; 0) = 0\,; \quad \dot{q}^{,\rho}_\alpha(h^d, b^0_\varrho; 0) = 0\,; \quad \tau \in [0, t]\,; \quad t \in [0, T] \\
&M^0_{\alpha\beta}(h^d, b^0_\varrho)\, \ddot{\lambda}^{,\rho}_\beta(h^d, b^0_\varrho; \tau) - C^0_{\alpha\beta}(h^d, b^0_\varrho)\, \dot{\lambda}^{,\rho}_\beta(h^d, b^0_\varrho; \tau) \\
&\qquad + K^0_{\alpha\beta}(h^d, b^0_\varrho)\, \lambda^{,\rho}_\beta(h^d, b^0_\varrho; \tau) \;=\; G^\rho_\alpha(h^d, b^0_\varrho; \tau, t) \\
&\lambda^{,\rho}_\alpha(h^d, b^0_\varrho; t) = 0\,; \quad \dot{\lambda}^{,\rho}_\alpha(h^d, b^0_\varrho; t) = 0\,; \quad \tau \in [0, t]\,; \quad t \in [0, T] \\
&\qquad\qquad \text{with} \quad \rho = 1, 2, \ldots, \tilde{N}
\end{aligned} \tag{7.33}$$

- Second-order (ϵ^2 terms, one pair of systems of N linear simultaneous ordinary differential equations for $q^{(2)}_\alpha(h^d, b^0_\varrho; \tau)$ and $\lambda^{(2)}_\alpha(h^d, b^0_\varrho; \tau)$, respectively)

$$\begin{aligned}
&M^0_{\alpha\beta}(h^d, b^0_\varrho)\, \ddot{q}^{(2)}_\beta(h^d, b^0_\varrho; \tau) + C^0_{\alpha\beta}(h^d, b^0_\varrho)\, \dot{q}^{(2)}_\beta(h^d, b^0_\varrho; \tau) \\
&\qquad + K^0_{\alpha\beta}(h^d, b^0_\varrho)\, q^{(2)}_\beta(h^d, b^0_\varrho; \tau) \;=\; Q^{(2)}_\alpha(h^d, b^0_\varrho; \tau) \\
&q^{(2)}_\alpha(h^d, b^0_\varrho; 0) = 0\,; \quad \dot{q}^{(2)}_\alpha(h^d, b^0_\varrho; 0) = 0\,; \quad \tau \in [0, t]\,; \quad t \in [0, T] \\
&M^0_{\alpha\beta}(h^d, b^0_\varrho)\, \ddot{\lambda}^{(2)}_\beta(h^d, b^0_\varrho; \tau) - C^0_{\alpha\beta}(h^d, b^0_\varrho)\, \dot{\lambda}^{(2)}_\beta(h^d, b^0_\varrho; \tau) \\
&\qquad + K^0_{\alpha\beta}(h^d, b^0_\varrho)\, \lambda^{(2)}_\beta(h^d, b^0_\varrho; \tau) \;=\; G^{(2)}_\alpha(h^d, b^0_\varrho; \tau, t) \\
&\lambda^{(2)}_\alpha(h^d, b^0_\varrho; t) = 0\,; \quad \dot{\lambda}^{(2)}_\alpha(h^d, b^0_\varrho; t) = 0\,; \quad \tau \in [0, t]\,; \quad t \in [0, T]
\end{aligned} \tag{7.34}$$

In Eqs. (7.33) and (7.34) $d = 1,2,\ldots,D$; $\alpha,\beta = 1,2,\ldots,N$; the second-order primary and adjoint displacement vectors $q_\alpha^{(2)}$ and $\lambda_\alpha^{(2)}$ are defined by Eqs. (7.20), while the first- and second-order primary generalized load vectors Q_α^{ρ} and $Q_\alpha^{(2)}$ are given in Eqs. $(7.21)_1$ and $(7.22)_1$, respectively. The following notation is employed for the first- and second-order adjoint generalized load vectors G_α^{ρ} and $G_\alpha^{(2)}$, cf. Eqs. $(7.21)_2$ and $(7.22)_2$

$$G_\alpha^{\rho}(h^d, b_\varrho^0; \tau, t) = G_{.\alpha}^{;\rho}(h^d, b_\varrho^0; t)\,\delta(t-\tau) - \Big[M_{\alpha\beta}^{;\rho}(h^d, b_\varrho^0)\,\ddot{\lambda}_\beta^0(h^d, b_\varrho^0; \tau) - C_{\alpha\beta}^{;\rho}(h^d, b_\varrho^0)\,\dot{\lambda}_\beta^0(h^d, b_\varrho^0; \tau) + K_{\alpha\beta}^{;\rho}(h^d, b_\varrho^0)\,\lambda_\beta^0(h^d, b_\varrho^0; \tau)\Big] \qquad (7.35)$$

$$\begin{aligned} G_\alpha^{(2)}(h^d, b_\varrho^0; \tau, t) = \Big\{ & G_{.\alpha}^{;\rho\sigma}(h^d, b_\varrho^0; t)\,\delta(t-\tau) - 2\Big[M_{\alpha\beta}^{;\rho}(h^d, b_\varrho^0)\,\ddot{\lambda}_\beta^{;\sigma}(h^d, b_\varrho^0; \tau) \\ & - C_{\alpha\beta}^{;\rho}(h^d, b_\varrho^0)\,\dot{\lambda}_\beta^{;\sigma}(h^d, b_\varrho^0; \tau) + K_{\alpha\beta}^{;\rho}(h^d, b_\varrho^0)\,\lambda_\beta^{;\sigma}(h^d, b_\varrho^0; \tau)\Big] \\ & - \Big[M_{\alpha\beta}^{;\rho\sigma}(h^d, b_\varrho^0)\,\ddot{\lambda}_\beta^0(h^d, b_\varrho^0; \tau) - C_{\alpha\beta}^{;\rho\sigma}(h^d, b_\varrho^0)\,\dot{\lambda}_\beta^0(h^d, b_\varrho^0; \tau) \\ & + K_{\alpha\beta}^{;\rho\sigma}(h^d, b_\varrho^0)\,\lambda_\beta^0(h^d, b_\varrho^0; \tau)\Big]\Big\}\, S_{\mathrm{b}}^{\rho\sigma} \end{aligned} \qquad (7.36)$$

Having solved the initial-terminal problem (7.32)–(7.34) for the zero-, first- and second-order primary and adjoint responses, the solution to the time instant SSDS problem can be obtained by following a similar process as for the time interval SSDS case. The expressions for the second-order accurate expectations for the sensitivity gradient coefficients at any time instant $\tau = t$ and their first-order accurate cross-covariances at $\xi_1 = (x_k^{(1)}, t_1)$ and $\xi_2 = (x_k^{(2)}, t_2)$ can be shown to have the following form

$$\begin{aligned} E[\mathcal{G}^{\cdot d}(t)] = {} & G^{0.d}(t) \\ & + \int_0^t \Big\{\Big[\lambda_\alpha^0(\tau) + \tfrac{1}{2}\lambda_\alpha^{(2)}(\tau)\Big]\mathcal{A}_\alpha^d(\tau) - \tfrac{1}{2}\lambda_\alpha^0(\tau)\,\mathcal{F}_\alpha^{d(2)}(\tau) \\ & + \tfrac{1}{2}\Big[G^{\cdot d,\rho\sigma}(\tau) + 2\lambda_\alpha^{;\rho}(\tau)\mathcal{B}_\alpha^{d\sigma}(\tau) + \lambda_\alpha^0(\tau)\,\mathcal{C}_\alpha^{d\rho\sigma}(\tau)\Big]S_{\mathrm{b}}^{\rho\sigma}\Big\}\,\mathrm{d}\tau \end{aligned} \qquad (7.37)$$

$$\begin{aligned} \mathrm{Cov}\Big(\mathcal{G}^{\cdot d}(t_1), \mathcal{G}^{\cdot e}(t_2)\Big) = {} & S_{\mathcal{G}^{\cdot}}^{de}(t_1, t_2) = \Big(G^{\cdot d,\rho}(t_1)\,G^{\cdot e,\sigma}(t_2) \\ & + G^{\cdot d,\rho}(t_1)\int_0^{t_2}\Big[\lambda_\alpha^{;\sigma}(\tau)\mathcal{A}_\alpha^e(\tau) + \lambda_\alpha^0(\tau)\,\mathcal{B}_\alpha^{e\sigma}(\tau)\Big]\,\mathrm{d}\tau \\ & + G^{\cdot e,\rho}(t_2)\int_0^{t_1}\Big[\lambda_\alpha^{;\sigma}(\tau)\mathcal{A}_\alpha^d(\tau) + \lambda_\alpha^0(\tau)\,\mathcal{B}_\alpha^{d\sigma}(\tau)\Big]\,\mathrm{d}\tau \\ & + \int_0^{t_1}\int_0^{t_2}\Big\{\lambda_\alpha^{;\rho}(\tau)\lambda_\beta^{;\sigma}(v)\,\mathcal{A}_\alpha^d(\tau)\,\mathcal{A}_\beta^e(v) + \lambda_\alpha^0(\tau)\lambda_\beta^0(v)\,\mathcal{B}_\alpha^{d\rho}(\tau)\,\mathcal{B}_\beta^{e\sigma}(v) \\ & + \lambda_\alpha^{;\rho}(\tau)\,\lambda_\beta^0(v)\Big[\mathcal{A}_\alpha^d(\tau)\,\mathcal{B}_\beta^{e\sigma}(v) + \mathcal{A}_\beta^e(\tau)\,\mathcal{B}_\alpha^{d\sigma}(v)\Big]\Big\}\,\mathrm{d}\tau\mathrm{d}v\Big)\,S_{\mathrm{b}}^{\rho\sigma} \end{aligned} \qquad (7.38)$$

with τ, v being integration (time) variables; $t, t_1, t_2 \in [0, T]$ (time instants); $d, e = 1,2,\ldots,D$ (design variable indices); $\rho,\sigma = 1,2,\ldots,\tilde{N}$ (random variable indices); $\alpha,\beta = 1,2,\ldots,N$ (degree-of-freedom numbers); and $\mathcal{A}_\alpha^d$, $\mathcal{B}_\alpha^{d\rho}$, $\mathcal{C}_\alpha^{d\rho\sigma}$, $\mathcal{D}_\alpha^d$, $\mathcal{E}_\alpha^{d\rho}$, $\mathcal{F}_\alpha^{d(2)}$, $\mathcal{H}_\alpha^{d\rho}$, $\mathcal{K}_\alpha^{d\rho\sigma}$, $\mathcal{L}_\alpha^{d\rho\sigma}$ being given in Eqs. (7.25). The symbol $S_{\mathrm{b}}^{\rho\sigma}$ stands for the covariance matrix of the random variables b_ρ.

7.2 Computational Aspects

We wish to point out again that all the FEM-based problems dealt with so far in this book can be treated as specific cases of the dynamic response SSDS analysis considered in this chapter. In view of this unification that we have managed to achieve, many computational considerations discussed before apply also to the wider class of problems presented now; thus appropriate repetitions will be avoided.

Formally, the computation process for both the time interval and time instant SSDS of systems with correlated random variables proceeds as follows:

(i) If the primary equations of motion are specified with nonhomogeneous initial conditions, the nonhomogeneous initial–terminal problem is converted into a corresponding homogeneous problem by using a similar procedure as in the deterministic analysis, cf. Section 2.8.2.4.

(ii) The primary equations are sequentially integrated with respect to time to determine the zeroth-, first- and second-order responses of the primary accelerations, velocities and displacements; the right-hand-side vector of the zeroth-order adjoint equation is formed.

(iii) By using the backward time variable $\bar{\tau} = T - \tau$, cf. Section 2.8.2.4, the zeroth-, first- and second-order adjoint systems are integrated sequentially in time by the same algorithm used for the initial problem, yielding a unique solution for the zeroth-, first- and second-order adjoint displacements.

(iv) Once the initial–terminal problem is solved the second-order accurate expectations and first-order accurate covariances (cross-covariances) of the sensitivity gradient coefficients are evaluated. In the case of the time instant SSDS this process has to be repeated for all the running discrete time instants t_i (a sampling interval Δt apart) used in the analysis.

The number of systems with N degrees of freedom to be solved equals $2L(\tilde{N}+2)$ in the time interval SSDS problem and $L(L+1)(\tilde{N}+2)$ in the time instant SSDS, L being the number of discrete time intervals (sampling length) used in the analysis. Taking into account the algebraic operations necessary to form the double sums $(.)^{\rho\sigma} S_{\mathrm{b}}^{\rho\sigma}$ in calculating the second-order load vectors and the first two moments for the sensitivity gradient coefficients, the computational cost turns out to be unacceptably high. Therefore, the use of the two-fold superposition approach in the dynamic response SSDS analysis appears inevitable. We shall thus focus our attention on the computational aspects of this effective technique.

Using the random variable transformation (4.149),(4.151) from the set of correlated random variables b_ρ in the system considered to the set of uncorrelated random variables $c_{\bar{\rho}}$, the 'correlated' initial–terminal problems described by Eqs. (7.17)–(7.19) for the time interval SSDS and by Eqs. (7.32)–(7.34) for the time instant SSDS can now be rewritten together in the transformed form as follows:

- Zeroth-order (ϵ^0 terms, one pair of systems of N linear simultaneous ordinary differential equations for $q^0_\alpha(h^d, c^0_{\check{\varrho}}; \tau)$ and $\lambda^0_\alpha(h^d, c^0_{\check{\varrho}}; \tau)$, respectively)

$$\begin{aligned} M^0_{\alpha\beta}(h^d, c^0_{\check{\varrho}})\, \ddot{q}^0_\beta(h^d, c^0_{\check{\varrho}}; \tau) + C^0_{\alpha\beta}(h^d, c^0_{\check{\varrho}})\, \dot{q}^0_\beta(h^d, c^0_{\check{\varrho}}; \tau) \qquad & \\ + K^0_{\alpha\beta}(h^d, c^0_{\check{\varrho}})\, q^0_\beta(h^d, c^0_{\check{\varrho}}; \tau) &= Q^0_\alpha(h^d, c^0_{\check{\varrho}}; \tau) \\ M^0_{\alpha\beta}(h^d, c^0_{\check{\varrho}})\, \ddot{\lambda}^0_\beta(h^d, c^0_{\check{\varrho}}; \tau) - C^0_{\alpha\beta}(h^d, c^0_{\check{\varrho}})\, \dot{\lambda}^0_\beta(h^d, c^0_{\check{\varrho}}; \tau) \qquad & \\ + K^0_{\alpha\beta}(h^d, c^0_{\check{\varrho}})\, \lambda^0_\beta(h^d, c^0_{\check{\varrho}}; \tau) &= G^{(0)}_\alpha(h^d, c^0_{\check{\varrho}}; .) \end{aligned} \tag{7.39}$$

- First-order (ϵ^1 terms, $\check{N}$ pairs of systems of N linear simultaneous ordinary differential equations for $q^{,\check{\rho}}_\alpha(h^d, c^0_{\check{\varrho}}; \tau)$ and $\lambda^{,\check{\rho}}_\alpha(h^d, c^0_{\check{\varrho}}; \tau)$, respectively)

$$\begin{aligned} M^0_{\alpha\beta}(h^d, c^0_{\check{\varrho}})\, \ddot{q}^{,\check{\rho}}_\beta(h^d, c^0_{\check{\varrho}}; \tau) + C^0_{\alpha\beta}(h^d, c^0_{\check{\varrho}})\, \dot{q}^{,\check{\rho}}_\beta(h^d, c^0_{\check{\varrho}}; \tau) \qquad & \\ + K^0_{\alpha\beta}(h^d, c^0_{\check{\varrho}})\, q^{,\check{\rho}}_\beta(h^d, c^0_{\check{\varrho}}; \tau) &= Q^{\check{\rho}}_\alpha(h^d, c^0_{\check{\varrho}}; \tau) \\ M^0_{\alpha\beta}(h^d, c^0_{\check{\varrho}})\, \ddot{\lambda}^{,\check{\rho}}_\beta(h^d, c^0_{\check{\varrho}}; \tau) - C^0_{\alpha\beta}(h^d, c^0_{\check{\varrho}})\, \dot{\lambda}^{,\check{\rho}}_\beta(h^d, c^0_{\check{\varrho}}; \tau) \qquad & \\ + K^0_{\alpha\beta}(h^d, c^0_{\check{\varrho}})\, \lambda^{,\check{\rho}}_\beta(h^d, c^0_{\check{\varrho}}; \tau) &= G^{\check{\rho}}_\alpha(h^d, c^0_{\check{\varrho}}; \tau, .) \\ \text{with} \quad \check{\rho} = 1, 2, \ldots, \check{N} \ll \tilde{N} & \end{aligned} \tag{7.40}$$

- Second-order (ϵ^2 terms, one pair of systems of N linear simultaneous ordinary differential equations for $q^{(2)}_\alpha(h^d, c^0_{\check{\varrho}}; \tau)$ and $\lambda^{(2)}_\alpha(h^d, c^0_{\check{\varrho}}; \tau)$, respectively)

$$\begin{aligned} M^0_{\alpha\beta}(h^d, c^0_{\check{\varrho}})\, \ddot{q}^{(2)}_\beta(h^d, c^0_{\check{\varrho}}; \tau) + C^0_{\alpha\beta}(h^d, c^0_{\check{\varrho}})\, \dot{q}^{(2)}_\beta(h^d, c^0_{\check{\varrho}}; \tau) \qquad & \\ + K^0_{\alpha\beta}(h^d, c^0_{\check{\varrho}})\, q^{(2)}_\beta(h^d, c^0_{\check{\varrho}}; \tau) &= Q^{(2)}_\alpha(h^d, c^0_{\check{\varrho}}; \tau) \\ M^0_{\alpha\beta}(h^d, c^0_{\check{\varrho}})\, \ddot{\lambda}^{(2)}_\beta(h^d, c^0_{\check{\varrho}}; \tau) - C^0_{\alpha\beta}(h^d, c^0_{\check{\varrho}})\, \dot{\lambda}^{(2)}_\beta(h^d, c^0_{\check{\varrho}}; \tau) \qquad & \\ + K^0_{\alpha\beta}(h^d, c^0_{\check{\varrho}})\, \lambda^{(2)}_\beta(h^d, c^0_{\check{\varrho}}; \tau) &= G^{(2)}_\alpha(h^d, c^0_{\check{\varrho}}; \tau, .) \end{aligned} \tag{7.41}$$

  with the corresponding initial conditions for the primary equations

$$\begin{aligned} q^0_\alpha(h^d, c^0_{\check{\varrho}}; 0) &= 0\,; & \dot{q}^0_\alpha(h^d, c^0_{\check{\varrho}}; 0) &= 0 \\ q^{,\check{\rho}}_\alpha(h^d, c^0_{\check{\varrho}}; 0) &= 0\,; & \dot{q}^{,\check{\rho}}_\alpha(h^d, c^0_{\check{\varrho}}; 0) &= 0 \\ q^{(2)}_\alpha(h^d, c^0_{\check{\varrho}}; 0) &= 0\,; & \dot{q}^{(2)}_\alpha(h^d, c^0_{\check{\varrho}}; 0) &= 0 \end{aligned} \tag{7.42}$$

  the terminal conditions for the time-interval adjoint equations

$$\begin{aligned} \lambda^0_\alpha(h^d, c^0_{\check{\varrho}}; T) &= 0\,; & \dot{\lambda}^0_\alpha(h^d, c^0_{\check{\varrho}}; T) &= 0 \\ \lambda^{,\check{\rho}}_\alpha(h^d, c^0_{\check{\varrho}}; T) &= 0\,; & \dot{\lambda}^{,\check{\rho}}_\alpha(h^d, c^0_{\check{\varrho}}; T) &= 0 \\ \lambda^{(2)}_\alpha(h^d, c^0_{\check{\varrho}}; T) &= 0\,; & \dot{\lambda}^{(2)}_\alpha(h^d, c^0_{\check{\varrho}}; T) &= 0 \end{aligned} \tag{7.43}$$

and the terminal conditions for the time-interval adjoint equations

$$\left.\begin{aligned} \lambda_{\alpha}^{0}(h^d, c_{\check{\varrho}}^{0}; t) = 0\ ; \quad \dot{\lambda}_{\alpha}^{0}(h^d, c_{\check{\varrho}}^{0}; t) = 0 \\ \lambda_{\alpha}^{,\check{\rho}}(h^d, c_{\check{\varrho}}^{0}; t) = 0\ ; \quad \dot{\lambda}_{\alpha}^{,\check{\rho}}(h^d, c_{\check{\varrho}}^{0}; t) = 0 \\ \lambda_{\alpha}^{(2)}(h^d, c_{\check{\varrho}}^{0}; t) = 0\ ; \quad \dot{\lambda}_{\alpha}^{(2)}(h^d, c_{\check{\varrho}}^{0}; t) = 0 \end{aligned}\right\} \quad \tau \in [0,t]\ ; \ t \in [0,T] \tag{7.44}$$

In Eqs. (7.39)–(7.41) the indices run over the following sequence: $d = 1, 2, \ldots, D$; $\check{\rho} = 1, 2, \ldots, \check{N}$; $\alpha, \beta = 1, 2, \ldots, N$. The second-order primary and adjoint displacement vectors are expressed in terms of the uncorrelated random variables $c_{\check{\rho}}$ as, cf. Eq. (4.158)

$$\begin{aligned} q_{\alpha}^{(2)}(h^d, c_{\check{\varrho}}^{0}; \tau) &= \sum_{\check{\rho}=1}^{\check{N}} q_{\alpha}^{,\check{\rho}\check{\rho}}(h^d, c_{\check{\varrho}}^{0}; \tau)\, S_{\mathrm{c}}^{\check{\rho}} \\ \lambda_{\alpha}^{(2)}(h^d, c_{\check{\varrho}}^{0}; \tau) &= \sum_{\check{\rho}=1}^{\check{N}} \lambda_{\alpha}^{,\check{\rho}\check{\rho}}(h^d, c_{\check{\varrho}}^{0}; \tau)\, S_{\mathrm{c}}^{\check{\rho}} \end{aligned} \tag{7.45}$$

The first- and second-order primary generalized load vectors are denoted, respectively, by

$$\begin{aligned} Q_{\alpha}^{\check{\rho}}(h^d, c_{\check{\varrho}}^{0}; \tau) &= Q_{\alpha}^{,\check{\rho}}(h^d, c_{\check{\varrho}}^{0}; \tau) - \Big[M_{\alpha\beta}^{,\check{\rho}}(h^d, c_{\check{\varrho}}^{0})\, \ddot{q}_{\beta}^{0}(h^d, c_{\check{\varrho}}^{0}; \tau) \\ &\quad + C_{\alpha\beta}^{,\check{\rho}}(h^d, c_{\check{\varrho}}^{0})\, \dot{q}_{\beta}^{0}(h^d, c_{\check{\varrho}}^{0}; \tau) + K_{\alpha\beta}^{,\check{\rho}}(h^d, c_{\check{\varrho}}^{0})\, q_{\beta}^{0}(h^d, c_{\check{\varrho}}^{0}; \tau)\Big] \\ Q_{\alpha}^{(2)}(h^d, c_{\check{\varrho}}^{0}; \tau) &= \sum_{\check{\rho}=1}^{\check{N}} \Big\{ Q_{\alpha}^{,\check{\rho}\check{\rho}}(h^d, c_{\check{\varrho}}^{0}; \tau) \\ &\quad - 2\Big[M_{\alpha\beta}^{,\check{\rho}}(h^d, c_{\check{\varrho}}^{0})\, \ddot{q}_{\beta}^{,\check{\rho}}(h^d, c_{\check{\varrho}}^{0}; \tau) + C_{\alpha\beta}^{,\check{\rho}}(h^d, c_{\check{\varrho}}^{0})\, \dot{q}_{\beta}^{,\check{\rho}}(h^d, c_{\check{\varrho}}^{0}; \tau) \\ &\quad + K_{\alpha\beta}^{,\check{\rho}}(h^d, c_{\check{\varrho}}^{0})\, q_{\beta}^{,\check{\rho}}(h^d, c_{\check{\varrho}}^{0}; \tau)\Big] - \Big[M_{\alpha\beta}^{,\check{\rho}\check{\rho}}(h^d, c_{\check{\varrho}}^{0})\, \ddot{q}_{\beta}^{0}(h^d, c_{\check{\varrho}}^{0}; \tau) \\ &\quad + C_{\alpha\beta}^{,\check{\rho}\check{\rho}}(h^d, c_{\check{\varrho}}^{0})\, \dot{q}_{\beta}^{0}(h^d, c_{\check{\varrho}}^{0}; \tau) + K_{\alpha\beta}^{,\check{\rho}\check{\rho}}(h^d, c_{\check{\varrho}}^{0})\, q_{\beta}^{0}(h^d, c_{\check{\varrho}}^{0}; \tau)\Big]\Big\} S_{\mathrm{c}}^{\check{\rho}} \end{aligned} \tag{7.46}$$

The zeroth-, first- and second-order adjoint generalized load vectors are defined for the time interval SSDS as

$$\begin{aligned} G_{\alpha}^{(0)}(h^d, c_{\check{\varrho}}^{0}; .) &= G_{.\alpha}^{0}(h^d, c_{\check{\varrho}}^{0}; \tau) \\ G_{\alpha}^{\check{\rho}}(h^d, c_{\check{\varrho}}^{0}; \tau, .) &= G_{.\alpha}^{,\check{\rho}}(h^d, c_{\check{\varrho}}^{0}; \tau) - \Big[M_{\alpha\beta}^{,\check{\rho}}(h^d, c_{\check{\varrho}}^{0})\, \ddot{\lambda}_{\beta}^{0}(h^d, c_{\check{\varrho}}^{0}; \tau) \\ &\quad - C_{\alpha\beta}^{,\check{\rho}}(h^d, c_{\check{\varrho}}^{0}) \dot{\lambda}_{\beta}^{0}(h^d, c_{\check{\varrho}}^{0}; \tau) + K_{\alpha\beta}^{,\check{\rho}}(h^d, c_{\check{\varrho}}^{0}) \lambda_{\beta}^{0}(h^d, c_{\check{\varrho}}^{0}; \tau)\Big] \\ G_{\alpha}^{(2)}(h^d, c_{\check{\varrho}}^{0}; \tau, .) &= \sum_{\check{\rho}=1}^{\check{N}} \Big\{ G_{.\alpha}^{,\check{\rho}\check{\rho}}(h^d, c_{\check{\varrho}}^{0}; \tau) \\ &\quad - 2\Big[M_{\alpha\beta}^{,\check{\rho}}(h^d, c_{\check{\varrho}}^{0}) \ddot{\lambda}_{\beta}^{,\check{\rho}}(h^d, c_{\check{\varrho}}^{0}; \tau) - C_{\alpha\beta}^{,\check{\rho}}(h^d, c_{\check{\varrho}}^{0}) \dot{\lambda}_{\beta}^{,\check{\rho}}(h^d, c_{\check{\varrho}}^{0}; \tau) \\ &\quad + K_{\alpha\beta}^{,\check{\rho}}(h^d, c_{\check{\varrho}}^{0}) \lambda_{\beta}^{,\check{\rho}}(h^d, c_{\check{\varrho}}^{0}; \tau)\Big] - \Big[M_{\alpha\beta}^{,\check{\rho}\check{\rho}}(h^d, c_{\check{\varrho}}^{0})\, \ddot{\lambda}_{\beta}^{0}(h^d, c_{\check{\varrho}}^{0}; \tau) \\ &\quad - C_{\alpha\beta}^{,\check{\rho}\check{\rho}}(h^d, c_{\check{\varrho}}^{0}) \dot{\lambda}_{\beta}^{0}(h^d, c_{\check{\varrho}}^{0}; \tau) + K_{\alpha\beta}^{,\check{\rho}\check{\rho}}(h^d, c_{\check{\varrho}}^{0}) \lambda_{\beta}^{0}(h^d, c_{\check{\varrho}}^{0}; \tau)\Big]\Big\} S_{\mathrm{c}}^{\check{\rho}} \end{aligned} \tag{7.47}$$

while for the time instant SSDS as

$$
\begin{aligned}
G^{(0)}_{\alpha}(h^d, c^0_{\check{\varrho}}; t) &= G^0_{.\alpha}(h^d, c^0_{\check{\varrho}}; t)\,\delta(t-\tau) \\
G^{(\check{\rho})}_{\alpha}(h^d, c^0_{\check{\varrho}}; \tau, t) &= G^{,\check{\rho}}_{.\alpha}(h^d, c^0_{\check{\varrho}}; t)\,\delta(t-\tau) - \Big[M^{,\check{\rho}}_{\alpha\beta}(h^d, c^0_{\check{\varrho}})\,\ddot{\lambda}^0_{\beta}(h^d, c^0_{\check{\varrho}}; \tau) \\
&\quad - C^{,\check{\rho}}_{\alpha\beta}(h^d, c^0_{\check{\varrho}})\dot{\lambda}^0_{\beta}(h^d, c^0_{\check{\varrho}}; \tau) + K^{,\check{\rho}}_{\alpha\beta}(h^d, c^0_{\check{\varrho}})\lambda^0_{\beta}(h^d, c^0_{\check{\varrho}}; \tau)\Big] \\
G^{(2)}_{\alpha}(h^d, c^0_{\check{\varrho}}; \tau, t) &= \sum_{\check{\rho}=1}^{\check{N}} \Big\{ G^{,\check{\rho}\check{\rho}}_{.\alpha}(h^d, c^0_{\check{\varrho}}; \tau)\,\delta(t-\tau) \\
&\quad - 2\Big[M^{,\check{\rho}}_{\alpha\beta}(h^d, c^0_{\check{\varrho}})\ddot{\lambda}^{,\check{\rho}}_{\beta}(h^d, c^0_{\check{\varrho}}; \tau) - C^{,\check{\rho}}_{\alpha\beta}(h^d, c^0_{\check{\varrho}})\dot{\lambda}^{,\check{\rho}}_{\beta}(h^d, c^0_{\check{\varrho}}; \tau) \\
&\quad + K^{,\check{\rho}}_{\alpha\beta}(h^d, c^0_{\check{\varrho}})\lambda^{,\check{\rho}}_{\beta}(h^d, c^0_{\check{\varrho}}; \tau)\Big] - \Big[M^{,\check{\rho}\check{\rho}}_{\alpha\beta}(h^d, c^0_{\check{\varrho}})\ddot{\lambda}^0_{\beta}(h^d, c^0_{\check{\varrho}}; \tau) \\
&\quad - C^{,\check{\rho}\check{\rho}}_{\alpha\beta}(h^d, c^0_{\check{\varrho}})\dot{\lambda}^0_{\beta}(h^d, c^0_{\check{\varrho}}; \tau) + K^{,\check{\rho}\check{\rho}}_{\alpha\beta}(h^d, c^0_{\check{\varrho}})\lambda^0_{\beta}(h^d, c^0_{\check{\varrho}}; \tau)\Big]\Big\} S^{\check{\rho}}_{c}
\end{aligned}
\tag{7.48}
$$

Recall that in Eqs. (7.40), (7.41) and (7.45)–(7.49) the symbols $(.)^{,\check{\rho}}$ and $(.)^{,\check{\rho}\check{\rho}}$ denote the first and second derivatives with respect to the uncorrelated random variables $c_{,\check{\rho}}$ and summation over the repeated index $\check{\rho}$ is not implied.

To decouple the initial–terminal equations the transformation (4.214) from the generalized finite element coordinates of the primary system q^0_{α}, $q^{,\check{\rho}}_{\alpha}$, $q^{(2)}_{\alpha}$ and of the adjoint system λ^0_{α}, $\lambda^{,\check{\rho}}_{\alpha}$, $\lambda^{(2)}_{\alpha}$ to the normalized coordinates $r^0_{\bar{\alpha}}$, $r^{,\check{\rho}}_{\bar{\alpha}}$, $r^{(2)}_{\bar{\alpha}}$ and $\vartheta^0_{\bar{\alpha}}$, $\vartheta^{,\check{\rho}}_{\bar{\alpha}}$, $\vartheta^{(2)}_{\bar{\alpha}}$ is employed. We have then, cf. Eqs. (4.214)

$$
\begin{aligned}
q^0_{\alpha}(h^d, c^0_{\check{\varrho}}; \tau) &= \phi^0_{\alpha\bar{\alpha}}(h^d, c^0_{\check{\varrho}})\, r^0_{\bar{\alpha}}(h^d, c^0_{\check{\varrho}}; \tau) \\
q^{,\check{\rho}}_{\alpha}(h^d, c^0_{\check{\varrho}}; \tau) &= \phi^0_{\alpha\bar{\alpha}}(h^d, c^0_{\check{\varrho}})\, r^{,\check{\rho}}_{\bar{\alpha}}(h^d, c^0_{\check{\varrho}}; \tau) \\
q^{(2)}_{\alpha}(h^d, c^0_{\check{\varrho}}; \tau) &= \phi^0_{\alpha\bar{\alpha}}(h^d, c^0_{\check{\varrho}})\, r^{(2)}_{\bar{\alpha}}(h^d, c^0_{\check{\varrho}}; \tau) \\
\lambda^0_{\alpha}(h^d, c^0_{\check{\varrho}}; \tau) &= \phi^0_{\alpha\bar{\alpha}}(h^d, c^0_{\check{\varrho}})\, \vartheta^0_{\bar{\alpha}}(h^d, c^0_{\check{\varrho}}; \tau) \\
\lambda^{,\check{\rho}}_{\alpha}(h^d, c^0_{\check{\varrho}}; \tau) &= \phi^0_{\alpha\bar{\alpha}}(h^d, c^0_{\check{\varrho}})\, \vartheta^{,\check{\rho}}_{\bar{\alpha}}(h^d, c^0_{\check{\varrho}}; \tau) \\
\lambda^{(2)}_{\alpha}(h^d, c^0_{\check{\varrho}}; \tau) &= \phi^0_{\alpha\bar{\alpha}}(h^d, c^0_{\check{\varrho}})\, \vartheta^{(2)}_{\bar{\alpha}}(h^d, c^0_{\check{\varrho}}; \tau)
\end{aligned}
\tag{7.49}
$$

$$\bar{\alpha} = 1, 2, \ldots, \bar{N} \ll N\,; \quad \alpha = 1, 2, \ldots, N\,; \quad (\text{no sum on } \check{\rho})$$

The mode shape matrix $\phi^0_{\alpha\bar{\alpha}}$ is obtained from the generalized eigensolution of the corresponding zeroth-order undamped free vibration problem (4.215) and the eigenpairs are the same for both the primary and adjoint systems. From Eqs. (7.39)–(7.41) we arrive at the uncoupled initial–terminal system in the following form:

- Zeroth-order (ϵ^0 terms, one pair of systems of $\bar{N}$ uncoupled linear ordinary differential equations for $r^0_{\alpha}(h^d, c^0_{\check{\varrho}}; \tau)$ and $\vartheta^0_{\alpha}(h^d, c^0_{\check{\varrho}}; \tau)$, respectively)

$$
\begin{aligned}
&\ddot{r}^0_{\bar{\alpha}}(h^d, c^0_{\check{\varrho}}; \tau) + 2\,\xi^0_{(\bar{\alpha})}(h^d, c^0_{\check{\varrho}})\,\omega^0_{(\bar{\alpha})}(h^d, c^0_{\check{\varrho}})\,\dot{r}^0_{\bar{\alpha}}(h^d, c^0_{\check{\varrho}}; \tau) \\
&\qquad + [\omega^0_{(\bar{\alpha})}(h^d, c^0_{\check{\varrho}})]^2\, r^0_{\bar{\alpha}}(h^d, c^0_{\check{\varrho}}; \tau) \;=\; R^0_{\bar{\alpha}}(h^d, c^0_{\check{\varrho}}; \tau) \\
&\ddot{\vartheta}^0_{\bar{\alpha}}(h^d, c^0_{\check{\varrho}}; \tau) - 2\,\xi^0_{(\bar{\alpha})}(h^d, c^0_{\check{\varrho}})\,\omega^0_{(\bar{\alpha})}(h^d, c^0_{\check{\varrho}})\,\dot{\vartheta}^0_{\bar{\alpha}}(h^d, c^0_{\check{\varrho}}; \tau) \\
&\qquad + [\omega^0_{(\bar{\alpha})}(h^d, c^0_{\check{\varrho}})]^2\, \vartheta^0_{\bar{\alpha}}(h^d, c^0_{\check{\varrho}}; \tau) \;=\; P^0_{\bar{\alpha}}(h^d, c^0_{\check{\varrho}}; \tau)
\end{aligned}
\tag{7.50}
$$

- First-order (ϵ^1 terms, $\check{N}$ pairs of systems of $\bar{N}$ uncoupled linear ordinary differential equations for $r_{\alpha}^{,\check{\rho}}(h^d, c_{\check{\varrho}}^0; \tau)$ and $\vartheta_{\alpha}^{,\check{\rho}}(h^d, c_{\check{\varrho}}^0; \tau)$, respectively)

$$
\begin{aligned}
\ddot{r}_{\bar{\alpha}}^{,\check{\rho}}(h^d, c_{\check{\varrho}}^0; \tau) + 2\, \xi_{(\bar{\alpha})}^0(h^d, c_{\check{\varrho}}^0)\, \omega_{(\bar{\alpha})}^0(h^d, c_{\check{\varrho}}^0)\, \dot{r}_{\bar{\alpha}}^{,\check{\rho}}(h^d, c_{\check{\varrho}}^0; \tau) & \\
+ [\omega_{(\bar{\alpha})}^0(h^d, c_{\check{\varrho}}^0)]^2\, r_{\bar{\alpha}}^{,\check{\rho}}(h^d, c_{\check{\varrho}}^0; \tau) &= R_{\bar{\alpha}}^{\check{\rho}}(h^d, c_{\check{\varrho}}^0; \tau) \\
\ddot{\vartheta}_{\bar{\alpha}}^{,\check{\rho}}(h^d, c_{\check{\varrho}}^0; \tau) - 2\, \xi_{(\bar{\alpha})}^0(h^d, c_{\check{\varrho}}^0)\, \omega_{(\bar{\alpha})}^0(h^d, c_{\check{\varrho}}^0)\, \dot{\vartheta}_{\bar{\alpha}}^{,\check{\rho}}(h^d, c_{\check{\varrho}}^0; \tau) & \\
+ [\omega_{(\bar{\alpha})}^0(h^d, c_{\check{\varrho}}^0)]^2\, \vartheta_{\bar{\alpha}}^{,\check{\rho}}(h^d, c_{\check{\varrho}}^0; \tau) &= P_{\bar{\alpha}}^{\check{\rho}}(h^d, c_{\check{\varrho}}^0; \tau)
\end{aligned}
\tag{7.51}
$$

- Second-order (ϵ^2 terms, one pair of systems of $\bar{N}$ uncoupled linear ordinary differential equations for $r_{\alpha}^{(2)}(h^d, c_{\check{\varrho}}^0; \tau)$ and $\vartheta_{\alpha}^{(2)}(h^d, c_{\check{\varrho}}^0; \tau)$, respectively)

$$
\begin{aligned}
\ddot{r}_{\bar{\alpha}}^{(2)}(h^d, c_{\check{\varrho}}^0; \tau) + 2\, \xi_{(\bar{\alpha})}^0(h^d, c_{\check{\varrho}}^0)\, \omega_{(\bar{\alpha})}^0(h^d, c_{\check{\varrho}}^0)\, \dot{r}_{\bar{\alpha}}^{(2)}(h^d, c_{\check{\varrho}}^0; \tau) & \\
+ [\omega_{(\bar{\alpha})}^0(h^d, c_{\check{\varrho}}^0)]^2\, r_{\bar{\alpha}}^{(2)}(h^d, c_{\check{\varrho}}^0; \tau) &= R_{\bar{\alpha}}^{(2)}(h^d, c_{\check{\varrho}}^0; \tau) \\
\ddot{\vartheta}_{\bar{\alpha}}^{(2)}(h^d, c_{\check{\varrho}}^0; \tau) - 2\, \xi_{(\bar{\alpha})}^0(h^d, c_{\check{\varrho}}^0)\, \omega_{(\bar{\alpha})}^0(h^d, c_{\check{\varrho}}^0)\, \dot{\vartheta}_{\bar{\alpha}}^{(2)}(h^d, c_{\check{\varrho}}^0; \tau) & \\
+ [\omega_{(\bar{\alpha})}^0(h^d, c_{\check{\varrho}}^0)]^2\, \vartheta_{\bar{\alpha}}^{(2)}(h^d, c_{\check{\varrho}}^0; \tau) &= P_{\bar{\alpha}}^{(2)}(h^d, c_{\check{\varrho}}^0; \tau)
\end{aligned}
\tag{7.52}
$$

Since the eigenvector matrix is orthogonal, the zeroth-, first-, second-order modal primary and adjoint displacements $r_{\bar{\alpha}}^0$, $r_{\bar{\alpha}}^{,\check{\rho}}$, $r_{\bar{\alpha}}^{(2)}$; $\vartheta_{\bar{\alpha}}^0$, $\vartheta_{\bar{\alpha}}^{,\check{\rho}}$, $\vartheta_{\bar{\alpha}}^{(2)}$ and loads $R_{\bar{\alpha}}^0$, $R_{\bar{\alpha}}^{\check{\rho}}$, $R_{\bar{\alpha}}^{(2)}$; $P_{\bar{\alpha}}^0$, $P_{\bar{\alpha}}^{\check{\rho}}$, $P_{\bar{\alpha}}^{(2)}$ can be written respectively as:

$$
\begin{aligned}
r_{\bar{\alpha}}^0(h^d, c_{\check{\varrho}}^0; \tau) &= \phi_{\bar{\alpha}\alpha}^{0^{\mathrm{T}}}(h^d, c_{\check{\varrho}}^0)\, q_{\alpha}^0(h^d, c_{\check{\varrho}}^0; \tau) \\
r_{\bar{\alpha}}^{,\check{\rho}}(h^d, c_{\check{\varrho}}^0; \tau) &= \phi_{\bar{\alpha}\alpha}^{0^{\mathrm{T}}}(h^d, c_{\check{\varrho}}^0)\, q_{\alpha}^{,\check{\rho}}(h^d, c_{\check{\varrho}}^0; \tau) \\
r_{\bar{\alpha}}^{(2)}(h^d, c_{\check{\varrho}}^0; \tau) &= \phi_{\bar{\alpha}\alpha}^{0^{\mathrm{T}}}(h^d, c_{\check{\varrho}}^0)\, q_{\alpha}^{(2)}(h^d, c_{\check{\varrho}}^0; \tau) \\
\vartheta_{\bar{\alpha}}^0(h^d, c_{\check{\varrho}}^0; \tau) &= \phi_{\bar{\alpha}\alpha}^{0^{\mathrm{T}}}(h^d, c_{\check{\varrho}}^0)\, \vartheta_{\alpha}^0(h^d, c_{\check{\varrho}}^0; \tau) \\
\vartheta_{\bar{\alpha}}^{,\check{\rho}}(h^d, c_{\check{\varrho}}^0; \tau) &= \phi_{\bar{\alpha}\alpha}^{0^{\mathrm{T}}}(h^d, c_{\check{\varrho}}^0)\, \vartheta_{\alpha}^{,\check{\rho}}(h^d, c_{\check{\varrho}}^0; \tau) \\
\vartheta_{\bar{\alpha}}^{(2)}(h^d, c_{\check{\varrho}}^0; \tau) &= \phi_{\bar{\alpha}\alpha}^{0^{\mathrm{T}}}(h^d, c_{\check{\varrho}}^0)\, \vartheta_{\alpha}^{(2)}(h^d, c_{\check{\varrho}}^0; \tau)
\end{aligned}
\tag{7.53}
$$

$$
\begin{aligned}
R_{\bar{\alpha}}^0(h^d, c_{\check{\varrho}}^0; \tau) &= \phi_{\bar{\alpha}\alpha}^{0^{\mathrm{T}}}(h^d, c_{\check{\varrho}}^0)\, Q_{\alpha}^0(h^d, c_{\check{\varrho}}^0; \tau) \\
R_{\bar{\alpha}}^{\check{\rho}}(h^d, c_{\check{\varrho}}^0; \tau) &= \phi_{\bar{\alpha}\alpha}^{0^{\mathrm{T}}}(h^d, c_{\check{\varrho}}^0)\, Q_{\alpha}^{\check{\rho}}(h^d, c_{\check{\varrho}}^0; \tau) \\
R_{\bar{\alpha}}^{(2)}(h^d, c_{\check{\varrho}}^0; \tau) &= \phi_{\bar{\alpha}\alpha}^{0^{\mathrm{T}}}(h^d, c_{\check{\varrho}}^0)\, Q_{\alpha}^{(2)}(h^d, c_{\check{\varrho}}^0; \tau) \\
P_{\bar{\alpha}}^0(h^d, c_{\check{\varrho}}^0; \tau) &= \phi_{\bar{\alpha}\alpha}^{0^{\mathrm{T}}}(h^d, c_{\check{\varrho}}^0)\, G_{\alpha}^0(h^d, c_{\check{\varrho}}^0; \tau, .) \\
P_{\bar{\alpha}}^{\check{\rho}}(h^d, c_{\check{\varrho}}^0; \tau) &= \phi_{\bar{\alpha}\alpha}^{0^{\mathrm{T}}}(h^d, c_{\check{\varrho}}^0)\, G_{\alpha}^{\check{\rho}}(h^d, c_{\check{\varrho}}^0; \tau, .) \\
P_{\bar{\alpha}}^{(2)}(h^d, c_{\check{\varrho}}^0; \tau) &= \phi_{\bar{\alpha}\alpha}^{0^{\mathrm{T}}}(h^d, c_{\check{\varrho}}^0)\, G_{\alpha}^{(2)}(h^d, c_{\check{\varrho}}^0; \tau, .)
\end{aligned}
\tag{7.54}
$$

We note that the initial–terminal conditions expressed in terms of the normalized coordinates for the modal initial–terminal equations are homogeneous, cf. Eqs. (7.42)–(7.44).

As mentioned at the beginning of this section, since the left-hand side of the normalized adjoint equations in the system (7.50)–(7.52) differs from that of the primary equations only by the algebraic sign at the damping terms, the terminal problem given by Eqs. $(7.50)_2$, $(7.51)_2$ and $(7.52)_2$ can be converted into an initial problem by using the backward time variable $\bar{\tau} = T - \tau$. The terminal conditions for the adjoint equations then become the initial conditions expressed in terms of the backward time variable $\bar{\tau}$, and the operators in the primary and adjoint systems assume exactly the same form. This observation is very useful since: (i) the eigenproblem has to be solved only once and the eigensolution can be used for either the primary or adjoint system, and (ii) the zeroth-, first- and second-order primary and adjoint modal displacements may be solved for 'in parallel' by the same algorithm.

Formally, to evaluate the first two moments for the sensitivity gradient coefficients the time instant SSDS requires integration over $L(L+1)/2$ time steps Δt, since at each time instant of interest the calculations proceed similarly as those in the single time interval SSDS problem. However, we observe that: (i) the initial–terminal equations are linear, (ii) the initial conditions for the primary equations are homogeneous, and (iii) at any time instant $\tau = t$ the running terminal conditions for the adjoint equations are also homogeneous. This suggests that we may develop a combined algorithm such that the zeroth-, first- and second-order primary and adjoint systems can be integrated *only once* over L time steps from 0 to T. The adjoint response to a Dirac-type forcing distribution can be treated as the unit impulse response (or the response to a unit 'terminal' velocity) of a normalized system, in which either the linear operator or terminal conditions are independent of the running terminal time t and the value of G at t. The procedure for the time instant deterministic SDS introduced in Section 2.8.2.4 can thus be directly employed for stochastic systems since all quantities involved in the linear operator of Eqs. (7.50)–(7.52) are deterministic.

Since the right-hand side of the (linear) modal adjoint equations involves two load functions: (i) the Dirac delta measure evaluated at t with the weighting factor $G(t)$, and (ii) a conventional forcing function of the backward time variable $\bar{\tau}$, by using the superposition principle the adjoint response can be computed as a linear combination of the response to the unit impulse excitation and the response to the time-dependent force.

As stressed in Chapter 4 the perturbation-based analysis may cause secularities in the first- and second-order solutions. The effect is even more significant in the SSDS analysis since the sensitivity expectations are bilinear functions of the primary and adjoint displacements and the covariances are quadratic functions of these variables. According to the technique described in Section 4.5.4, to eliminate secular terms the first- and second-order forcing sequences are composed and treated as complex-valued sequences. The fast Fourier analysis is performed on each modal force sequence to obtain its discrete Fourier spectra. To remove the resonant part from the Fourier spectra the coefficients of each frequency series which lie within a specified range close to the natural frequency are weighted, and the fast Fourier synthesis is carried out on the weighted frequency series to recover the modal force sequence with no resonant excitation. In any algorithm for SSDS problems the procedure for the elimination of secularity should necessarily be implemented.

Having solved and superposed the primary and adjoint responses, the second-order accurate expectations and first-order accurate covariances for the sensitivity gradient coefficients can be calculated. For the time interval SSDS the expectations and cross-covariances are obtained as

$$\begin{aligned} E[\mathcal{G}^{\cdot d}] = \int_0^T \Big\{ & G^{0.d}(\tau) \\ & + \Big[\lambda^0_\alpha(\tau) + \tfrac{1}{2}\lambda^{(2)}_\alpha(\tau)\Big]\,\mathcal{A}^d_\alpha(\tau) - \tfrac{1}{2}\,\lambda^0_\alpha(\tau)\,\mathcal{F}^{d(2)}_\alpha(\tau) \\ & + \tfrac{1}{2}\sum_{\check\rho=1}^{\check N}\Big[G^{\cdot d,\check\rho\check\rho}(\tau) + 2\lambda^{,\check\rho}_\alpha(\tau)\,\mathcal{B}^{d\check\rho}_\alpha(\tau) + \lambda^0_\alpha(\tau)\,\mathcal{C}^{d\check\rho\check\rho}_\alpha(\tau)\Big] S^{\check\rho}_{\mathrm{c}}\Big\}\,\mathrm{d}\tau \end{aligned} \tag{7.55}$$

$$\begin{aligned} \mathrm{Cov}(\mathcal{G}^{\cdot d},\mathcal{G}^{\cdot e}) = {} & S^{de}_{\mathcal{G}\cdot} \\ = {} & \sum_{\check\rho=1}^{\check N}\Big(\int_0^T\!\!\int_0^T \Big\{G^{\cdot d,\check\rho}(\tau)\Big[G^{\cdot e,\check\rho}(\upsilon) + \lambda^{,\check\rho}_\alpha(\upsilon)\mathcal{A}^e_\alpha(\upsilon) + \lambda^0_\alpha(\upsilon)\mathcal{B}^{e\check\rho}_\alpha(\upsilon)\Big] \\ & + G^{\cdot e,\check\rho}(\tau)\,\Big[\lambda^{,\check\rho}_\alpha(\upsilon)\,\mathcal{A}^d_\alpha(\upsilon) + \lambda^0_\alpha(\upsilon)\,\mathcal{B}^{d\check\rho}_\alpha(\upsilon)\Big] \\ & + \lambda^{,\check\rho}_\alpha(\tau)\lambda^{,\check\rho}_\beta(\upsilon)\mathcal{A}^d_\alpha(\tau)\mathcal{A}^e_\beta(\upsilon) + \lambda^0_\alpha(\tau)\lambda^0_\beta(\upsilon)\,\mathcal{B}^{d\check\rho}_\alpha(\tau)\,\mathcal{B}^{e\check\rho}_\beta(\upsilon) \\ & + \lambda^{,\check\rho}_\alpha(\tau)\lambda^0_\beta(\upsilon)\Big[\mathcal{A}^d_\alpha(\tau)\,\mathcal{B}^{e\check\rho}_\beta(\upsilon) + \mathcal{A}^e_\beta(\tau)\,\mathcal{B}^{d\check\rho}_\alpha(\upsilon)\Big]\Big\}\,\mathrm{d}\tau\mathrm{d}\upsilon\Big) S^{\check\rho}_{\mathrm{c}} \end{aligned} \tag{7.56}$$

For the case of the time instant SSDS the expectations of the sensitivity gradient coefficients at any time instant $\tau = t$ and their cross-covariances at $\xi_1 = (x_k^{(1)}, t_1)$ and $\xi_2 = (x_k^{(2)}, t_2)$ can be written as

$$\begin{aligned} E[\mathcal{G}^{\cdot d}(t)] = {} & G^{0.d}(t) \\ & + \int_0^t \Big\{\Big[\lambda^0_\alpha(\tau) + \tfrac{1}{2}\lambda^{(2)}_\alpha(\tau)\Big]\mathcal{A}^d_\alpha(\tau) - \tfrac{1}{2}\lambda^0_\alpha(\tau)\,\mathcal{F}^{d(2)}_\alpha(\tau) \\ & + \tfrac{1}{2}\sum_{\check\rho=1}^{\check N}\Big[G^{\cdot d,\check\rho\check\rho}(\tau) + 2\lambda^{,\check\rho}_\alpha(\tau)\,\mathcal{B}^{d\check\rho}_\alpha(\tau) + \lambda^0_\alpha(\tau)\,\mathcal{C}^{d\check\rho\check\rho}_\alpha(\tau)\Big] S^{\check\rho}_{\mathrm{c}}\Big\}\,\mathrm{d}\tau \end{aligned} \tag{7.57}$$

$$\begin{aligned} \mathrm{Cov}\Big(\mathcal{G}^{\cdot d}(t_1),\mathcal{G}^{\cdot e}(t_2)\Big) = {} & S^{de}_{\mathcal{G}\cdot}(t_1,t_2) = \sum_{\check\rho=1}^{\check N}\Big(G^{\cdot d,\check\rho}(t_1)\,G^{\cdot e,\check\rho}(t_2) \\ & + G^{\cdot d,\check\rho}(t_1)\int_0^{t_2}\Big[\lambda^{,\check\rho}_\alpha(\tau)\mathcal{A}^e_\alpha(\tau) + \lambda^0_\alpha(\tau)\,\mathcal{B}^{e\check\rho}_\alpha(\tau)\Big]\,\mathrm{d}\tau \\ & + G^{\cdot e,\check\rho}(t_2)\int_0^{t_1}\Big[\lambda^{,\check\rho}_\alpha(\tau)\mathcal{A}^d_\alpha(\tau) + \lambda^0_\alpha(\tau)\,\mathcal{B}^{d\check\rho}_\alpha(\tau)\Big]\,\mathrm{d}\tau \\ & + \int_0^{t_1}\!\!\int_0^{t_2}\Big\{\lambda^{,\check\rho}_\alpha(\tau)\lambda^{,\check\rho}_\beta(\upsilon)\,\mathcal{A}^d_\alpha(\tau)\,\mathcal{A}^e_\beta(\upsilon) + \lambda^0_\alpha(\tau)\lambda^0_\beta(\upsilon)\,\mathcal{B}^{d\check\rho}_\alpha(\tau)\,\mathcal{B}^{e\check\rho}_\beta(\upsilon) \\ & + \lambda^{,\check\rho}_\alpha(\tau)\,\lambda^0_\beta(\upsilon)\,\Big[\mathcal{A}^d_\alpha(\tau)\,\mathcal{B}^{e\check\rho}_\beta(\upsilon) + \mathcal{A}^e_\beta(\tau)\,\mathcal{B}^{d\check\rho}_\alpha(\upsilon)\Big]\Big\}\,\mathrm{d}\tau\mathrm{d}\upsilon\Big)\,S^{\check\rho}_{\mathrm{c}} \end{aligned} \tag{7.58}$$

where, as before, τ and υ are dummy (time) variables of integration, and the following notation is employed:

$$
\begin{aligned}
\mathcal{A}_{\alpha}^{d}(\tau) &= Q_{\alpha}^{0.d}(\tau) - \mathcal{D}_{\alpha}^{d}(\tau) \\
\mathcal{B}_{\alpha}^{d\check{p}}(\tau) &= Q_{\alpha}^{\cdot d,\check{p}}(\tau) - \mathcal{E}_{\alpha}^{d\check{p}}(\tau) - \mathcal{H}_{\alpha}^{d\check{p}}(\tau) \\
\mathcal{C}_{\alpha}^{d\check{p}\check{p}}(\tau) &= Q_{\alpha}^{\cdot d,\check{p}\check{p}}(\tau) - 2\,\mathcal{K}_{\alpha}^{d\check{p}\check{p}}(\tau) - \mathcal{L}_{\alpha}^{d\check{p}\check{p}}(\tau) \\
\mathcal{D}_{\alpha}^{d}(\tau) &= M_{\alpha\beta}^{0.d}\,\ddot{q}_{\beta}^{0}(\tau) + C_{\alpha\beta}^{0.d}\,\dot{q}_{\beta}^{0}(\tau) + K_{\alpha\beta}^{0.d}\,q_{\beta}^{0}(\tau) \\
\mathcal{E}_{\alpha}^{d\check{p}}(\tau) &= M_{\alpha\beta}^{0.d}\,\ddot{q}_{\beta}^{,\check{p}}(\tau) + C_{\alpha\beta}^{0.d}\,\dot{q}_{\beta}^{,\check{p}}(\tau) + K_{\alpha\beta}^{0.d}\,q_{\beta}^{,\check{p}}(\tau) \\
\mathcal{F}_{\alpha}^{d(2)}(\tau) &= M_{\alpha\beta}^{0.d}\,\ddot{q}_{\beta}^{(2)}(\tau) + C_{\alpha\beta}^{0.d}\,\dot{q}_{\beta}^{(2)}(\tau) + K_{\alpha\beta}^{0.d}\,q_{\beta}^{(2)}(\tau) \\
\mathcal{H}_{\alpha}^{d\check{p}}(\tau) &= M_{\alpha\beta}^{\cdot d,\check{p}}\,\ddot{q}_{\beta}^{0}(\tau) + C_{\alpha\beta}^{\cdot d,\check{p}}\,\dot{q}_{\beta}^{0}(\tau) + K_{\alpha\beta}^{\cdot d,\check{p}}\,q_{\beta}^{0}(\tau) \\
\mathcal{K}_{\alpha}^{d\check{p}\check{p}}(\tau) &= M_{\alpha\beta}^{\cdot d,\check{p}}\,\ddot{q}_{\beta}^{,\check{p}}(\tau) + C_{\alpha\beta}^{\cdot d,\check{p}}\,\dot{q}_{\beta}^{,\check{p}}(\tau) + K_{\alpha\beta}^{\cdot d,\check{p}}\,q_{\beta}^{,\check{p}}(\tau) \\
\mathcal{L}_{\alpha}^{d\check{p}\check{p}}(\tau) &= M_{\alpha\beta}^{\cdot d,\check{p}\check{p}}\,\ddot{q}_{\beta}^{0}(\tau) + C_{\alpha\beta}^{\cdot d,\check{p}\check{p}}\,\dot{q}_{\beta}^{0}(\tau) + K_{\alpha\beta}^{\cdot d,\check{p}\check{p}}\,q_{\beta}^{0}(\tau)
\end{aligned}
\tag{7.59}
$$

with $\tau \in [0, T]$; $d, e = 1, 2, \ldots, D$; $\check{p} = 1, 2, \ldots, \check{N}$; $\alpha, \beta = 1, 2, \ldots, N$; summation over the index $\check{p}$ is not implied.

It is pointed out in closing that:

(i) The sampling interval Δt must be selected smaller than that for the conventional analysis to eliminate the effects of amplitude decay and period elongation appearing in the numerical integration.

(ii) Since a local variation in some parameters may cause a nonlinear change in the structural functional, it is necessary to employ the Taylor expansion at least up to the second order to cover such nonlinearity. (If a third-order perturbation is taken, the computation cost will greatly increase as the third-order rates of change must be included, and the first six moments of the random variables need to be known.)

(iii) It can be observed that the time response of the first two sensitivity moments is damped away more rapidly than the time response of the first two displacement moments. In addition, the first two sensitivity moments are found to have finite values even when the displacement response appears to have a tendency to exhibit chaotic characteristics. The same conclusion was arrived at analytically in [112,113] when considering the stochastic sensitivity function of a distributed-parameter dynamic control system.

7.3 Numerical Illustrations

Example 7.1 Let us now consider the time-instant sensitivity response of the structural system of Fig. 4.7. The structural response functional is defined as

$$\mathcal{G}(\tau) = \frac{[q_y(\tau)]^2}{(q_y^{(A)})^2} - 1 \leq 0$$

where $q_y(\tau)$ is the vertical displacement at the apex A and $q_y^{(A)}$ is an admissible displacement value taken as $q_y^{(A)} = 0.012$. Deterministic and random input data are assumed to be the same as in Example 4.7; the cross-sectional areas are assumed to be random design variables. To solve the initial–terminal-value problem the two-fold superposition technique is used with the 10 lowest structural eigenvalues and the 10 highest uncorrelated random variables. The equations are integrated with respect to time using 512 time steps with the sampling interval $\Delta t = 0.001$. The secularity effects are eliminated using the frequency range factor $r = 0.15$ and 1024 Fourier terms. Fig. 7.1 shows the time distribution of the spatial expectations, variances and covariances of the sensitivity gradient at points A and B; the deterministic solutions are also plotted.

Example 7.2 This example is concerned with the time distribution of the sensitivity gradient for a cylindrical shell clamped at the boundaries and subjected to a concentrated time-varying load at the midpoint A as shown in Fig. 4.9, Example 4.8. The stochastic finite element mesh again includes 60 constant thickness rectangular elements; the random design variables are defined as the element thicknesses $t^\rho(x,y)$, $\rho = 1, 2, \ldots, 60$. The structural response functional is assumed to have the same form as that of Example 7.1 with y replaced by z; $q_z(\tau)$ is the vertical displacement at the midpoint A and $q_z^{(A)}$ is an allowable displacement value set to 1.0 in this analysis. Deterministic and random data are defined similarly as in Example 4.8. To solve the primary and adjoint systems of the SFEM equations of motion the mode superposition technique is used with the 10 lowest modes. The set of 60 correlated random variables is transformed to a set of uncorrelated variables, out of which only the 10 highest modes are retained. The equations are integrated with respect to time using 512 time steps ($\Delta t = 0.001$). The secular terms are eliminated using the frequency range factor $r = 0.15$ and 1024 Fourier terms. The time distributions of the expectations, variances and covariances for the sensitivity gradient coefficients at points A and B are displayed in Fig. 7.2. In Fig. 7.3 the solutions obtained by the Wilson θ-integration ($\theta = 1.4$) and Newmark ($\delta = 0.5$, $\alpha = 0.25$) integration schemes are compared. It is seen that the results compare very well for both integration techniques; the Newmark scheme yields small perturbations around time response peaks in the second moment sensitivity gradient coefficients, though.

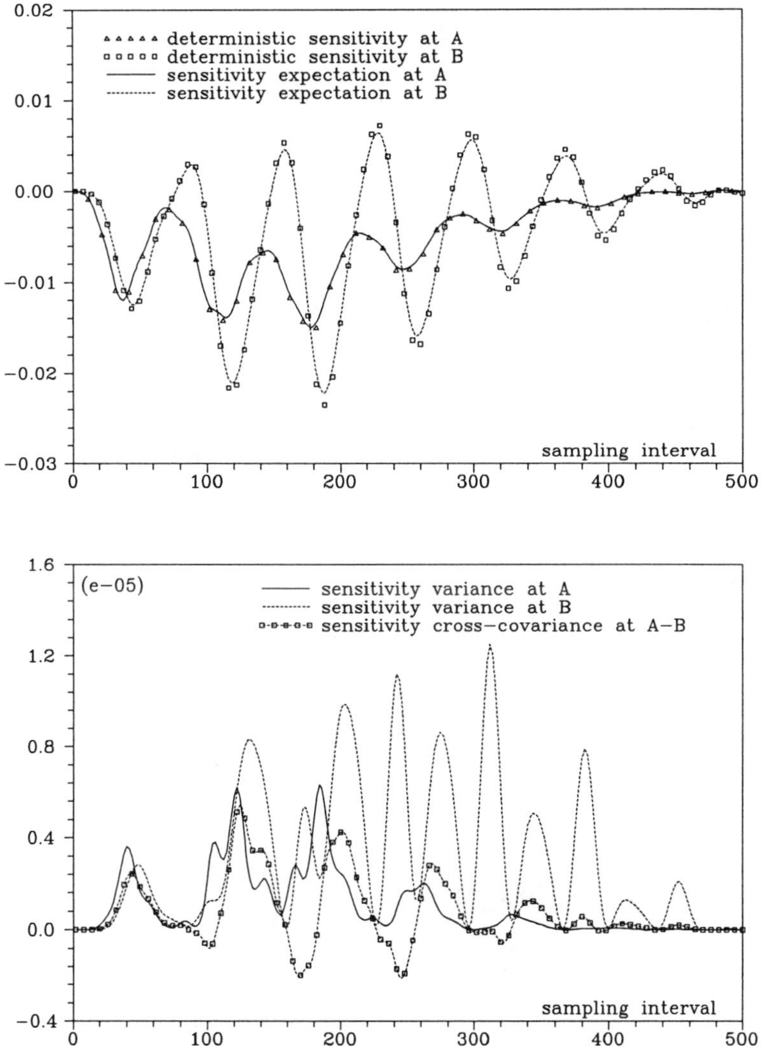

Figure 7.1 100-element frame. Time response of stochastic sensitivity.

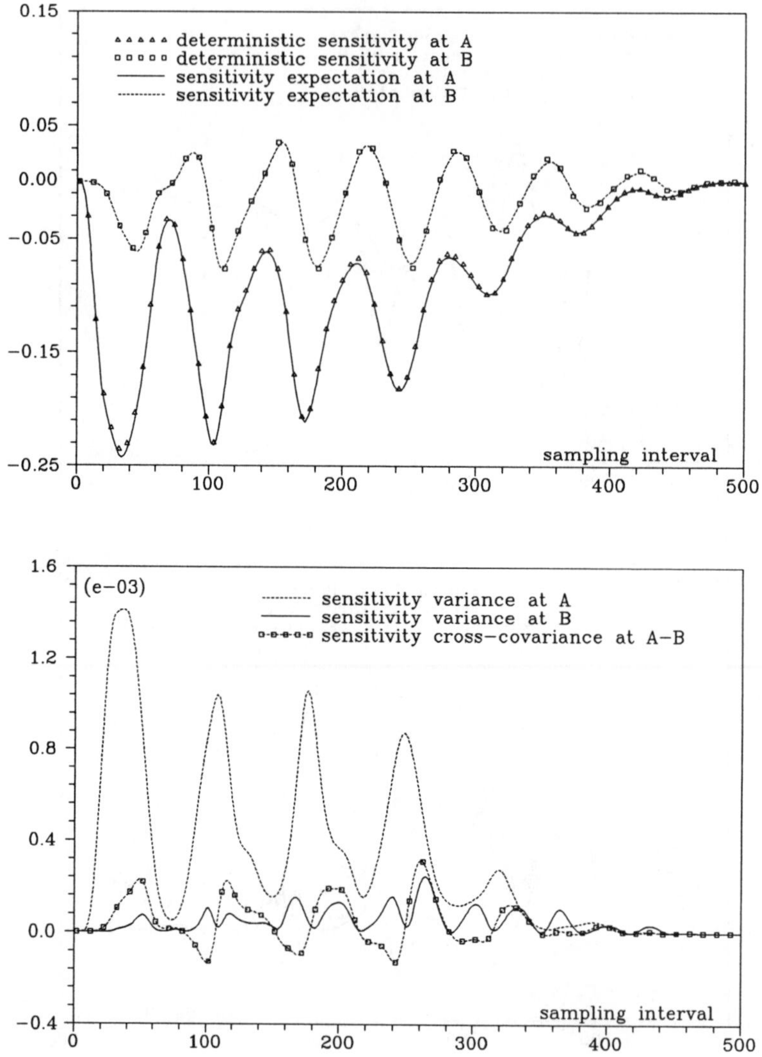

Figure 7.2 60-element shell. Time response of stochastic sensitivity.

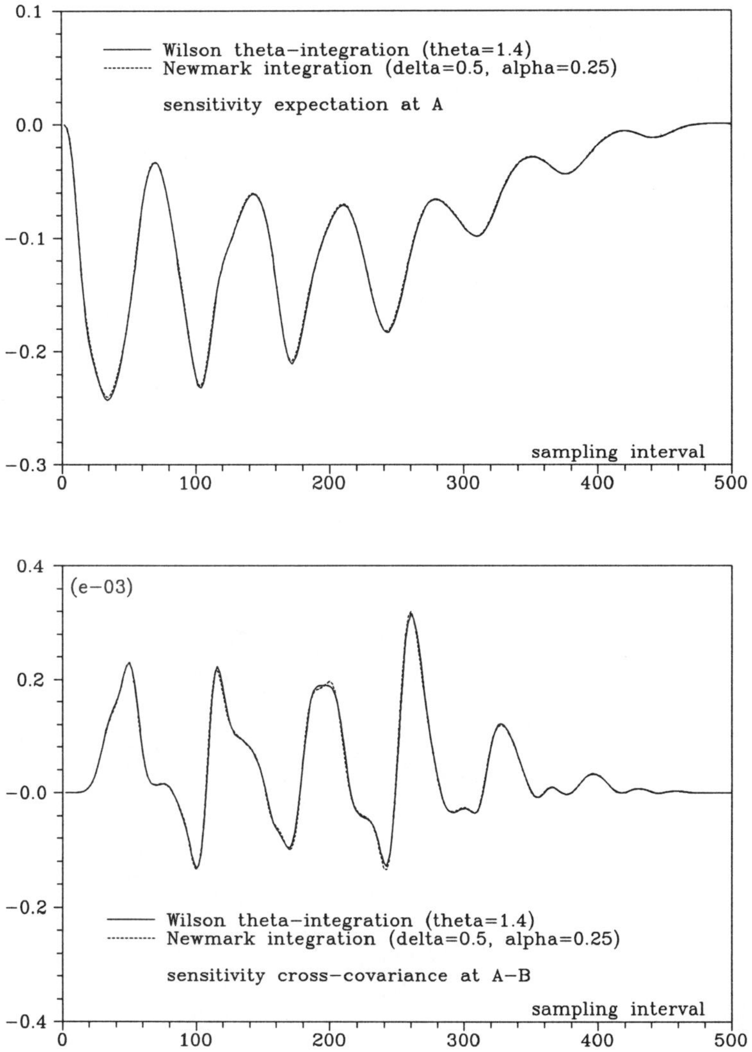

Figure 7.3 60-element shell. Response integrated by Wilson θ- and Newmark schemes.

Chapter 8

Program SFEDYN

A CODE FOR THE DETERMINISTIC AND STOCHASTIC ANALYSIS OF DYNAMICS AND DYNAMIC SENSITIVITY OF 3D FRAMES

8.1 Data Preparation

On the basis of the developments in Chapters 2, 4 and 7 a Fortran code called SFEDYN has been worked out to deal with dynamic problems of beam-type structures with parameters described deterministically and/or stochastically.

Similarly as in Chapter 6, the program is described here to further familiarize the reader with different aspects of programming the stochastic finite element method. Only a part of the source code, which deals directly with stochastic aspects, is included in the text; the explanations provided should make it possible to use the routines as an extension to any standard FEM program in order to cover the stochastic analysis option. As mentioned in Preface, the source code of the complete program SFEDYN can be obtained from the Publisher.

The computer program SFEDYN has been designed for educational, research and application environments to deal with dynamic deterministic and stochastic problems of medium- and large-scale three-dimensional frames. The system has been written for IBM compatible personal computers in Fortran 77 extended.[1] Minimum hardware requirements are 640KB of high-speed memory and a hard disk drive of 10MB.

For deterministic systems SFEDYN is capable of solving problems of free and forced vibrations[2] and dynamic response sensitivities. In the DSDS analysis a class of response functionals typical of dynamic control systems, namely the displacement-squared-type functionals, can be considered; response functionals of arbitrary form may be incorporated into the program with no difficulty. The design variables considered are cross-sectional area, length, Young's modulus and mass density in any of the structural members of which the system is made up.

[1]It includes *do while, do enddo, block if, on-line comments*, etc. The MS Fortran Compiler 5.0 (and later versions) and its UNIX and XENIX compatible versions are recommended.

[2]These two analysis options are based on a considerably modified version of the SAP-IV program [11].

In stochastic analysis uncertainties in geometry and material parameters are incorporated through the combination of the FEM and the second-order perturbation method. In the version given below, each random field variable representing the field value of an element is approximated by its spatial average over the element. The probabilistic distributions (i.e. spatial expectations and cross-covariances) for the displacement response can be evaluated at any time instant. For the SSDS analysis the adjoint variable technique is used. The response functionals and design variables are defined similarly as in the deterministic case; the design variables assumed to be possibly random are cross-sectional area, length, Young's modulus and mass density of the structural members. The transformation of correlated random variables to the set of uncorrelated variables is used and the fast Fourier transform may be employed to eliminate secularities. The time response of the expectations and cross-covariances for the design sensitivity gradients (and the displacements and stresses, if necessary) can be obtained on output.

Similarly as in Chapter 6, to appreciate the material being presented in this chapter the reader should be familiar with some basic aspects of Fortran programming.

In this section we sketch the major characteristics of SFEDYN through a description of several representative `format` statements; in programming practice reading format statements and their associated **read, write** commands is a simple and fast way to familiarize oneself with a new program provided it is well documented otherwise.

The present version of SFEDYN includes a main module and 84 routines. Modules developed in the code are grouped so that they (i) are functionally independent, (ii) interact with the global data base by means of a database control system included in the main module, and (iii) may be conveniently compiled, linked and run in the overlay mode. The main module reads in the control parameters, sets up the low- and high-speed storage required by the FEM (SFEM) solution process, selects the appropriate solution scheme (single- or multi-block mode depending on the operating memory and the model size) and monitors the macro process of setting up and solving the equations. Six analysis options are available in SFEDYN as defined by the analysis type parameter `ndyn` in the following output `format` statement

```
2000 format (1h ,20a4///
    *'P R O B L E M   C O N T R O L   I N F O R M A T I O N'  //4x,
    *' NUMBER OF NODAL POINTS (numnp).......................',i4/4x,
    *' NUMBER OF FINITE ELEMENT GROUPS (neltyp).............',i4/4x,
    *' NUMBER OF REQUIRED FREQUENCIES (nf)..................',i4/4x,
    *' ANALYSIS TYPE (ndyn).................................',i4/4x,
    *'   ndyn=1 - DETERMINISTIC EIGENPROBLEM                '   /4x,
    *'        2 - DETERMINISTIC FORCED VIBRATION            '   /4x,
    *'        3 - DETERMINISTIC EIGENVALUE SENSITIVITY      '   /4x,
    *'        4 - DETERMINISTIC DYNAMICS SENSITIVITY        '   /4x,
    *'        5 - STOCHASTIC FORCED VIBRATION               '   /4x,
    *'        6 - STOCHASTIC DYNAMICS SENSITIVITY           '   /4x,
```

```
* ' EXECUTION MODE (modex)...............................',i4/4x,
* '  modex=0 - PROBLEM EXECUTION                         '   /4x,
* '        1 - DATA CHECK ONLY                           '   /4x,
* ' NUMBER OF SUBSPACE ITERATION VECTORS (nad)............',i4/4x,
* ' NUMBER OF EQUATIONS PER BLOCK (keqb)..................',i4   )
```

Besides the analysis type parameter **ndyn** and the character field **(20a4)** reserved for printing the alphanumeric information (**title**) which identifies the finite element problem considered, other control variables (formal parameters, lower-case letter expressions in parentheses) are defined as follows:

- **numnp** – overall number of nodal points in the finite element model considered; each of the **numnp** nodes has six degrees of freedom,
- **neltyp** – number of different finite element groups in the model; **neltyp** is effectively used to define various types of random and design variables for a single SFEM model and for output-data processing of element data (the solvers implemented in SFEDYN are all of the Gauss-elimination type),
- **nf** – number of normalized coordinates (normal modes) used to change the basis of the displacement-type generalized coordinates and employed to integrate primary–adjoint equations of motion by the mode superposition technique,
- **nad** – size of the subspace employed in the subspace iteration scheme to solve for the smallest zeroth-order eigenpairs; **nad** can be selected by the user to change the number of iteration vectors from that determined automatically by the program (**nad=0** or default),
- **keqb** – number of degrees of freedom per storage block; for large-scale problems or small computers, by setting **keqb=0** (or default) SFEDYN automatically divides the system mass and stiffness matrices and load vectors into submatrix blocks and solves the equation system in the multi-block mode; setting a nonzero value for **keqb** may also be used to control the number of blocks or to test an extended version of SFEDYN,
- **modex** – operating mode; if **modex=1** SFEDYN runs in the data-check mode, verifying and printing all the data input as generated by the input routines; it skips high-cost operations in the assembly and solution phase of the execution mode invoked when **modex=0** (or default).

Similarly as in SFESTA, labels of the input and output **format** statements in the entire program SFEDYN are given as four-digit numbers; the first digit of an input format label is the number one (1) while an output **format** label begins with the number two (2).

For the sake of simplicity and consistency with the design sensitivity formulation, in this version of the program random fields are discretized by using the spatial averaging approach, i.e. the element random variable is defined as the spatial average of the random field over the element domain. Different random and design variable types available in SFEDYN can be seen from an output `format` statement in the routine `elbeam` used to compute element stiffness matrices together with their derivatives with respect to random and design variables and to transform these quantities into the system (global) generalized coordinates. This output `format` statement provides the control information of element input data for each element group out of the set of `neltyp` groups existing in the finite element model considered (the maximum value of `neltyp` is 30) and has the following form

```
2005 format (///          ' 3 / D   B E A M   E L E M E N T S '  //4x,
     *  ' NUMBER OF ELEMENTS (nbeam)...........................',i5/4x,
     *  ' NUMBER OF GEOMETRIC PROPERTY SETS (numetp)...........',i5/4x,
     *  ' NUMBER OF MATERIAL SETS (nummat).....................',i5/4x,
     *  ' FLAG FOR RANDOM OR DESIGN VAR. DEFINITION (isens).....',i5/4x,
     *  '   isens=0 - NONE OF THESE                             '   /4x,
     *  '           1 - ONE OR TWO OF THESE                     '   /4x,
     *  ' RANDOM OR DESIGN VARIABLE TYPE (istype)..............',i5/4x,
     *  '  istype=0 - IF isens=0                                '   /4x,
     *  '           1 - CROSS-SECTIONAL AREA                    '   /4x,
     *  '           2 - YOUNG'S MODULUS                         '   /4x,
     *  '           3 - MASS DENSITY                            '   /4x,
     *  '           4 - LENGTH                                  '     )
```

The fact that elements in a FEM model may be assembled into separate groups and elements in each group may have different cross-sectional areas or material properties enables one to define different types of random and design variables. The zero value of `isens` assigned to an element group means that all the elements in this group are described deterministically and no design variables are defined. For `isens=1` random and/or design variables are defined for all the elements in this group by setting `istype` equal to one of the integer values 1,2,3,4; the analysis is then defined in accordance with the analysis type `ndyn` (`ndyn`$\geq$ 3) entered on the problem control information line.

The way in which the element control data are entered suggests that: (i) the definition of the random and design variables has an 'element' character; for instance, design variables can be assigned to one or several element groups of interest, and the design variable 'mesh' need not necessarily coincide with the finite element mesh, and (ii) the computations of the first and second derivatives of the system matrices and vectors can be carried out at the element level.

The technique for handling the input covariance matrix of random variables plays an important role in SFEM analysis. For (very) small SFEM problems the user may enter the value and row–column indices for nonzero entries of the covariance matrix directly in the common input data file in which the expectations of random variables are stored. However, the direct way may become inefficient when dealing with larger systems because of the matrix size and scattered population of its entries. In this case it is more efficient to create the covariance matrix prior to the solution and save it (in binary form) in an unformatted file. The following `format` statement

```
2000 format (///    ' C O N T R O L   D A T A   F O R   I N P U T ',
     *                   ' C O V A R I A N C E   M A T R I X ' ///
     * ' FLAG FOR COVARIANCE INPUT (intype)........................',i5/
     * '   intype=0 - FROM COMMON INPUT DATA FILE                  '  /
     * '          1 - FROM UNFORMATTED FILE                        '  /
     * ' NUMBER OF REQUIRED TRANSFORMED RANDOM VARIABLES (nvar)....',i5)
```

defined in the routine `rcovin` used for data processing of the input covariances and the standard normal transformation of correlated random variables, demonstrates that both the input options are available in SFEDYN. In this format `nvar` is the number of random variables in the uncorrelated variable set used to approximate the input set of correlated random variables. In stochastic analysis the set of `nrand` correlated random variables (`nrand` is determined by SFEDYN) is transformed to a set of `nvar` uncorrelated random variables. The routine `rcovin` will be given a detailed description in the next section.

To become familiar with the analysis options for deterministic forced vibrations or time instant dynamic response sensitivity (`ndyn`=2 or 4) and stochastic forced vibrations or stochastic dynamics sensitivity (`ndyn`=5 or 6, respectively) the user may begin by reading the output `format` statement in the routine `modelsup`

```
2000 format (///        ' C O N T R O L   D A T A   F O R ',
     *                  ' D Y N A M I C   R E S P O N S E '        ///4x,
     * ' NUMBER OF TIME FORCE FUNCTIONS (nfn)..............',i5     /4x,
     * ' FLAG FOR GROUND MOTION LOADING (ngm)..............',i5     /4x,
     * '      ngm=0 - NO                                   '        /4x,
     * '          1 - YES                                  '        /4x,
     * ' NUMBER OF ARRIVAL TIMES (nat).....................',i5     /4x,
     * ' NUMBER OF TIME STEPS (SAMPLE LENGTH) (nt).........',i5     /4x,
     * ' OUTPUT PRINTING INTERVAL (not)....................',i5     /4x,
     * ' TIME INCREMENT (SAMPLING INTERVAL) (dt)...........',1pd11.3/4x,
     * ' DAMPING FACTOR (xsi)..............................',1pd11.3/4x,
     * ' FLAG FOR INTEGRATION SCHEME USED (integr).........',i5     /4x,
     * '   integr=0 - WILSON THETA-METHOD                  '        /4x,
     * '          1 - NEWMARK METHOD                       '        /4x,
     * ' FLAG FOR SECULAR ELIMINATION (isecular)...........',i5     /4x,
     * '  isecular=0 - NO                                  '        /4x,
     * '          1 - YES                                  '        /4x,
     * ' RANGE FACTOR FOR SECULAR ELIMINATION (range)......',1pd11.3/4x,
     * '  (taken with respect to first natural frequency)  '          )
```

The procedure `modelsup` is used to read in the control information for dynamic response analyses and to monitor procedures for generating the time sequence of load functions and those used for the zeroth-order response calculation. The dynamic response can be evaluated for any time-dependent load corresponding to any degree of freedom and for a ground acceleration in any (or all) of the three spatial directions. Initial time instants defined for the time-dependent loads may be different and arbitrarily taken. Integration of uncoupled equations of motion may be performed alternatively by using the Wilson θ-method or Newmark method. The fast Fourier transform may be used to eliminate secular terms; for `isecular`=1 the frequency range factor `range` must be specified in accordance with the first natural frequency. For `ndyn`$\geq$ 5 and `isecular`=1 the sampling length `nt` should be chosen as greater than (or equal to) 8 and be divisible by 4; we note that if the length `nt` equals 2^n, n being a natural number, the fast Fourier transform is used the most effectively.

8.2 Fortran Routines

In this section we provide some routines which are representative of SFEDYN and can readily be adapted to fit into existing conventional FEM-based packages. This should extend the range of applicability of the latter systems to cover free vibration and time instant sensitivity problems for deterministic systems as well as stochastic forced vibration and time instant SSDS problems. The routines are designed not only to illustrate the SFEM formulations and implementation discussed in Chapters 2, 4 and 7 but also with a view towards increased computational efficiency of the program in engineering practice. In general, the routines are self-explanatory and employ suggestive notation for global and local variables and off-line and on-line comments; a brief synopsis will be additionally given at the beginning of each subroutine. We note that in the `common` statements variables or formal parameters which are locally inactive are frequently denoted by the variable names with the suffix `nul`. Also, in SFEDYN some parameters are identified by more-than-six-character names only for the sake of clarity of the program's commands. This is not restrictive also when using earlier versions of Fortran compilers; the variable name ambiguity does not occur as the first six characters of any multi-character name are never repetitive.

8.2.1 Subroutine `elstiff`

The subroutine `elstiff` forms the element mass and stiffness matrices together with their derivatives with respect to random and design variables and the matrix relations between generalized element stresses (i.e. nodal forces, including axial and shear forces and bending and torque moments) expressed in the local coordinates (`x,y,z`) and nodal displacement-type variables expressed in the global coordinates (`X,Y,Z`). The element masses (and their derivatives with respect to random and design variables) are assumed to be concentrated at two edge nodal points in the three-dimensional beam element so that the corresponding matrices have a diagonal form. In the analysis of sensitivity and in stochastic problems the shear effects dependent on the element cross-sectional area are neglected and variations in the cross-sectional dimensions are assumed proportional (i.e. the dimension ratio is kept fixed) so that the moments of inertia are proportional to the cross-sectional area squared.

```
      subroutine elstiff (e,g,rho,coprop,dl,numetp,meltyp,mattyp,nummat)

*     To form 3D beam element mass and stiffness matrices,
*     their derivatives with respect to random and design variables
*     and stress-displacement relation arrays

*     ON INPUT:
*       isens         Flag for random or design variable identification
*                     (0, no; 1, yes)
*       istype        Random and/or design variable type
*                     (inactive if isens.eq.0; 1, cross-sectional area;
*                     2, Young's modulus; 3, mass density; 4, el. length)
*       mattyp        Material set number
*       meltyp        Cross-section set number
*       nummat        Number of material sets
*       nummetp       Number of cross-section sets
*       lm(.)         Equation numbers of element d.o.f.'s
*       coprop(.,1)   Cross-sectional (x-axial) areas (mean values)
*       coprop(.,2)   y-axial shear areas
*       coprop(.,3)   z-axial shear areas
*       coprop(.,4)   Torsional moments of inertia
*       coprop(.,5)   Moments of inertia about y-axis
*       coprop(.,6)   Moments of inertia about z-axis
*       dl            Element length (mean values)
*       e(.)          Young's moduli (mean values)
*       g(.)          Poisson ratios
*       rho(.)        Mass densities (mean values)
*       t(.)          Coordinate transformation matrix

*     ON OUTPUT:
*       nc            Number of element stress components
*       nd            Number of element d.o.f.'s
*       asa(.,.)      Stiffness matrix, increased in size to (24,24)
*                     if slave d.o.f.'s are specified
*       dkdb(.,.)     Stiffness derivative matrix
*       dmdb(.)       Mass derivative diagonal matrix (vector)
*       sa(.,.)       Stress-displacement transformation matrix
*       xm(.)         Mass diagonal matrix (vector)

      implicit        real*8 (a-h,o-z)

      common /sstiv/ isens,istype,nnul(31)
      common /sfems/ nd,ns,lm(24),asa(24,24),xm(24),sa(12,24),
     *               dkdb(24,24),dmdb(24),dkdbl(12,12),
     *               sas(12,12),t(3,3),tsnul(4),s(12,12)
      dimension       e(nummat),g(nummat),rho(nummat),coprop(numetp,6)

*     Set element geometry and material properties

      ax=coprop(meltyp,1)                          ! Cross-sectional area
```

```
      ay=coprop(meltyp,2)                        ! Shear area in y-direction
      az=coprop(meltyp,3)                        ! Shear area in z-direction
      if(isens.eq.1) then                          ! If sensitivity analysis
        ay=0.0d0
        az=0.0d0
      endif

      aax=coprop(meltyp,4)                  ! Moments of inertia about x-axis
      aay=coprop(meltyp,5)                  ! Moments of inertia about y-axis
      aaz=coprop(meltyp,6)                  ! Moments of inertia about z-axis
      shfy=0.0d0
      shfz=0.0d0

      zy=e(mattyp)/(dl*dl)
      eiy=zy*aay
      eiz=zy*aaz
      if(ay.ne.0.0d0) shfy=6.d0*eiz/(g(mattyp)*ay)
      if(az.ne.0.0d0) shfz=6.d0*eiy/(g(mattyp)*az)
      commy=eiy/(1.d0+shfz+shfz)
      commz=eiz/(1.d0+shfy+shfy)

*     Form element stiffness in local coordinates

      do j=1,12
        do i=1,12
          s(i,j)=0.0d0
        enddo
      enddo

      s(1,1)= e(mattyp)* ax/dl
      s(4,4)= g(mattyp)*aax/dl
      s(2,2)= 12.d0*commz/dl
      s(3,3)= 12.d0*commy/dl
      s(5,5)=  4.d0*commy*dl*(1.d0+0.5d0*shfz)
      s(6,6)=  4.d0*commz*dl*(1.d0+0.5d0*shfy)
      s(2,6)=  6.d0*commz
      s(3,5)= -6.d0*commy
      do i=1,6
        j=i+6
        s(j,j)=s(i, i)
      enddo
      do i=1,4
        j=i+6
        s(i,j)=-s(i,i)
      enddo
      s(6,12)= s(6,6)*(1.d0-shfy)/(2.d0+shfy)
      s(5,11)= s(5,5)*(1.d0-shfz)/(2.d0+shfz)
      s(2,12)= s(2,6)
      s(6, 8)=-s(2,6)
      s(8,12)=-s(2,6)
      s(3,11)= s(3,5)
      s(5, 9)=-s(3,5)
      s(9,11)=-s(3,5)
```

```
      do i=2,12                               ! Update lower triangular part
        k=i-1
        do j=1,k
          s(i,j)=s(j,i)
        enddo
      enddo

*     Form stiffness derivatives in local coordinates

      if(isens.ne.1) goto 100
      do i=1,24
        do j=1,24
          dkdb (i,j)=0.0d0
        enddo
        dmdb(i)=0.0d0
      enddo
      do i=1,12
        do j=1,12
          dkdbl(i,j)=0.0d0
        enddo
      enddo

      if(istype.lt.4) then                      ! Design or random variable
        if(istype.eq.1) then                   ! ... as cross-sectional area
          tmp=2.0d0/ax
          dkdbl(1,1)=s(1,1)/ax
        elseif(istype.eq.2) then                    ! ... as Young's modulus
          tmp=1.0d0/e(mattyp)
          dkdbl(1,1)=s(1,1)*tmp
        else                                           ! ... as mass density
          tmp=0.0d0
          dkdbl(1,1)=0.0d0
        endif
        do i=2,6
          dkdbl(i,i)=s(i,i)*tmp
        enddo
        dkdbl(2,6)=s(2,6)*tmp
        dkdbl(3,5)=s(3,5)*tmp
      else                                            ! ... as element length
        tmp=-1.0d0/dl
        dkdbl(1,1)=s(1,1)*tmp
        dkdbl(4,4)=s(4,4)*tmp
        dkdbl(5,5)=s(5,5)*tmp
        dkdbl(6,6)=s(6,6)*tmp
        tmp=tmp+tmp
        dkdbl(2,6)=s(2,6)*tmp
        dkdbl(3,5)=s(3,5)*tmp
        tmp=1.5d0*tmp
        dkdbl(2,2)=s(2,2)*tmp
        dkdbl(3,3)=s(3,3)*tmp
      endif
```

```
      do i=1,6
        j=i+6
        dkdbl(j,j)= dkdbl(i,i)
      enddo
      do i=1,4
        j=i+6
        dkdbl(i,j)=-dkdbl(i,i)
      enddo
      dkdbl(5,11)= 0.5d0*dkdbl(5,5)
      dkdbl(6,12)= 0.5d0*dkdbl(6,6)
      tmp=dkdbl(2,6)
      dkdbl(2,12)= tmp
      dkdbl(6, 8)=-tmp
      dkdbl(8,12)=-tmp
      tmp=dkdbl(3,5)
      dkdbl(3,11)= tmp
      dkdbl(5, 9)=-tmp
      dkdbl(9,11)=-tmp

      do i=2,12                                ! Update lower triangular part
        k=i-1
        do j=1,k
          dkdbl(i,j)=dkdbl(j,i)
        enddo
      enddo

*     Form stress-displacement transformation and stiffness derivatives
*     in global coordinates

  100 do j=1,24
        do i=1,12
          sa(i,j)=0.0d0
        enddo
      enddo
      do j=1,12
        do i=1,12
          sas(i,j)=0.0d0
        enddo
      enddo

      do la=1,10,3
        lb=la+2
        do ma=1,10,3
          mb=ma-1
          do i=la,lb
            do jm=1,3
              j=jm+mb
              if(isens.eq.1) then
                tmp=0.0d0
                do k=1,3
                  tmp=tmp+dkdbl(i,k+mb)*t(k,jm)
                enddo
```

```
            sas(i,j)=tmp
          endif
          tmp=0.0d0
          do k=1,3
            tmp=tmp+s(i,k+mb)*t(k,jm)
          enddo
          sa(i,j)=tmp
        enddo
      enddo
    enddo
  enddo

*     Update stiffness matrix and stiffness derivatives
*     in global coordinates

      do i1=1,24
        do i=1,24
          asa(i1,i)=0.0d0
        enddo
      enddo
      do la=1,10,3
        lb=la-1
        do ma=1,10,3
          mb=ma+2
          do il=1,3
            i=il+lb
            do j=ma,mb
              if(isens.eq.1) then
                tmp=0.0d0
                do k=1,3
                  tmp=tmp+t(k,il)*sas(k+lb,j)
                enddo
                dkdb(i,j)=tmp
              endif
              tmp=0.0d0
              do k=1,3
                tmp=tmp+t(k,il)*sa(k+lb,j)
              enddo
              asa(i,j)=tmp
            enddo
          enddo
        enddo
      enddo

*     Form mass (diagonal) matrix and its derivatives

      tmp=0.5d0*rho(mattyp)*ax*dl
      do m=1,3
        xm(m)=tmp
        xm(m+3)=0.0d0
        xm(m+9)=0.0d0
        xm(m+6)=tmp
      enddo
```

```
      if(isens.eq.1) then                       ! Design or random variable
        if(istype.eq.1) then                   ! ... as cross-sectional area
          tmp=1.0d0/ax
        elseif(istype.eq.2) then                     ! ... as Young's modulus
          tmp=0.0d0
        elseif(istype.eq.3) then                        ! ... as mass density
          tmp=1.0d0/rho(mattyp)
        else
          tmp=1.0d0/dl                                ! ... as element length
        endif
        do l=1,12
          dmdb(l)=xm(l)*tmp
        enddo
      endif

      end
```

8.2.2 Subroutine eigsens

This routine deals with the deterministic sensitivity problem of structural eigenvalues (natural frequencies). On the problem control data line the integer number 3 is entered for the analysis type **ndyn**, while the number of **nfreq** smallest eigenvalues required in the sensitivity analysis is specified by the parameter **nf** on the problem control data line, cf. Section 8.2.1. As shown in Section 2.8.2.3, having solved the generalized eigenproblem for the **nfreq** smallest eigenvalues and associated eigenvectors, the sensitivity of the $\bar{\alpha}$-th eigenvalue with respect to the variation of design parameters can be evaluated directly as, cf. Eq. (2.155)

$$\omega_{(\bar{\alpha})}^{,d} = \phi_{\alpha(\bar{\alpha})}^{\mathrm{T}} \left(K_{\alpha\beta}^{,d} - \omega_{(\bar{\alpha})} M_{\alpha\beta}^{,d} \right) \phi_{\beta(\bar{\alpha})} \qquad \begin{array}{l} d = 1,2,\ldots,D \\ \bar{\alpha} = 1,2,\ldots,\bar{N} \\ \alpha,\beta = 1,2,\ldots,N \end{array} \tag{8.1}$$

Recall that the symbol $(.)^{,d}$ denotes derivative with respect to the design variable h^d, D is the total number of design variables collected from the element groups, while N is the number of degrees of freedom in the system and $\bar{N}$ is the number of smallest eigenvalues required for sensitivity analysis. In SFEDYN the parameter D is set up at the 'element group' level since the design variables are defined separately in accordance with the element groups, while N is automatically determined by the program and $\bar{N}$ is given by the user; N and $\bar{N}$ are specified by the formal parameters **neq** and **nfreq**, respectively.

In the case of large equation systems the eigenvector matrix computed is saved in the multi-block mode in the reverse order of low-speed blocks because of the nature of the back-substitution process. Thus, prior to the first step of the eigenvalue sensitivity computation all the **nfreq** eigenvectors associated with the eigenvalues required are entered into the operating memory. The flow of data for evaluation and printing the sensitivity gradient coefficients is clearly commented upon inside the routine.

```
      subroutine eigsens (freq,lm,zl,dkdh,dmdh,domegadh,
     *                    nblock,neltyp,nt,neqb,neqblk,nfreq,npmax,nd)

*     To evaluate deterministic eigenvalue sensitivities

*     ON INPUT:
*       nblock        Number of core blocks used to save system matrices
*       nd            Number of element d.o.f.'s
*       neltyp        Number of design variable groups
*       neqb          Number of equations per block
*       neqblk        Number equal to nblock*neqb
*       nfreq         Number of smallest eigenvalues required
*       npmax         Max. number of design variables in a group
*       nt            Scratch file number (used to save eigensolution)
*       lm(.)         Equation numbers of element d.o.f.'s
*       mpar(.)       Number of design variables in each group
*       dkdh(.,.)     Stiffness derivative matrix
*       dmdh(.)       Mass derivative matrix (vector)
*       freq(.)       Required eigenvalues
*       zl(.,.)       Required eigenvectors

*     ON OUTPUT:
*       domegadh(.,.) Eigenvalue sensitivity matrix

      implicit        real*8 (a-h,o-z)

      common /sstiv/ nnul(3),mpar(30)
      dimension       freq(nfreq),lm(nd),zl(neqblk,nfreq),
     *                dkdh(nd,nd),dmdh(nd),domegadh(nfreq,npmax)

      write (0,'(1h+, ''* Evaluate sensitivity gradient'')')
      write (0,'('''')')

*     Read in eigensolution and assemble eigenvector blocks in core

      rewind nt
      read  (nt) freq                                   ! Eigenvalues
      ne=neqblk
      ns=ne+1-neqb
      do n=1,nblock
        read (nt) ((zl(i,l),i=ns,ne),l=1,nfreq)         ! Eigenvectors
        ns=ns-neqb
        ne=ne-neqb
      enddo

*     Computation and printout for each design variable group

      n6=(nfreq-1)/6+1                ! Number of 6-column print pages
      rewind 14

      do 10 m=1,neltyp               ! Loop over design variable groups
      write (6,2000) m
      nes=mpar(m)
```

```
      do n=1,nes                          ! Loop over variables in each group

        write (0,'(1h+,14x,''variable'',i5,4x,''group'',i4)') n,m
        read (14) nd,(lm(j),j=1,nd),((dkdh(j,k),j=1,nd),k=1,nd),
     *                                 (dmdh(j),j=1,nd)

        do l=1,nfreq                        ! Loop over required eigenvalues
          temp1=0.0d0
          temp2=0.0d0
          do k=1,nd                        ! Outer loop over element d.o.f.'s
            kk=lm(k)
            if(kk.gt.0) then
              temp3=0.0d0
              do j=1,nd                    ! Inner loop over element d.o.f.'s
                jj=lm(j)
                if(jj.gt.0) temp3=temp3+dkdh(k,j)*zl(jj,l)
              enddo
              zlkkl=zl(kk,l)
              temp1=temp1+zlkkl*temp3
              temp2=temp2+zlkkl*zlkkl*dmdh(k)
            endif
          enddo
          domegadh(l,n) = temp1-temp2*freq(l)
        enddo
      enddo

      do in=1,n6                                      ! Print 6-column pages
        l1=6*in-5
        l6=l1+5
        if(l6.gt.nfreq) l6=nfreq
        write (6,2001) (l,l=l1,l6)
        write (6,'('' '')')
        do n=1,nes
          write (6,2002) n,(domegadh(l,n),l=l1,l6)
        enddo
      enddo
   10 continue

 2000 format (///    'D E S I G N   S E N S I T I V I T Y',
     *  ' G R A D I E N T  -  V A R I A B L E   G R O U P', i5)
 2001 format (//1x,7h DESIGN,6(:7x,10hEIGENVALUE,i3))
 2002 format (i8,1p6d20.10)

      end
```

8.2.3 Subroutine rcovinp

As indicated in Section 8.1 the input data for the covariance matrix of random variables may be entered alternatively in the 'small data' or 'large data' mode. Running the program with the former option, SFEDYN reads in the matrix from a unformatted (binary) file created by the user prior to the solution; the file specification (filename with extension) is expected to be entered when a prompt shows up on the screen during the session. By using the latter option only nonzero covariance entries are typed in the common input data file which involves the problem control parameters, nodal point and element data (including the expectations of random variables), discrete time signals of the loading functions, additional concentrated masses, etc.; in this case each covariance entry must be specified by its value and row and column addresses.

Since the covariance matrix `Cov(nrand,nrand)` of the input random variables is symmetric, `nrand` being the overall number of random variables in the system considered, only the part above (and including) the diagonal needs to be entered. Entries are saved in compact form[1] in a one-dimensional array `a(.)` of length `nrand*(nrand+1)/2` and ordered as follows:

```
a(1) := Cov(1,1),
a(2) := Cov(2,2), a(3) := Cov(1,2),
a(4) := Cov(3,3), a(5) := Cov(2,3), ...
...
a(nrand*(nrand+1)/2-nrand+1) := Cov(nrand,nrand),
...
a(nrand*(nrand+1)/2) := Cov(1,nrand)
```

In Fortran this is carried out by the following commands

```
k=0
do j=1,nrand
  do i=j,1,-1
    k=k+1
    a(k)=Cov(i,j)
  enddo
enddo
```

Note that if the covariance matrix is created by the user himself using the 'large data' option, then in order to be consistent with the SFEDYN programming style the arrays `a(.)` and `Cov(.,.)` should be declared as double precision variables and the scratch file used to save `a(.)` opened in the unformatted form. In other words, the declaration `form='unformatted'` should be included in the `open` command in the user's subroutine for preparation of the covariance matrix.

The subroutine `rcovinp` is written to process the input covariance data in the 'small data' case. In this routine the data processing for the covariance entries is controlled by the parameter `intype` specified for the input option. If the 'large data' option is chosen (`intype=1`), `rcovin` does nothing but expects the file specification of the covariance file prepared a priori so that it becomes accessible for the standard normal transformation of correlated random variables. In the 'small data' mode `rcovin` reads in the value and address of all the nonzero covariance entries, generates and prints the input data and saves the generated matrix in compact form.

[1]In SFEDYN the *from-diagonals storage* mode is employed, cf. Section 2.4.2.

```
      subroutine rcovin (irowcol,cov,covar,covinp,nrand,nvar,nwm)

*     To read in and arrange input covariances for random variables

*     ON INPUT:
*       intype     Flag for covariance input
*                  (0, from common input data file;
*                   1, from unformatted file)
*       modex      Execution mode (0, solution; 1, data check)
*       n          Row index of input covariance entries
*       nrand      Overall number of random variables
*       nvar       Number of required uncorrelated variables
*       nwm        Number equal to nrand*(nrand+1)/2
*       itmp(.)    Column indices of input covariance entries
*       covtmp(.)  Values of input covariance entries

*     ON OUTPUT:
*       covar(.)   Covariance matrix saved in compact form

*       irowcol(.,.), cov(.), covinp(.)  Working arrays

      implicit       real*8 (a-h,o-z)

      character*64   filename
      common /exmod/ modex,inul
      dimension      irowcol(nwm,2),cov(nwm),covar(nwm),
     *               covinp(nrand),itmp(5),covtmp(5)

      read  (5,1000) intype,nvar                       ! Control parameters
      write (6,2000) intype,nvar
                                                ! Read in covariance entries
      if(intype.eq.1) then                       ! ... from unformatted file
        if(modex.eq.0) then
          write (0,'(15x,a\)') ' file specification: '
    1     read  (0,'(a64)') filename
          open  (2,file=filename,status='old',form='unformatted',
     *               iostat=ierror)
          if(ierror.ne.0) then
            write (0,'(15x,a\)') ' ERROR!  type again: '
            goto 1
          endif
          write (6,2001) filename
          read  (2) covar
          close (2,status='keep')
          open  (2,status='unknown',form='unformatted')
        endif
        return
      endif

      ncov=0                                    ! ... from common input file
    3 read (5,1000) n,(itmp(j),covtmp(j),j=1,5)
      if(n.eq.0.and.itmp(1).eq.0) goto 7
      if(n.ne.0) ntmp=n
```

```
      do j=1,5                    ! Loop over 5 entry pairs on input line
        if(itmp(j).eq.0) goto 3
        ncov=ncov+1
        irowcol(ncov,1)=ntmp
        irowcol(ncov,2)=itmp(j)
        cov(ncov)=covtmp(j)
      enddo
      goto 3
    7 write (6,2002)                              ! Print input covariances
      n20=(ncov+19)/20                 ! Number of 20-entry printing blocks
      do n=1,n20                         ! Processing print 4-column pages
        ne=20*n
        ns=ne-19
        if(ne.gt.ncov) ne=ncov
        write (6,2003) ((irowcol(i,j),j=1,2),cov(i),i=ns,ne)
        write (6,'('''')')
      enddo
      if(modex.eq.1) return                      ! Return if data-check mode

      jj=0                      ! Save covariances matrix in compact form
      do i=1,nrand
        do j=1,nrand
          covinp(j)=0.0d0
        enddo
        do k=1,ncov
          ir=irowcol(k,1)
          ic=irowcol(k,2)
          if(ir.lt.i) then
             if(ic.eq.i) covinp(ir)=cov(k)
          else if(ir.eq.i) then
             covinp(ic)=cov(k)
          else
             goto 9
          endif
        enddo
    9   do j=i,1,-1
          jj=jj+1
          covar(jj)=covinp(j)
        enddo
      enddo

 1000 format (i5,5(i5,d10.0))
 2000 format (///    ' C O N T R O L   D A T A   F O R   I N P U T',
     *                    ' C O V A R I A N C E   M A T R I X' ///
     * ' FLAG FOR COVARIANCE INPUT (intype).......................',i5/
     * '   intype=0 - FROM COMMON INPUT DATA FILE                  '/
     * '          1 - FROM UNFORMATTED FILE                        '/
     * ' NUMBER OF REQUIRED TRANSFORMED RANDOM VARIABLES (nvar)....',i5)
 2001 format (///' INPUT COVARIANCES ARE READ IN FROM FILE  ',a64)
 2002 format (///' I  N  P  U  T    C  O  V  A  R  I  A  N  C  E  S'//)
 2003 format (4(:1x,4hCov(,i3,1h,,i3,4h) = ,1pd11.5,4x))

      end
```

8.2.4 Subroutine rhs1ord

Some of the most typical procedures in the computer implementation of the stochastic finite element method are those written to form the right-hand sides of the first- and second-order equations of motion. The reason is that the system considered is modelled so that all the random characteristics are transformed entirely into first- and second-order forcing functions. The presentation of both the routines which are essentially similar in terms of their structure and algorithm goes beyond the scope of the book. We therefore focus our attention only on the first of them, called **rhs1ord**, which is used to establish the right-hand side of the first-order equations. In accordance with the two-fold superposition technique the discrete time signal sequence describing the first-order excitation is computed in the normalized coordinate system. It is assumed in SFEDYN that the damping coefficients are linear functions of the corresponding mass and stiffness entries and the modal damping factors are the same for all the modes.

```
      subroutine rhs1ord (accel,displ,ftdmcf,ftdkcf,rhs1,nt,nf,nvar,i12)

*     To form right-hand side of first-order uncoupled equations

*     ON INPUT:
*       integr        Flag for integration scheme used in analysis
*                     (0, Wilson theta-method; 1, Newmark method)
*       isecular      Flag for secularity elimination (0, no; 1, yes)
*       istyp14       Random variable is cross-sectional area or length
*       i12           Control parameter for response do-loops
*                     (1, primary response only;
*                      2, primary and adjoint responses)
*       nf            Number of required normalized coordinates
*       nt            Sampling length (number of time intervals)
*       nvar          Number of required uncorrelated coordinates
*       accel(.,.)    Modal accelerations
*       displ(.,.)    Modal displacements
*       ftdkcf(.,.,.) Modal stiff. derivatives in uncorr. coords
*       ftdmcf(.,.,.) Modal mass derivatives in uncorr. coords

*     ON OUTPUT:
*       rhs1(.,.)     First-order modal forcing sequence

*       nfile1,nfile2 Scratch file unit numbers
*     (equivalence (accel,displ) and (ftdmcf,ftdkcf) if .not.istyp14)

      implicit       real*8 (a-h,o-z)

      logical        istyp14
      common /istyp/ istyp14,inul(2)
      common /fftdt/ isecular,xnul(2),integr,nfile1,nfile2
      dimension      accel(nf,nt),displ(nf,nt),
     *               ftdmcf(nf,nf,nvar),ftdkcf(nf,nf,nvar),rhs1(nt,nvar)
```

```
      if(isecular.eq.1) then
        nfile1=9
        nfile2=10
      else
        nfile1=10
        nfile2=9
        write (0,'('''')')
      endif

      write (0,'(1h+,8x,''form 1st-order r.h.s.'',40(1h )/)')

      rewind 9
      rewind 10
      rewind 15                                       ! Random variable
      if(istyp14) then           ! ... as cross-sectional area or length
        read (15) ftdmcf
        read (15) ftdkcf
      else
         read (15) ftdkcf        ! ... as Young's modulus or mass density
      endif

*     Form first-order primary loads (mm=1) and adjoint load (mm=2)

      do 200 mm=1,i12                                 ! Random variable
      if(istyp14) then                ! as cross-sectional area or length
        read (nfile1) accel
        read (nfile1) displ

        if(integr.eq.0) then       ! Forcing sequence in Wilson t-scheme
          call rwilson (accel,nf,nt)
          call rwilson (displ,nf,nt)
        endif
        do j=1,nf
          do k=1,nvar
            write (0,2000) k,j,mm
            do i=1,nt
              tmp=0.0d0
              do n=1,nf
               tmp=tmp+ftdmcf(j,n,k)*accel(n,i)+ftdkcf(j,n,k)*displ(n,i)
              enddo
              rhs1(i,k)=-tmp
            enddo
          enddo
          write (nfile2) rhs1
        enddo                                         ! Random variable
      else                            ! as Young's modulus or mass density
        read (nfile1) displ
        if(integr.eq.0) call rwilson (displ,nf,nt)
        do j=1,nf
          do k=1,nvar
            write (0,2000) k,j,mm
            do i=1,nt
              tmp=0.0d0
```

```
               do n=1,nf
                 tmp=tmp+ftdkcf(j,n,k)*displ(n,i)
               enddo
               rhs1(i,k)=-tmp
             enddo
           enddo
           write (nfile2) rhs1                     ! Save first-order r.h.s.
         enddo
       endif
  200 continue

 2000 format (1h+,13x,'uncorrelated mode',i4,
     *        ', structural/adjoint mode',i4,3x,'(',i1,')')

      end
*********************************************************************
      subroutine rwilson (rhs,nf,nt)

*     Kernel of routines rhs1ord and rhs2ord
*     To scale time forcing sequence for Wilson theta-integration

      real*8 rhs(nf,nt),rold,rnew

      do n=1,nf
        rold=rhs(n,1)
        rhs(n,1)=0.6d0*rold+0.4d0*rhs(n,2)
        do i=2,nt
          rnew=rhs(n,i)
          rhs(n,i)=1.4d0*rnew-0.4d0*rold
          rold=rnew
        enddo
      enddo

      end
```

The subroutine `rwilson` forms the kernel of `rhs1ord` and `rhs2ord` (computation of the second-order right-hand sides); it is used to process the first- and second-order forcing sequences to obtain equivalent discrete time signals used in turn in the Wilson θ-scheme for numerical integration of the equation of motion. Clearly, by choosing the Newmark scheme of integration (`integr=1` entered on the control data line for dynamic response, cf. Section8.1) the execution of `rwilson` is skipped.

8.2.5 Subroutines `respo1` and `rcovprt`

To complement the discussion of the routine `rhs1ord` given above, in this section we provide the subroutine `respo1` designed to: (i) evaluate the first-order primary and adjoint modal responses by the direct step-by-step integration technique and their superposition, and (ii) compute the time variation of the cross-covariances for selected displacement components.

Assuming that the solution at time instant t is known, the system response at time $\tau = t+\Delta t$ can be determined by the recursive expression [7]

$$\begin{bmatrix} \ddot{x}(t+\Delta t) \\ \dot{x}(t+\Delta t) \\ x(t+\Delta t) \end{bmatrix} = A_{3\times 3} \begin{bmatrix} \ddot{x}(t) \\ \dot{x}(t) \\ x(t) \end{bmatrix} + B_{3\times 1}\, y(t+\theta\Delta t) \tag{8.2}$$

The symbols x and y represent the normal coordinates and modal excitations in the primary and/or adjoint systems. The direct integration approximation operator A and load operator B are functions of the natural frequencies $\omega_{(\bar{\alpha})}$, modal damping factors $\psi_{(\bar{\alpha})}$ and sampling time interval Δt; they are thus constant in the time interval considered. The explicit form of A and B for the Newmark algorithm and Wilson θ-algorithm is given in [7]. In SFEDYN the value of θ is taken as 1.4 for the Wilson θ-scheme and as 1 for the Newmark scheme.

```
      subroutine respo1 (rhs1,accel1,displ1,d1st,r1st,rcov,
     *                   nt,nf,nvar,nds,ndofpr,i12)

*     To evaluate first-order modal responses

*     ON INPUT:
*       iocov         Flag for displacement cross-covariance printing
*       istyp2        Random variable is Young's modulus
*       istyp3        Random variable is mass density
*       istyp14       Random variable is cross-sectional area or length
*       i12           Control parameter for response do-loops
*                     (1, primary resp.; 2, primary and adjoint resp.)
*       ndofpr        Number of required displ. components on output
*       nds           Number of printing time instants (nt/not)
*       nf            Number of required normalized coordinates
*       not           Printing interval
*       nrec?         Record addresses in direct access mode
*       nt            Sampling length (number of time intervals)
*       numset        Number of 6-column page for expect. printing
*       nvar          Number of required uncorrelated coordinates
*       kd(.,.)       Node, nodal d.o.f. and equation numbers
*       a(.,.)        Direct integration approximation operator
*       b(.)          Modal load operator
*       f(.,.)        Eigenvectors required in analysis
*       rhs1(.,.)     First-order modal forcing sequence
*       var(.)        Variances for uncorrelated random variables

*     ON OUTPUT:
*       covmax        Maximum amplitude of cross-covariance response
*       timax         Maximum cross-covariance time instant
*       accel1(.,.)   First-order modal accelerations
*       displ1(.,.)   First-order modal displacements
*       rcov(.)       Displacement cross-covariance time response
*       r1st(.,.,.)   First-order uncorrelated displacements

      implicit        real*8 (a-h,o-z)
```

```
      logical        istyp14,istyp2,istyp3,meq1,m1ioc
      common /istyp/ istyp14,istyp2,istyp3
      common /fftdt/ inul1(6),nfile1,nfile2
      common /acode/ nnul(2),numset,not,inul2(3),dtnot,knul
      common /sfems/ a(3,3),b(3),xold(3),xnew(3),kd(3,6),f(6,50),
     *               znul(1619)
      common /vartm/ var(48),tnul(14)
      common /inout/ ionul,iocov,iosnul,nrec0,nrec1,nrec2,nrec3,nrec4
      dimension      rhs1(nt,nvar),accel1(nt,nvar),displ1(nt,nvar),
     *               d1st(nds,nvar),r1st(ndofpr,nds,nvar),rcov(nds)

      write (0,'(1h+,8x,
     *       ''integrate first-order equations'',32(1h )/)')

      if(iocov.eq.1) then            ! For displ. cross-covariance printing
        do k=1,nvar
          do i=1,nds
            do m=1,ndofpr
              r1st(m,i,k)=0.0d0
            enddo
          enddo
        enddo
      endif

      rewind 9
      rewind 10
      if(i12.eq.2) nrec2=nrec1

*     Do initial-terminal loop over normalized coordinates

      do 500 mm=1,i12                               ! Initial-terminal loop

      meq1=mm.eq.1
      m1ioc=meq1.and.iocov.eq.1
      if(.not.meq1) nrec3=nrec2
      rewind 13

      do 300 n=1,nf                                ! Normal coordinate loop

      read (nfile2) rhs1
      read (13) a,b

      do k=1,nvar
        write (0,'(1h+,13x,''uncorrelated mode'',i4,
     *     '', structural/adjoint mode'',i4,3x,''('',i1,'')'')') k,n,mm
        xold(1)=rhs1(1,k)                      ! xold(1), acceleration at t
        xold(2)=0.0d0                              ! xold(2), velocity at t
        xold(3)=0.0d0                          ! xold(3), displacement at t
        xnew(1)=0.0d0                       ! xnew(1), acceleration at t+dt
        xnew(2)=0.0d0                           ! xnew(2), velocity at t+dt
        xnew(3)=0.0d0                       ! xnew(3), displacement at t+dt
        accel1(1,k)=xold(1)
        displ1(1,k)=0.0d0
```

```
        ii=0
        nout=not+1
        do i=2,nt
          do j=1,3
            xnew(j)=b(j)*rhs1(i,k)
            do m=1,3
              xnew(j)=xnew(j)+a(j,m)*xold(m)
            enddo
          enddo
          do j=1,3
            xold(j)=xnew(j)
          enddo
          accel1(i,k)=xnew(1)
          displ1(i,k)=xnew(3)
          if(i.eq.nout.and.m1ioc) then
            ii=ii+1
            d1st(ii,k)=xnew(3)       ! First-order modal displ. response
            nout=nout+not
          endif
        enddo
      enddo

*     Save first-order responses for second-order calculation
                                            ! Save modal primary response
      if(istyp3) then                       ! Random var. as mass density
        write (nfile1) accel1
      elseif(istyp2) then                ! Random var. as Young's modulus
        write (nfile1) displ1
      else                          ! Random var. as cross-section or length
        write (nfile1) accel1
        write (nfile1) displ1
      endif
      if(i12.eq.2) then      ! Save modal adjoint response to unit impulse
        if(meq1) then
          do k=1,nvar
            nrec2=nrec2+1
            if(istyp3) then                 ! Design var. as mass density
               write (18,rec=nrec2) (accel1(i,k),i=1,nt)
            elseif(istyp2) then          ! Design var. as Young's modulus
               write (18,rec=nrec2) (displ1(i,k),i=1,nt)
            else                    ! Design var. as cross-section or length
               write (18,rec=nrec2) (accel1(i,k),i=1,nt)
               nrec2=nrec2+1
               write (18,rec=nrec2) (displ1(i,k),i=1,nt)
            endif
          enddo
        else
          do k=1,nvar
            nrec3=nrec3+1
            write (18,rec=nrec3) (displ1(i,k),i=1,nt)
          enddo
        endif
      endif
```

```
*     Perform superposition for first-order modal displacements

      if(m1ioc) then
        rewind 3
        m=0
        do ns=1,numset
          read (3) lcol,kd,((f(l,j),l=1,6),j=1,nf)
          do l=1,lcol
            m=m+1
            tmp=f(l,n)
            do k=1,nvar
              do i=1,nds
                           ! First-order uncorrelated displ. time response
                r1st(m,i,k)=r1st(m,i,k)+tmp*d1st(i,k)
              enddo
            enddo
          enddo
        enddo
      endif

  300 continue

      if(.not.m1ioc) goto 500

      write (0,'(1h+,8x,
     *  ''calculate displacement cross-covariance response'',15(1h ))')
      rewind 14

      do m=1,ndofpr
        do j=m,ndofpr
          covmax=0.0d0
          do i=1,nds
            tmp=0.0d0
            do k=1,nvar
              tmp=tmp+r1st(m,i,k)*r1st(j,i,k)*var(k)
            enddo
            abstmp=dabs(tmp)
            if(abstmp.gt.covmax) then
              covmax=abstmp            ! Max. cross-covariance amplitude
              timax=dtnot*dfloat(i) ! Max. cross-covariance time instant
            endif
            rcov(i)=tmp                 ! cross-covariance time response
          enddo

          write (14) m,j,rcov,covmax,timax

        enddo
      enddo

  500 continue

      end
```

In **respo1** cross-covariances for the required displacement components at any time instant $\tau = t$ are computed using the formula, cf. Eq. (4.192)

$$S_{\mathbf{q}}^{\alpha\beta}(t) = \sum_{\check{\rho}=1}^{\check{N}} q_{\alpha}^{;\check{\rho}}(t)\, q_{\beta}^{;\check{\rho}}(t)\, S_{\mathrm{c}}^{\check{\rho}} \qquad t \in [0, T] \tag{8.3}$$

with $\check{N}$ and $S_{\mathrm{c}}^{\check{\rho}}$ specified by the formal parameters **nvar** and **var(.)**, respectively.

Since SFEDYN is intended to be useful also in the analysis of large-scale systems, only the displacement components suggested by the user are processed and the time variation of the first two statistical moments is printed at required time instants. In the routine **covprt** given below the displacement component addresses are stored in the array **kd(.,.)** and the corresponding response values are arranged into six-column pages for printing. In order to obtain the second-order accurate displacement expectations the second-order response has to be evaluated while only the first-order response needs to be known to determine the first-order accurate cross-covariances. That is why the second moment of the nodal displacements is computed at this stage of the analysis. Owing to symmetry only the upper triangular part of the three-dimensional cross-covariance matrix is computed; absolute values of maximum amplitudes and corresponding time instants are printed at the end of each column. Calculation and printout of nodal displacement expectations and cross-covariances is optional; in the stochastic sensitivity analysis this step may be skipped.

```
      subroutine rcovprt (noddof,rcov,time,nds)

*     To print out time response of displacement cross-covariances

*       ndofpr        Number of required displ. components on output
*       nds           Number of printing time instants (nt/not)
*       not           Printing interval
*       nt            Sampling length (number of time intervals)
*       numset        Number of 6-column pages for expect. printing
*       kd(.,.)       Node, nodal d.o.f. and equation numbers
*       covmax(.)     Maximum amplitudes of cross-covariance response
*       rcov(.)       Displacement cross-covariance time response
*       timax(.)      Maximum cross-covariance time instants
*       time(.)       Sampling points (required time instants)

      implicit        real*8(a-h,o-z)

      common /acode/ nnul,nt,numset,not,ndofpr,dnul,dtnot,knul
      common /sfems/ kd(3,6),kk(2,6),jj(2,6),covmax(6),timax(6),
     *               xnul(1913)
      dimension       noddof(2,nds),rcov(nds,6),time(nds)
      character*22    rfmt
      character*1     fmt(15),digit(6)
      equivalence     (fmt,rfmt)
      data rfmt       /'(0pf12.5,2x,1p6d19.10)'/
      data digit      /'1','2','3','4','5','6'/
```

```
      rewind 3
      k=0
      do nb=1,numset                              ! Loop over printing sets
        read (3) lcol,kd
        do i=1,lcol                               ! Loop over page columns
          k=k+1
          noddof(1,k)=kd(1,i)                         ! Nodal point number
          noddof(2,k)=kd(2,i)                         ! Nodal d.o.f. number
        enddo
      enddo
      do n=1,nds
        time(n)=dtnot*dfloat(n)                           ! Time instants
      enddo

      rewind 14
      nblsum=ndofpr*(ndofpr+1)/2          ! Total number of printing columns
      n6=(nblsum+5)/6                            ! Number of 6-column pages
      do i=1,n6                                  ! Loop over 6-column pages
        if(i.eq.n6) then
          i6=nblsum-(n6-1)*6
          fmt(15)=digit(i6)
        else
          i6=6
          fmt(15)=digit(6)
        endif
        do m=1,i6                                 ! Loop over page columns
          read (14) k,j,(rcov(n,m),n=1,nds),covmax(m),timax(m)
          kk(1,m)=noddof(1,k)
          kk(2,m)=noddof(2,k)
          jj(1,m)=noddof(1,j)
          jj(2,m)=noddof(2,j)
        enddo
        write (6,2001) (kk(1,m),kk(2,m),jj(1,m),jj(2,m),m=1,i6)   ! Head
        write (6,'('' '')')
                                          ! Time instants and amplitudes
        write (6,rfmt) (time(n),(rcov(n,m),m=1,i6),n=1,nds)
        write (6,2002) (covmax(m),m=1,i6)                 ! Max. amplitude
        write (6,2003) (timax(m),m=1,i6)                    ! Time instant
      enddo

 2001 format(///            'D I S P L A C E M E N T',
     *              '  A U T O C O V A R I A N C E',
     *                                  '  R E S P O N S E'//
     *         'P R I N T   O F   U P P E R   T R I A N G L E',
     *                                  ' C O M P O N E N T S ',
     * ' C o v (Ith NODE - Kth D.O.F., Jth NODE - Lth D.O.F.)'///
     * (5x,7hT I M E,2x,6(:3x,4hCov(,i3,1h-,i1,1h,,i3,1h-,i1,1h)))
 2002 format(/
     *'M A X I M U M   A B S O L U T E   C O V A R I A N C E S'//
     *  3x,9hV A L U E,2x,1p6d19.10)
 2003 format(5x,7hT I M E,2x,1p6e19.5)

      end
```

8.2.6 Subroutine rsecular

It was shown in Chapter 4 that the second-order perturbation technique employed in stochastic finite element modelling can cause secularities owing to the inherently secular character of discrete time signal sequences describing the first- and second-order generalized forcing functions. To maintain validity of solutions for a relatively long response duration, which is frequently required in stochastic analysis (in particular, for stochastic sensitivity problems), the elimination of secular terms involved on the right-hand side of the first- and second-order initial and terminal equations is strongly recommended, the case of systems with structural damping being no exception. The concepts and comprehensive aspects of the computer implementation for the secularity problem and the fast Fourier transform were discussed in detail in Section 4.5.4. Fast Fourier transform algorithms have been extensively investigated in the literature and effective FFT Fortran procedures may be found in many software libraries. It should be emphasized that with a good FFT algorithm available the computation effort taken by the direct and inverse Fourier transform for sampling records of several thousand sampling points (time intervals) is only a very small fraction of the total computational cost.

The routine **rsecular** has been written to eliminate the resonant part (secular terms) involved on the right-hand side of the first- and second-order equations of the initial as well as initial–terminal problem. To filtrate the secular effects from the n-th discrete time sequence describing the right-hand side of the n-th modal equation the following steps have to be accomplished: (i) the discrete finite-range Fourier analysis is carried out for the time sequence to obtain its discrete Fourier spectra; (ii) in the resulting frequency sequence the coefficients at frequencies within the domain extending from $\omega_n - \Delta\Omega$ to $\omega_n + \Delta\Omega$, $\Delta\Omega$ being a specified frequency range, are weighted by using a data window; and (iii) the discrete finite-range Fourier synthesis is performed for the weighted frequency sequence to recover the time sequence with no resonant part. In SFEDYN the frequency range $\Delta\Omega$ is specified by the user through a frequency range factor r (identified by **range** entered on the control data line for dynamic response, cf. Section 8.1) and defined with respect to the smallest natural frequency ω_1, i.e. $\Delta\Omega = r\,\omega_1$, while the data window employed is of the cosine-squared type, cf. Eq. $(4.226)_3$.

```
      subroutine rsecular (a,accel,displ,freq,sample,
     *                     randdt,nf,nt,ntp1,mtot,n5,iflag,i12)

*     To eliminate secular terms by fast Fourier transform

*     ON INPUT:
*       iflag        Flag for controlling dynamic response process
*       istyp2       Random variable is Young's modulus
*       istyp3       Random variable is mass density
*       istyp14      Random variable is cross-sectional area or length
*       i12          Control parameter for response do-loops
*                    (1, primary resp.; 2, primary and adjoint resp.)
```

```
*       mtot          RAM-checking parameter ('Mother' array length)
*       nf            Number of required normalized coordinates
*       not           Printing interval
*       nt            Sampling length (number of time intervals)
*       ntp1          Extended sampling length
*       n5            Tail address of working array a(.)
*       randdt        Sampling time interval
*       range         Frequency range factor for secularity elimination
*       sample(.)     Input time real-valued sequence
*       freq(.,.)     Natural frequencies

*     ON OUTPUT:
*       accel(.,.)    Weighted acceleration-type time sequence
*       displ(.,.)    Weighted displacement-type time sequence

*       a(.)          'Mother' (working) array
*     (sample of length nt is extended with sample(nt+1)=sample(nt))

      implicit        real*8 (a-h,o-z)
      logical         istyp14,istyp2,istyp3,iflag0
      common /istyp/  istyp14,istyp2,istyp3
      common /fftdt/  inul,range,i001(5)
      dimension       a(1)
      dimension       accel(nf,nt),displ(nf,nt),freq(nf),sample(ntp1)
      data            pi/3.14159265358979d0/,one/0.999999999999999d0/

      write (0,'(1h+,8x,
     * ''eliminate secularities by fast Fourier transform'',20(1h )/)')

      iflag0=iflag.eq.0
      ntm1=nt-1                     ! Check if sampling length is satisfied
      if(iflag0) then                       ! (nt-1.ge.8 and divisible by 4)
        if(ntm1.lt.8.or.ntm1.ne.(ntm1/4*4)) then
          write (6,2001) ntm1
          STOP 'E R R O R !  Bad time-interval number'
        endif
      endif

      nyquist=ntm1/2
      mtwo=2
      do i=1,32                        ! Max. sampling length 2**32 allowed
        if(nyquist.le.mtwo) goto 15
        mtwo=mtwo+mtwo
      enddo
   15 if(nyquist.ne.mtwo) then
        n6=n5+ntm1                       ! Set FFT memory if ntm1.ne.2**m
        if(n6.gt.mtot) call error (n6-mtot)      ! Check if memory enough
      endif

      rewind 7
      read  (7) freq                           ! Read in natural frequencies
      domega=range*freq(1)                         ! Secular frequency range
      pidr=pi/domega
```

```
      fnyquist=pi/randdt                       ! Nyquist cutoff frequency
      deltaf=fnyquist/dfloat(nyquist)       ! Sampling frequency interval

      if(iflag0) then                                ! Set do-loop counters
        if(istyp3) then                 ! Random variable as mass density
          jstart=1
          jend=1
        elseif(istyp2) then           ! Random variable as Young's modulus
          jstart=2
          jend=2
        else                   ! Random variable as cross-section and length
          jstart=1
          jend=2
        endif
      else
        jstart=3
        jend=3
      endif
      rewind 9
      rewind 10

*     Process primary (mm=1) and adjoint (mm=2) loop over nf frequencies

      do 200 mm=1,i12                                 ! Primary-adjoint loop

      do 100 n=1,nf                             ! Normalized coordinate loop

      freqn=freq(n)                                ! n-th natural frequency
      ffirst=freqn-domega                            ! Left-bound frequency
      if(ffirst.le.0.0d0) ffirst=0.0d0
      flast=freqn+domega                            ! Right-bound frequency
      if(flast.gt.fnyquist) flast=fnyquist
      k=(ffirst+one*deltaf)/deltaf                      ! Frequency pointer
      fkm1=deltaf*dfloat(k-1)

      do 100 j=jstart,jend

      if(j.eq.1) then
        write (0,'(1h+,15x,
     *        ''acceleration-type mode'',i4,3x,''('',i1,'')'')') n,mm
      elseif(j.eq.2) then
        write (0,'(1h+,15x,
     *        ''displacement-type mode'',i4,3x,''('',i1,'')'')') n,mm
      else
        write (0,'(1h+,17x,
     *          ''structural-type mode'',i4,3x,''('',i1,'')'')') n,mm
      endif

      read (10) (sample(i),i=1,ntm1)                 ! Read in time sequence
      sample(nt)=0.0d0
      sample(ntp1)=0.0d0
                                              ! Calculating Fourier spectra
      call rfanaly (sample,a(n5),ntm1,-1)                     ! by direct FFT
```

```
                    ! Eliminating secular terms from frequency sequence
      fk=fkm1
      i=k+k-3
   30 fk=fk+deltaf
      if(fk.gt.flast) goto 40
      i=i+2
      coeff=0.5d0-0.5d0*dcos(pidr*(freqn-fk))
      sample(i) =sample(i) *coeff
      sample(i+1)=sample(i+1)*coeff
      goto 30
                                     ! Recovering (weighted) time sequence
   40 call rfsynth (sample,a(n5),ntm1,+1)              ! by inverse FFT

                                                   ! and saving results
      if(j.eq.1) then
         do i=1,ntm1
           accel(n,i)=sample(i)/ntm1
         enddo
         accel(n,nt)=accel(n,ntm1)
      else
         do i=1,ntm1
           displ(n,i)=sample(i)/ntm1
         enddo
         displ(n,nt)=displ(n,ntm1)
      endif

  100 continue

      if(iflag0) then
        if(istyp14) then
          write (9) accel
          write (9) displ
        else
          write (9) displ
        endif
      else
        write (9) displ
      endif

  200 continue

 2001 format(///
     * ' E R R O R ! ',
     * ' NUMBER OF TIME INTERVALS (',i5,') IS LESS THAN 8',
     * ' OR NOT DIVISIBLE BY 4. SECULAR SOLUTION TERMINATED'///)

      end
```

The subroutine `error` is used to check whether the rest of the operating memory is sufficient to continue the solution process. If the value of the last address (`n5` in this case) required in the analysis is larger than the maximum available size `mtot` reserved for the 'mother' array `a(.)` in the main procedure, then the solution is terminated.

The routine **rfanaly** assembles the real-valued time sequence into a complex-valued sequence of half the sampling length and performs the (fast) direct Fourier transform of this complex-valued sequence. Once the secular terms are eliminated from the resulting frequency sequence, the routine **rfsynth** carries out the (fast) inverse Fourier transform of the weighted complex-valued sequence and recovers the real-valued time sequence. It is noted that the pre- and post-processing of data for the Fourier analysis and synthesis procedures, i.e. conversion from a real-valued sequence to a complex-valued sequence and vice versa, have purely numerical significance.

8.2.7 Subroutines resen1 and resen2

In SFEDYN, the analysis type **ndyn=4** specified on the problem control information line, cf. Section 8.1, monitors the time instant sensitivity solution for deterministic structural systems.

The finite element formulation and computer implementation for the deterministic time instant sensitivity problem have been discussed in Sections 2.8.2.2 and 2.8.2.4. Since (i) the adjoint system is excited by the Dirac-type load and (ii) in SFEDYN the uncoupled (modal) equations of the initial–terminal problem can be integrated alternatively by the Wilson θ- or Newmark technique, it is worth pointing out that the unit impulse can be simulated as the discrete time sequences shown in Fig. 8.1.

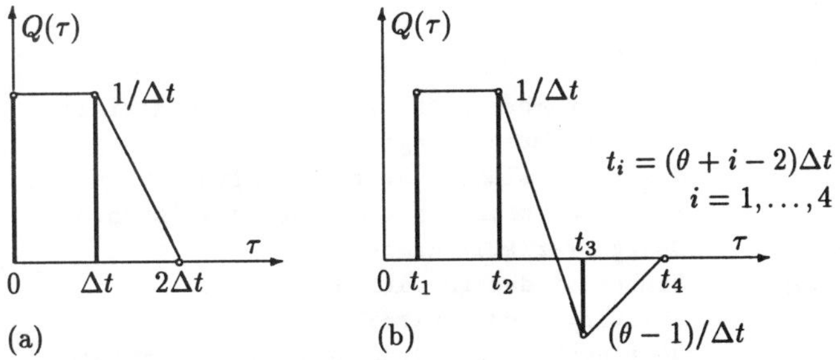

Figure 8.1 Simulation of the unit impulse in Newmark (a) and Wilson (b) schemes.

To elaborate, let us assume that the structural response functional $\mathcal{G}(\tau)$, $\tau \in [0, T]$, is defined implicitly with respect to the design variable h^d and the time-dependent load vector is a function independent of h^d. As an example, the functional used to control the displacement amplitude $q_\alpha(\tau)$ at any point in a structural system subjected to dynamic loads may be defined as

$$\mathcal{G}[q_\alpha(h^d;t)] = \frac{1}{q^2_{(A)}} \int_0^t q^2_\alpha(h^d;\tau)\,\delta(t-\tau)\,\mathrm{d}\tau - 1 \leq 0 \qquad t \in [0,T] \quad (8.4)$$

$q_{(A)}$ being a prescribed admissible value. The expression for the time-dependent sensitivity gradient for this class of functionals with respect to the design variables h^d at any time instant $\tau = t$ can be written as, cf. Eq. (2.151)

$$\mathcal{G}^{\cdot d}(t) \;=\; -\int_0^t \lambda_\alpha(\tau)\, \mathcal{D}^d_\alpha(\tau)\, \mathrm{d}\tau \qquad\qquad t \in [0,T] \tag{8.5}$$

where the solution to the terminal problem $\lambda_\alpha(\tau)$ is a displacement-type vector variable for the adjoint system, while the first-order nodal force vectors $\mathcal{D}^d_\alpha(\tau)$ are expressed as

$$\mathcal{D}^d_\alpha(\tau) \;=\; M^{\cdot d}_{\alpha\beta}\, \ddot{q}_\beta(\tau) + C^{\cdot d}_{\alpha\beta}\, \dot{q}_\beta(\tau) + K^{\cdot d}_{\alpha\beta}\, q_\beta(\tau) \qquad\qquad \tau \in [0,t] \tag{8.6}$$

The primary and adjoint modal time responses are evaluated by the procedure used to solve for the zeroth-order response of stochastic systems. The flow of data in this procedure is similar to that presented in Section 8.2.5 for the subroutine `respo1`.

The subroutine `resen1` accomplishes the last step in the algorithm described in Section 2.8.2.4, i.e. it computes the sensitivity gradient coefficients $\mathcal{G}^{\cdot d}(\tau)$ when the primary responses $\ddot{q}_\alpha(\tau)$, $\dot{q}_\alpha(\tau)$ and $q_\alpha(\tau)$ (and consequently, $\mathcal{D}^d_\alpha(\tau)$) and adjoint displacements $\lambda_\alpha(\tau)$ have already been found and saved on scratch files. The computation process and printout of the final results are restricted to the design variables required. The subroutine `resen2`, which forms the kernel of `resen1`, computes the time variation of $\mathcal{G}^{\cdot d}$ for each design variable at required time instants (a printing time interval apart).

```
      subroutine resen1 (a,dsens,dsmax,timax,neprt,
     *                   neltyp,nt,not,nds,npmax,mtot,n5)

*     To evaluate time instant sensitivity response

*     ON INPUT:
*       icsens      D.o.f. number with displacement constraint
*       mtot        'Mother' array length
*       nd          Number of element degrees of freedom
*       nds         Number of printing time instants (nt/not)
*       nel         Design variable number
*       neltyp      Number of design variable groups
*       not         Printing time interval
*       npsens      Node number with displacement constraint
*       npmax       Max. number of design variables in a group
*       nt          Sampling length (number of time intervals)
*       n5          Working array first address
*       randdt      Sampling time interval
*       lm(.)       Equation numbers of element d.o.f.'s
*       mpar(.)     Number of design variables in groups

*     ON OUTPUT:
*       neprt(.)    Design variable numbers
*       dsens(.,.)  Time response sensitivity amplitudes
*       dsmax(.)    Maximum (absolute) sensitivity amplitudes
*       timax(.)    Max. amplitude time instants

*       a(.)        'Mother' (working) array
```

```
      implicit       real*8 (a-h,o-z)

      common /sstiv/ inul1(3),mpar(30)
      common /idonc/ npsens,icsens,inul2(98)
      common /fftdt/ inul3(3),randdt,inul4(3)
      common /sfems/ nel,nd,lm(12),xnul(1939)
      dimension      a(1),dsens(nds,npmax),dsmax(npmax),timax(npmax),
     *               neprt(npmax)

      write(0,'(1h+,8x,''time instant sensitivity response'',28(1h )/)')

      rewind 2
      rewind 10
      rewind 18
      dtnot=dfloat(not)*randdt                ! Required printing interval

      do 10 nn=1,neltyp                 ! Loop over design variable groups

      nelout=mpar(nn)
      if(nelout.eq.0) goto 10
      write (6,2000) npsens,icsens,nn

      do n=1,nelout              ! Loop over required variables in a group
        read (2) nel,nd,lm
        n6=n5+nd*nt
        n7=n6+nd*nt
                                              ! Time response sensitivity
                                   ! with respect to each design variable
        call resen2 (dsens,a(n5),a(n6),
     *               dtnot,dsemax,timmax,nd,nt,not,nds,npmax,n,nn)
        neprt(n)=nel                                   ! Variable number
        dsmax(n)=dsemax                                 ! Max. amplitude
        timax(n)=timmax                     ! Max. amplitude time instant
      enddo

      i6=(nelout+5)/6                 ! Number of 6-columns printing pages
      do i=1,i6                                      ! Loop over columns
        ns=6*i-5
        ne=ns+5
        if(ne.gt.nelout) ne=nelout
        write (6,2100) (neprt(n),n=ns,ne)
        write (6,'('''')')
        time=0.0d0
        do m=1,nds                    ! Loop over printing time intervals
          time=time+dtnot                                 ! Time instant
          write (6,2101) time,(dsens(m,n),n=ns,ne)
        enddo
        write (6,2102) (dsmax(n),n=ns,ne)
        write (6,2103) (timax(n),n=ns,ne)
      enddo

   10 continue
```

```
 2000 format(///
     *'D E S I G N   S E N S I T I V I T Y   R E S P O N S E ',
     *' (DISPLACEMENT CONSTRAINT AT NODE',i5,',  D.O.F.',i3,')'//
     *                                ' VARIABLE GROUP',i3)
 2100 format (//5x,7hT I M E,2x,6(:4x,11hDESIGN VAR.,i4))
 2101 format (f12.5,2x,1p6d19.9)
 2102 format (/'M A X I M U M   A B S O L U T E   V A L U E S'//
     *                           3x,9hV A L U E,2x,1p6d19.9)
 2103 format (5x,7hT I M E,2x,1p6e19.5)

      end
***********************************************************************
      subroutine resen2 (dsens,xlamb0,dalfd,
     *                  dtnot,dsemax,timmax,nd,nt,not,nds,npmax,n,nn)

*     Kernel of subroutine resen1
*     To compute sensitivity with respect to each design variable

*       n              Design variable number
*       nn             Design variable group number
*       dsemax         Maximum (absolute) sensitivity amplitude
*       dtnot          Required printing interval
*       timmax         Maximum amplitude time instant
*       dalfd(.,.)     First-order nodal forces
*       gaster(.)      Magnitude of adjoint load (weighting numbers)
*       xlamb0(.,.)    Adjoint response to unit impulse excitation

      implicit       real*8 (a-h,o-z)

      common /sfems/ nel,inul,lm(12),xnul(756),gaster(1183)
      dimension      dsens(nds,npmax),xlamb0(nd,nt),dalfd(nd,nt)

      read (18) xlamb0
      read (10) dalfd
      dsemax=0.0d0
      ntout=1
                                    ! Loop over required time instants
      do 10 m=1,nds
      write (0,'(1h+,15x,''interval'',i5,'', variable'',i5,
     *                               '', group'',i4)') m,nel,nn
      dsensn=0.0d0
      ntout=ntout+not
                        ! Loop for sensitivity response at each instant
      do ks=1,ntout
        ke=ntout+1-ks
        tmp1=0.0d0
              ! Loop over element (design variable) degrees of freedom
        do i=1,nd
          if(lm(i).gt.0) tmp1=tmp1+xlamb0(i,ke)*dalfd(i,ks)
        enddo
        dsensn=dsensn+tmp1
      enddo
```

```
                                ! Multiply by magnitude and find maxima
      dsens(m,n)=gaster(m)*dsensn
      absds=dabs(dsens(m,n))
      if(absds.gt.dsemax) then
        dsemax=absds
        timmax=dfloat(m)*dtnot
      endif

   10 continue

      end
```

8.2.8 Subroutines rsenmean and rsentime

The time instant sensitivity analysis of stochastic structural systems by the second-order SFEM method is controlled in SFEDYN by the analysis type `ndyn=6` entered on the problem control information line, cf. Section 8.1. We will focus our attention on structural response functionals defined in the form similar to that employed in the deterministic case, cf. Section 8.2.7, i.e. we consider the class of functionals which do not explicitly depend on the design variables h^d. Furthermore, we assume that:

- all the design variables are described stochastically and all the components of both the random and design variable vectors coincide,
- the structural load vector is defined deterministically and assumed independent of design parameters,
- the element mass, damping and stiffness matrices are explicit in terms of random and design parameters, i.e. the corresponding derivative matrices with respect to random and design variables can be generated explicitly by direct differentiation,
- the first two statistical moments of interest are the spatial expectations and cross-covariances for the sensitivity gradient coefficients at any time instant $\tau = t \in [0, T]$.

Under these assumptions, by following the two-fold superposition technique for the time instant SSDS problem of three-dimensional beam linear elastic systems we may arrive at the expression for the second-order accurate expectations for the time instant sensitivity gradient with respect to d-th random design variable in the form, cf. Eqs. (7.57) and (7.59)

$$E[\mathcal{G}^{,d}(t)] = -\int_0^t \Big\{ \left[\lambda^0_\alpha(\tau) + \tfrac{1}{2} \lambda^{(2)}_\alpha(\tau) \right] \mathcal{D}^d_\alpha(\tau) + \tfrac{1}{2}\, \lambda^0_\alpha(\tau)\, \mathcal{F}^{d(2)}_\alpha(\tau) + \sum_{\check{\rho}=1}^{\check{N}} \lambda^{,\check{\rho}}_\alpha(\tau)\, \mathcal{E}^{d\check{\rho}}_\alpha(\tau)\, S^{\check{\rho}}_{c} \Big\}\, \mathrm{d}\tau \tag{8.7}$$

with τ being a dummy variable of integration; $\check{\rho} = 1, 2, \ldots, \check{N}$ (number of uncorrelated random variables used in the analysis); $\alpha = 1, 2, \ldots, N$ (number of degrees of freedom in the system).

The subroutine `rsenmean` and its kernel part `rsentime` are designed to control the process of computing the sensitivity gradient expectations expressed by Eq. (8.7) at required time instants and for required random design variables. Similarly as in the deterministic case summation over the system degrees of freedom is performed at the element level.

```
      subroutine rsenmean (a,ransen,neprt,ssmax,timax,senprt,
     *                     neltyp,nvar,nt,not,nds,npmax,mtot,n5)

*     To evaluate expectations of stochastic sensitivity time response

*     ON INPUT:
*       icsens          D.o.f. number with displacement constraint
*       mtot            'Mother' array length
*       nd              Number of element degrees of freedom
*       nds             Number of printing time instants (nt/not)
*       nel             Random design variable number
*       neltyp          Number of random design variable groups
*       not             Printing time interval
*       npsens          Node number with displacement constraint
*       npmax           Max. number of design variables in a group
*       nt              Sampling length (number of time intervals)
*       nvar            Number of uncorrelated random variables
*       n5              Working array first address
*       randdt          Sampling time interval
*       lm(.)           Equation numbers of element d.o.f.'s
*       mpar(.)         Number of design variables in groups
*       gaster(.)       Magnitude of adjoint load (weighting numbers)
*       var(.)          Variances of uncorrelated random variables

*     ON OUTPUT:
*       ransen(.,.,.)   Time instant sensitivity expectations
*       senprt(.,.)     Sensitivity array for printing
*       ssmax(.)        Maximum (absolute) sensitivity amplitudes
*       timax(.)        Max. amplitude time instants

*       a(.)            'Mother' (working) array
*       nrev0,nrev1,nrev2,nrev3,nrev4  Direct access record numbers

      implicit        real*8 (a-h,o-z)

      common /sstiv/ inul1(3),mpar(30)
      common /idonc/ npsens,icsens,inul2(98)
      common /fftdt/ inul3(3),randdt,inul4(3)
      common /sfems/ nel,nd,lm(12),vnul(756),gaster(1183)
      common /vartm/ var(48),tnul(14)
      common /nrvar/ nrev0,nrev1,nrev2,nrev3,nrev4
      dimension      a(1),ransen(nds,npmax,neltyp),neprt(npmax),
     *               timax(npmax),ssmax(npmax),senprt(nds,npmax)
```

```
      write (0,'(1h+,8x,
     *      ''update sensitivity expectations'',35(1h )/)')

      dtnot=dfloat(not)*randdt                     ! Printing time interval
      do nn=1,neltyp                                  ! Zero working array
        do n=1,npmax
          do m=1,nds
            ransen(m,n,nn)=0.0d0
          enddo
        enddo
      enddo

*     Do loop over terms involved in expectation expression

      do 20 mm=1,3
                                  ! Find first address of direct access records
      if(mm.eq.1) then     ! ... for first term (lambda0+0.5lambda2)*Dalfd
        nrevl=nrev1
        nrevd=nrev2
        n1var=1
      elseif(mm.eq.2) then          ! ... for second term 0.5lambda0*Fad2
        nrevl=0
        nrevd=nrev4
        n1var=1
      else                 ! ... for third term Sum(lambdarho*Ead*Var(brho))
        nrevl=nrev0
        nrevd=nrev3
        n1var=nvar
      endif

      do 20 nv=1,n1var             ! (nv=nvar for third term, =1 otherwise)
      if(n1var.ne.1) then
        varnv=var(nv)
      else
        varnv=1.0d0
      endif
      ndelta=(nv-1)*nrev0            ! Current addresses of scratch records
      nrevll=nrevl+ndelta
      nrevdd=nrevd+ndelta
      rewind 2

      do 20 nn=1,neltyp                   ! Loop over design variable groups
      nelout=mpar(nn)                   ! Number of design variables in group
      if(nelout.ne.0) then
        do n=1,nelout                   ! Loop over required design variables
          read (2) nel,nd,lm
          write (0,'(1h+,29x,6hdesign,i4,7h, group,i3,
     *          14h, uncorr. var.,i3,6h, term,i3)') nel,nn,nv,mm
          icount=0
          do i=1,nd                  ! Count unconstrained element d.o.f.'s
            if(lm(i).gt.0) icount=icount+1
          enddo
```

```
          ntic=nt*icount
          n6=n5+ntic
          n7=n6+ntic
          if(n7.gt.mtot) call error (n7-mtot)       ! Check total memory
          do i=1,icount        ! Loop over unconstrained element d.o.f.'s
            im1nt=(i-1)*nt
            ns=n5+im1nt
            ne=ns+nt-1
            nrevll=nrevll+1
            read (17,rec=nrevll) (a(j),j=ns,ne)
            ns=n6+im1nt
            ne=ns+nt-1
            nrevdd=nrevdd+1
            read (17,rec=nrevdd) (a(j),j=ns,ne)
          enddo
                                  ! Required time response for each variable
          call rsentime (ransen,a(n5),a(n6),
     *                  varnv,1,neltyp,icount,nt,not,nds,npmax,n,nn)
        enddo
      endif
   20 continue

*     Update and print time sensitivity expectations

      rewind 2

      do 50 nn=1,neltyp                    ! Loop over design variable groups
      nelout=mpar(nn)                ! Number of printing variables in group
      if(nelout.eq.0) goto 50
      write (6,2000) npsens,icsens,nn
                                     ! Multiply by magnitude and find maxima
      do n=1,nelout                    ! Loop over required design variables
        ssemax=0.0d0
        do m=1,nds                          ! Loop over printing time interval
          senprt(m,n)=ransen(m,n,nn)*gaster(m)
          abstmp=dabs(senprt(m,n))
          if(abstmp.gt.ssemax) then
            ssemax=abstmp              ! Maximum amplitude and time instant
            timmax=m
          endif
        enddo
        read (2) nel
        neprt(n)=nel                                       ! Save for printing
        ssmax(n)=ssemax
        timax(n)=timmax*dtnot
      enddo

      i6=(nelout+5)/6           ! Number of 6-columns pages for each group
      do i=1,i6                                     ! Loop over printing pages
        ns=6*i-5
        ne=ns+5
        if(ne.gt.nelout) ne=nelout
        write (6,2100) (neprt(n),n=ns,ne)                  ! Title and indices
```

```
      write (6,'('''')')
      time=0.0d0
      do m=1,nds                  ! Loop over required printing intervals
        time=time+dtnot
                                ! Time instant and sensitivity amplitudes
        write (6,2101) time,(senprt(m,n),n=ns,ne)
      enddo
      write (6,2102) (ssmax(n),n=ns,ne)        ! Max. sens. amplitudes
      write (6,2103) (timax(n),n=ns,ne)      ! Max. ampl. time instant
    enddo
 50 continue

2000 format (///               ' E X P E C T A T I O N S   O F ',
   * ' D E S I G N   S E N S I T I V I T Y   R E S P O N S E'//
   *  ' (DISPLACEMENT CONSTRAINT AT NODE',i5,',  D.O.F.',i3,')'//
   *  ' ELEMENT GROUP',i4)
2100 format (//5x,7hT I M E,2x,6(:4x,11hDESIGN VAR.,i4))
2101 format (f12.5,2x,1p6d19.9)
2102 format (/' M A X I M U M   A B S O L U T E   V A L U E S'//
   *        3x,9hV A L U E,2x,1p6d19.9)
2103 format (5x,7hT I M E,2x,1p6e19.5)

    end
**********************************************************************
    subroutine rsentime (ransen,xlambda,dkmdbq,
   *                     varnv,nv,neltyp,icount,nt,not,nds,npmax,n,nn)

*     Kernel of subroutine rsenmean
*     To compute sensitivity with respect to each design variable
*     (time response for each uncorrelated random variable)

*       icount          Number of unconstrained d.o.f.'s in element
*       n               Random design variable number
*       nn              Random design variable group number
*       nv              nv=1 for expectation evaluation
*       timmax          Maximum amplitude time instant
*       dkmdbq(.,.)     Derivatives times primary response
*       gaster(.)       Magnitude of adjoint load (weighting numbers)
*       ransen(.,.,.)   Sensitivity amplitudes
*       xlambda(.,.)    Adjoint response to unit impulse excitation

      implicit       real*8 (a-h,o-z)

      common /sfems/ inul(2),lm(12),vnul(1939)
      dimension      ransen(nds,npmax,neltyp,1),
     *               xlambda(nt,icount),dkmdbq(nt,icount)

      ntout=1
                                  ! Loop over required printing interval
      do m=1,nds
        write (0,'(1h+,15x,8hinterval,i4,2h, )') m
        sensit=0.0d0
        ntout=ntout+not
```

```
      do ks=1,ntout               ! Loop for sensitivity at each instant
        ke=ntout+1-ks
        do i=1,icount                  ! Loop over unconstrained d.o.f.'s
          sensit=sensit+xlambda(ke,i)*dkmdbq(ks,i)
        enddo
      enddo
      ransen(m,n,nn,nv)=ransen(m,n,nn,nv)+sensit*varnv
    enddo

    end
```

8.2.9 Subroutine rsenc1v

This subroutine is designed to accomplish the last step of the evaluation of the cross-covariances for the time instant sensitivity gradient coefficients; it computes and prints the time variation of sensitivity gradient cross-covariances for required random design variables in each variable group at required printing time intervals. With the assumptions adopted in Section 8.2.8 for structural response functionals and three-dimensional beam systems the first-order accurate cross-covariance for the sensitivity gradient coefficients associated with the d-th and e-th components of the random design variable vector at any time instant $\tau = t \in [0, T]$ can be expressed as, cf. Eqs. (7.58) and (7.59)

$$\mathrm{Cov}\Big(\mathcal{G}^{\cdot d}(t), \mathcal{G}^{\cdot e}(t)\Big) \;=\; \sum_{\check{\rho}=1}^{\check{N}} R^{d\check{\rho}}(t)\, R^{e\check{\rho}}(t)\, S_{\mathrm{c}}^{\check{\rho}} \tag{8.8}$$

where

$$R^{d\check{\rho}}(t) \;=\; \int_0^t \Big[\lambda_\alpha^{,\check{\rho}}(\tau)\mathcal{D}_\alpha^d(\tau) + \lambda_\alpha^0(\tau)\,\mathcal{E}_\alpha^{d\check{\rho}}(\tau)\Big]\,\mathrm{d}\tau \tag{8.9}$$

Since the adjoint displacement variations previously calculated are the response of a linear system excited by the unit impulse, the response of the system subjected to the adjoint load can be obtained by using the superposition principle, i.e. by multiplying the unit impulse response by the weighting numbers of the response functional. Similarly as in the sensitivity expectation case, cf. Section 8.2.8, this step is carried out in the final phase of the computation process.

```
      subroutine rsenc1v (ransen,covsen,covmax,timmax,neprt,ncprt,dtnot,
     *                    nvar,neltyp,nds,npmax,nelout,ncol,nn)

*     To update and print cross-covariances of sensitivities
*     for each random design variable group

*     ON INPUT:
*       ncol              Number of printing columns (nelout(nelout+1)/2)
*       icsens            D.o.f. number with displacement constraint
*       nds               Number of printing time instants (nt/not)
*       nelout            Number of design variables required in group
*       neltyp            Number of design variable groups
```

```
*       nn                 Random design variable group number
*       npsens             Node number with displacement constraint
*       npmax              Max. number of design variables in a group
*       nvar               Number of uncorrelated random variables
*       dtnot              Printing time interval
*       gaster(.)          Magnitude of adjoint load (weighting numbers)
*       ransen(.,.,.,.)    R_d_rho(t) in Eq. (8.9)
*       var(.)             Variances of uncorrelated random variables

*     ON OUTPUT:
*       covsen(.,.,.)      Time instant sensitivity cross-covariances
*       ncprt(.,.)         Cross-covariance component indices
*       neprt(.,.)         Required design variable numbers
*       covmax(.)          Maximum (absolute) sensitivity amplitudes
*       timmax(.)          Maximum amplitude time instants

      implicit       real*8 (a-h,o-z)

      common /sfems/ xnul(763),gaster(1183)
      common /vartm/ var(48),tnul(14)
      common /idonc/ npsens,icsens,i002(98)
      dimension      ransen(nds,npmax,neltyp,nvar),
     *               covsen(nds,ncol),covmax(ncol),timmax(ncol),
     *               neprt(nelout),ncprt(ncol,2)

      write (6,2000) npsens,icsens,nn                 ! Head line printing
      do n=1,nelout
        read (2) neprt(n)          ! Read in required design var. numbers
      enddo
      k=0
      do n=1,nelout              ! Two loops over required design variables
        do j=n,nelout                   ! (nelout*(nelout+1)/2) components
          k=k+1
          cmax=0.0d0
          do m=1,nds             ! Loop over required printing time instants
            tmp=0.0d0
            do nv=1,nvar           ! Loop over uncorrelated random variables
              tmp=tmp+ransen(m,n,nn,nv)*ransen(m,j,nn,nv)*var(nv)
            enddo
                            ! Multiply 'unit' response by weighting number
            covsen(m,k)=gaster(m)*gaster(m)*tmp
            abscov=dabs(covsen(m,k))                         ! Find maxima
            if(abscov.gt.cmax) then
              cmax=abscov
              time=m
            endif
          enddo
          covmax(k)=cmax                       ! Maximum cross-covariances
          timmax(k)=time*dtnot               ! Max. cross-cov. time instant
          ncprt(k,1)=neprt(n)                    ! First index of component
          ncprt(k,2)=neprt(j)                   ! Second index of component
        enddo
      enddo
```

```
      i6=(ncol+5)/6                               ! Number of 6-columns pages

      do i=1,i6                                   ! Loop over printing columns
        ns=6*i-5
        ne=ns+5
        if(ne.gt.ncol) ne=ncol
        write (6,2011) (ncprt(k,1),ncprt(k,2),k=ns,ne)
        write (6,'('''')')
        time=0.0d0

        do m=1,nds              ! Process print at required time instants
          time=time+dtnot
          write (6,2012) time,(covsen(m,k),k=ns,ne)
        enddo
        write (6,2013) (covmax(k),k=ns,ne)
        write (6,2014) (timmax(k),k=ns,ne)
      enddo

 2000 format (///         ' A U T O C O V A R I A N C E S   O F ',
     * ' D E S I G N   S E N S I T I V I T Y   R E S P O N S E '//
     * ' (DISPLACEMENT CONSTRAINT AT NODE',i5,',  D.O.F.',i3,')'//
     * ' ELEMENT GROUP',i4)
 2011 format (//5x,7hT I M E,2x,6(:7x,4hCov(,i3,1h-,i3,1h)))
 2012 format (f12.5,2x,1p6d19.9)
 2013 format (/' M A X I M U M   A B S O L U T E   V A L U E S'//
     *      3x,9hV A L U E,2x,1p6d19.9)
 2014 format (5x,7hT I M E,2x,1p6e19.5)

      end
```

We note that when the sensitivity gradient coefficients with respect to `nelout` random design variables are required to be computed in each variable group, the number of components (time response columns) required at printing equals `nelout(nelout+1)/2` as the cross-covariance matrix `Cov(nelout,nelout)` computed at each time instant is symmetric.

8.3 Source Files and Overlay Linking

As seen from the above discussion the code SFEDYN has been designed not only to illustrate the SFEM formulation but also to be useful in an educational research and application environment. In order to make the program as efficient as possible we note that: (i) subroutines should be appropriately assembled into source files such that they are independent in terms of specific solution phases and analysis options, and (ii) response files for linking should be created in the overlay mode[1] in an optimal way for each of the analysis options. These aspects will be discussed below.

[1]In the overlay mode a subsequent binary segment is loaded to perform its own task and is unloaded after fulfilling this task.

The subroutines in SFEDYN may best be stored in nine source files; the names and functions of each file are as follows:

`SFEMDYN.for` Depending on the problem to be solved, this file determines its size and sets up the operating and back-up storage, monitors the process of modelling and solution, checks whether there is enough of the high-speed memory at each computation step[1] and counts the time lapse after each computation phase. In a segmented executable file this is the only part which necessarily remains resident; other segments may be overlaid.

`FEMMODEL.for` This file opens scratch files required for the specific analysis, generates nodal point data and updates dependent ('slave') degrees of freedom, forms element stiffness matrices and their derivatives with respect to random and design variables, forms the stress–displacement transformation matrices, computes the bandwidth of the equation system, assembles the global stiffness matrix and inputs additional concentrated masses. The object code of this file may be linked in the overlay mode since it remains inactive at the next stages of the computation process.

`GENEIGEN.for` This file deals with the generalized eigenproblem of the (zeroth-order) system. The smallest eigenpairs required are computed by the subspace iterations; the Sturm sequence check may be employed to verify if any eigenvalue is omitted in the frequency domain computed. The deterministic solution for the eigenvalue sensitivity problem is found. The object code of this file may be overlaid at linking.

`STDEIGEN.for` This file inputs and processes the covariance matrix for correlated random variables. The standard eigenproblem for this matrix is solved for the largest eigenvalues (variances for uncorrelated variables) and associated eigenvectors. The mass and stiffness derivative matrices are transformed into the standard normal space of the uncorrelated coordinates. The object code of this file is of the overlay type.

`MODESUP1.for` This file reads in the dynamic response control parameters and generates time sequences of forcing functions (concentrated loads and/or ground accelerations). The time sequences are updated to fit into the integration scheme used in the analysis. The zeroth-order response of the primary system (and adjoint response to the unit impulse, if a sensitivity option is invoked) can be evaluated by using alternatively the Wilson θ- or Newmark scheme. The associated object code is of the overlay type.

`MODESUP2.for` This file is the second part for the forced vibration and sensitivity control. Output requests for displacement components (or expectations and cross-covariances) are assembled in six-column printing pages. In

---

[1] If the required operating storage is exceeded the program is terminated with an error message. The user should increase the size of the 'mother' array declared in the main module.

the sensitivity analysis the constraint for the response functional is read in; the design variables required for computation of the sensitivity gradients are assembled also in six-column printing pages. The solution process of the forced vibration and sensitivity analysis of stochastic systems is monitored. Combinations of the object code of `MODESUP2.for` with those of `DETSENS.for` or `RANDDISP.for` and `RANDSENS.for` (described below) form overlay units.

`RANDDISP.for` This file forms the time sequences of the first- and second-order generalized loads. Secularity elimination may be used to remove the resonant part from the sequences (of length assumed to be divisible by 4 and not larger than 2^{32}); to do this the fast direct and inverse Fourier transforms of complex-valued sequences are employed. The first- and second-order primary and adjoint responses are determined and the expectations and cross-covariances for the required displacement components are evaluated and printed.

`DETSENS.for` This file evaluates the time instant sensitivity gradient coefficients for deterministic systems by using the adjoint variable technique. The results are output at the required time printing intervals and for the design variable groups required.

`RANDSENS.for` This file solves the time instant sensitivity problem for stochastic systems. Using the adjoint variables and stochastic finite elements we obtain the expectations and cross-covariances for the sensitivity gradient coefficients. Time response values are output at the required time printing intervals and for the selected random design variable groups.

To create executable (running) files corresponding to the analysis options available in SFEDYN on PC-type computers with less than 1MB operating memory the use of overlay linkage is strongly recommended. The assembly of SFEDYN's routines into functionally independent modules as described above plays an important role in linking the object-code files. We shall focus our discussion on creating effective response files for linking each from the six analysis options, assuming that all the source files have been compiled; the DOS-type commands are used in the presentation.

When the analysis type number 1 or 3 is entered for `ndyn` on the problem control data line, cf. Section 8.1, i.e. we look for the smallest structural eigenpairs or eigenvalue sensitivity gradient coefficients for a deterministic system, the running file for both the analysis options `ndyn=1` and `ndyn=3` may be created by the command

```
link /e /se:256 @SFEDY1_3.lnk
```

Without going into details of the DOS-system commands we recall that this statement monitors linking of the object-code files that are specified on the response file with the file specification SFEDY1_3.lnk. To obtain the executable file by name, say SFEDY1_3.exe, the response file SFEDY1_3.lnk can be written in the overlay mode as

```
SFEMDYN+
(FEMMODEL)+
(GENEIGEN)+
(STDEIG_0)+
(MODESU_0)
c:\sfem\exe\SFEDY1_3
```

It is seen from this response file that while running the file SFEMD1_3.exe the only part resident in the operating storage in the whole solution process is the binary segment associated with the object-code module SFEMDYN.obj; the next four object modules are overlaid. The last two object-code files correspond to the so-called blank files; they contain one or more subroutines that do nothing else but: (i) preserve the minimum size of the running file maintaining correctness of the Fortran syntax, and (ii) provide an error message on the screen if the control parameter ndyn is not appropriately entered. The source file MODESU_0.for, for instance, has the following Fortran commands:

```
subroutine mod1sup (a)
real*8 a(1)
STOP 'The analysis option is not implemented in this running file'
end

subroutine mod2sup (a)
real*8 a(1)
end
```

It is apparent that to treat the structural eigenproblem and/or eigenvalue sensitivity analysis for deterministic systems only the first three object-code files are needed in the solution process; other files do not necessarily appear in the executable file in complete form and may be replaced by a blank file which includes the blank kernel(s) corresponding to the calling subroutine(s). Running by mistake this executable file with the input data file specified for the forced vibration analysis (ndyn=2) or time instant sensitivity analysis (ndyn=4), for instance, yields the above error message on the monitor screen.

In view of these observations the following blank source files corresponding to their 'active' files can be created:

STDEIG_0.for (associated with STDEIGEN.for)
MODESU_0.for (associated with MODESUP1.for and MODESUP2.for)
DETSEN_0.for (associated with DETSENS.for)
RANDDI_0.for (associated with RANDDISP.for)
RANDSE_0.for (associated with RANDSENS.for)

as the object-code files of SFEMDYN.for, FEMMODEL.for and GENEIGEN.for appear on all the executable files for the six analysis options. We can now create the response files which may be effectively used for linking all the analysis options available in SFEDYN as follows:

file `SFEDYN_2.lnk`, deterministic analysis of forced vibrations (`ndyn=2`)

```
SFEMDYN+
(FEMMODEL)+
(GENEIGEN)+
(STDEIG_0)+
(MODESUP1)+
(MODESUP2+DETSEN_0+RANDDI_0+RANDSE_0)
c:\sfem\exe\SFEDYN_2
```

file `SFEDYN_4.lnk`, deterministic analysis of time instant sensitivity (`ndyn=4`)

```
SFEMDYN+
(FEMMODEL)+
(GENEIGEN)+
(STDEIG_0)+
(MODESUP1)+
(MODESUP2+DETSENS+RANDDI_0+RANDSE_0)
c:\sfem\exe\SFEDYN_4
```

file `SFEDYN_5.lnk`, stochastic analysis of forced vibrations (`ndyn=5`)

```
SFEMDYN+
(FEMMODEL)+
(GENEIGEN)+
(STDEIGEN)+
(MODESUP1)+
(MODESUP2+DETSEN_0+RANDDISP+RANDSE_0)
c:\sfem\exe\SFEDYN_5
```

file `SFEDYN_6.lnk`, stochastic analysis of time instant sensitivity (`ndyn=6`)

```
SFEMDYN+
(FEMMODEL)+
(GENEIGEN)+
(STDEIGEN)+
(MODESUP1)+
(MODESUP2+DETSEN_0+RANDDISP+RANDSENS)
c:\sfem\exe\SFEDYN_6
```

The assembly of subroutines into appropriate groups and the use of overlay linkage and of blank files enable one to run various versions of SFEDYN on small computers and considerably reduce the computational time. It should be pointed out that in the stochastic analysis of forced vibrations and time instant sensitivity the use of overlay linkage seems to be the only way to run SFEDYN on a computer with 640KB of RAM.

Since the complete presentation of this medium-scale program with 85 modules of total length of about 6500 Fortran lines goes beyond the scope of this book, the discussion of this chapter should be understood as an introduction to the code SFEDYN and a description of the computer implementation of the stochastic finite element method. On the other hand, the reader-programmer may use the above discussion and the set of Fortran routines given in Section 8.2 as the basis for extension of his/her finite element program to make it capable in a computationally efficient way of sophisticated stochastic finite element analysis.

The authors believe that the complete programs SFESTA (cf. Chapter 6) and SFEDYN available with this book provide a powerful solution software to assist the interested reader in performing stochastic structural mechanics computations. While the programs have been tested on numerous test examples, it is likely that some errors still exist within the program modules – any information about these errors (and other issues involved) will help the authors to improve the presentation in the future.

Chapter 9

SFEM in Nonlinear Mechanics

9.1 Finite Element Formulation

In Section 2.6 we developed equations which account for both the geometrical and material nonlinearities in structural systems subject to static or dynamic loads. This book is essentially concerned with the numerical analysis of stochastic problems for linear structural systems. However, the analysis of nonlinear effects has now become a crucial aspect of any attempt to realistically assess performance of structural configurations under complex load histories; accounting for parameter uncertainties by using techniques of stochastic analysis should therefore become in the near future an essential ingredient in any sophisticated computer simulation and design method used for such systems. Therefore we conclude the book with an outline of how the methodology advocated here can be extended towards an analysis of nonlinear structural behaviour.

An additional word of caution seems to be necessary as this stage. As emphasized several times throughout the book, the second-order perturbation technique can safely be used for the class of problems at hand provided the uncertainties are not too large. This limitation may become particularly crucial in the analysis of nonlinear structural problems; a rule of thumb could be that the maximum coefficient of variation of the random variables should not exceed 10% of their mean values, respectively.

In general, the way the spatial randomness in structural sizing, material and load parameters is incorporated into the SFEM formulation for nonlinear structural systems is similar to that discussed for linear systems in Chapter 4. Using the variational formulation we may start with a variational principle describing the response of deterministic nonlinear systems, such as the principle of minimum potential energy, cf. Eq. (2.9), the multi-field principle of the type (2.23), or the Hamilton principle (2.29). As a result, we may arrive at governing equations describing static or dynamic stochastic response using one or more fields represented independently.

As an alternative to the variational method the second-moment approach may be applied directly to equations describing nonlinear discretized structural problems with random parameters. The latter method will be discussed below by considering the incremental FEM equations in the form (2.100) expressed now for a typical time interval $[t, t+\Delta t]$ as

$$M_{\alpha\beta}(b_\rho)\Delta\ddot{q}_\beta(b_\rho) + C_{\alpha\beta}(b_\rho)\,\Delta\dot{q}_\beta(b_\rho) + K^{(\mathrm{T})}_{\alpha\beta}(b_\rho)\Delta q_\beta(b_\rho) = \Delta Q_\alpha(b_\rho)$$
$$\rho = 1,2,\ldots,\tilde{N}\ ;\quad \alpha,\beta = 1,2,\ldots,N \qquad \text{(no sum on } \rho) \qquad (9.1)$$

where $M_{\alpha\beta}$, $C_{\alpha\beta}$ and $K^{(\mathrm{T})}_{\alpha\beta}$ are the system mass, damping and tangent stiffness matrices, respectively; Δq_α and ΔQ_α are the vectors of nodal generalized incremental displacements and loads; b_ρ is the vector of nodal random variables, cf. Eq. (4.1); $\tilde{N}$ is the total number of nodal random variables; and N is the total number of degrees of freedom in the system. We note that Δq_α is an implicit function of b_ρ, $M_{\alpha\beta}$ and (possibly) $C_{\alpha\beta}$ are known, explicit functions[1] of b_ρ, while $K^{(\mathrm{T})}_{\alpha\beta}$ (and possibly $C_{\alpha\beta}$ if it is deformation dependent) are generally implicit in b_ρ but can be determined at each time instant provided we know the full solution at this moment, as will be discussed below. For the sake of brevity of notation, the dependence on time in Eq. (9.1) is not indicated explicitly; it is clear, however, that ΔQ_α and consequently Δq_α, $K^{(\mathrm{T})}_{\alpha\beta}$ and possibly $C_{\alpha\beta}$ are all time-dependent quantities. Also, no specific time instant is suggested for setting up the tangent stiffness matrix $K^{(\mathrm{T})}_{\alpha\beta}$ – any explicit or implicit time integration may be considered at this stage.

An alternative formulation of the spatially discretized equations of motion based on Eq. (2.104) rather than on (9.1) can be employed as well. Apart from some numerical time integration errors which are of no concern here both formulations are fully equivalent. Eq. (9.1) has been adopted in this chapter since it seems more consistent with the incremental methodology advocated in the book.

The build-up of the tangent stiffness matrix and its dependence on b_ρ requires additional explanation. We may recall the incremental constitutive equation given by Eq. (2.92); it is now conveniently rewritten in a very general form typical of rate-independent inelastic materials as

$$\Delta\sigma_{ij} \;=\; C_{ijkl}[x_m; \sigma_{pr}(b_\rho); \alpha_v(b_\rho); b_\rho]\,\Delta\varepsilon_{kl}(x_m; b_\rho) \qquad (9.2)$$

in which the vector α_v, $v = 1,2,\ldots,V$, represents the set of internal variables describing the kinematic and/or isotropic hardening of the material, for which the incremental form of the evolution equation is postulated as

$$\Delta\alpha_v \;=\; A_{vkl}[x_m; \sigma_{ij}(b_\rho); \alpha_v(b_\rho)]\,\Delta\varepsilon_{kl}(x_m; b_\rho) \qquad (9.3)$$

C_{ijkl} and A_{vkl} being known functions of their arguments. The dependence of σ_{ij} and α_v on b_ρ is implicit. We note that in order to limit the proliferation of symbols, plastic strains are formally treated on this level of generality as included in the vector α_v; therefore in what follows no separate expressions for them need be displayed. An analytical treatment of other general forms of the evolution law (such as those typical of rate-dependent plasticity) is quite similar to the derivation based on Eqs. (9.2) and (9.3) which we shall present below.

[1]The issue of numerically integrated matrices is not taken up in this section to better address the very basic aspects of stochastic analysis for nonlinear systems.

The incremental displacement–nodal displacement and strain–nodal displacement relations are presented as

$$\begin{aligned} \Delta u_i &= \varphi_{i\alpha}(x_m; b_\rho)\,\Delta q_\alpha(b_\rho) \\ \Delta \varepsilon_{ij} &= B_{ij\alpha}(x_m; b_\rho)\,\Delta q_\alpha(b_\rho) \end{aligned} \tag{9.4}$$

with the operator matrices $\varphi_{i\alpha}$ and $B_{ij\alpha}$ assumed possibly to depend on b_ρ in an explicit way.

Since the definition of the tangent stiffness matrix $K^{(\mathrm{T})}_{\alpha\beta}$ involves the constitutive moduli C_{ijkl}, stresses σ_{ij} and geometric operator matrices $\varphi_{i\alpha}$ and $B_{ij\alpha}$, cf. Eqs. (2.101) and (2.102), we may symbolically present $K^{(\mathrm{T})}_{\alpha\beta}$ as a known function of its arguments in the form

$$K^{(\mathrm{T})}_{\alpha\beta} = K^{(\mathrm{T})}_{\alpha\beta}[\sigma_{ij}(b_\rho); \alpha_v(b_\rho); b_\rho] \tag{9.5}$$

where the symbols σ_{ij} and α_v now stand collectively for stress and internal state variable components at all integration points of the finite element system. More precisely, Eq. (9.5) may be presented as

$$\begin{aligned} K^{(\mathrm{T})}_{\alpha\beta} = K^{(\mathrm{con})}_{\alpha\beta} + K^{(\sigma)}_{\alpha\beta} &= \int_\Omega C_{ijkl}[\sigma_{mn}(b_\rho); \alpha_v(b_\rho); b_\rho]\, B_{ij\alpha}(b_\rho) B_{kl\beta}(b_\rho)\,\mathrm{d}\Omega \\ &+ \int_\Omega \sigma_{ij}(b_\rho)\,\varphi_{k\alpha,i}(b_\rho)\,\varphi_{k\beta,j}(b_\rho)\,\mathrm{d}\Omega \end{aligned} \tag{9.6}$$

which is more adequate for further discussion than the compact notation used in Eq. (9.1). By employing the second-order perturbation approach we now expand the matrices $M_{\alpha\beta}$, $C_{\alpha\beta}$ and $K^{(\mathrm{T})}_{\alpha\beta}$ about the spatial expectations b^0_ρ of the nodal random variables b_ρ; retaining terms up to the second-order leads to

$$\begin{aligned} M_{\alpha\beta}(b_\varrho) &= M^0_{\alpha\beta}(b^0_\varrho) + \epsilon\, M^{;\rho}_{\alpha\beta}(b^0_\varrho)\,\Delta b_\rho + \frac{1}{2}\epsilon^2\, M^{;\rho\sigma}_{\alpha\beta}(b^0_\varrho)\,\Delta b_\rho \Delta b_\sigma \\ C_{\alpha\beta}(b_\varrho) &= C^0_{\alpha\beta}(b^0_\varrho) + \epsilon\, C^{;\rho}_{\alpha\beta}(b^0_\varrho)\,\Delta b_\rho + \frac{1}{2}\epsilon^2\, C^{;\rho\sigma}_{\alpha\beta}(b^0_\varrho)\,\Delta b_\rho \Delta b_\sigma \\ K^{(\mathrm{T})}_{\alpha\beta}(b_\varrho) &= K^{(\mathrm{T})0}_{\alpha\beta}(b^0_\varrho) + \epsilon\, K^{(\mathrm{T});\rho}_{\alpha\beta}(b^0_\varrho)\,\Delta b_\rho + \frac{1}{2}\epsilon^2\, K^{(\mathrm{T});\rho\sigma}_{\alpha\beta}(b^0_\varrho)\,\Delta b_\rho \Delta b_\sigma \end{aligned} \tag{9.7}$$

where the symbol $(.)^{;\rho}$ stands for the total derivative with respect to b_ρ so that

$$\begin{aligned} K^{(\mathrm{T})0}_{\alpha\beta} &= \int_\Omega C^0_{ijkl} B^0_{ij\alpha} B^0_{kl\beta}\,\mathrm{d}\Omega + \int_\Omega \sigma^0_{ij}\,\varphi^0_{k\alpha,i}\,\varphi^0_{k\beta,j}\,\mathrm{d}\Omega \\ K^{(\mathrm{T});\rho}_{\alpha\beta} &= \int_\Omega \Big(C^{;\rho}_{ijkl} B_{ij\alpha} B_{kl\beta} + C^0_{ijkl} B^{;\rho}_{ij\alpha} B^0_{kl\beta} + C^0_{ijkl} B^0_{ij\alpha} B^{;\rho}_{kl\beta} \Big)\mathrm{d}\Omega \\ &+ \int_\Omega \Big(\sigma^{;\rho}_{ij}\,\varphi^0_{k\alpha,i}\,\varphi^0_{k\beta,j} + \sigma^0_{ij}\,\varphi^{;\rho}_{k\alpha,i}\,\varphi^0_{k\beta,j} + \sigma^0_{ij}\,\varphi^0_{k\alpha,i}\,\varphi^{;\rho}_{k\beta,j} \Big)\mathrm{d}\Omega \\ K^{(\mathrm{T});\rho\sigma}_{\alpha\beta} &= \int_\Omega \Big(C^{;\rho\sigma}_{ijkl} B^0_{ij\alpha} B^0_{kl\beta} + C^{;\rho}_{ijkl} B^{;\sigma}_{ij\alpha} B^0_{kl\beta} + C^{;\rho}_{ijkl} B^0_{ij\alpha} B^{;\sigma}_{kl\beta} \\ &+ C^{;\sigma}_{ijkl} B^{;\rho}_{ij\alpha} B^0_{kl\beta} + C^{;\sigma}_{ijkl} B^0_{ij\alpha} B^{;\rho}_{kl\beta} + C^0_{ijkl} B^{;\rho}_{ij\alpha} B^{;\sigma}_{kl\beta} \\ &+ C^0_{ijkl} B^{;\sigma}_{ij\alpha} B^{;\rho}_{kl\beta} + C^0_{ijkl} B^{;\rho\sigma}_{ij\alpha} B^0_{kl\beta} + C^0_{ijkl} B^0_{ij\alpha} B^{;\rho\sigma}_{kl\beta} \Big)\mathrm{d}\Omega \\ &+ \int_\Omega \Big(\sigma^{;\rho\sigma}_{ij}\,\varphi^0_{k\alpha,i}\,\varphi^0_{k\beta,j} + \sigma^{;\rho}_{ij}\,\varphi^{;\sigma}_{k\alpha,i}\,\varphi^0_{k\beta,j} + \sigma^{;\rho}_{ij}\,\varphi^0_{k\alpha,i}\,\varphi^{;\sigma}_{k\beta,j} \\ &+ \sigma^{;\sigma}_{ij}\,\varphi^{;\rho}_{k\alpha,i}\,\varphi^0_{k\beta,j} + \sigma^{;\sigma}_{ij}\,\varphi^0_{k\alpha,i}\,\varphi^{;\rho}_{k\beta,j} + \sigma^0_{ij}\,\varphi^{;\rho}_{k\alpha,i}\,\varphi^{;\sigma}_{k\beta,j} \\ &+ \sigma^0_{ij}\,\varphi^{;\sigma}_{k\alpha,i}\,\varphi^{;\rho}_{k\beta,j} + \sigma^0_{ij}\,\varphi^{;\rho\sigma}_{k\alpha,i}\,\varphi^0_{k\beta,j} + \sigma^0_{ij}\,\varphi^0_{k\alpha,i}\,\varphi^{;\rho\sigma}_{k\beta,j} \Big)\mathrm{d}\Omega \end{aligned} \tag{9.8}$$

all functions of b_ρ being evaluated at b^0_ρ. The derivatives $C^{;\rho}_{ijkl}$ and $C^{;\rho\sigma}_{ijkl}$ will be given below in a more explicit form.

For the deformation-dependent damping matrix $C_{\alpha\beta}$ a similar interpretation of the expansion $(9.7)_2$ may be necessary.

The vectors ΔQ_α and Δq_α are expanded as

$$\begin{aligned} \Delta Q_\alpha(b_\varrho) &= \Delta Q^0_\alpha(b^0_\varrho) + \epsilon\,\Delta Q^{;\rho}_\alpha(b^0_\varrho)\,\Delta b_\rho + \frac{1}{2}\epsilon^2 \Delta Q^{;\rho\sigma}_\alpha(b^0_\varrho)\,\Delta b_\rho \Delta b_\sigma \\ \Delta q_\alpha(b_\varrho) &= \Delta q^0_\alpha(b^0_\varrho) + \epsilon\,\Delta q^{;\rho}_\alpha(b^0_\varrho)\,\Delta b_\rho + \frac{1}{2}\epsilon^2 \Delta q^{;\rho\sigma}_\alpha(b^0_\varrho)\,\Delta b_\rho \Delta b_\sigma \end{aligned} \tag{9.9}$$

Substituting Eqs. (9.2) and (9.3) into Eq. (9.1), multiplying all the second-order terms in the resulting equation by the $\tilde{N}$-variate probability density function $p_{\tilde{N}}(b_1, b_2, \ldots, b_{\tilde{N}})$ and integrating over the domain of b_ρ, and collecting terms of equal order with respect to the small parameter ϵ, we obtain the zeroth-, first- and second-order equations for the stochastic, nonlinear dynamics problem in the following form, cf. Eqs. (4.17)–(4.20):

- Zeroth-order (ϵ^0 terms, one system of N linear simultaneous ordinary differential incremental equations for $\Delta q^0_\alpha(b^0_\varrho)$, $\alpha = 1, 2, \ldots, N$)

$$M^0_{\alpha\beta}(b^0_\varrho)\Delta\ddot{q}^0_\beta(b^0_\varrho) + C^0_{\alpha\beta}(b^0_\varrho)\Delta\dot{q}^0_\beta(b^0_\varrho) + K^{(\mathrm{T})0}_{\alpha\beta}(b^0_\varrho)\,\Delta q^0_\beta(b^0_\varrho) = \Delta Q^0_\alpha(b^0_\varrho) \tag{9.10}$$

- First-order (ϵ^1 terms, $\tilde{N}$ systems of N linear simultaneous ordinary differential incremental equations for $\Delta q^{;\rho}_\alpha(b^0_\varrho)$, $\rho = 1, 2, \ldots, \tilde{N}$, $\alpha = 1, 2, \ldots, N$)

$$\begin{aligned} &M^0_{\alpha\beta}(b^0_\varrho)\Delta\ddot{q}^{;\rho}_\beta(b^0_\varrho) + C^0_{\alpha\beta}(b^0_\varrho)\Delta\dot{q}^{;\rho}_\beta(b^0_\varrho) + K^{(\mathrm{T})0}_{\alpha\beta}(b^0_\varrho)\Delta q^{;\rho}_\beta(b^0_\varrho) = \Delta Q^{;\rho}_\alpha(b^0_\varrho) \\ &\quad - \left[M^{;\rho}_{\alpha\beta}(b^0_\varrho)\,\Delta\ddot{q}^0_\beta(b^0_\varrho) + C^{;\rho}_{\alpha\beta}(b^0_\varrho)\,\Delta\dot{q}^0_\beta(b^0_\varrho) + K^{(\mathrm{T});\rho}_{\alpha\beta}(b^0_\varrho)\,\Delta q^0_\beta(b^0_\varrho)\right] \end{aligned} \tag{9.11}$$

- Second-order (ϵ^2 terms, one system of N linear simultaneous ordinary differential incremental equations for $\Delta q^{(2)}_\alpha(b^0_\varrho)$, $\alpha = 1, 2, \ldots, N$)

$$\begin{aligned} &M^0_{\alpha\beta}(b^0_\varrho)\Delta\ddot{q}^{(2)}_\beta(b^0_\varrho) + C^0_{\alpha\beta}(b^0_\varrho)\Delta\dot{q}^{(2)}_\beta(b^0_\varrho) + K^{(\mathrm{T})0}_{\alpha\beta}(b^0_\varrho)\Delta q^{(2)}_\beta(b^0_\varrho) = \Big\{\Delta Q^{;\rho\sigma}_\alpha(b^0_\varrho) \\ &\quad - 2\left[M^{;\rho}_{\alpha\beta}(b^0_\varrho)\Delta\ddot{q}^{;\sigma}_\beta(b^0_\varrho) + C^{;\rho}_{\alpha\beta}(b^0_\varrho)\Delta\dot{q}^{;\sigma}_\beta(b^0_\varrho) + K^{(\mathrm{T});\rho}_{\alpha\beta}(b^0_\varrho)\,\Delta q^{;\sigma}_\beta(b^0_\varrho)\right] \\ &\quad - \left[M^{;\rho\sigma}_{\alpha\beta}(b^0_\varrho)\Delta\ddot{q}^0_\beta(b^0_\varrho) + C^{;\rho\sigma}_{\alpha\beta}(b^0_\varrho)\Delta\dot{q}^0_\beta(b^0_\varrho) + K^{(\mathrm{T});\rho\sigma}_{\alpha\beta}(b^0_\varrho)\Delta q^0_\beta(b^0_\varrho)\right]\Big\} S^{\rho\sigma}_{\mathrm{b}} \end{aligned} \tag{9.12}$$

where

$$\Delta q^{(2)}_\alpha(b^0_\varrho) = \Delta q^{;\rho\sigma}_\alpha(b^0_\varrho)\, S^{\rho\sigma}_{\mathrm{b}} \tag{9.13}$$

and $S^{\rho\sigma}_{\mathrm{b}}$ is, as before, the covariance matrix of nodal random parameters; we note that $S^{\rho\sigma}_{\mathrm{b}}$ is assumed time-independent.

Quite similarly as for linear systems, in the static case the SFEM differential equations of motion (9.10)–(9.12) reduce to the equilibrium equations which read:

- Zeroth-order (ϵ^0 terms, one system of N linear simultaneous algebraic incremental equations for $\Delta q^0_\alpha(b^0_\varrho)$, $\alpha = 1, 2, \ldots, N$)

$$K^{(\mathrm{T})0}_{\alpha\beta}(b^0_\varrho)\,\Delta q^0_\beta(b^0_\varrho) = \Delta Q^0_\alpha(b^0_\varrho) \tag{9.14}$$

- First-order (ϵ^1 terms, $\tilde{N}$ systems of N linear simultaneous algebraic incremental equations for $\Delta q_\alpha^{,\rho}(b_\varrho^0)$, $\rho = 1, 2, \ldots, \tilde{N}$, $\alpha = 1, 2, \ldots, N$)

$$K_{\alpha\beta}^{(\mathrm{T})0}(b_\varrho^0)\,\Delta q_\beta^{,\rho}(b_\varrho^0) \;=\; \Delta Q_\alpha^{,\rho}(b_\varrho^0) - K_{\alpha\beta}^{(\mathrm{T});\rho}(b_\varrho^0)\,\Delta q_\beta^0(b_\varrho^0) \tag{9.15}$$

- Second-order (ϵ^2 terms, one system of N linear simultaneous algebraic incremental equations for $\Delta q_\alpha^{(2)}(b_\varrho^0)$, $\alpha = 1, 2, \ldots, N$)

$$\begin{aligned} K_{\alpha\beta}^{(\mathrm{T})0}(b_\varrho^0)\Delta q_\beta^{(2)}(b_\varrho^0) \;=\; & \Big[\Delta Q_\alpha^{,\rho\sigma}(b_\varrho^0) - 2\,K_{\alpha\beta}^{(\mathrm{T});\rho}(b_\varrho^0)\,\Delta q_\beta^{,\sigma}(b_\varrho^0) \\ & - K_{\alpha\beta}^{(\mathrm{T});\rho\sigma}(b_\varrho^0)\,\Delta q_\beta^0(b_\varrho^0)\Big]\, S_{\mathrm{b}}^{\rho\sigma} \end{aligned} \tag{9.16}$$

The terms $K_{\alpha\beta}^{(\mathrm{T});\rho}$ and $K_{\alpha\beta}^{(\mathrm{T});\rho\sigma}$ in Eq. (9.7)$_3$ may be considered known provided the appropriate derivatives of C_{ijkl} and σ_{ij} required in Eqs. (9.8) are known. These derivatives must be computed incrementally at every time step and accumulated along the whole deformation history. To derive explicitly relevant expressions we note that, cf. Eq. (9.4)$_2$

$$\begin{aligned} \Delta\varepsilon_{ij}(x_m) = & \Big[B_{ij\alpha}^0(x_m; b_\varrho^0) + B_{ij\alpha}^{,\rho}(x_m; b_\varrho^0)\Delta b_\rho + \tfrac{1}{2} B_{ij\alpha}^{,\rho\sigma}(x_m; b_\varrho^0)\Delta b_\rho \Delta b_\sigma\Big] \\ & \times \Big[\Delta q_\alpha^0(b_\varrho^0) + \Delta q_\alpha^{,\varsigma}(b_\varrho^0)\,\Delta b_\varsigma + \tfrac{1}{2}\Delta q_\alpha^{,\varsigma\omega}(b_\varrho^0)\,\Delta b_\varsigma \Delta b_\omega\Big] \end{aligned} \tag{9.17}$$

or, retaining terms up to second order

$$\Delta\varepsilon_{ij} \;=\; \Delta\varepsilon_{ij}^0 + \Delta\varepsilon_{ij}^{,\rho}\,\Delta b_\rho + \tfrac{1}{2}\Delta\varepsilon_{ij}^{,\rho\sigma}\,\Delta b_\rho \Delta b_\sigma \tag{9.18}$$

with

$$\begin{aligned} \Delta\varepsilon_{ij}^0 \;&=\; B_{ij\alpha}^0\,\Delta q_\alpha^0 \\ \Delta\varepsilon_{ij}^{,\rho} \;&=\; B_{ij\alpha}^{,\rho}\,\Delta q_\alpha^0 + B_{kl\alpha}^0\,\Delta q_\alpha^{,\rho} \\ \Delta\varepsilon_{ij}^{,\rho\sigma} \;&=\; B_{ij\alpha}^{,\rho\sigma}\,\Delta q_\alpha^0 + B_{ij\alpha}^{,\rho}\,\Delta q_\alpha^{,\sigma} + B_{ij\alpha}^{,\sigma}\,\Delta q_\alpha^{,\rho} + B_{ij\alpha}^0\,\Delta q_\alpha^{,\rho\sigma} \end{aligned} \tag{9.19}$$

Furthermore

$$\begin{aligned} \Delta\sigma_{ij} \;=\; & \Big(C_{ijkl}^0 + C_{ijkl}^{;\rho}\Delta b_\rho + \tfrac{1}{2} C_{ijkl}^{;\rho\sigma}\Delta b_\rho \Delta b_\sigma\Big) \\ & \times \Big(\Delta\varepsilon_{kl}^0(b_\varrho^0) + \Delta\varepsilon_{kl}^{,\varsigma}(b_\varrho^0)\,\Delta b_\varsigma + \tfrac{1}{2}\Delta\varepsilon_{kl}^{,\varsigma\omega}(b_\varrho^0)\,\Delta b_\varsigma \Delta b_\omega\Big) \end{aligned} \tag{9.20}$$

with

$$\begin{aligned} C_{ijkl}^{0} \;=\; & C_{ijkl}^0[x_m; \sigma_{mn}(b_\varrho^0); \alpha_v(b_\varrho^0); b_\varrho^0] \\ C_{ijkl}^{;\rho} \;=\; & \frac{\partial C_{ijkl}}{\partial \sigma_{mn}}\,\sigma_{mn}^{,\rho} + \frac{\partial C_{ijkl}}{\partial \alpha_v}\,\alpha_v^{,\rho} + \frac{\partial C_{ijkl}}{\partial b_\rho} \\ C_{ijkl}^{;\rho\sigma} \;=\; & \left(\frac{\partial^2 C_{ijkl}}{\partial\sigma_{mn}\partial\sigma_{pr}}\sigma_{pr}^{,\sigma} + \frac{\partial^2 C_{ijkl}}{\partial\sigma_{mn}\partial\alpha_v}\alpha_v^{,\sigma} + \frac{\partial^2 C_{ijkl}}{\partial\sigma_{mn}\partial b_\sigma}\right)\sigma_{mn}^{,\rho} + \frac{\partial C_{ijkl}}{\partial\sigma_{mn}}\sigma_{mn}^{,\rho\sigma} \\ & + \left(\frac{\partial^2 C_{ijkl}}{\partial\alpha_v\partial\sigma_{mn}}\sigma_{mn}^{,\sigma} + \frac{\partial^2 C_{ijkl}}{\partial\alpha_v\partial\alpha_s}\alpha_s^{,\sigma} + \frac{\partial^2 C_{ijkl}}{\partial\alpha_v\partial b_\sigma}\right)\alpha_v^{,\rho} + \frac{\partial C_{ijkl}}{\partial\alpha_v}\alpha_v^{,\rho\sigma} \\ & + \frac{\partial^2 C_{ijkl}}{\partial b_\rho\partial\sigma_{mn}}\,\sigma_{mn}^{,\sigma} + \frac{\partial^2 C_{ijkl}}{\partial b_\rho\partial\alpha_v}\,\alpha_v^{,\sigma} + \frac{\partial^2 C_{ijkl}}{\partial b_\rho\partial b_\sigma} \end{aligned} \tag{9.21}$$

all functions of b_ρ being evaluated at b_ϱ^0. Upon defining:

$$
\begin{aligned}
\Delta\sigma^0_{ij} &= C^0_{ijkl}\,\Delta\varepsilon^0_{kl} \\
\Delta\sigma^{;\rho}_{ij} &= C^{;\rho}_{ijkl}\,\Delta\varepsilon^0_{kl} + C^0_{ijkl}\,\Delta\varepsilon^{;\rho}_{kl} \\
\Delta\sigma^{;\rho\sigma}_{ij} &= C^{;\rho\sigma}_{ijkl}\,\Delta\varepsilon^0_{kl} + C^{;\rho}_{ijkl}\,\Delta\varepsilon^{;\sigma}_{kl} + C^{;\sigma}_{ijkl}\,\Delta\varepsilon^{;\rho}_{kl} + C^0_{ijkl}\,\Delta\varepsilon^{;\rho\sigma}_{kl}
\end{aligned}
\tag{9.22}
$$

we then have expansion for the stress increment in the form

$$
\Delta\sigma_{ij} = \Delta\sigma^0_{ij} + \Delta\sigma^{;\rho}_{ij}\,\Delta b_\rho + \frac{1}{2}\Delta\sigma^{;\rho\sigma}_{ij}\,\Delta b_\rho \Delta b_\sigma \tag{9.23}
$$

Quite similarly we may obtain the explicit form of the state variable evolution law (9.3) as

$$
\Delta\alpha_v = \Delta\alpha^0_v + \Delta\alpha^{;\rho}_v\,\Delta b_\rho + \frac{1}{2}\Delta\alpha^{;\rho\sigma}_v\,\Delta b_\rho \Delta b_\sigma \tag{9.24}
$$

in which

$$
\begin{aligned}
\Delta\alpha^0_v &= A^0_{vkl}\,\Delta\varepsilon^0_{kl} \\
\Delta\alpha^{;\rho}_v &= A^{;\rho}_{vkl}\,\Delta\varepsilon^0_{kl} + A^0_{vkl}\,\Delta\varepsilon^{;\rho}_{kl} \\
\Delta\alpha^{;\rho\sigma}_v &= A^{;\rho\sigma}_{vkl}\,\Delta\varepsilon^0_{kl} + A^{;\rho}_{vkl}\,\Delta\varepsilon^{;\sigma}_{kl} + A^{;\sigma}_{vkl}\,\Delta\varepsilon^{;\rho}_{kl} + A^0_{vkl}\,\Delta\varepsilon^{;\rho\sigma}_{kl}
\end{aligned}
\tag{9.25}
$$

and

$$
\begin{aligned}
A^0_{vkl} &= A_{vkl}[x_m; \sigma_{ij}(b^0_\varrho); \alpha_v(b^0_\varrho)] \\
A^{;\rho}_{vkl} &= \frac{\partial A_{vkl}}{\partial \sigma_{ij}}\,\sigma^{;\rho}_{ij} + \frac{\partial A_{vkl}}{\partial \alpha_w}\,\alpha^{;\rho}_w \\
A^{;\rho\sigma}_{vkl} &= \left(\frac{\partial^2 A_{vkl}}{\partial\sigma_{ij}\partial\sigma_{mn}}\,\sigma^{;\sigma}_{mn} + \frac{\partial^2 A_{vkl}}{\partial\sigma_{ij}\partial\alpha_w}\,\alpha^{;\sigma}_w\right)\sigma^{;\rho}_{ij} + \frac{\partial A_{vkl}}{\partial\sigma_{ij}}\,\sigma^{;\rho\sigma}_{ij} \\
&\quad + \left(\frac{\partial^2 A_{vkl}}{\partial\alpha_w\partial\sigma_{ij}}\,\sigma^{;\sigma}_{ij} + \frac{\partial^2 A_{vkl}}{\partial\alpha_w\partial\alpha_u}\,\alpha^{;\sigma}_u\right)\alpha^{;\rho}_w + \frac{\partial A_{vkl}}{\partial\alpha_w}\,\alpha^{;\rho\sigma}_w
\end{aligned}
\tag{9.26}
$$

The updating procedure for the different functions involved in the above derivation follows the standard pattern and has the form

$$
\begin{aligned}
q^0_\alpha(t+\Delta t) &= q^0_\alpha(t) + \Delta q^0_\alpha \\
q^{;\rho}_\alpha(t+\Delta t) &= q^{;\rho}_\alpha(t) + \Delta q^{;\rho}_\alpha \\
q^{;\rho\sigma}_\alpha(t+\Delta t) &= q^{;\rho\sigma}_\alpha(t) + \Delta q^{;\rho\sigma}_\alpha \qquad \text{etc.}
\end{aligned}
\tag{9.27}
$$

We should observe that all the general expressions derived in this section are likely to get significantly simplified in problems involving only limited types of nonlinearities, such as large displacement linear elastic problems, for instance.

9.2 Probabilistic Distribution Output

Assume for the given time step $[t, t+\Delta t]$ that we are given the current values of all the functions necessary to set up Eqs. (9.10)–(9.12). In other words we are given the values of $M^0_{\alpha\beta}$, $M^{;\rho}_{\alpha\beta}$, $M^{;\rho\sigma}_{\alpha\beta}S^{\rho\sigma}_{\mathrm{b}}$, $C^0_{\alpha\beta}$, $C^{;\rho}_{\alpha\beta}$, $C^{;\rho\sigma}_{\alpha\beta}S^{\rho\sigma}_{\mathrm{b}}$ and $K^{(\mathrm{T})0}_{\alpha\beta}$, $K^{(\mathrm{T});\rho}_{\alpha\beta}$, $K^{(\mathrm{T});\rho\sigma}_{\alpha\beta}S^{\rho\sigma}_{\mathrm{b}}$; we may observe that similarly as in the linear analysis the second derivatives of $M_{\alpha\beta}$, $C_{\alpha\beta}$ and $K^{(\mathrm{T})}_{\alpha\beta}$ with respect to b_ρ never appear as such – instead, we employ the values of $M^{,\rho\sigma}_{\alpha\beta} S^{\rho\sigma}_{\mathrm{b}}$, $C^{,\rho\sigma}_{\alpha\beta} S^{\rho\sigma}_{\mathrm{b}}$ and $K^{(\mathrm{T});\rho\sigma}_{\alpha\beta} S^{\rho\sigma}_{\mathrm{b}}$ each having substantially fewer entries. Having solved Eqs. (9.10)–(9.12) for Δq^0_α, $\Delta q^{,\rho}_\alpha$ and $\Delta q^{(2)}_\alpha$ and their time derivatives (or Eqs. (9.14)–(9.16) for Δq^0_α, $\Delta q^{,\rho}_\alpha$ and $\Delta q^{(2)}_\alpha$ if a static problem is considered) we may evaluate the first two statistical moments for the nodal displacements (velocities, accelerations) and element strains and stresses. As before, the formal solution can be obtained by setting $\epsilon = 1$ in the perturbation expansion, i.e. for instance

$$\Delta q_\alpha(b_\varrho) = \Delta q^0_\alpha(b^0_\varrho) + \Delta q^{,\rho}_\alpha(b^0_\varrho)\,\Delta b_\rho + \frac{1}{2}\Delta q^{,\rho\sigma}_\alpha(b^0_\varrho)\,\Delta b_\rho \Delta b_\sigma \tag{9.28}$$

with all the expectations determined up to second-order accuracy while cross-covariances (autocovariances) up to first-order accuracy. Various types of second moments can be generated; for the nodal displacements we may talk of $\mathrm{Cov}(\Delta q_\alpha, \Delta q_\beta)$, $\mathrm{Cov}(q_\alpha, \Delta q_\beta)$ or $\mathrm{Cov}(q_\alpha, q_\beta)$, for instance.

Following a procedure similar to that used for linear systems, cf. Section 4.2, and using the random variable linear transformation, cf. Section 1.1.3, the expressions for the expectations of the incremental and total displacements are directly obtained as

$$\begin{aligned} E[\Delta q_\alpha] &= \Delta q^0_\alpha + \frac{1}{2}\Delta q^{(2)}_\alpha \\ E[q_\alpha(t+\Delta t)] &= E[q_\alpha(t)] + E[\Delta q_\alpha] \end{aligned} \tag{9.29}$$

whereas the displacement cross-covariances evaluated at $\xi_1 = (x^{(1)}_m, t_1)$ and $\xi_2 = (x^{(2)}_m, t_2)$ are

$$\mathrm{Cov}\Big(q_\alpha(t_1), q_\beta(t_2)\Big) = q^{,\rho}_\alpha(t_1)\, q^{,\sigma}_\beta(t_2)\, S^{\rho\sigma}_{\mathrm{b}} \tag{9.30}$$

and, similarly

$$\begin{aligned} \mathrm{Cov}\Big(q_\alpha(t), \Delta q_\beta\Big) &= q^{,\rho}_\alpha(t)\, \Delta q^{,\sigma}_\beta\, S^{\rho\sigma}_{\mathrm{b}} \\ \mathrm{Cov}\Big(\Delta q_\alpha, \Delta q_\beta\Big) &= \Delta q^{,\rho}_\alpha\, \Delta q^{,\sigma}_\beta\, S^{\rho\sigma}_{\mathrm{b}} \qquad \text{etc.} \end{aligned} \tag{9.31}$$

Using Eq. (9.17) the expectations for the incremental and total strain components are

$$\begin{aligned} E[\Delta \varepsilon_{ij}] &= \Delta \varepsilon^0_{ij} + \frac{1}{2}\Delta \varepsilon^{(2)}_{ij} \\ E[\varepsilon_{ij}(t+\Delta t)] &= E[\varepsilon_{ij}(t)] + E[\Delta \varepsilon_{ij}] \end{aligned} \tag{9.32}$$

where $\Delta\varepsilon^0_{ij}$ is given in Eq. $(9.19)_1$ and

$$\begin{aligned} \Delta \varepsilon^{(2)}_{ij} &= \Delta \varepsilon^{,\rho\sigma}_{ij}\, S^{\rho\sigma}_{\mathrm{b}} = B^0_{ij\alpha}\, \Delta q^{(2)}_\alpha + B^{(2)}_{ij\alpha}\, \Delta q^0_\alpha + 2\, B^{,\rho}_{ij\alpha}\, \Delta q^{,\sigma}_\alpha\, S^{\rho\sigma}_{\mathrm{b}} \\ B^{(2)}_{ij\alpha} &= B^{,\rho\sigma}_{ij\alpha}\, S^{\rho\sigma}_{\mathrm{b}} \end{aligned} \tag{9.33}$$

cf. Eq. $(9.19)_3$. The strain cross-covariances at any two space–time points $\xi_1 = (x^{(1)}_k, t_1)$ and $\xi_2 = (x^{(2)}_k, t_2)$ can be presented as

$$\mathrm{Cov}\Big(\varepsilon_{ij}(x^{(1)}_m, t_1), \varepsilon_{kl}(x^{(2)}_m, t_2)\Big) = \varepsilon^{,\rho}_{ij}(x^{(1)}_m, t_1)\, \varepsilon^{,\sigma}_{kl}(x^{(2)}_m, t_2)\, S^{\rho\sigma}_{\mathrm{b}} \tag{9.34}$$

and

$$\mathrm{Cov}\Big(\varepsilon_{ij}(x^{(1)}_m, t), \Delta\varepsilon_{kl}(x^{(2)}_m)\Big) = \varepsilon^{,\rho}_{ij}(x^{(1)}_m, t)\, \Delta\varepsilon^{,\sigma}_{kl}(x^{(2)}_m)\, S^{\rho\sigma}_{\mathrm{b}} \tag{9.35}$$

with the fields $\varepsilon_{ij}^{;\rho}(t)$ defined using the accumulation rules of Eq. (9.27) and the definition of the first-order incremental strains given in Eq. $(9.19)_2$.

The evaluation of the first two moments for stress is carried out in a similar way. Using Eq. (9.23) the expectations for the incremental and total stress components are generated as

$$\begin{aligned} E[\Delta\sigma_{ij}] &= \Delta\sigma_{ij}^0 + \frac{1}{2}\Delta\sigma_{ij}^{(2)} \\ E[\sigma_{ij}(t+\Delta t)] &= E[\sigma_{ij}(t)] + E[\Delta\sigma_{ij}] \end{aligned} \tag{9.36}$$

with $\Delta\sigma_{ij}^0$ given in Eq. $(9.22)_1$ while, cf. Eq. $(9.22)_3$

$$\begin{aligned} \Delta\sigma_{ij}^{(2)} &= \Delta\sigma_{ij}^{;\rho\sigma} S_{\mathrm{b}}^{\rho\sigma} = C_{ijkl}^0 \, \Delta\varepsilon_{kl}^{(2)} + C_{ijkl}^{(2)} \, \Delta\varepsilon_{kl}^0 + 2\, C_{ijkl}^{;\rho} \, \Delta\varepsilon_{kl}^{;\sigma} \, S_{\mathrm{b}}^{\rho\sigma} \\ C_{ijkl}^{(2)} &= C_{ijkl}^{;\rho\sigma} \, S_{\mathrm{b}}^{\rho\sigma} \end{aligned} \tag{9.37}$$

The stress cross-covariances at $\xi_1 = (x_k^{(1)}, t_u)$ and $\xi_2 = (x_k^{(2)}, t_v)$ read

$$\mathrm{Cov}\Big(\sigma_{ij}(x_m^{(1)}, t_1), \sigma_{kl}(x_m^{(2)}, t_2)\Big) = \sigma_{ij}^{;\rho}(x_m^{(1)}, t_1)\, \sigma_{kl}^{;\sigma}(x_m^{(2)}, t_2)\, S_{\mathrm{b}}^{\rho\sigma} \tag{9.38}$$

and

$$\mathrm{Cov}\Big(\sigma_{ij}(x_m^{(1)}, t), \Delta\sigma_{kl}(x_m^{(2)})\Big) = \sigma_{ij}^{;\rho}(x_m^{(1)}, t)\, \Delta\sigma_{kl}^{;\sigma}(x_m^{(2)})\, S_{\mathrm{b}}^{\rho\sigma} \tag{9.39}$$

Finding the statistical moments for the internal state variables follows the same pattern.

In order to proceed after using Eqs. (9.10)–(9.12) for Δq_α^0, $\Delta q_\alpha^{;\rho}$ and $\Delta q_\alpha^{(2)}$ the following steps have to be carried out to set up the equations for the next step calculations:

(a) Eqs. $(9.19)_{1,2}$ and (9.33) are used to find $\Delta\varepsilon_{ij}^0$, $\Delta\varepsilon_{ij}^{;\rho}$ and $\Delta\varepsilon_{ij}^{(2)}$, respectively,

(b) Eqs. $(9.22)_{1,2}$ and (9.37) are used to find $\Delta\sigma_{ij}^0$, $\Delta\sigma_{ij}^{;\rho}$ and $\Delta\sigma_{ij}^{(2)}$, respectively,

(c) Eqs. $(9.26)_{1,2}$ are used to find A_{vkl}^0 and $A_{vkl}^{;\rho}$, i.e. by Eqs. $(9.25)_{1,2}$ and $(9.19)_{1,2}$ the values of $\Delta\alpha_v^0$ and $\Delta\alpha_v^{;\rho}$; the value of $\Delta\alpha_v^{(2)}$ is found using the expression $\Delta\alpha_v^{(2)} = \Delta\alpha_v^{;\rho\sigma} S_{\mathrm{b}}^{\rho\sigma}$ and Eqs. $(9.25)_3$, (9.26) and (9.19),

(d) an updating procedure is performed using Eqs. (9.27),

(e) an appropriate stress transformation is carried out to change the reference configuration from the one at time t to the next one at time $t+\Delta t$; this step is necessary for large deformation problems only,

(f) Eqs. (9.21) are used to find updated values of C_{ijkl}^0, $C_{ijkl}^{;\rho}$ and $C_{ijkl}^{(2)} = C_{ijkl}^{;\rho\sigma} S_{\mathrm{b}}^{\rho\sigma}$,

(g) Eqs. (9.6) are used to find update values of $K_{\alpha\beta}^{(\mathrm{T})0}$, $K_{\alpha\beta}^{(\mathrm{T});\rho}$ and $K_{\alpha\beta}^{(\mathrm{T})(2)} = K_{\alpha\beta}^{(\mathrm{T});\rho\sigma} S_{\mathrm{b}}^{\rho\sigma}$.

By examining the above derivations we may conclude that no second derivatives of σ_{ij} and α_{ij} (and, in fact, of any other field implicitly dependent on b_ρ) need be computed at all; what is required instead are only the values of $\sigma_{ij}^{(2)}$ and $\alpha_{ij}^{(2)}$, etc.

All the above considerations apply to static stochastic analysis as well.

9.3 Some Computational Aspects

For an implicit integration scheme the system configuration at $t+\Delta t$ which is required for setting up the incremental 'equilibrium' conditions is unknown prior to the solution. Therefore, the zeroth-order perturbation equations of motion may require iterations in order to improve the solution accuracy within the time step, cf. Eqs. (2.112) and (2.114). In contrast to the zeroth-order equations, the first- and second-order equations are linear.[1] This observation makes it possible to work out an efficient algorithm for the implicit integration of SFEM equations, in which the zeroth-, first- and second-order systems are not treated uniformly at the solution stage as in the case of linear-elastic systems. Instead, the forward–backward substitution process for the first- and second-order equations is carried out once the convergence of the zeroth-order solution has been reached, and not performed simultaneously (with a phase shift) as for linear systems.

To reduce the computational cost a transformation of the full covariance matrix $S_{\mathrm{b}}^{\rho\sigma}$ of the input correlated random variables b_ρ to a diagonal variance matrix $S_{\mathrm{c}}^{\check\rho}$ of the uncorrelated random variables $c_{\check\rho}$, cf. Eqs. (4.149)–(4.152), can be employed. The procedure for the standard normal transformation follows similar lines as presented in Section 4.5.2.2; for the equivalent system described in the space of uncorrelated random variables we thus obtain $\check N + 2$ equations in the following form, cf. Eqs. (4.155)–(4.157):

- Zeroth-order (ϵ^0 terms, one system of N linear simultaneous ordinary differential incremental equations for $\Delta q_\alpha^0(c_{\check\varrho}^0)$, $\alpha = 1, 2, \ldots, N$)

$$M_{\alpha\beta}^0(c_{\check\varrho}^0)\,\Delta\ddot q_\beta^0(c_{\check\varrho}^0) + C_{\alpha\beta}^0(c_{\check\varrho}^0)\,\Delta\dot q_\beta^0(c_{\check\varrho}^0) + K_{\alpha\beta}^{(\mathrm{T})0}(c_{\check\varrho}^0)\,\Delta q_\beta^0(c_{\check\varrho}^0) \;=\; \Delta Q_\alpha^0(c_{\check\varrho}^0) \tag{9.40}$$

- First-order (ϵ^1 terms, $\check N$ systems of N linear simultaneous ordinary differential incremental equations for $\Delta q_\alpha^{,\check\rho}(c_{\check\varrho}^0)$, $\check\rho = 1, 2, \ldots, \check N$, $\alpha = 1, 2, \ldots, N$)

$$M_{\alpha\beta}^0(c_{\check\varrho}^0)\,\Delta\ddot q_\beta^{,\check\rho}(c_{\check\varrho}^0) + C_{\alpha\beta}^0(c_{\check\varrho}^0)\,\Delta\dot q_\beta^{,\check\rho}(c_{\check\varrho}^0) + K_{\alpha\beta}^{(\mathrm{T})0}(c_{\check\varrho}^0)\,\Delta q_\beta^{,\check\rho}(c_{\check\varrho}^0) \;=\; \Delta Q_\alpha^{\check\rho}(c_{\check\varrho}^0) \tag{9.41}$$

- Second-order (ϵ^2 terms, one system of N linear simultaneous ordinary differential incremental equations for $\Delta q_\alpha^{(2)}(c_{\check\varrho}^0)$, $\alpha = 1, 2, \ldots, N$)

$$M_{\alpha\beta}^0(c_{\check\varrho}^0)\,\Delta\ddot q_\beta^{(2)}(c_{\check\varrho}^0) + C_{\alpha\beta}^0(c_{\check\varrho}^0)\,\Delta\dot q_\beta^{(2)}(c_{\check\varrho}^0) + K_{\alpha\beta}^{(\mathrm{T})0}(c_{\check\varrho}^0)\,\Delta q_\beta^{(2)}(c_{\check\varrho}^0) \;=\; \Delta Q_\alpha^{(2)}(c_{\check\varrho}^0) \tag{9.42}$$

where the uncorrelated second-order displacement increments are defined as

$$\Delta q_\alpha^{(2)}(c_{\check\varrho}^0) \;=\; \sum_{\check\rho=1}^{\check N} \Delta q_\alpha^{,\check\rho\check\rho}(c_{\check\varrho}^0)\,S_{\mathrm{c}}^{\check\rho} \tag{9.43}$$

[1]This is so also in the adjoint approach discussed in Section 4.5.5; the equations describing response of the adjoint structure are always linear irrespective of the primary system.

while the transformed first- and second-order incremental load functions are

$$\begin{aligned}
\Delta Q^{\check{\rho}}_{\alpha}(c^{0}_{\check{\varrho}}) &= \Delta Q^{,\check{\rho}}_{\alpha}(c^{0}_{\check{\varrho}}) \\
&\quad - \Big[M^{,\check{\rho}}_{\alpha\beta}(c^{0}_{\check{\varrho}})\Delta\ddot{q}^{0}_{\beta}(c^{0}_{\check{\varrho}}) + C^{,\check{\rho}}_{\alpha\beta}(c^{0}_{\check{\varrho}})\Delta\dot{q}^{0}_{\beta}(c^{0}_{\check{\varrho}}) + K^{(\mathrm{T});\check{\rho}}_{\alpha\beta}(c^{0}_{\check{\varrho}})\Delta q^{0}_{\beta}(c^{0}_{\check{\varrho}})\Big] \\
\Delta Q^{(2)}_{\alpha}(c^{0}_{\check{\varrho}}) &= \sum_{\check{\rho}=1}^{\check{N}}\Big\{\Delta Q^{,\check{\rho}\check{\rho}}_{\alpha}(c^{0}_{\check{\varrho}}) \\
&\quad - 2\Big[M^{,\check{\rho}}_{\alpha\beta}(c^{0}_{\check{\varrho}})\Delta\ddot{q}^{,\check{\rho}}_{\beta}(c^{0}_{\check{\varrho}}) + C^{,\check{\rho}}_{\alpha\beta}(c^{0}_{\check{\varrho}})\Delta\dot{q}^{,\check{\rho}}_{\beta}(c^{0}_{\check{\varrho}}) + K^{(\mathrm{T});\check{\rho}}_{\alpha\beta}(c^{0}_{\check{\varrho}})\Delta q^{,\check{\rho}}_{\beta}(c^{0}_{\check{\varrho}})\Big] \\
&\quad - \Big[M^{,\check{\rho}\check{\rho}}_{\alpha\beta}(c^{0}_{\check{\varrho}})\Delta\ddot{q}^{0}_{\beta}(c^{0}_{\check{\varrho}}) + C^{,\check{\rho}\check{\rho}}_{\alpha\beta}(c^{0}_{\check{\varrho}})\Delta\dot{q}^{0}_{\beta}(c^{0}_{\check{\varrho}}) + K^{(\mathrm{T});\check{\rho}\check{\rho}}_{\alpha\beta}(c^{0}_{\check{\varrho}})\Delta q^{0}_{\beta}(c^{0}_{\check{\varrho}})\Big]\Big\} S^{\check{\rho}}_{\mathrm{c}}
\end{aligned} \tag{9.44}$$

In the static analysis of nonlinear structural systems the SFEM equations expressed in terms of the uncorrelated random variables take the following form, cf. Eqs. (4.172)–(4.174):

- Zeroth-order (ϵ^0 terms, one system of N linear simultaneous algebraic equations for $\Delta q^{0}_{\alpha}(c^{0}_{\check{\varrho}})$, $\alpha = 1, 2, \ldots, N$)

$$K^{(\mathrm{T})0}_{\alpha\beta}(c^{0}_{\check{\varrho}})\,\Delta q^{0}_{\beta}(c^{0}_{\check{\varrho}}) = \Delta Q^{0}_{\alpha}(c^{0}_{\check{\varrho}}) \tag{9.45}$$

- First-order (ϵ^1 terms, $\check{N}$ systems of N linear simultaneous algebraic equations for $\Delta q^{,\check{\rho}}_{\alpha}(c^{0}_{\check{\varrho}})$, $\check{\rho} = 1, 2, \ldots, \check{N}$, $\alpha = 1, 2, \ldots, N$)

$$K^{(\mathrm{T})0}_{\alpha\beta}(c^{0}_{\check{\varrho}})\,\Delta q^{,\check{\rho}}_{\beta}(c^{0}_{\check{\varrho}}) = \Delta Q^{,\check{\rho}}_{\alpha}(c^{0}_{\check{\varrho}}) - K^{(\mathrm{T});\check{\rho}}_{\alpha\beta}(c^{0}_{\check{\varrho}})\,\Delta q^{0}_{\beta}(c^{0}_{\check{\varrho}}) \tag{9.46}$$

- Second-order (ϵ^2 terms, one system of N linear simultaneous algebraic equations for $\Delta q^{(2)}_{\alpha}(c^{0}_{\check{\varrho}})$, $\alpha = 1, 2, \ldots, N$)

$$\begin{aligned}
K^{(\mathrm{T})0}_{\alpha\beta}(c^{0}_{\check{\varrho}})\,\Delta q^{(2)}_{\beta}(c^{0}_{\check{\varrho}}) &= \sum_{\check{\rho}=1}^{\check{N}}\Big[\Delta Q^{,\check{\rho}\check{\rho}}_{\alpha}(c^{0}_{\check{\varrho}}) \\
&\quad - 2\,K^{(\mathrm{T});\check{\rho}}_{\alpha\beta}(c^{0}_{\check{\varrho}})\,\Delta q^{,\check{\rho}}_{\beta}(c^{0}_{\check{\varrho}}) - K^{(\mathrm{T});\check{\rho}\check{\rho}}_{\alpha\beta}(c^{0}_{\check{\varrho}})\,\Delta q^{0}_{\beta}(c^{0}_{\check{\varrho}})\Big]\,S^{\check{\rho}}_{\mathrm{c}}
\end{aligned} \tag{9.47}$$

where

$$\Delta q^{(2)}_{\alpha}(c^{0}_{\check{\varrho}}) = \sum_{\check{\rho}=1}^{\check{N}} \Delta q^{,\check{\rho}\check{\rho}}_{\alpha}(c^{0}_{\check{\varrho}})\,S^{\check{\rho}}_{\mathrm{c}} \tag{9.48}$$

The evaluation of the expectations and first-order accurate cross-covariances for the nodal displacements and element strains and stresses is processed identically as in the case of the correlated random variables. The expectations for the incremental and total displacement components are evaluated by using Eq. (9.29) with $\Delta q^{(2)}_{\alpha}$ given by Eq. (9.43), while their cross-covariances at $\xi_1 = (x^{(1)}_m, t_1)$ and $\xi_2 = (x^{(2)}_m, t_2)$ read, cf. Eq. (9.30)

$$\mathrm{Cov}\Big(q_\alpha(t_1), q_\beta(t_2)\Big) = \sum_{\check{\rho}=1}^{\check{N}} q^{,\check{\rho}}_{\alpha}(t_1)\,q^{,\check{\rho}}_{\beta}(t_2)\,S^{\check{\rho}}_{\mathrm{c}} \tag{9.49}$$

and, similarly, cf. Eq. (9.31)

$$\begin{aligned}\mathrm{Cov}\Big(q_\alpha(t),\Delta q_\beta\Big) &= \sum_{\check{\rho}=1}^{\check{N}} q_\alpha^{,\check{\rho}}(t)\,\Delta q_\beta^{,\check{\rho}}\,S_{\mathrm{c}}^{\check{\rho}}\\ \mathrm{Cov}\Big(\Delta q_\alpha,\Delta q_\beta\Big) &= \sum_{\check{\rho}=1}^{\check{N}} \Delta q_\alpha^{,\check{\rho}}\,\Delta q_\beta^{,\check{\rho}}\,S_{\mathrm{c}}^{\check{\rho}}\end{aligned} \tag{9.50}$$

with

$$q_\alpha^{,\check{\rho}}(t+\Delta t) = q_\alpha^{,\check{\rho}}(t) + \Delta q_\alpha^{,\check{\rho}} \tag{9.51}$$

being the first partial derivatives of the total displacements $q_\alpha(t+\Delta t)$ with respect to the uncorrelated random variables $c_{\check{\rho}}$.

The expressions for the expectations of the incremental and total strains have the form identical to Eq. (9.32) with, cf. Eqs. (9.33)

$$\begin{aligned}\Delta\varepsilon_{ij}^{(2)} &= B_{ij\alpha}^{0}\,\Delta q_\alpha^{(2)} + B_{ij\alpha}^{(2)}\,\Delta q_\alpha^{0} + 2\sum_{\check{\rho}=1}^{\check{N}} B_{ij\alpha}^{,\check{\rho}}\,\Delta q_\alpha^{,\check{\rho}}\,S_{\mathrm{c}}^{\check{\rho}}\\ B_{ij\alpha}^{(2)} &= \sum_{\check{\rho}=1}^{\check{N}} B_{ij\alpha}^{,\check{\rho}\check{\rho}}\,S_{\mathrm{c}}^{\check{\rho}}\end{aligned} \tag{9.52}$$

while the strain cross-covariances at $\xi_1 = (x_k^{(1)}, t_1)$ and $\xi_2 = (x_k^{(2)}, t_2)$ can be written as, cf. Eq. (9.34)

$$\mathrm{Cov}\Big(\varepsilon_{ij}(x_m^{(1)},t_1),\varepsilon_{kl}(x_m^{(2)},t_2)\Big) = \sum_{\check{\rho}=1}^{\check{N}} \varepsilon_{ij}^{,\check{\rho}}(x_m^{(1)},t_1)\,\varepsilon_{kl}^{,\check{\rho}}(x_m^{(2)},t_2)\,S_{\mathrm{c}}^{\check{\rho}} \tag{9.53}$$

and, cf. Eq. (9.35)

$$\mathrm{Cov}\Big(\varepsilon_{ij}(x_m^{(1)},t),\Delta\varepsilon_{kl}(x_m^{(2)})\Big) = \sum_{\check{\rho}=1}^{\check{N}} \varepsilon_{ij}^{,\check{\rho}}(x_m^{(1)},t)\,\Delta\varepsilon_{kl}^{,\check{\rho}}(x_m^{(2)})\,S_{\mathrm{c}}^{\check{\rho}} \tag{9.54}$$

with

$$\begin{aligned}\Delta\varepsilon_{ij}^{,\check{\rho}} &= B_{ij\alpha}^{,\check{\rho}}\,\Delta q_\alpha^{0} + B_{kl\alpha}^{0}\,\Delta q_\alpha^{,\check{\rho}}\\ \varepsilon_{ij}^{,\check{\rho}}(t+\Delta t) &= \varepsilon_{ij}^{,\check{\rho}}(t) + \Delta\varepsilon_{ij}^{,\check{\rho}}\end{aligned} \tag{9.55}$$

The expectations for the incremental and total stress components have form similar to Eq. (9.36) in which, cf. Eq. (9.37)

$$\begin{aligned}\Delta\sigma_{ij}^{(2)} &= C_{ijkl}^{0}\,\Delta\varepsilon_{kl}^{(2)} + C_{ijkl}^{(2)}\,\Delta\varepsilon_{kl}^{0} + 2\sum_{\check{\rho}=1}^{\check{N}} C_{ijkl}^{,\check{\rho}}\,\Delta\varepsilon_{kl}^{,\check{\rho}}\,S_{\mathrm{c}}^{\check{\rho}}\\ C_{ijkl}^{(2)} &= \sum_{\check{\rho}=1}^{\check{N}} C_{ijkl}^{,\check{\rho}\check{\rho}}\,S_{\mathrm{c}}^{\check{\rho}}\end{aligned} \tag{9.56}$$

The expressions for the stress cross-covariances at $\xi_1 = (x_k^{(1)}, t_1)$ and $\xi_2 = (x_k^{(2)}, t_2)$ read, cf. Eq. (9.38)

$$\mathrm{Cov}\Big(\sigma_{ij}(x_m^{(1)}, t_1), \sigma_{kl}(x_m^{(2)}, t_2)\Big) = \sigma_{ij}^{;\bar{\rho}}(x_m^{(1)}, t_1)\, \sigma_{kl}^{;\bar{\rho}}(x_m^{(2)}, t_2)\, S_{\mathrm{c}}^{\bar{\rho}} \tag{9.57}$$

and, cf. Eq. (9.39)

$$\mathrm{Cov}\Big(\sigma_{ij}(x_m^{(1)}, t), \Delta\sigma_{kl}(x_m^{(2)})\Big) = \sigma_{ij}^{;\bar{\rho}}(x_m^{(1)}, t)\, \Delta\sigma_{kl}^{;\bar{\rho}}(x_m^{(2)})\, S_{\mathrm{c}}^{\bar{\rho}} \tag{9.58}$$

with

$$\begin{aligned} \Delta\sigma_{ij}^{;\bar{\rho}} &= C_{ijkl}^{;\bar{\rho}}\, \Delta\varepsilon_{kl}^{0} + C_{ijkl}^{0}\, \Delta\varepsilon_{kl}^{;\bar{\rho}} \\ \sigma_{ij}^{;\bar{\rho}}(t+\Delta t) &= \sigma_{ij}^{;\bar{\rho}}(t) + \Delta\sigma_{ij}^{;\bar{\rho}} \end{aligned} \tag{9.59}$$

The fact that only a few highest uncorrelated modes may be needed to describe many random systems has a tremendous significance when dealing with nonlinear structural systems by using the incremental approach. This is because the computational time needed for updating the system's tangent stiffness matrix, assembling derivative matrices and first- and second-order right-hand-side vectors and evaluating the second statistical moments at each (or some) time step(s) constitute the most costly part of all computations.

It should be pointed out in closing that by employing the second-order perturbation technique the inherent nonlinearity in both structural material and element geometry randomness may be taken into account. The covariance matrix of nonlinear random variables, however, may be changing in time depending on the actual configuration. Thus, the use of a constant-in-time covariance matrix associated with the initial configuration and, in particular, of the uncorrelated random variable transformation implies that the time-dependent effects have to be small to maintain an acceptable accuracy of the perturbation solution. This is similar to the analysis of nonlinear structural systems by the mode superposition approach; the constant-in-time eigenpairs corresponding to the initial configuration may be efficiently used in the analysis of deterministic systems with only small nonlinearities [8,67]. In terms of linear algebra the changes of the base for correlated random variables and for finite element displacement-type variables are quite similar, since the normalized coordinates of these variables are obtained through an eigenproblem. In view of these observations a combination of the second-order perturbation method and two-fold superposition technique can be effectively applied to stochastic finite element systems with only small nonlinearities. Although future work is required to overcome the numerical limitations of the mode superposition procedure applied to nonlinear mechanics as a whole (cf. [88,92], for instance), the two-fold superposition strategy has certainly a great potential as a general tool for the analysis of finite element stochastic nonlinear structural problems.

Appendix A

SFESTA – User's Manual

Below we give a complete description of the input data necessary to run the program SFESTA which can be ordered from the Publisher. Before trying to use the program the user is encouraged to read at least Chapter 6 of this book to become familiar with the capabilities offered by the program and the programming techniques employed in it. Sample input data corresponding to some of the example problems discussed in the book are provided on the diskettes.

The input data have the following format:[1]

A.1 Heading Line (20a4)

columns	*variable*	*entry*
1–80	`title(20)`	Alphanumeric heading information

A.2 Problem Control Line (7i5)

comment	*columns*	*variable*	*entry*
(1)	1–5	`numnp`	Number of nodal points
(2)	6–10	`neltyp`	Number of element groups
(3)	11–15	`ll`	Number of load cases
	16–20	`nsta`	Analysis type 1 – deterministic statics 2 – deterministic static sensitivity 3 – stochastic statics 4 – stochastic static sensitivity
(4)	21–25	`modex`	Operating mode 0 – problem solution 1 – data-check mode
(5)	26–30	`keqb`	Number of equations (degrees of freedom) per block of storage 0 – set by program
(6)	31–35	`nc`	Number of displacement constraints

[1]The example of input data preparation for SAP [11] is closely followed.

Comments

(1) Nodes are numbered from 1 to `numnp`. Data for nodal points are described in Section A.3.

(2) Elements in each group are numbered from 1 to the number of elements in this group. For stochastic problems, random and/or design variables are numbered uniformly in all element groups in the order corresponding to the group numbering. Data for element groups are described in Section A.4.

(3) At least one load condition (`ll`≥`1`) must be specified; `ll` is automatically set equal to one (1) if `nsta=3` or `4`. The program always processes data of Section A.5 (concentrated loads) and Section A.6 (element load multipliers); it thus expects to find some data there.

(4) In the data-check mode (`modex=1`), SFESTA produces the complete data input as generated by the program, skipping calculations required at the solution stage which correspond to `modex=0` (or blank).

(5) By setting `keqb=0` (or blank) SFESTA automatically divides the stiffness matrices and load vectors into submatrices and solves the equation system in the multi-block mode; a nonzero value of `keqb` may also be used to control the number of blocks in stochastic analysis or to test an extended version of SFESTA.

(6) The parameter `nc` denotes the number of displacement constraints used in structural design sensitivity analysis; this parameter is active for `nsta=2` or `4`. SFESTA expects to read the data in Section A.5.

A.3 Nodal Point Data (i5, a1, i4, 2i5, 3d10.0, i5, d10.0)

comment	*columns*	*variable*	*entry*
(1)	2–5	`n`	Node number
	6	`ipr`	Flag for nodal data printing blank – full printing A – suppress printing of generated data B – suppress printing of equation mapping array C – both A and B
(2)	7–10	`ix(n,1)`	X-translation (1-st nodal d.o.f.) code
	11–15	`ix(n,2)`	Y-translation (2-nd nodal d.o.f.) code
	16–20	`ix(n,3)`	Z-translation (3-rd nodal d.o.f.) code
	21–30	`x(n)`	X-ordinate
	31–40	`y(n)`	Y-ordinate
	41–50	`z(n)`	Z-ordinate
(3)	51–55	`k`	Generating increment for node numbers blank if no generation
(4)	56–65	`t(n)`	Nodal temperature

Comments

(1) Nodal data must be defined for all `numnp` nodes. These input data may be read in for each node on a separate line or generated if this option is applicable. Printing flags are typed in upper case letters.

(2) The boundary condition code is defined in the global coordinate system $\{XYZ\}$ for each d.o.f. as

`ix(n,m)=0` – free d.o.f.
`ix(n,m)=1` – fixed d.o.f.

Fixed d.o.f.'s are removed from the final system equations; any load applied at these d.o.f.'s is ignored. Nodal d.o.f.'s associated with undefined stiffness (out-of-plane components in a two-dimensional model, for instance) should be fixed to reduce the size of the system equations.

(3) Nodal point lines need not be input in the node order. Data for a sequence of nodes (n_1, n_1+k_2, n_1+2k_2, ..., n_2-k_2) may be generated by two lines

Line 1	n_1	$ix(n_1,1)$	...	$x(n_1)$	...	k_1	$t(n_1)$
Line 2	n_2	$ix(n_2,1)$	...	$x(n_2)$	...	k_2	$t(n_2)$

The generating parameter k_2 is given on the second line (k_1 may be a generating parameter for a previous nodal point sequence). The node difference n_2-n_1 must be divisible by k_2. Coordinates of the intermediate nodes between n_1 and n_2 are determined by linear interpolation. D.o.f.'s boundary codes `ix(n,m)` for the generated data are set equal to those entered on the first line.

(4) Nodal temperatures describe the actual (physical) temperature distribution in the structure. Average element temperatures resulting from those of the nodal values are used to select material properties and to compute thermal strains in the model.

A.4 3D Truss Element Data

A.4.1 *Control line* (3i5)

columns 6–10	Number of truss elements in the group
11–15	Number of material-geometry property sets
16–20	Flag for sensitivity or stochastic analysis (`isens`) 0 – none of these 1 – one or two of these
21–25	Design and/or random variable type inactive if `isens=0` 1 – cross-sectional area 2 – Young's modulus 3 – mass density 4 – length

A.4.2 *Material-geometry property data* (i5, 5d10.0)

One line for each material-geometry property set.

columns	1–5	Material identification number
	6–15	Young's modulus
	16–25	Coefficient of thermal expansion
	26–35	Mass density
	36–45	Cross-sectional area
	46–55	Weight density (used to calculate gravity loads)

A.4.3 *Element load factors* (4d10.0); Four lines

Line 1 Multipliers of gravity load in the global $+X$ direction.

columns	1–10	Element load case A
	11–20	Element load case B
	21–30	Element load case C
	31–40	Element load case D

Line 2 As above for gravity load in the global $+Y$ direction.

Line 3 As above for gravity load in the global $+Z$ direction.

Line 4 Fraction of the thermal load to be added to each of the element load cases A, B, C and D, respectively.

A.4.4 *Element data* (4i5, d10.0, i5)

One line per element in increasing order starting with one (1).

columns	1–5	Element number
	6–10	Node number i
	11–15	Node number j
	16–20	Material-geometry property number
	21–30	Reference temperature for zero stress
	31–35	Generating parameter **k**

If a sequence of elements exist such that for the element number n_i greater by one than the previous element number (i.e. $n_i=n_{i-1}+1$) the nodal point numbers can be presented as $i_i=i_{i-1}+k$ and $j_i=j_{i-1}+k$, then only the first element in the sequence need be provided. The element property identification number and the temperature for the generated elements are set equal to the values on the first line. If **k** (given on the first line) is input as zero, it is set to 1 by SFESTA. The element data line for the last element must always be given.

The element temperature increment ΔT used to calculate thermal loads is given by $\Delta T = (T_i + T_j - 2T_{(r)})/2$, where the symbols T_i and T_j represent the nodal temperatures specified on the nodal point data lines for nodes `i` and `j` and $T_{(r)}$ is the reference temperature specified on the element line. It is generally more convenient to set $T_i = T_j = 0.0$ so that $\Delta T = -T_{(r)}$ is used as the thermal load generator. Some other types of element loadings can be specified using an equivalent ΔT. For instance, if a truss element has an initial length imperfection by an amount Δl (positive if too long) then the equivalent temperature increment should be taken as $\Delta T = \Delta l/(\alpha l)$. If an initial pre-stress force P (positive if tensile) is applied to the element end that is released after the element is connected to the rest of the structure then $\Delta T = -P/(\alpha AE)$, where A is the cross-sectional area, l is the element length and α is the coefficient of thermal expansion.

A.5 Concentrated Loads. Design Sensitivity Constraints

A.5.1 *Concentrated load data* (2i5, 3d10.0)

comment	*columns*	*variable*	*entry*
(1)	1–5	`n`	Nodal point number
(2)	6–10	`l`	Load case number
	11–20	`fx(n,l)`	Global X-direction force
	21–30	`fy(n,l)`	Global Y-direction force
	31–40	`fz(n,l)`	Global Z-direction force

Comments

(1) One line is required for each nodal point (`n`) at which a nonzero concentrated force is applied. All load cases must be grouped together for node `n` before data are entered for the next node at which loads are also applied. Only the load cases for which node `n` is loaded need be given but the load case numbers (`l`) which are referenced must be supplied in ascending order. Nodal loadings must be defined in increasing node number order, but again only those nodes actually loaded are required on input. The static loads defined in this section act on the structure exactly as input and are not scaled, factored, etc. by the element load case (A, B, C, D) multipliers (Section A.6 below). Nodal forces arising from element loadings are additively combined with any concentrated loads given in this section. Applied force vectors are positive in accordance with the global coordinate directions. The program stops reading load data when a zero (or blank) node number (`n`) is encountered, i.e. this section should be terminated by inputting a blank line.

(2) If `nsta=1` or `2` the load case numbers should range from one (1) to the total number of load cases required on the problem control line, i.e. `l=1,2,...,ll`.

A.5.2 *Displacement constraints for sensitivity analysis* (2i5, 6d10.0)

Skip this line if `nsta`≠2 or `nsta`≠4.

comment	*columns*	*variable*	*entry*
(1)	1–5	`n`	Nodal point number
	6–10	`l`	Sensitivity constraint case number
	11–20	`qx(n,l)`	Allowable X-translation
	21–30	`qy(n,l)`	Allowable Y-translation
	31–40	`qz(n,l)`	Allowable Z-translation

Comments

(1) For the deterministic sensitivity analysis (`nsta=2`) one line is required for each degree of freedom constrained by imposing an admissible value on its displacement. Only the constrained degrees of freedom need to be entered. All cases of the sensitivity constraints must be grouped together for node `n` before data are entered for the next node at which the displacement components are also bounded. Nodes must be typed in increasing order. The allowable displacements or rotations must be positive. SFESTA stops reading constraint data when a blank line is encountered. In the stochastic sensitivity analysis (`nsta=4`) only one displacement constraint is allowed.

A.6 Element Load Multipliers (4d10.0)

comment	*columns*	*variable*	*entry*
(1)	1–10	`em(1)`	Multiplier for element load case A
	11–20	`em(2)`	Multiplier for element load case B
	21–30	`em(3)`	Multiplier for element load case C
	31–40	`em(4)`	Multiplier for element load case D

Comments

(1) One line must be given for each load out of `ll` load cases specified on the problem control line. The lines must refer to the load case numbers in ascending order. The four element load sets A, B, C, D, if given in Section A.4 above, are combined with the concentrated loads specified in Section A.5 for the load cases. For example, if seven static loading conditions are considered (i.e. `ll=7`), then SFESTA expects to read seven lines in this section. Furthermore, if line number three in this section contains the entries

[`em(1)` `em(2)` `em(3)` `em(4)`] = [−3.0 0.0 2.0 0.0]

then the load case No. 3 will be constructed by using the concentrated loads specified in Section A.5 minus the loads from the element set A multiplied by 3.0, plus the loads from the element set C multiplied by 2.0. Load sets B and D will not be applied in this load case. Element load sets may be referred to any number of times in order to construct different structure loading conditions. Element-based loads (gravity, thermal, etc.) can only be applied to the structure by means of the data entries in this section.

> *Input data for deterministic analysis of static response and static sensitivity (*`nsta=1` *or* 2*) are now complete.*

A.7 Covariances of Random Variables

Skip this section if `nsta<3`.

A.7.1 *Control flags for data input and printout* (4i5)

columns 1–5	`intype`	Flag for covariance input 0 – from the common input data file 1 – from an unformatted file
6–10	`iodis`	Printing of expectations for nodal displacements
11–15	`iocov`	Printing of cross-covariances for nodal displacements
16–20	`iostr`	Printing of stresses at mean displacements 0 – no 1 – yes

Comments

There are two options for data input of the covariance matrix of random variables. If `intype=1` SFESTA expects to read the covariance matrix from an unformatted file whose specification (file name) is given by the user at a screen prompt during the input session. Since the matrix is symmetric only the part above and including the diagonal (upper triangle) is stored in compact form as a one-dimensional array `a(.)` of length `nrand×(nrand+1)/2` so that

```
a(1):=cov(1,1),
a(2):=cov(2,2), a(3):=cov(1,2),
a(4):=cov(3,3), ...
...
a(nrand×(nrand+1)/2-nrand+1):=cov(nrand,nrand),
...
a(nrand×(nrand+1)/2):=cov(1,nrand)
```

where `nrand` denotes the number of random variables defined by SFESTA in the finite element model considered. The entries of the vector `a(.)` must be declared in double precision format. For `intype=0` the data format is specified below.

A.7.2 *Covariance matrix data* (i5, 5(i5, d10.0)).

Skip these lines if `intype=1`.

columns	1–5	`n1`	Row number
	6–10	`m1`	Column number
	11–20	`cov(n1,m1)`	Covariance matrix entry at `(n1,m1)` location
	26–30	`m2`	Column number
	31–35	`cov(n1,m2)`	Covariance matrix entry at `(n1,m2)` location
	...	...	...

Comments

Only nonzero elements of the upper triangle including the diagonal of the covariance matrix, i.e. entries of `cov(n,m)` such that `m`≥`n`, need be entered. Input data for row–column indices must be entered in increasing order; the entries skipped are set to zero. If the number of components in the `n`-th row is larger than five, they should be typed on new line(s) in the format of 5(i5,d10.0) starting from columns 6–10 without repeating the row number (columns 1–5 left blank). Each row of the covariance matrix starts with the diagonal element `cov(n,n)` and must begin on a new line. Data input for the covariance matrix is terminated with a blank line.

Input data for stochastic statics and stochastic static sensitivity are now complete.

Appendix B

SFEDYN – User's Manual

Below we give a complete description of the input data necessary to run the program SFEDYN which can be ordered from the Publisher. Before trying to use the program the user is encouraged to read at least Chapter 8 of this book to become familiar with the capabilities offered by the program and the programming techniques employed in it. Sample input data corresponding to some of the example problems discussed in the book are provided on the diskettes.

The input data have the following format:[1]

B.1 Heading Line (20a4)

columns	*variable*	*entry*
1–80	`title(20)`	Alphanumeric heading information

B.2 Problem Control Line (7i5)

comment	*columns*	*variable*	*entry*
(1)	1–5	`numnp`	Number of nodal points
(2)	6–10	`neltyp`	Number of element groups
(3)	16–20	`nf`	Number of required eigenpairs
(3)	21–25	`ndyn`	Analysis type
			1 – deterministic eigenproblem
			2 – deterministic forced vibration
			3 – deterministic eigenvalue sensitivity
			4 – deterministic dynamics sensitivity
			5 – stochastic forced vibration
			6 – stochastic dynamics sensitivity
(4)	26–30	`modex`	Operating mode
			0 – problem solution
			1 – data-check mode

[1]The example of input data preparation for SAP [11] is closely followed.

(5)	31–35	`nad`	Number of vectors used in a subspace iteration solution for eigenproblem 0 – set to `min(2nf,nf+8)`
(6)	36–40	`keqb`	Number of equations (degrees of freedom) per block of storage 0 – set by program

Comments

(1) Nodes are numbered from 1 to `numnp`. Data for nodal points are described in Section B.3.

(2) For each element type a new element group may be defined. Elements in each group are numbered from 1 to the number of elements in this group. Data for element groups are described in Section B.4.

(3) If `ndyn=1` the lowest `nf` eigenpairs are computed and printed. If `ndyn>1` SFEDYN first solves the eigenproblem and then carries out the dynamic response solution; SFEDYN expects to read the control line governing the eigenproblem (Section B.6) before reading data in Sections B.7 and B.8, except for `ndyn=3`. For the analysis options with `ndyn`$\geq$`4` the number of required modes should not be greater than 48.

(4) In the data-check mode (`modex=1`), SFEDYN produces all data input in its completely generated form, skipping calculations required in the solution process (`modex=0` or blank).

(5) Since SFEDYN solves for eigenpairs using the subspace iteration algorithm, `nad` may be entered to set the total number of iteration vectors instead of using the default value (minimum of `2nf` and `nf+8`). A greater `nad` may accelerate convergence in the eigenproblem solution process.

(6) The parameter `keqb` makes the program use the multi-block mode for equation solving and some other operations. It is useful when solving a large-scale problem on a computer with a small operating memory.

B.3 Nodal Point Data (i5, a1, i4, 5i5, 3d10.0, i5)

comment	*columns*	*variable*	*entry*
(1)	2–5	`n`	Node number
	6	`ipr`	Flag for nodal data printing blank – full printing A – suppress printing of generated data B – suppress printing of equation mapping array C – both A and B

(2)	7–10	`ix(n,1)`	X-translation (1-st nodal d.o.f.) code
	11–15	`ix(n,2)`	Y-translation (2-nd nodal d.o.f.) code
	16–20	`ix(n,3)`	Z-translation (3-rd nodal d.o.f.) code
	21–25	`ix(n,4)`	X-rotation (4-th nodal d.o.f.) code
	26–30	`ix(n,5)`	Y-rotation (5-th nodal d.o.f.) code
	31–35	`ix(n,6)`	Z-rotation (6-th nodal d.o.f.) code 0 – free 1 – fixed (no load acting) >1 – master node number
(3)	36–45	`x(n)`	X-ordinate
	46–55	`y(n)`	Y-ordinate
	56–65	`z(n)`	Z-ordinate
(3)	66–70	`kn`	Generating increment for node numbers.

Comments

(1) Nodal data must be defined for all `numnp` nodes. These data may be read in for each node on a separate line, or generated if this option is applicable. Printing flags are typed in upper case letters.

(2) The boundary condition code is defined in the global coordinate system $\{XYZ\}$ for each d.o.f. as

`ix(n,m)=0` – free d.o.f.
`ix(n,m)=1` – fixed d.o.f.
`ix(n,m)=k` – `m`-th d.o.f. at node `n` is slave to `m`-th d.o.f. at master node `k`, $1<\mathtt{k}\leq\mathtt{numnp}$ and $\mathtt{k}\neq\mathtt{n}$

Fixed d.o.f.'s are removed from the final system equations; any load applied at these d.o.f.'s is ignored. Nodal d.o.f.'s associated with undefined stiffness (out-of-plane components in a two-dimensional model, for instance) should be fixed to reduce the scale of the system equations. Rigid links can be modelled by using the master–slave option; `ix(n,m)=k` means that the `m`-th displacement at node `n` is equal to the same `m`-th displacement at node `k`. No actual beam needs to run from `k` to `n`. Slave displacements are not recovered for printing; and zero values appear as output for these d.o.f.'s. The following restrictions hold: (i) $\mathtt{k}\neq 1$, (ii) a node `n` can be slave to only one master node `k` (however, multiple nodes can be slave to the same master), (iii) if the beam running from `n` to `k` is to be a rigid link arbitrarily oriented in the $\{XYZ\}$ space, then all six d.o.f.'s at `n` must be assigned to be slave to `k`.

(3) Nodal point lines need not be input in node order. Data for a sequence of nodes (n_1, n_1+kn_2, n_1+2kn_2, …, n_2-kn_2) may be generated by two lines

Line 1	n_1	$ix(n_1,1)$	…	$x(n_1)$	…	kn_1
Line 2	n_2	$ix(n_2,1)$	…	$x(n_2)$	…	kn_2

The generating parameter kn_2 is given on the second line (kn_1 may be a generating parameter for a previous nodal point sequence). The node difference n_2-n_1 must be divisible by kn_2. Coordinates of the intermediate nodes between n_1 and n_2 are determined by linear interpolation. D.o.f.'s boundary codes `ix(n,m)` for the generated data are set equal to those entered on the first line.

B.4 3D Beam Element Data

The coordinate system $\{xyz\}$ for each element `i–j` is defined with the x-axis directed along the beam axis from node `i` to node `j` and the y- and z-axes as the cross-sectional axes of inertia; the coordinate axes are directed so as to form the system which is right-handed.

B.4.1 *Control line* (5x, 2i5, 5x, i5, 5x, 2i5)

columns	6–10	Number of beam elements in the group
	11–15	Number of geometry sets
	21–25	Number of material sets
	31–35	Flag for sensitivity and/or stochastic analysis (`isens`) 0 – none of these 1 – one or two of these
	21–25	Design and/or random variable type inactive if `isens=0` 1 – cross-sectional area 2 – Young's modulus 3 – mass density 4 – length

B.4.2 *Material set lines* (i5, 5d10.0)

columns	1–5	Material set number
	6–15	Young's modulus
	16–25	Poisson's ratio
	26–35	Mass density
	36–45	Weight density

B.4.3 *Geometry set lines* (i5, 6d10.0)

columns	1–5	Geometry set number
	6–15	Axial area
	16–25	Shear area associated with shear forces along local y-axis
	26–35	Shear area associated with shear forces along local z-axis
	36–45	Torsional inertia (about local x-axis)
	46–55	Flexural inertia about local y-axis
	56–65	Flexural inertia about local z-axis

Comments

One line is required for each set. Shear areas need be specified only if shear deformations are to be included.

B.4.4 *Element lines* (7i5)

columns	1–5	Element number
	6–10	i-node number
	11–15	j-node number.
	16–20	k-node number (used for fixing the local y-axis)
	21–25	Material set number
	26–30	Geometry set number
	31–35	Generating increment `ke` for element data

Comments

Nodal point `k` is used to define the y-axis and, consequently, the z-axis of the local coordinate system; the y-axis lies on the plane formed by the node `k` and the x-axis and it is directed towards `k`. The node `k` does not have to be assigned to any beam element in which case all its six degrees of freedom should be deleted. For a sequence of elements for which $n_i = n_{i-1}+1$ only the element data line for the first element in the sequence need be read in provided that: (i) the nodal point numbers can be specified as $i_i = i_{i-1}+ke$ and $j_i = j_{i-1}+ke$, and (ii) material set number, geometry set number and direction of local y-axis are the same for each element. The default value of `ke` is set to be 1. The element data line for the last element must always be given. Internal forces and moments are printed out for each beam in its local coordinate system $\{xyz\}$. When successive beam elements have the same material and geometry properties and the same orientation, SFEDYN automatically skips recomputation of the element stiffness.

B.5 Lumped Added Mass (i5, 5x, 6d10.0)

comment	*columns*	*variable*	*entry*
(1)	1–5	`n`	Nodal point number
	11–20	`fx(n)`	Global X-translational mass coefficient
	21–30	`fy(n)`	Global Y-translational mass coefficient
	31–40	`fz(n)`	Global Z-translational mass coefficient
	41–50	`mx(n)`	Global X-rotational inertia
	51–60	`my(n)`	Global Y-rotational inertia
	61–70	`mz(n)`	Global Z-rotational inertia

Comments

(1) The mass coefficients of this section are added to the system mass matrix. The program terminates reading added mass data when a blank line is encountered. With no added mass applied, enter a blank line.

B.6 Deterministic Analysis of Eigenpairs and/or Eigenvalue Sensitivity (3i5, 2d10.0, i5)

comment	*columns*	*variable*	*entry*
(1)	1–5	`ifpr`	Flag for additional printing 0 – no; 1 – yes
(2)	6–10	`ifss`	Flag for the Sturm sequence check 0 – yes; 1 – no
(2)	11–15	`nitem`	Maximum number of iterations blank – set to 16
(3)	16–25	`rtol`	Convergence tolerance for the highest `nf`-th eigenvalue blank – set to 10^{-5}
(4)	26–35	`cofq`	Cutoff frequency (cycles per second) blank – `nf` eigenpairs are calculated >0 – only those with eigenvalues below `cofq` are calculated
(1)	36–40	`iprint`	Flag for eigenvector printing 0 – yes; 1 – no

Comments

(1) The values of intermediate matrices, convergence norms, etc. at each iteration may be printed. Standard output includes printing of `nf` eigenvalues and corresponding eigenvectors.

(2) The Sturm sequence check can be applied to verify if any eigenvalue is missing. The factorization of the system matrix is performed at a shift just to the right of the `nf`-th eigenvalue. If during the subspace iteration the `nf`-th eigenvalue fails to converge within the tolerance of `rtol` in `nitem` iterations then the Sturm sequence check is automatically ignored and the values of eigenpairs are accepted as those after the `nitem`-th iteration. If a rigid body mode occurs SFEDYN will stop with an error message.

(3) Lower modes are determined more precisely with accuracy decreasing with increasing mode number until the highest `nf`-th mode is accurate within `rtol`. Iteration is terminated after k-th iteration if the `nf`-th eigenvalue λ satisfies the condition $|\lambda_k - \lambda_{k-1}| / \lambda_{k-1} \leq$ `rtol`.

(4) Computations will terminate if all eigenvalues below `cofq` are found. If only `n` (`n<nf`) eigenvalues smaller than `cofq` are determined, the Sturm sequence check is performed using the `n`-th eigenvalue.

Input data for deterministic analysis of eigenpairs and eigenvalue sensitivity (`ndyn=1` or `3`) are now complete.

B.7 Covariances of Random Variables

Skip this section if `ndyn<5`.

B.7.1 *Control line* (2i5)

comment	*columns*	*variable*	*entry*
(1)	1–5	`intype`	Flag for covariance input 0 – from the common input data file 1 – from an unformatted file
(2)	6–10	`nvar`	Number of required transformed (uncorrelated) random variables

Comments

(1) When `intype=0`, data are entered as shown in Section B.7.2. If `intype=1` SFEDYN reads the covariance matrix from an unformatted (binary) file; the user will be prompted to give the file name during the input session. Since the matrix is symmetric, only the part above and including the diagonal needs to be saved in a one-dimensional array `a(.)` of length `nrand×(nrand+1)/2` according to the following scheme

```
a(1):=cov(1,1),
a(2):=cov(2,2), a(3):=cov(1,2),
a(4):=cov(3,3), ...
...
a(nrand×(nrand+1)/2-nrand+1):=cov(nrand,nrand),
...
a(nrand×(nrand+1)/2):=cov(1,nrand)
```

where `nrand` is the number of random variables determined by SFEDYN. This matrix must be positive definite; its entries are declared in the double precision.

(2) When `ndyn=5` or `6`, SFEDYN performs the transformation of the set of `nrand` correlated random variables onto a set of `nvar` uncorrelated ones through the standard eigenproblem. In analogy with the modal analysis, only a few (*highest!*) modes are required to capture the major characteristics of probabilistic distributions. The value of `nvar=0` is illegal.

B.7.2 *Covariance matrix data* (i5, 5(i5, d10.0))

Skip these lines if `intype=1`.

columns	1–5	`n1`	Row number
	6–10	`m1`	Column number
	11–20	`cov(n1,m1)`	Covariance matrix entry at `(n1,m1)` location
	26–30	`m2`	Column number
	31–35	`cov(n1,m2)`	Covariance matrix entry at `(n1,m2)` location
	...	...	...

Comments

Only nonzero entries of the upper triangle, i.e. entries of `cov(n,m)` such that `m`≥`n` need be entered. Data input is terminated with a blank line. Row and column numbers must be entered in increasing order, but not necessarily in successive order; the entries skipped are set to zero. When the number of entries in the `n`-th row is larger than 5, they should be typed on the new line(s) in the format of 5(i5, d10.0) starting from columns 6–10, without necessarily repeating the row number (columns 1–5 are omitted by SFEDYN). Each new (say, `n`-th) row must be typed on a new line and begun by entering the diagonal entry `cov(n,n)`.

B.8 Deterministic and Stochastic Analysis of Dynamic Response by Mode Superposition

B.8.1 *Two control lines*

Line 1 (5i5, 2d10.0, 5x, 2i5, d10.0)

comment	*columns*	*variable*	*entry*
(1)	1–5	`nfn`	Number of different forcing functions `nfn`≥1
(1)	6–10	`ngm`	Flag for ground motion 0 – no 1 – yes
(2)	11–15	`nat`	Number of different arrival instants for the forcing functions 0 (or blank) – set all instants to zero
(3)	16–20	`nt`	Number of solution time steps
(4)	21–25	`not`	Printing interval ≥`1` and ≤`nt`
(3)	26–35	`dt`	Time step Δt
(5)	36–45	`damp`	Damping factor (fraction of critical damping)
(3)	51–55	`integr`	Flag for integrating decoupled equations 0 – Wilson θ-algorithm 1 – Newmark algorithm
(3)	56–60	`isecular`	Flag for eliminating secularities 0 – no 1 – yes
(3)	61–70	`range`	Frequency range factor for secularity elimination

Line 2. Flags for stochastic sensitivity output (2i5).

Skip this line if `ndyn≠6`.

comment	*columns*	*variable*	*entry*
(4)	1–5	`iodis`	Printout of expectations for nodal displacement response 0 – no 1 – yes
	6–10	`iocov`	Printout of cross-covariances for nodal displacement response 0 – no 1 – yes

Comments

(1) At least one forcing function must be defined. Dynamic response can be evaluated with two function types: (i) time-dependent loads applied at any d.o.f., except for the slave ones, and (ii) ground acceleration in any (or all) of the three global X-, Y- and Z-directions.

(2) If no arrival instants are input, all forcing functions are set to act at the zero instant. The same arrival instant may be referred to by different forcing functions. The parameter `nat` determines the number of nonzero arrival time entries.

(3) Integration of uncoupled equations of motion may be performed alternatively by using the Wilson θ-method with θ=1.4 or by the Newmark method with δ=0.5, α=0.25. If `ndyn=5` or `6` the set of `nrand` correlated random variables is transformed to a set of `nvar` uncorrelated ones by the standard eigensolution. The fast Fourier transform may be used (`isecular=1`) to eliminate secular terms. When `isecular=1` the frequency range factor `range` corresponding to the first natural frequency must be specified. The entry `range` is active only when `isecular=1` and it should be chosen to be between 0.1 and 0.3. The value of `range=0.0` is illegal. For `ndyn=5` or `6` and `isecular=1` the number of time intervals `nt` should be chosen as greater than or equal to 8 and be divisible by 4. Moreover, the fast Fourier transform is the most effective when `nt` is of the form 2^β, β being an integer.

(4) Printout of the nodal displacements, element stresses, sensitivity gradient coefficients or their expectations and cross-covariances is carried out only after time steps controlled by `not`. For the stochastic sensitivity analysis (`ndyn=6`), printout of displacement expectations and cross-covariances is optional.

(5) The damping factor (`damp`) is applied to all `nf` modes. The allowable range for `damp` is between 0.0 (undamped) and 1.0 (critical viscous damping).

B.8.2 *Time-dependent load lines* (4i5, d10.0)

comment	*columns*	*variable*	*entry*
(1)	1–5	`np`	Nodal point number where load component(s) is applied
(1)	6–10	`idof`	d.o.f.'s number $q_x=1$, $q_y=2$, $q_z=3$, $\vartheta_x=4$, $\vartheta_y=5$, $\vartheta_z=6$
(2)	11–15	`ifn`	Forcing function number ≥1 and ≤`nfn`
(3)	16–20	`iat`	Arrival instant number 0 – load applied at the zero instant ≥1 – nonzero arrival instant(s)
(4)	21–30	`p`	Load scale factor 0.0 – no load applied

Comments

(1) Each of the forcing functions is defined by a set of discrete points $[Q(t_i), t_i]$, $i = 1, 2, \ldots, k$, a scalar factor `p` and additional identification parameters `np`, `idof`, `ifn` and `iat`. One line is required for each 'loaded' d.o.f. The parameters' `idof`'s are arranged in ascending d.o.f. order at any given node. This sequence is input in ascending node order and terminated with a blank line, which must also be provided even if no loads are applied.

(2) Input data for the forcing functions are described in Section B.8.5. Function values between the preset input instants are linearly interpolated.

(3) If `iat` is set to zero or the default value, the forcing function is set to act at the zero instant. If 1≤`iat`≤`nat` the forcing function begins acting when the solution reaches the `iat`-th arrival instant (defined in Section B.8.4).

(4) The actual value of nodal forces acting at time t_i equals the magnitude of the `ifn`-th function at t_i times `p`.

B.8.3 *Ground motion control line* (6i5)

Skip this line if `ngm=0`.

comment	*columns*	*variable*	*entry*
(1)	1–5	`nfnx`	Forcing function number specifying ground acceleration in the global X-direction
	6–10	`nfny`	Forcing function number specifying ground acceleration in the global Y-direction
	11–15	`nfnz`	Forcing function number specifying ground acceleration in the global Z-direction
(2)	16–20	`natx`	Arrival instant number, X-direction
	21–25	`naty`	Arrival instant number, Y-direction
	26–30	`natz`	Arrival instant number, Z-direction

Comments

(1) A zero time function number indicates that no ground motion is applied in that particular direction.

(2) A zero (blank) arrival instant means that the ground acceleration begins acting on the structure at the zero instant.

B.8.4 *Arrival instant data*

Line 1 (8d10.0)

comment	*columns*	*variable*	*entry*
(1)	1–10	`at(1)`	Arrival time number 1
	11–20	`at(2)`	Arrival time number 2
	...	...	...
	71–80	`at(8)`	Arrival time number 8

Line(s) 2 (if `nat`>8) (8d10.0)

comment	*columns*	*variable*	*entry*
(1)	1–10	`at(9)`	Arrival time number 9
	...	...	...

Comments

(1) Input as many lines as are required to define `nat` different arrival times, eight entries per line. Skip this section and leave a blank line if `nat=0`.

B.8.5 *Forcing function data*

Line 1 (i5, d10.0, 12a5)

comment	*columns*	*variable*	*entry*
(1)	1–5	`nlp`	Number of function definition points, ≥2
(2)	6–15	`sftr`	Scale factor for the $Q(t)$ values 0.0 or blank – set to 1.0
	16–75	`hed(12)`	Alphanumeric information describing this forcing function

Line(s) 2 (8d10.0)

comment	*columns*	*variable*	*entry*
(3)	1–10	`t(1)`	Time instant value at point 1, t_1
	11–20	`Q(1)`	Function value at t_1, $Q(t_1)$
	21–30	`t(2)`	Time instant value at point 2, t_2
	31–40	`Q(2)`	Function value at t_2, $Q(t_2)$
	...	...	...

Comments

(1) One set of *Line 1* and *Line(s) 2* must be read in for each of the `nfn` forcing functions entered in columns 1–5 of the control line (Section B.8.1). The line sets are input in ascending function number order. At least two points (two pairs of $[Q(t), t]$) must be entered to describe each function.

(2) The factor `sftr` is used to scale the function values only; input arrival instant values are not changed. If the scale factor is omitted, `sftr` is reset by SFEDYN to 1.0.

(3) Input as many *Lines 2* as are required to define `nlp` pairs $[t_i, Q(t_i)]$, four pairs per line. These pairs must be entered in ascending order of time instant values. The highest input instant value (t_{NLP}) should be at least equal to the value of the final time instant in the analysis.

B.8.6 *Output information data*

B.8.6.1 *Displacement output*

Output control line (2i5)

comment	*columns*	*variable*	*entry*
(1)	1–5	`kkk`	Flag for output printing 3 – print maxima only ≠3 – print history and maxima

Comments

(1) The type of output applies to all requests in both the deterministic and stochastic solutions.

Displacement output data (7i5)

comment	*columns*	*variable*	*entry*
(1)	1–5	`np`	Nodal point number
(2)	6–10	`idof(1)`	Displacement component, enter 1
	11–15	`idof(2)`	Displacement component, enter 2
	16–20	`idof(3)`	Displacement component, enter 3
	21–25	`idof(4)`	Displacement component, enter 4
	26–30	`idof(5)`	Displacement component, enter 5
	31–35	`idof(6)`	Displacement component, enter 6

Comments

(1) Only displacement components at nodes at which output is to be produced are entered. Lines must be input in ascending node number order. Node numbers may not be repeated. These lines are terminated with a blank line.

(2) Displacement component requests `idof(n)` range from 1 to 6, with q_x=1, q_y=2, q_z=3, ϑ_x=4, ϑ_y=5, ϑ_z=6. The first zero (or blank) value encountered terminates information for the line; d.o.f.'s components at the node may be required in any order. For instance, if q_y, ϑ_z and ϑ_x at node 1001 are to be printed, the line could be entered as (1001, 2, 6, 4, 0). For the stochastic analysis (`ndyn=5` or `6`), only the upper triangle of the symmetric displacement cross-covariance matrix needs to be computed and printed at given time instants. In other words, SFEDYN calculates and prints N history tables of the displacement expectations and $N(N+1)/2$ history tables of the displacement cross-covariances, with N being the number of required displacement components.

B.8.6.2 *Sensitivity response data* (2i5, d10.0)

Skip this line if `ndyn≠4` and `ndyn≠6`.

comment	*columns*	*variable*	*entry*
(1)	1–5	`npsens`	Nodal point number
	6–10	`idof`	Nodal d.o.f. number ≥1 and ≤6
	11–20	`qallow`	Displacement constraint value

Comments

(1) Only one displacement constraint component is permitted.

B.8.6.3 *Stress and sensitivity output*

Output control line (2i5)

comment	*columns*	*variable*	*entry*
(1)	1–5	`kkk`	Flag for output printing 3 – print maxima only ≠3 – print history and maxima

Comments

(1) See Section B.8.6.1.

Stress and sensitivity data (13i5)

comment	*columns*	*variable*	*entry*
(1)	1–5	`nel`	Element number for output
(2)	6–10	`is(1)`	Stress component number, enter 1
	11–15	`is(1)`	Stress component number, enter 2
	...	...	...
	61–65	`is(12)`	Stress component number, enter 12

Comments

(1) Requests are entered by element groups in the same order as the Element Data Input (Section B.4). Element numbers within each group must be typed in ascending order; they may be omitted but not repeated. Terminate each element output group with a blank line, which is also used for element groups whose output is not to be printed.

For the deterministic sensitivity analysis (**ndyn=4**) the request list is treated as the list of required elements for stress printing as well as required design variables for sensitivity output. If **ndyn=5** stresses are evaluated at mean displacements. In stochastic sensitivity analysis (**ndyn=6**) the stress field is not computed, the request list is treated as that of required random design variables for printing of the time-dependent expectations and cross-covariances for the sensitivity gradients; and the stress component numbers `is(n)` are inactive. Because of symmetry only the upper triangle of the sensitivity cross-covariance matrix is computed and printed. That is, SFEDYN calculates and prints N history tables of sensitivity expectations and $N(N+1)/2$ history tables of sensitivity cross-covariances, where N denotes the total number of required design random variables.

(2) The first zero (or blank) request encountered while reading `is(1)`, ..., `is(12)` terminates information for the line. The table below lists the stress component numbers, symbols and descriptions:

number	*symbol*	*description*
(1)	`P1(i)`	x-force at node i
(2)	`V2(i)`	y-shear at node i
(3)	`V3(i)`	z-shear at node i
(4)	`T1(i)`	x-torque at node i
(5)	`M2(i)`	y-moment at node i
(6)	`M3(i)`	z-moment at node i
(7)	`P1(j)`	x-force at node j
(8)	`V2(j)`	y-shear at node j
(9)	`V3(j)`	z-shear at node j
(10)	`T1(j)`	x-torque at node j
(11)	`M2(j)`	y-moment at node j
(12)	`M3(j)`	z-moment at node j

Input data for deterministic and stochastic analysis of forced vibrations and dynamic sensitivity (`ndyn=2` *or* `4` *or* `5` *or* `6`*) are now complete.*

References

[1] G. Adomian. *Stochastic Systems.* Academic Press, 1983.

[2] G. Arfken. *Mathematical Methods for Physicists.* Academic Press, 1985.

[3] J.H. Argyris and S. Kelsey. *Energy Theorems and Structural Analysis.* Butterworth, 1960.

[4] J.H. Argyris and M. Kleiber. Incremental formulation in nonlinear mechanics and finite strain elasto-plasticity – Natural approach. Part 1. *Comput. Meth. Appl. Mech. Eng.*, 11:125–157, 1977.

[5] J.H. Argyris, M. Kleiber, and J.S. Doltsinis. Incremental formulation in nonlinear mechanics and finite strain elasto-plasticity – Natural approach. Part 2. *Comput. Meth. Appl. Mech. Eng.*, 14:259–292, 1978.

[6] G. Augusti, A. Baratta, and F. Casciati. *Probabilistic Methods in Structural Engineering.* Chapman and Hall, 1984.

[7] K.J. Bathe. *Finite Element Procedures in Engineering Analysis.* Prentice-Hall, 1982.

[8] K.J. Bathe and S. Gracewski. On nonlinear dynamic analysis using substructuring and mode superposition. *Comput. Struct.*, 13:699–707, 1981.

[9] K.J. Bathe, E. Ramm, and E.L. Wilson. Finite element formulations for large deformation dynamic analysis. *Int. J. Num. Meth. Eng.*, 9:353–386, 1975.

[10] K.J. Bathe and E.L. Wilson. *Numerical Methods in Finite Element Analysis.* Prentice-Hall, 1976.

[11] K.J. Bathe, E.L. Wilson, and F.E. Peterson. *SAP IV - A Structural Analysis Program for Static and Dynamic Response of Linear Systems.* Technical Report, College of Engineering Univ. California, June 1973.

[12] T. Belytschko and R. Mullen. Stability of explicit-implicit mesh partitions in time integration. *Int. J. Num. Meth. Eng.*, 12:1575–1586, 1979.

[13] H. Benaroya and M. Rehark. Finite element methods in probabilistic structural analysis: A selective review. *Appl. Mech. Rev.*, 41:201–213, 1988.

[14] J.S. Bendat. *Principles and Applications of Random Noise Theory.* Wiley, 1958.

[15] J.S. Bendat and A.G. Piersol. *Engineering Applications of Correlation and Spectral Analysis.* Wiley-Interscience, 1980.

[16] J.S. Bendat and A.G. Piersol. *Random Data: Analysis and Measurement Procedures.* Wiley-Interscience, 1971.

[17] A.T. Bharrucha-Reid. On random operator equations in Banach space. *Bull. Pol. Acad. Sci. Ser. Sci. Math. Astr. Phys.*, 7:561–564, 1959.

[18] N.M. Bogoliubov. *Asymptotic Methods in the Theory of Non-Linear Oscillators.* Gordon and Breach, 1961.

[19] R.N. Bracewell. *The Fourier Transform and Its Applications.* McGraw-Hill, 1978.

[20] E.O. Brigham. *The Fast Fourier Transform.* Prentice-Hall, 1974.

[21] P.G. Ciarlet. *The Finite Element Method for Elliptic Problems.* North-Holland, 1978.

[22] R.W. Clough and J. Penzien. *Dynamics of Structures.* McGraw-Hill, 1975.

[23] E.A. Coddington and N. Levinson. *Theory of Ordinary Differential Equations.* McGraw-Hill, 1955.

[24] J.D. Collins and W.T. Thomson. The eigenvalue problem for structural systems with statistical properties. *AIAA J.*, 7(4):642–648, 1969.

[25] R.D. Cook. *Concepts and Applications of Finite Element Analysis.* Wiley, 1974.

[26] J.W. Cooley and J.W. Tukey. An algorithm for machine calculation of complex Fourier series. *Math. Comput.*, 19(4):297–301, 1965.

[27] H. Crammer. *The Elements of Probability Theory.* Wiley, 1955.

[28] S.H. Crandall and W.D. Mark. *Random Vibration in Mechanical Systems.* Academic Press, 1963.

[29] G. Dahlquist and A. Björck. *Numerical Methods. Series in Automatic Computation.* Prentice-Hall, 1969 (translated by N. Anderson).

[30] W.B. Davenport and W.L. Root. *An Introduction to the Theory of Random Signals and Noise.* McGraw-Hill, 1958.

[31] H.F. Davis. *Fourier Series and Orthogonal Functions.* Allyn and Bacon, 1963.

[32] P.J. Davis and P. Rabinowitz. *Methods of Numerical Integration. Series in Comput. Sci. Appl. Math.* Academic Press, 1975.

[33] H. Dym and H.P. McKean. *Fourier Series and Integrals.* Academic Press, 1972.

[34] V.N. Faddeeva. *Computational Methods of Linear Algebra.* Dover, 1959.

[35] C.A. Felippa. Solution of linear equations with skyline-stored symmetric matrix. *Comput. Struct.*, 7:13–29, 1975.

[36] G.E. Forsythe and C.B. Moler. *Computer Solution of Linear Algebraic Systems.* Prentice-Hall, Englewood Cliffs, 1967.

[37] R.L. Fox and M.P. Kapoor. Rate of change of eigenvalues and eigenvectors. *AIAA J.*, 6(12):2426–2429, 1968.

[38] R.H. Gallagher. *Finite Element Analysis: Fundamentals.* Wiley, 1975.

[39] R. Ghanem and P. Spanos. Polynomial chaos in stochastic finite elements. *Appl. Mech., ASME*, 57:197–202, 1990.

[40] C. Goffman. *Calculus of Several Variables.* Harper-Row, 1965.

[41] H. Goldstein. *Classical Mechanics.* Addison-Wesley, 1953.

[42] I. Halperin. *Introduction to the Theory of Distribution.* Univ. Toronto Press, 1952.

[43] G.C. Hart. *Uncertainty Analysis, Load, and Safety in Structural Engineering.* Prentice-Hall, 1982.

[44] E.J. Haug, K.K. Choi, and V. Komkov. *Design Sensitivity Analysis of Structural Systems. Series in Math. Sci. Eng.* Academic Press, 1986.

[45] H.D. Hibbitt, P.V. Marcal, and J.R. Rice. A finite element formulation for problems of large strain and large displacement. *Solids Struct.*, 6:1069–1086, 1970.

[46] T.D. Hien. *Deterministic and Stochastic Sensitivity in Computational Structural Mechanics.* Habilitation thesis, *IFTR-Report Pol. Acad. Sci.*, No. 46, 1990.

[47] T.D. Hien and M. Kleiber. Finite element analysis based on stochastic Hamilton variational principle. *Comput. Struct.*, 37(6):893–902, 1990.

[48] T.D. Hien and M. Kleiber. Stochastic design sensitivity in structural dynamics. *Int. J. Num. Meth. Eng.*, 32:1247–1265, 1991.

[49] T.D. Hien and M. Kleiber. Stochastic structural design sensitivity of static response. *Comput. Struct.*, 38(5/6):659–667, 1991.

[50] E. Hinton and D.R.J. Owen. *An Introduction to Finite Element Computations.* Pineridge Press, 1979.

[51] T. Hisada and S. Nakagiri. Role of stochastic finite element method in structural safety and reliability. In *Proc. 4th Int. Conf. on Struct. Safety and Reliability*, Kobe, Japan, Vol. I, pages 385–394, 1985.

[52] T. Hisada and S. Nakagiri. Stochastic finite element method developed for structural safety and reliability. In *Proc. 3rd Int. Conf. on Struct. Safety and Reliability*, pages 395–402, 1981.

[53] T. Hisada, S. Nakagiri, and M. Mashimo. A note on stochastic finite element method (Part 10) – 1D invariance of advanced first-order second-moment reliability index in analysis of continuum. *Inst. of Industrial Science, Univ. Tokyo*, 37(3):111–114, 1985.

[54] S.M. Holzer. *Computer Analysis of Structures*. Elsevier, 1985.

[55] L.P. Huelsman and P.E. Allen. *Introduction to the Theory and Design of Active Filters*. McGraw-Hill, 1980.

[56] T.J.R. Hughes and W.K. Liu. Implicit-explicit finite elements in transient analysis: stability theory. *Appl. Mech.*, 45:371–374, 1978.

[57] R.A. Ibrahim. Structural dynamics with parameter uncertainties. *Appl. Mech. Rev.*, 40(3):309–328, 1987.

[58] B. Irons and S. Ahmad. *Techniques of Finite Elements*. Wiley, 1980.

[59] D. Jackson. *The Theory of Approximation*. Amer. Math. Soc., New York, 1930.

[60] D.E. Johnson. *Introduction to Filter*. Prentice-Hall, Englewood Cliffs, 1976.

[61] A. Der Kiureghian. Measures of structural safety under imperfect states of knowledge. *J. Struct. Eng.*, 115(5):1119–1140, 1989.

[62] A. Der Kiureghian and Jyh-Bin Ke. The stochastic finite element method in structural reliability. *Prob. Eng. Mechanics*, 3(2):83–91, 1988.

[63] A. Der Kiureghian and P.L. Liu. First- and second-order finite element reliability methods in structural reliability. In W.K. Liu and T. Belytschko, editors, *Computational Mechanics of Probabilistic and Reliability Analysis*, pages 281–298, Elmepress Int., 1989.

[64] M. Kleiber. *Incremental Finite Element Modelling in Non-Linear Solid Mechanics*. Ellis Horwood – PWN, 1989.

[65] M. Kleiber and P. Breitkopf. *Finite Elements in Structural Mechanics: An Introduction with Turbo Pascal Programs for Microcomputers*. Ellis Horwood – PWN, 1992.

[66] M. Kleiber, T.D. Hien, and E. Postek. Incremental finite element analysis of nonlinear structural design sensitivity problems. In E. Oñate, J. Periaux, and A. Samuelsson, editors, *The Finite Element Method in the 1990's*. A book dedicated to O.C. Zienkiewicz, pages 241–247, Springer-Verlag, 1991.

[67] M. Kleiber and Cz. Woźniak. *Nonlinear Mechanics of Structures*. Kluwer – PWN, 1991.

[68] K. Knopp. *Theory and Application of Infinite Series.* Blackie, London, 1951.

[69] V. Kumar, S.J. Lee, and M.D. German. Finite element design sensitivity analysis and its integration with numerical optimization techniques for structural design. *Comput. Struct.*, 32(3-4):883–897, 1989.

[70] C. Lanczos. *The Variational Principles of Mechanics.* Univ. Toronto Press, 1964.

[71] J.H. Laning and L.H. Battin. *Random Processes in Automatic Control.* McGraw-Hill, 1956.

[72] M.A. Lawrence. Basic random variables in finite element analysis. *Int. J. Num. Meth. Eng.*, 24:1849–1863, 1987.

[73] M.A. Lawrence. A finite element solution technique for plates of random thickness. In T.J.R. Hughes and E. Hinton, editors, *Finite Element Method for Plate and Shell Structures. Vol. 2: Formulations and Algorithms*, pages 213–228, Pineridge Press, 1986.

[74] M.J. Lighthili. *Introduction to Fourier Analysis and Generalized Functions.* Cambridge Univ. Press, 1958.

[75] Y.K. Lin. *Probabilistic Theory of Structural Dynamics.* McGraw-Hill, 1967.

[76] H. Lippmann. *Extremum and Variational Principles in Mechanics.* Springer, 1972.

[77] P.L. Liu. Size effect of random field elements on finite-element reliability methods. In *Proc. 7th Int. Conf. on Struct. Saf. Reliab.*, pages 223–239, 1991.

[78] W.K. Liu, T. Belytschko, and G.H. Besterfield. A variational principle for probabilistic mechanics. In T.J.R. Hughes and E. Hinton, editors, *Finite Element Method for Plate and Shell Structures. Vol. 2: Formulations and Algorithms*, pages 285–311, Pineridge Press, 1986.

[79] W.K. Liu, T. Belytschko, and A. Mani. Probabilistic finite elements for nonlinear structural dynamics. *Comput. Meth. Appl. Mech. Eng.*, 56:61–81, 1986.

[80] W.K. Liu, T. Belytschko, and A. Mani. Random field finite elements. *Int. J. Num. Meth. Eng.*, 23:1831–1845, 1986.

[81] W.K. Liu, G.H. Besterfield, and T. Belytschko. Transient probabilistic systems. *Comput. Meth. Appl. Mech. Eng.*, 67:27–54, 1988.

[82] W.K. Liu, G.H. Besterfield, and T. Belytschko. Variational approach to probabilistic finite elements. *Eng. Mech.*, 114(12):2115–2133, 1988.

[83] M. Loeve. *Probability Theory.* G. Van Nostrand Co., Princeton, 1960.

[84] H.C. Martin and G.F. Carey. *Introduction to Finite Element Analysis.* McGraw-Hill, 1969.

[85] E.F. Masur and Z. Mróz. Singular solutions in structural optimization problems. In S. Nemat-Nasser, editor, *Variational Methods in Mechanics of Solids*, pages 337–343, Pergamon Press, 1986.

[86] L. Meirowitch. *Elements of Vibration Analysis.* McGraw-Hill, 1986.

[87] S.G. Mikhlin. *Integral Equations.* Pergamon Press, 1957.

[88] N.F. Morris. The use of modal superposition in nonlinear dynamics. *Comput. Struct.*, 7:65–72, 1977.

[89] A. Myslinsky. Bimodal optimal design of vibrating plates using theory and methods of nondifferentiable optimization. *J.O.T.A.*, 46(2):187–203, 1985.

[90] A.H. Nayfeh. *Perturbation Methods.* Wiley, 1973.

[91] R.B. Nelson. Simplified calculation of eigenvector derivatives. *AIAA J.*, 14(9):1201–1205, 1976.

[92] A. Nickell. Nonlinear dynamics by mode superposition. *Comput. Meth. Appl. Mech. Eng.*, 7:107–129, 1976.

[93] B. Noble. *Applied Linear Algebra.* Prentice-Hall, Englewood Cliffs, 1969.

[94] J.T. Oden. *Finite Element of Nonlinear Continua.* McGraw-Hill, 1972.

[95] N. Olhoff and S.H. Rasmussen. On single and bimodal optimum buckling loads of clamped columns. *Solids Struct.*, 13:605–614, 1977.

[96] A.V. Oppenheim and R.W. Schafer. *Digital Signal Processing.* Prentice-Hall, Englewood Cliffs, 1975.

[97] A.V. Oppenheim, A.S. Willsky, and I.T. Young. *Signals and Systems. Series in Signal Processing.* Prentice-Hall, 1983.

[98] R.K. Otnes and L. Enochson. *Digital Time Series Analysis.* Wiley, 1972.

[99] D.R.J. Owen and E. Hinton. *Finite Element in Plasticity.* Pineridge Press, 1980.

[100] A. Papoulis. *The Fourier Integral and Its Applications.* McGraw-Hill, 1962.

[101] S. Prager and W. Prager. A note on optimal design of columns. *Mech. Sci.*, 21:249–251, 1979.

[102] W. Prager. *Variational Principles of Linear Elastostatics for Discontinuous Displacements, Strains and Stresses.* Folk Odquist Volume, 1967.

[103] J.S. Przemieniecki. Matrix structural analysis of substructures. *AIAA J.*, 1:138–147, 1963.

[104] J.S. Przemieniecki. *Theory of Matrix Structural Analysis.* McGraw-Hill, 1968.

[105] L.R. Rabiner and B. Gold. *Theory and Application of Digital Signal Processing.* Prentice-Hall, Englewood Cliffs, 1975.

[106] M. Shinozuka. Stochastic fields and their digital simulation. In G.I. Schuëller and M. Shinozuka, editors, *Stochastic Methods in Structural Dynamics*, pages 93–133, Martinus Nijhoff, 1987.

[107] M. Shinozuka and T. Nomoto. *Response Variability due to Spatial Randomness of Material Properties.* Technical Report, Dept. Civ. Eng., Columbia Univ., 1980.

[108] I.N. Sneddon. *Fourier Transform.* McGraw-Hill, 1951.

[109] T.T. Soong. *Random Differential Equations in Science and Engineering.* Academic Press, 1973.

[110] P.D. Spanos and R. Ghanem. Stochastic finite element expansion for random media. *Eng. Mech., ASCE*, 115:1035–1053, 1989.

[111] J.F. Steffesen. *Interpolation.* Chelsea, 1950.

[112] J. Szopa. Application of stochastic sensitivity functions to chaotic systems. *Sound Vib.*, 104(1):176–178, 1986.

[113] J. Szopa. Sensitivity of stochastic systems to initial conditions. *Sound Vib.*, 97(4):645–649, 1984.

[114] I. Tadjbakhsh and J.B. Keller. Strongest columns and isoparametric inequalities for eigenvalues. *Appl. Mech.*, 29:159–164, 1977.

[115] J.M.T. Thomson and G.W. Hunt. Dangers of structural optimization. *Eng. Optim.*, 2:99–110, 1974.

[116] K.D. Tocher. *The Art of Simulation.* McGraw-Hill, 1968.

[117] P. Tong and J.N. Rossettos. *Finite Element Method. Basic Technique and Implementation.* The MIT Press, 1978.

[118] E.H. Vanmarcke. *Random Fields: Analysis and Synthesis.* The MIT Press, 1983.

[119] E.H. Vanmarcke and M. Grigoriu. Stochastic finite element analysis of simple beams. *J. Eng. Mech., ASCE*, 109(5):1203–1214, 1983.

[120] K. Washizu. *Variational Methods in Elasticity and Plasticity.* Pergamon Press, 1975.

[121] E.L. Wilson. The static condensation algorithm. *Int. J. Num. Meth. Eng.*, 8:199–103, 1974.

[122] E.L. Wilson. Structural analysis of axisymmetric solids. *AIAA J.*, 3:2269–2274, 1965.

[123] F. Yamazaki, M. Shinozuka, and G. Dasgupta. *Neumann Expansion for Stochastic Finite Element Analysis.* Technical Report, Dept. Civ. Eng., Columbia Univ., 1985.

[124] O.C. Zienkiewicz. *The Finite Element Method in Engineering Science.* McGraw-Hill, second edition, 1971.

[125] O.C. Zienkiewicz, R.W. Lewis, and K.G. Stagg (Ed.). *Numerical Methods in Offshore Engineering.* Wiley, 1978.

[126] O.C. Zienkiewicz and R.C. Taylor. *The Finite Element Method.* McGraw-Hill, fourth edition, 1989.

Index

Adaptive formulation, 170
Adjoint system, 69, 73, 77, 168–169, 188, 202 , 205, 207, 243, 254
Adjoint variable, 69, 72, 143–144, 164–165, 201
Adjoint variable method (AVM), 69, 142–144, 145, 168, 201, 224
d'Alembert principle, 48, 55, 93
Aliasing, 29
Amplitude decay, 217
Autocorrelation, 16, 17
Autocovariance, 16
Average, *see* Expectation

Back-up memory, *see* Low-speed storage
Backward substitution, 53, 124–125, 168, 234, 277
Backward time, 75, 76, 215
Backward-time integration, 73
Banded-storage technique, 55
Base-two fast Fourier transform, 30
Basic random variable, 97
Bifurcation load, *see* Buckling load
Bifurcation point, 67, 114
Blank source file, 267–268
Body force, 41, 43, 99
Boundary condition, 41
Boundary traction, 41, 43 ,99
Boundary-value problem, 43, 67
Buckling load, 66, 75, 114, 115, 119–121, 134–135
Buckling load factor, 115, 119, 134–135

Calculus of variations, 34–37
Cauchy stress tensor, 62
Chain rule, *see* Leibnitz rule
Change of basis, 59, 145
Chebyshev inequality, 8, 17
Cholesky factorization, 53, 54
Coefficient of variation, 7, 86, 103, 271
Compatible displacement model, 50, 53
Compiler (Fortran), 177, 180, 228
Complex sequence length, *see* Fourier transform degrees
Consistent mass matrix, 59
Constitutive relations, 41
Constitutive stiffness matrix, 63, 114
Continuous frequency signal, 27
Continuous random process, 18
Continuous random sequence, 18
Continuous time signal, 27
Cooley-Tukey algorithm, 32
Correlated random variable, 98, 124, 125, 127, 129, 136, 145, 169, 198, 213, 227, 278, 281
Correlation coefficient, 8, 98
Correlation function, 16, 86
Correlation length, 98, 128, 146
Correlation matrix, 17, 98, 146
Coupled system, 124
Covariance, 8, 17, 86, 97, 131, 166–167
Covariance matrix, 8, 10, 11, 12, 13, 99, 116, 125, 146, 168, 192, 236–239, 265, 281
Cramer's rule, 53
Cross-correlation function, 7, 16
Cross-covariance, 16, 89, 90, 95–97, 104, 105, 120–122, 124, 132, 139, 247, 262, 264, 266, 275–276, 279–280
Cumulative distribution function (CDF), 4

Damping factor, 60, 101, 137
Damping force, 55
Damping matrix, 56, 102, 106, 200, 272
Database control system, 180, 224
Data-check mode, 179, 225
Data processing, 192, 227
Data window, 141, 249
Design variable, 67, 75, 164, 168, 180, 195, 200–201, 223
Deterministic quantity, 3
Dirac delta distribution, 19–21, 24, 74, 77, 151, 206, 215
Direct access file, 195
Direct differentiation method (DDM), 69
Direct Fourier transform, *see* Fourier analysis
Direct integration, 56, 76

Direct integration operator, 243
Discrete Fourier transform, 22–29, 140, 249
Displacement-rate formulation, 62
Discrete random process, 18
Discrete random sequence, 18
Discrete signal, 21, 27, 140, 237, 242
Discrete variable, 21
Discretized system, 50, 97, 140, 272
Displacement vector, 40
Duhamel's convolution, 60, 151
Dynamic control system, 217

Effective 'load' vector, 57
Effective 'stiffness' matrix, 56
Eigenpair, 67
Eigenvalue, 59, 106, 116
Eigenvalue sensitivity, 75, 234, 265, 266
Eigenvector, 59, 106, 116
Element level, 70, 180
Element stiffness matrix, 52
Equations of motion, 41, 56, 101, 202, 272, 273
Equilibrium equations, 43, 164, 273
Equilibrium iteration, 273
Euler beam, 39
Euler equation, 36, 37, 46
Executable file, 265, 267
Expansion coefficient, 22
Expansion method, 98
Expectation, 6, 15, 17, 86, 89, 92, 97, 99, 102, 104–105, 113–114, 123–124, 223, 225, 266, 271, 274–276, 279, 280
Expected value, *see* Expectation

Fast Fourier transform (FFT), 30–34, 140–142, 215–216, 228, 249, 253, 266
Field equations, 42
File specification, 180, 236
Finite difference scheme, 70, 165, 277
Finite element, 50
Finite element coordinates, 59
Finite-range Fourier transform, 27, 28
First density function, 15
First-order estimate, 89
First-order perturbation, 143
Folding, 29
Forced-vibration sensitivity, 79
Fortran, 30, 177, 178, 180, 188, 223, 237, 267, 269
Forward substitution (reduction), 54, 124–125, 168, 277
Forward iteration, 116, 117
Forward time, 76
Fourier analysis, *see* Fourier transform
Fourier integral, 23, 24, 151
Fourier series, 22, 151
Fourier synthesis, *see* Fourier transform
Fourier transform, 25, 56, 141, 151, 253
Fourier transform degree, 140, 218
Fourier transform pair, 25
Free vibration problem, 59, 75, 106–107
Frequency-domain integration, 30, 56
Frequency range, 141, 249
Frequency range factor, 141, 249
Frequency weighting function, 141
From-diagonals storage mode, 55
Frontal solution, 54, 179
Fundamental elastic solution, 66
Fundamental frequency, 27
Fundamental matrix, 125

Galerkin projection, 84
Gaussian distributed random variable, 11
Gaussian distribution, 17
Gaussian elimination, 53, 70, 179, 225
Gaussian process, 17
Gaussian random variable, 11
Gauss–Ostrogradski theorem, 44, 45, 92
Generalized coordinates, 56, 131, 139, 225, 226
Generalized eigenproblem, 59, 67, 75, 106–107, 139, 234
Generalized Hooke's law, 41
Geometric stiffness matrix, 63, 114
Global stiffness matrix, 52

Hamilton variational principle, 47
High-speed storage, 178, 180, 224, 252, 265
Hu–Washizu functional, 45, 92, 271

Implicit design derivative, 70
Implicit–explicit integration, 65
Implicit function theorem, 68
Implicit integration, 57, 276, 277
Incremental constitutive equation, 61, 274
Incremental equations of motion, 61, 272, 278
Incremental formulation, 61, 272, 281
Incremental problem, 61
Incremental strain–displacement relation, 61, 274
Incremental virtual work, 63
Index set, 14
Infinite-range Fourier transform, 26
Inherent variability, 83

Initial-boundary value problem, 42
Initial conditions, 41, 205
Initial stresses, 61, 277
Initial stress stiffness matrix, 63, 66
Initial–terminal-value problem, 73, 76, 200, 207, 209, 210, 218, 249
Initial-value problem, 73, 76, 200, 207, 209, 210
Input signal, 21, 140
Integral function of distribution, 14
Integral transform, 25
Internal nodal force, 64
Interpolation method, 98
Inverse Fourier transform, *see* Fourier synthesis
Inverse iteration, 117
Iterative method, 53

Joint statistical moment, 16

Karkunen–Loeve decomposition, 84
Kinematic boundary condition, 62
Kinetic energy, 47
Kirchhoff plate, 39
Kronecker delta, 21

Lagrange multiplier, 45
Laplace transform, 30, 56
Least squares fit method, 165, 277
Leibnitz differentiation rule, 68, 72, 164, 207
Limit theorem, 8
Linear acceleration method, 57
Linear algebraic equations, 53
Linear transformation, 9, 274
Linearized buckling, 65, 130
Load operator, 243
Local formulation, 42
Lower bound, 146
Lower triangular matrix, 53
Low-speed storage, 178, 180, 188, 224, 265
Lumped mass matrix, 59

Macro process, 180
Marginal CDF, 6
Marginal PDF, 5
Massless degree of freedom, 59
Mass matrix, 56, 102, 106, 200, 272
Material density, 41
Matrix decomposition, 30, 54
Mean, *see* Expectation
Mean of stochastic process, 15
Mean square, 6, 15
Mean-square treatment, 85
Mean value, *see* Expectation
Mean vector, 8
Midpoint method, 98
Mixed direct–adjoint method (DDM), 145
Modal analysis, 111
Modal damping, 60
Modal superposition, *see* Mode superposition
Mode shape matrix, 59, 77, 127, 137, 213
Mode superposition, 58, 60, 65, 76, 131, 138, 225, 281
Monte Carlo simulation, 83, 123
m-th moment, 7, 15
m-th central moment, 7, 15
Multi-block mode, 54, 180, 188, 224, 234
Multi-dimensional CDF, 15
Multi-dimensional PDF, 14–15
Multi-dimensional random field, 18, 96
Multi-dimensional vector random process, 17
Multi-field principle, 45–46
Multi-field variational statement, 45, 271
Multi-variate CDF, 5, 15, 19, 85
Multi-variate PDF, 4, 86, 88, 92, 93, 166, 272
Multi-variate normal PDF, 11
Multi-variate random field, 18, 96
Multi-variate vector random field, 18
Mutually independent random variable, 6

Natural frequency, 60, 139, 234
Neumann expansion, 84
Newmark method, 57, 218, 228, 242, 242, 243, 253, 265
Newton–Raphson method, 65
Newton-type iteration, 65
Nodal acceleration, 58, 123, 202
Nodal displacement, 50, 123, 202
Nodal velocity, 58, 123, 202
Normal strain, 40
Normal stress, 40
Normalized autocovariance, 16
Normalized coordinates, 59, 60, 131, 137, 213, 225, 243, 281
Normalized force, 60
Normally distributed random variable, 11
Nyquist cutoff frequency, 28, 141

Object-code file, 265–266
Off-line comment, 180, 223
One-dimensional PDF, 14
One-dimensional random field, 18

One-field variational statement, 45
One-parameter load, 65
On-line comment, 180, 223
Operating memory, *see* High-speed storage
Orthogonal transformation, 59, 145
Orthogonality condition, 60, 109, 115, 118
Orthonormality condition, 60, 75, 108–110, 115, 117, 118
Output signal, 21
Overlay linkage, 228, 264, 266, 268
Overlay mode, 228, 264, 265
Overlay unit, 265–266

Parameter set, 14
Parameter space, 19
Period elongation, 217
Perturbation approach, 83, 91, 101
Piece-wise linearization, 61–62
Piece-wise regular, 22
Piola–Kirchhoff stress tensors, 62
Plane strain, 39
Plane stress, 39
Poisson's ratio, 41
Polynomial chaos expansion, 84
Postbuckling shape, 67, 122
Potential energy, 43
Potential energy-based approach, 49
Potential energy functional, 44
Prebuckling deformation, 66
Primary system, 168–169, 188, 202, 213, 243
Principle of minimum potential energy, 43–44
Principle of stationary potential energy, 44
Principle of virtual work, 44, 93
Probabilistic distribution, 89, 103, 106, 113, 119, 131–135, 166, 168, 199, 205
Probability density function (PDF), 3
Probability of the event, 3
Proportionality factor, 55

Qualitative analysis, 85
Quantitative analysis, 85

Random differential equations, 84
Random displacement field, 89
Random field, 18, 19, 86, 98, 99, 123, 146
Random field variable, 87, 91, 97, 99, 123, 146
Random function, 14
Random numbers, 83
Random phenomenon, 3
Random process, 14
Random quantity, 3, 26
Random variable, 3, 97, 99, 102, 164, 180, 192, 272
Random vibration, 84
Randomness, 83
Randomness superposition, 136, 139
Rate-formulation, 62
Rayleigh–Ritz method, 49
Real line, 3
Rectangular shaped function, 19
Reduced stiffness matrix, 53
Reference configuration, 63
Reference load, 66
Reference state, 61
Repeated eigenvalue, 75, 79
Response file for linking, 264, 266–268
Rigid body motion, 43
Root mean square value, 6
Running file, *see* Executable file
Running terminal time, 74, 207
Running time instant, 76

Sample length, 27, 140, 228, 253
Sampling frequency interval, 27
Sampling (time) interval, 21, 27, 74, 77, 140, 206, 217, 274, 276, 277
Sampling point, 21
Sampling time signal, 74
Scratch file, 180, 195, 237, 254, 265
Second central moment, 7
Second density function, 16
Second-moment analysis, 86, 91, 97, 272
Second-order estimate, 89
Second-order expansion, 89, 93, 119, 139, 205
Second-order perturbation, 84, 87, 106, 115, 139–140, 144, 203, 207, 223, 249, 271, 272, 281
Secular terms, 125, 139, 228, 249, 252
Secularity elimination, 125, 139–142, 215, 216, 228, 249, 266
Sensitivity gradient, 68, 69, 73, 74, 75, 139, 163, 164, 166–168, 188, 192, 195, 198, 201, 209, 215–218, 224, 234, 254, 258, 262, 264, 266
Shape function, 50, 98, 99, 146
Shear strain, 40
Shear stress, 40
Shifting in phase, 124, 277
Single-block mode, 188
Skyline-storage technique, 55
Small parameter, 89, 104, 106, 115, 116, 165, 202, 209, 272, 273

Solution path, 61
Source file, 264, 265
Space–time moment, 19
Spatial averaging method, 98, 179, 226
Spatially discretized system, 56
Standard deviation, 7, 192
Standard eigenproblem, 59, 115, 125, 138, 265
Standard normal transformation, 98, 124–130, 131, 138, 146, 168, 198, 224, 227, 237, 265, 278
Standardized normal PDF, 11
Static boundary equations, 43
Static condensation, 54
Stationary condition, 46
Stationary value, 36
Statistical analysis, 84
Statistical average, 6
Statistical moments, 86
Stiffness matrix, 52, 102, 106, 195, 200
Stochastic adjoint analysis, 125, 142–144
Stochastic Hamilton principle, 93, 271
Stochastic multi-field variational principle, 91, 271
Stochastic potential energy principle, 86, 88, 271
Stochastic process, 14
Stochastic initial–terminal problem, 202
Stochastic terminal problem, 202
Strain–displacement relationship, 40, 265
Strain-rate, 62
Strain tensor, 40
Stress boundary condition, 62
Stress-rate, 62
Stress tensor, 62
Strongly homogeneous random process, 17
Strongly stationary random process, 17
Structural buckling, 67
Structural design sensitivity (SDS), 67
Structural response functional, 68, 71, 74, 75, 163, 168, 195, 198, 200, 206, 218, 223, 224, 253, 257, 266
Sturm–Liouville theory, 22
Sturm sequence check, 265
Subspace iteration, 117
Substructure analysis, 54
Surface traction, 43
System frequency, 59

Tangent stiffness matrix, 63, 272, 277, 281
Taylor expansion, 36, 83, 87, 99, 106, 131, 137, 139, 143, 199, 202, 217
Tensor of elastic moduli, 41
Terminal time, 71, 199, 200, 202
Terminal-time condition, 71, 199, 202, 205, 209, 212
Terminal-value problem, 73, 74, 76, 203, 207, 254
Time-domain integration, 30, 56
Time interval sensitivity, 71–73, 199, 200–206
Time instant sensitivity, 73–74, 199, 206–209, 227–228, 253, 257, 262, 266
Time-invariant randomness, 84, 86, 96
Time-stepping algorithm, 56, 60
Time-varying randomness, 84
To-diagonals storage mode, 55
Transformation matrix, 59
Triangular algebraic system, 53
Two-fold superposition, 125, 142–144, 136, 145, 218, 240, 257, 281

Unconditional stable, 58
Uncorrelated random variable, 8, 98, 124, 125, 127, 129, 136, 143, 145, 167, 169, 198, 212–213, 227, 258, 265, 278, 281
Uncoupled equations, 60, 77, 124, 138–139
Undamped system, 60
Unformatted file, 193, 227, 236, 237
Updated Lagrangian description, 61
Unit impulse, 21, 74, 77, 207, 253, 262
Uni-variate random field, 18
Upper bound, 146
Upper triangular matrix, 53
Upper triangular system, 53

Variance, 6, 16, 86, 103, 125, 265
Variance of random process, 16
Variational formulation, 42, 43, 271
Variational problem, 34
Variational statement, 42, 97, 101, 103
Variation of the actual path, 34
Varying-in-space random field, 84
Vibration mode, 59
Viscous damping, 55
Viscous stresses, 55

Weakly homogeneous random process, 17
Weakly stationary random process, 17
Weighting number, 144, 262
Wilson θ-method, 57–58, 64, 218, 228, 242, 243, 253, 265

Zero-order hold, 21

Glossary of Symbols

a, a_i	sample points on the real line
a_n	Fourier expansion coefficients
$a^{(e)}_{\alpha\xi}$	finite element coordinate transformation matrix
$\mathbf{A}$	coefficient matrix of algebraic equations
$A_{3\times 3}$	direct integration approximation operator
$A_{i\bar{\alpha}\alpha}$, $A_{r\bar{\alpha}\alpha}$	Boolean transformation matrices
A_ρ	cross-sectional areas
b, b_i	sample points on the real line
b_r	vector of continuous random variables
$b_{r\bar{\alpha}}$	matrix of random parameter nodal variables
$\mathbf{b}$, b_ρ	vector of nodal random variables
$B_{3\times 1}$	load operator in the direct integration scheme
$B_{ij\alpha}$	strain–nodal displacement matrix
c_ρ, $c_{\breve{\rho}}$	uncorrelated (normalized) random variables
C	number of constraints in sensitivity analysis
C_{ijkl}	constitutive tensor
$\mathbf{C}$, $C_{\alpha\beta}$	damping matrix
Cov_{XX}	autocovariance function of random variable (process) X
Cov_{XY}	cross-covariance function of random variables (processes) X, Y
$\mathrm{Cov}(b_r, b_s)$	covariance function of continuous random variables
$\mathrm{Cov}(b_\rho, b_\sigma)$	covariance matrix of nodal random variables
D	overall number of design variables
E	total number of finite elements in the system
	Young's modulus
$E[\cdot]$	expectation function of random variables
f	circular frequency
f_i	vector of body forces
f_j	discretized cycle frequency
$\mathbf{F}^{(\mathrm{eff})}$	vector of internal nodal forces
F	circular Nyquist frequency
G	structural response function
$\mathcal{G}$	structural response functional
$\mathbf{h}$, h^d	vector of design variables
I_x, I_y, I_z	moments of inertia
$\mathcal{J}$	potential energy functional
$k_{\xi\zeta}$	element stiffness matrix
$\mathbf{K}$, $K_{\alpha\beta}$	system stiffness matrix

$K^{(\mathrm{con})}_{\alpha\beta}$	constitutive stiffness matrix
$K^{(\mathrm{eff})}_{\alpha\beta}$	effective 'stiffness' matrix
$K^{(\mathrm{T})}_{\alpha\beta}$	tangent stiffness matrix
$K^{(\sigma)}_{\alpha\beta}$	initial stress (geometric) stiffness matrix
l, l_ρ	structural member length
L	number of load cases
	sampling length
$\mathbf{L}$	lower triangular matrix
$\mathbf{M}$, $M_{\alpha\beta}$	system mass matrix
N	overall number of system degrees of freedom
$\bar{N}$	overall number of system nodal points
$\tilde{N}$	overall number of nodal random variables
$\check{N}$	overall number of nodal uncorrelated random variables
$p(.)$	probability density function of a random variable
$p_n(.)$	n-dimensional probability density function of a random process
$P(.)$	cumulative distribution function of a random variable
$P_n(.)$	cumulative distribution function of a random process
$\Pr(.)$	probability of an event
q_α	vector of nodal displacement-type variables
Q_α	vector of nodal external loads
Q^*_α	vector of reference loads
r	frequency range factor for secularity elimination
$r_{\bar{\alpha}}$	normal displacement-type coordinates
R	total number of continuous random variables
$R_{\bar{\alpha}}$	nodal normal loads
$\mathbf{S}$	covariance matrix
S^{rs}_{b}	covariance function of continuous random variables
$S^{\rho\sigma}_{\mathrm{b}}$	covariance matrix of nodal random variables
$S^{de}_{\mathcal{G}\cdot}$	second moment matrix of sensitivity gradient coefficients
$S^{\alpha\beta}_{\mathrm{q}}$	second moment matrix of nodal displacement components
$S^{\alpha\hat{\alpha}\beta\hat{\beta}}_{Q}$	second moment matrix of bifurcation load components
S^{ijkl}_{ε}	second moment matrix of strain components
$S^{\hat{\alpha}\hat{\beta}}_{\lambda}$	second moment matrix of buckling load factors
S^{ijkl}_{σ}	second moment matrix of stress components
$S^{\alpha\hat{\alpha}\beta\hat{\beta}}_{\phi}$	second moment matrix of eigenvector components
$S^{\hat{\alpha}\hat{\beta}}_{\omega}$	second moment matrix of eigenvalues
t	element thickness
	running time variable
$\hat{t}_i$	vector of boundary tractions
T	terminal time instant
$\mathcal{T}$	kinetic energy
	terminal time condition function
$\mathbf{u}$, u_i	vector of continuous displacement variables
$u_{i\bar{\alpha}}$	matrix of displacement variable nodal values

$\mathbf{U}$	upper triangular matrix
$\mathcal{U}$	potential energy
v_α, $v_{\alpha(\hat{\alpha})}$	buckling mode vector
$V_{\alpha\beta}$	buckling mode matrix
Var(.)	variance function
x_k, x_m	spatial variables
X	continuous random variable (process)
α	proportional damping coefficient (mass term)
α, α_X	coefficient of variation
	first interpolating coefficient in Newmark integration
α_v	internal variable vector
β	proportional damping coefficient (stiffness term)
δ	second interpolating coefficient in Newmark integration
$\delta(.)$	Dirac delta distribution
δ_{ij}	Kronecker delta
Δt	sampling time interval (time increment)
$\Delta\Omega$	secular frequency range
ϵ	small parameter
$\boldsymbol{\varepsilon}$, ε_{ij}	strain tensor
θ	extrapolating coefficient of Wilson θ-integration
$\vartheta_{\bar{\alpha}}$	adjoint normal coordinates
$\Theta_{(\rho)}$	diagonal matrix of transformed variances
λ, $\lambda_{(i)}$	buckling load factors
λ_α	vector of adjoint displacement-type variables
$\Lambda_{(\alpha)}$	diagonal matrix of buckling load factors
$\bar{\mu}_{kl}$	cross-correlation function
μ_{XX}	autocorrelation function of random variable (process) X
μ_X^m	m-th statistical moment of random variable (process) X
μ_X^{mn}	joint statistical moment of random variable (process) X
$\bar{\mu}_X^m$	m-th central moment of random variable (process) X
ν	Poisson's ratio
ξ	modal damping factor
ϱ	mass density
ϱ_{XY}	autocorrelation-coefficient function of random variables X, Y
$\boldsymbol{\sigma}$, σ_{ij}	Cauchy stress tensor
σ_{ij}^{I}	first Piola–Kirchhoff stress tensor
$\sigma_{ij}^{\mathrm{II}}$	second Piola–Kirchhoff stress tensor
σ_X	standard deviation of random variable (process) X
τ	(forward) time variable
$\bar{\tau}$	backward time variable
$\Phi_{N\times N}$	mode shape matrix
$\phi_{\alpha\beta}$, $\phi_{\bar{\alpha}\bar{\beta}}$	eigenvector matrix
$\phi_{\alpha(\bar{\alpha})}$	eigenvector
$\varphi_{i\alpha}^{(e)}$	element shape function matrix

$\varphi_{i\alpha}$	system shape function matrix
$\varphi_{\bar{\alpha}}$	random field shape function vector
$\psi_{\rho(\breve{\rho})}$	fundamental vectors of the standard normal transformation
$\psi_{\rho\sigma}$	fundamental matrix of the standard normal transformation
ω_j	discretized natural frequencies
Ω	angular Nyquist frequency
	solid volume
$\Omega_{N\times N}$, $\Omega_{(\alpha)}$	diagonal matrix of system eigenvalues
$\partial\Omega$	solid boundary
$\partial\Omega_u$	solid boundary with prescribed displacement conditions
$\partial\Omega_\sigma$	solid boundary with prescribed stress conditions
$(.)^0$	zeroth-order quantities, taken at means of random variables
$(.)^{(2)}$	second-order quantities, taken at means of random variables
$(.)_{\text{binary}}$	base-two number representation
$(.)_{\text{decimal}}$	base-ten number representation
$(.)^T$	vector or matrix transposed
$(.)^{.d}$	first partial derivatives with respect to (w.r.t.) design variables
$(.)^{,r}$	first partial derivatives w.r.t. continuous random variables
$(.)^{,rs}$	second partial derivatives w.r.t. continuous random variables
$(.)^{,\rho}$	first partial derivatives w.r.t. nodal random variables
$(.)^{,\rho\sigma}$	second partial derivatives w.r.t. nodal random variables
$(.)^{,\breve{\rho}}$	first partial derivatives w.r.t. nodal uncorrelated random variables
$(.)^{,\breve{\rho}\breve{\rho}}$	second partial derivatives w.r.t. nodal uncorrelated random variables
$(.)^{;\rho}$	first total derivatives w.r.t. nodal random variables
$(.)^{;\rho\sigma}$	second total derivatives w.r.t. nodal random variables
$(.)_{.\alpha}$	first partial derivatives w.r.t. nodal displacements
$0(.)$	order estimate
$\dot{f}$, $\ddot{f}$	second time derivative of function f
mod(.)	remainder in integer-division operation
AVM	Adjoint Variable Method
CDF	Cumulative Distribution Function
DDM	Direct Differentiation Method
DSDS	Deterministic Structural Design Sensitivity
	Dynamic Structural Design Sensitivity
FEM	Finite Element Method
FFT	Fast Fourier Transform
PCM	Probabilistic Computational Mechanics
PDF	Probability Density Function
RAM	Random Access Memory
SDS	Structural Design Sensitivity
SFEM	Stochastic Finite Element Method
SSDS	Stochastic Structural Design Sensitivity

ML

UNIVERSITY OF STRATHCLYDE
30125 00470974 6

ANDERSONIAN LIBRARY
★
WITHDRAWN
FROM
LIBRARY
STOCK
★
UNIVERSITY OF STRATHCLYDE

Books are to be returned on or befo[re]
the last date below.

31/7/02
30/9/03
31/8/02
DUE
15 OCT 2002
25 Nov 2
MAY
17 JAN 20
28 Feb 2